SHAPE MEMORY
EFFECTS IN ALLOYS

The Metallurgical Society of AIME Proceedings
published by Plenum Press

1968–Refractory Metal Alloys: Metallurgy and Technology
Edited by I. Machlin, R. T. Begley, and E. D. Weisert

1969–Research in Dental and Medical Materials
Edited by Edward Korostoff

1969–Developments in the Structural Chemistry of Alloy Phases
Edited by B. C. Giessen

1970–Corrosion by Liquid Metals
Edited by J. E. Draley and J. R. Weeks

1971–Metal Forming: Interaction Between Theory and Practice
Edited by A. L. Hoffmanner

1972–The Nature and Behavior of Grain Boundaries
Edited by Hsun Hu

1973–Titanium Science and Technology (4 Volumes)
Edited by R. I. Jaffee and H. M. Burte

1973–Metallurgical Effects at High Strain Rates
Edited by R. W. Rohde, B. M. Butcher,
J. R. Holland, and C. H. Karnes

1975–Physics of Solid Solution Strengthening
Edited by E. W. Collings and H. L. Gegel

1975–Shape Memory Effects in Alloys
Edited by Jeff Perkins

A Publication of The Metallurgical Society of AIME

SHAPE MEMORY EFFECTS IN ALLOYS

Edited by

Jeff Perkins

Naval Postgraduate School
Monterey, California

PLENUM PRESS·NEW YORK AND LONDON

Library of Congress Cataloging in Publication Data

International Symposium on Shape Memory Effects and Applications, Toronto, 1975.
Shape memory effects in alloys.

"A publication of the Metallurgical Society of AIME."
Sponsored by the Physical Metallurgy Committee, Institute of Metals Division,
Metallurgical Society of AIME.
Includes bibliographical references and index.
1. Alloys—Congresses. 2. Physical metallurgy—Congresses. I. Perkins, Jeff. II. Metal-
lurgical Society of AIME. Institute of Metals Division Physical Metallurgy Committee.
III. Title.

TN689.2.I59 1975	669'.94	75-31928
ISBN 978-1-4684-2213-9	ISBN 978-1-4684-2211-5 (eBook)	
DOI 10.1007/978-1-4684-2211-5		

Proceedings of the International Symposium on Shape Memory Effects and Ap-
plications held in Toronto, Ontario, Canada, May 19-22, 1975 and sponsored by the
Physical Metallurgy Committee, Institute of Metals Division, The Metallurgical
Society of AIME

©1975 Plenum Press, New York
Softcover reprint of the hardcover 1st edition 1975
A Division of Plenum Publishing Corporation
227 West 17th Street, New York, N.Y. 10011

United Kingdom edition published by Plenum Press, London
A Division of Plenum Publishing Company, Ltd.
Davis House (4th Floor), 8 Scrubs Lane, Harlesden, London, NW10 6SE, England

Preface

The International Symposium on Shape Memory Effects and Applications was held at the University of Toronto on May 19-20, 1975, in four sessions over two days, as part of the regular 1975 Spring Meeting of The Metallurgical Society of AIME, sponsored by the Physical Metallurgy Committee of The Metallurgical Society. This was the first symposium on the subject, the only previous meeting at all related being the 1968 NOL Symposium on TiNi and Associated Compounds.

One of the major intentions of this Symposium was to provide a forum for cross-communication between workers in the diverse metallurgical areas pertinent to shape memory effects, areas such as martensitic transformation, crystallography and thermodynamics, mechanical behavior, stress-induced transformation, lattice stability, and alloy development. Authors were encouraged to place an emphasis on delineation of general controlling factors and mechanisms, and on comparison of shape memory effect alloy systems with systems not exhibiting SME.

Ten authors were invited to present thorough treatments in various areas of particular significance; these invited papers are grouped at the beginning of these Proceedings, although at the actual Symposium they were dispersed throughout the program. Twenty-two additional papers were contributed; these are grouped in the second and third sections, five being arbitrarily separated into the third section for papers dealing with applications which have reached a relatively advanced stage of development. Many of these contributed papers were specifically solicited by the Symposium coordinator because of their importance and interest, but because of time and space limitations were allowed slightly less presentation time at the meeting and slightly less space in these Proceedings. Professor Walter S. Owen was invited to overview the entire symposium; his paper is included in the first section. Papers 17, 22, and 27, and some written discussions were contributed after the meeting. Paper 20 was scheduled for presentation but not presented at the meeting. Dr. R.A. Vandermeer presented a paper at the meeting, but it was not available for publication. The symposium was well attended and provided much interesting oral discussion, which unfortunately could not be recorded. The meeting was truly inter-national, with non-USA origin for fifteen of the thirty-three pa-pers and authors from ten different countries.

An additional, impromptu session was informally organized on the evening of the final day of the Symposium, at which time Luc Delaey projected and narrated several of his excellent films on martensitic transformations in beta-brass type alloys (available through Encyclopedia Cinematographie), and Dave Lieberman regaled the group with demonstrations and descriptions of the remarkable behavior and mechanisms of ferroelastic effects such as in AuCd. The esthetic high point of this evening, and the symposium as a whole, however, occurred when Ridgeway Banks fired up his mechani-cally amazing Nitinol heat engine, which was assembled and observed, because of the nonavailability of a convenient room with hot and cold water that evening, in the second floor men's washroom facil-ity of Sidney Smith Hall at the University. One participant was inspired to speculate whether, in the historical parade of mile-stones in energy conversion, the occasion will rank with the gather-ing under the west stand at Stagg Field on December 2, 1942.

The papers and discussion presented at this symposium would seem to have significantly advanced both the theoretical and empir-ical aspects of SME phenomena. Also, the meetings served in them-selves as an excellent conference on martensitic transformation, and the papers published here constitute an excellent coverage of this area. The symposium coordinator and editor expresses his appreciation to the many enthusiastic authors, and especially to the four session chairmen (Horace Pops, Marv Wayman, Bill Buehler, Jack Harrison), all of whom made the symposium a great success.

Jeff Perkins
May 1975

Contents

DEFORMATION, MECHANISMS AND OTHER CHARACTERISTICS OF SHAPE MEMORY ALLOYS

C. M. Wayman

Department of Metallurgy and Mining Engineering and
Materials Research Laboratory, University of Illinois
at Urbana-Champaign, Urbana, Illinois 61801

I. INTRODUCTION

Although only a curiosity a few years ago, the shape memory effect ("marmem" (1) = martensite memory effect) is now a field of intense investigation which has reached a maturity level high enough to warrant an international conference devoted to the subject alone. At one time, the "uncoiling wire experiment" was used to amuse various audiences, small and large, and I, and many colleagues as well, have had fun with "Nitinol" in front of prospective metallurgy students and others who once thought that the primary function of a metallurgist was in the steel mill. In the past few years, we have seen the list of memory materials getting longer and longer, patents being issued, and useful devices fashioned. There is now thought about using the shape memory effect (SME) for energy conversion, and several "engine" prototypes have been built.

In 1971, Professor Shimizu and I, in putting together a cooperative research program, surveyed the state of the art of memory alloys (1). It then appeared that memory martensites were ordered, thermoelastic, and internally twinned. The first two of these "selection rules" still appear to be applicable but the matter concerning the martensite substructure requires additional clarification as will be discussed later. More recently, Delaey, Warlimont, Krishnan and Tass (2) have updated the field with an extensive three-part review, including a number of recent unpublished results. It will not be the purpose here to resurvey this material, but rather an attempt will be made to concentrate on certain problem areas and present some recent findings.

It is hoped that among other things, this conference might lead to certain ground rules on an acceptable working definition of the shape memory effect. This seems to be important because there are various reports of the SME some of which, compared say to the well known behavior in "Nitinol" should more properly be considered as cases exhibiting a partial memory, i.e., specimens which exhibit an imperfect shape recovery after a "light" deformation of only one or two percent. It is also hoped that as a result of this conference additional mechanistic insights on the actual deformation and re-covery processes will come to surface. At this stage it is clear that the detailed processes of martensite deformation and conse-quent shape recovery are many and complex. Consequently, the SME is still incompletely understood.

The following sections of this paper will be devoted to some recent thermodynamic analyses (with particular emphasis on thermo-elastic martensites and the "engine" efficiency of the shape memory effect), a presentation of recent results obtained from polycrystal-line Cu-Zn alloys (with emphasis on deformation modes, the deleteri-ous effects of grain boundaries, similarities and differences be-tween athermal and stress induced martensite (SIM), and stress-induced transformation of deformed martensite), some observations on the shape memory vs. martensite substructure, and finally a general discussion of some salient problems.

II. THERMODYNAMIC ASPECTS OF THE SHAPE MEMORY EFFECT

A. Thermoelastic Martensites

It is now generally recognized that the shape memory effect is associated with a thermoelastic martensitic transformation. The character of this type of transformation requires a revision in the traditional thermodynamic approach (3) because the reverse martens-ite-parent transformation is biased by elastic energy stored during the forward parent-martensite transformation. In other words, the reverse transformation occurs "prematurely" because stored elastic energy assists the chemical driving force gained by heating. As a consequence, the A_s temperature for the reverse transformation fre-quently lies below the M_s temperature marking the inception of transformation on cooling.

In ordinary martensitic transformations, e.g., Fe-Ni alloys (4), it is customary to deal with a thermodynamic equilibrium temperature, T_0, where the chemical free energies of the two phases (parent and martensite) are equal. For Fe-Ni alloys, T_0 was estimated as being midway between the M_d and A_d temperatures where, respectively, mar-tensite and the parent phase form by deformation. It is a good approximation that T_0 lies roughly halfway between M_s and A_s, i.e., $T_0 = \frac{1}{2}(M_s + A_s)$, and this is the assumption usually made (4). Thus on cooling, M_s lies below T_0 and on heating A_s lies above T_0; supercooling and superheating, respectively, are necessary to override the non-chemical energies.

The above situation cannot apply to certain thermoelastic transformations for which $A_s < M_s$ (designated as class II thermoelastic transformations (5)) and it has been suggested that in effect two T_0 temperatures are involved (3,5). Clearly M_s must lie below T_0 as usually defined, and similarly A_s should lie above T_0. But this cannot be the case when $A_s < M_s$, and consequently one must envision a modified equilibrium temperature, T_0', which lies below A_s applicable to the reverse transformation (3). As before, T_0 is where $G^M = G^P$ or $\Delta G^{P \rightarrow M} = 0$, but T_0' is the temperature where $\Delta G^{M \rightarrow P} + \Delta G_{nc}^{M \rightarrow P} = 0$. In the previous, $\Delta G_{nc}^{M \rightarrow P}$ is the non-chemical free energy difference between the martensite and parent, and for thermoelastic transformations consists primarily of stored elastic energy.

The relationships between M_s, A_s, T_0 and T_0' are shown in Fig. 1 (3). Arguments have been presented to suggest that for thermoelastic martensitic transformations

$$T_0 = \frac{1}{2}(M_s + A_f) \qquad (1)$$

$$T_0' = \frac{1}{2}(M_f + A_s) \qquad (2)$$

and in Class II transformations

$$A_f > T_0 > M_s > A_s > T_0' > M_f \qquad (3)$$

Equation (1) is also applicable to non-thermoelastic transformations and is probably a better estimate than $T_0 = \frac{1}{2}(M_s + A_s)$ (3).

It can be shown that $(T_0 - T_0')/T_0 \approx \Delta G_{nc}/\Delta Q$, where ΔQ is the transformation latent heat and ΔG_{nc} is the stored non-chemical free energy difference (6).

The above discussion indicates that a single T_0 equilibrium temperature is not applicable in general, and in particular, modifications are required when the reverse transformation is considered.

B. Use of the Shape Memory Effect for Energy Conversion

As is well known, and as will be further demonstrated by a number of papers presented during this symposium, shape memory alloys are rather easily deformed below the M_f temperature, and provided the extent of deformation is not too large, the original, undeformed state can be recovered by heating to the A_f temperature to effect

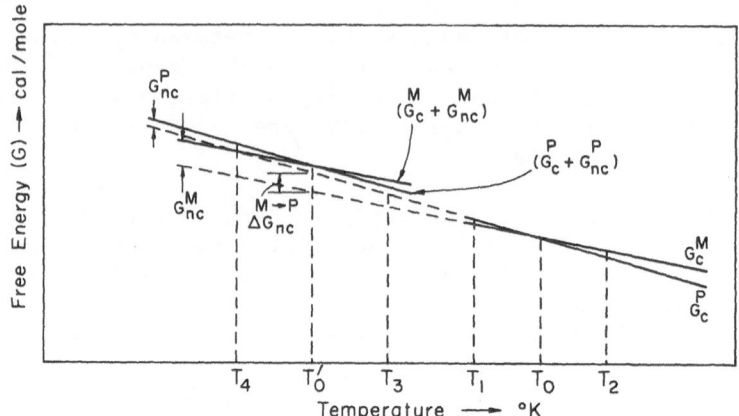

1. Schematic representation of free energy-temperature rela-
    tionship for a thermoelastic martensitic transformation,
    showing two characteristic temperatures, T_O and T_O'. The low
    temperature regime is displaced to a higher level because of
    the stored elastic energy $\Delta G_{nc}^{M \rightarrow P}$ (3).

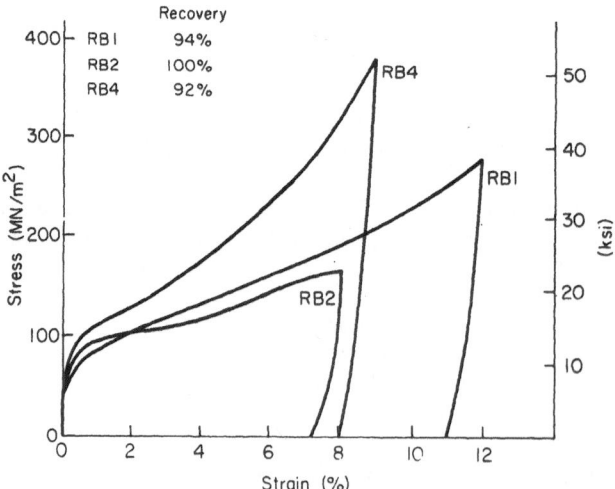

2. Stress-strain curves for bamboo type (RB) Cu-39.8 wt%Zn
    specimens deformed below M_f (-120°C). Strain recoveries
    after heating to room temperature are indicated (12).

the reverse transformation. It is also significant that during the heating process, substantial stresses are generated when a deformed specimen "springs back" to its original shape (7). Thus one can envision a generator or "engine" operating between two temperature limits. In effect a "small" stress is applied below M_f to deform the martensite and a "large" stress is given off by heating to a higher temperature (A_f). Of course thermal energy must be supplied to obtain the mechanical energy during heating. The general idea expressed above is not new. As far back as 1957 (and perhaps farther) the late Professor T. A. Read and colleagues constructed a cyclic weight lifting device from a Au-47.5 at% Cd alloy for exhibition at the 1958 World's Fair in Brussels. And more recently, workers at the University of California (Berkeley) (8) have fashioned a more engineering-wise sophisticated version in the form of a wheel turning device (to be described and shown during this symposium). However, only within the present year have detailed analyses (7,9) appeared which consider quantitatively the efficiency of the above energy conversion process. These treatments are equivalent (10) and one of the formulations (7) is briefly outlined here.

Basically a small stress σ_1 is applied below M_f to deform the martensite and a load giving an opposition stress σ_2 is applied at the same temperature. The deformed specimen expands against the stress σ_2 and the net work done is

$$W_{net} = (\sigma_2 - \sigma_1)V_o \ln(1 + \varepsilon) \qquad (4)$$

where $\varepsilon = (\ell - \ell_0)/\ell_o$ and $V_o = A_o \ell_o$. In general there will be a shift in the equilibrium temperature under stress which is

$$\Delta T_o = T_o(\sigma) - T_o(0) = \frac{T_o(0)V_o \varepsilon \sigma}{\Delta Q} \qquad (5)$$

where ΔQ is the transformation latent heat. Assuming $\sigma_1 << \sigma_2$, the net work becomes (7)

$$W_{net} = \Delta Q \left(\frac{\Delta T_o}{T_o(0)} \right)\left(\frac{\ln(1 + \varepsilon)}{\varepsilon} \right) \qquad (6)$$

The efficiency is the work done divided by the heat input which is, after some simplification,

$$\eta = \left(\frac{\ln(1 + \varepsilon)}{\varepsilon} \right)\left(\frac{1}{T_o(0)} \right)\left(\frac{1}{1/\Delta T_o + \bar{C}_p/\Delta Q} \right) \qquad (7)$$

where $\bar{C}_p$ is an average specific heat. This efficiency is lower than the Carnot efficiency

$$\eta_c = 1 - \frac{M_f(0)}{A_f(\sigma_2)} \qquad (8)$$

Examination of Eq. (7) shows that a high efficiency is favored by a large latent heat, ΔQ, a high recoverable strain ε, a low T_0, and a high parent phase flow stress, which will maximize ΔT_0. Calculated efficiencies vary from 5-21%, depending upon the material properties substituted in Eq. (7). Work is now in progress to compare theoretical and measured efficiencies.

Although somewhat preliminary in form, the calculations to date (7,9) indicate that certain of the shape memory materials may be highly promising for energy converting devices.

III. THE SHAPE MEMORY AND PSEUDOELASTIC BEHAVIOR IN Cu-Zn ALLOYS(12,13)

A.   Deformation Below M_f and Grain Size Effects

Following an initial report (11) of the shape memory effect in Cu-Zn alloys, investigations were undertaken to elucidate mechanical behavior in general (12), and microstructural aspects of martensite formation, deformation and reversal in polycrystalline materials (13). Alloys containing 38.5 to 40.0 wt%Zn were investigated with particular emphasis on grain size effects and the influence of grain boundaries since preliminary work indicated the inhibiting effect of grain boundaries on the magnitude of the shape recovery and the degree of recovery of stress-induced martensite. Fig. 2 shows three typical stress-strain curves obtained from cylindrical specimens deformed at 77°K. In the discussion which follows, the results apply to a Cu-39.8%Zn alloy with an M_s temperature of -120°C. The specimens were 3.4 mm in diameter with a typical grain size of 3.0 mm. As Fig. 2 shows the stress-strain curves were widely different although a general correlation was found between the shape recovery on subsequent heating to room temperature and the magnitude of the strain at 77°K. The observed variation in flow stress is attributed to different grain orientations considering that the grain size and specimen diameters were essentially the same. The fracture strain was typically above 15% in such round "bamboo" (RB) specimens.

Similar stress-strain curves are shown in Fig. 3 for flat large grain size (FLG, 1.2 mm), flat small grain size (FSG, 0.7 mm) and round small grain size (RPC, 0.25 mm) specimens. These specimens were also deformed in tension at 77°K but differed from those of Fig. 2 by having numerous grains extending across the width of the test section. As before, the extent of the shape recovery on heating was inversely proportional to the strain at low temperature, but strains to fracture were typically only 6-9% for the FLG, FSG and RPC specimens. The recoverability of the RPC specimens was the lowest for all series tested, ranging from 100% at 2.6% strain to only 87% following a 3.6% strain at low temperature, in marked

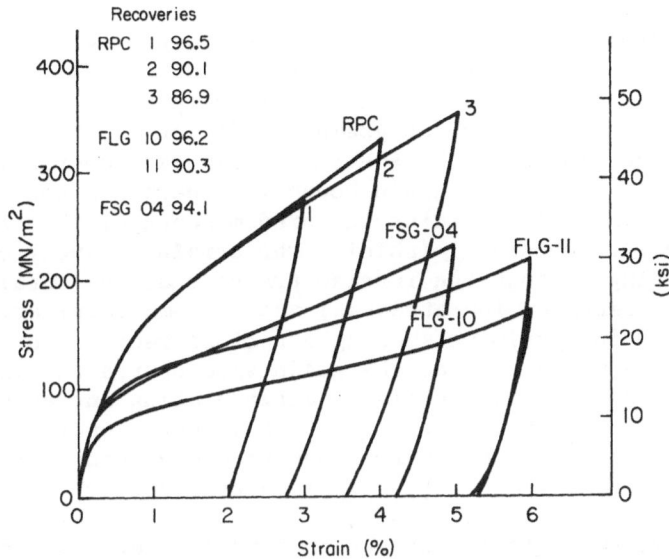

3. Same as Fig. 2, but for RPC, FLG and FSG (see text) specimens (12).

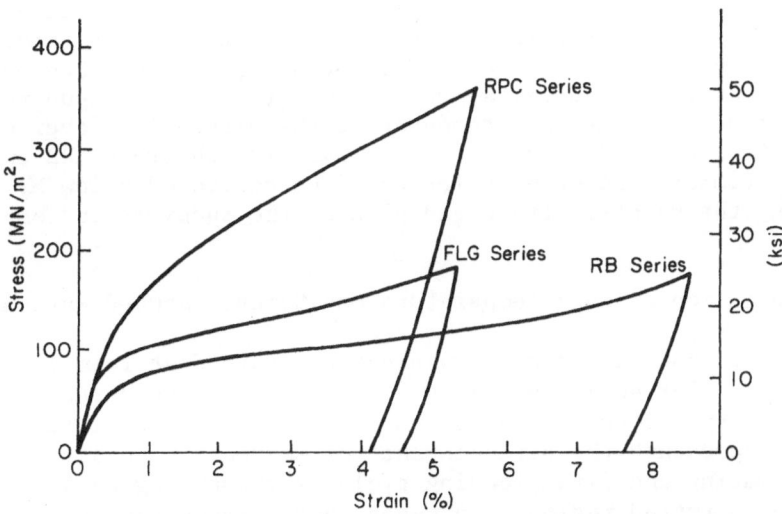

4. Typical characteristic stress-strain curves for polycrystalline Cu-39.8 wt%Zn specimens deformed in tension below M_f (12).

8

C.M. WAYMAN

contrast to a recoverable strain of 10.85% for one RB specimen.

In general the effect of a decrease grain size is to raise the
martensite flow stress and suppress the "easy glide" region of the
stress strain curve, i.e., the work hardening slope is increased.
Irrecoverable plastic flow at grain boundaries and triple points is
also more predominant as the specimen grain size decreases. Such
deformation gives rise to grain boundary separation and premature
fracture. Even at low strains the predominance of this type of
deformation (irreversible) inhibits the strain recovery during subsequent heating through the reverse transformation. Grain boundary
deformation starts at low stresses, and in some cases small amounts
were observed in specimens showing a nominal recovery of 100%. In
general it was noted that as the grain size decreases, the amount
of permanent deformation remaining after heating increases for the
same amount of recoverable strain. The amount of permanent deformation observed at grain boundaries and triple points increased with
lower shape recovery and larger applied stresses.

The difference in stress-strain behavior for three different
specimen types is summarized in Fig. 4, and Fig. 5 shows that the
percentage of strain recovered decreases with increasing stress.

Since the effect of increasing polycrystallinity is to inhibit
the degree of recoverable strain, the latter is expected to be even
greater in single crystal specimens.

Two additional features displayed by the polycrystalline Cu-Zn
specimens (12) are as follows. Fig. 6 shows the stress-strain curve
of an RPC specimen loaded in tension below M_f and then heated under
stress to above the normal A_f temperature (Point A). Upon releasing
the applied stress at room temperature the martensite reverted and
the strain recovered to point B. It was also observed (RB specimen)
that low temperature deformation could be recovered below M_f by
releasing the tensile stress and placing the specimen in compression.

B. Deformation at Room Temperature and Stress-Induced Martensite

Fig. 7 shows a stress-strain curve obtained at room temperature
(M_s= -120°C) for an RPC specimen. The flat, serrated region of the
curve corresponds to SIM formation, and the "reverse yield" on unloading represents the disappearance of martensite on stress relaxation. During the first loading cycle, a smooth region A-B precedes the serrated region. In cases where large strains were employed, strains to fracture ranging from 9-14% were observed.

The inhibiting effect of grain size is also noted in the room
temperature stress-strain curves. Fig. 8 shows stress strain
curves for FLG and FSG specimens at room temperature. Apart from

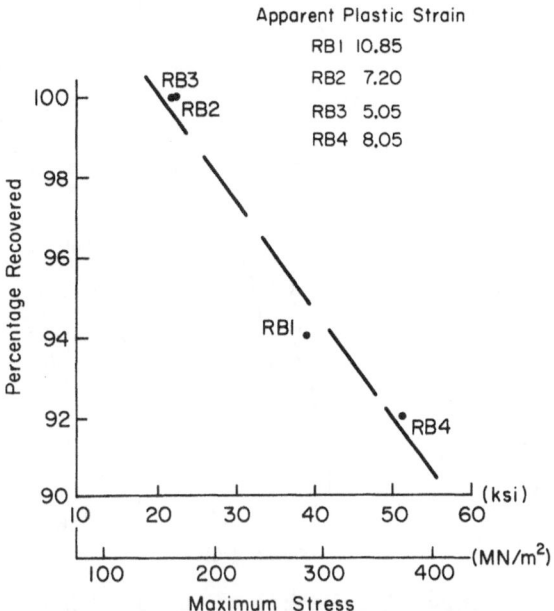

5. Relationship between shape recovery on heating vs. maximum
    applied stress below M_f for bamboo type Cu-39.8 wt%Zn
    specimens (12).

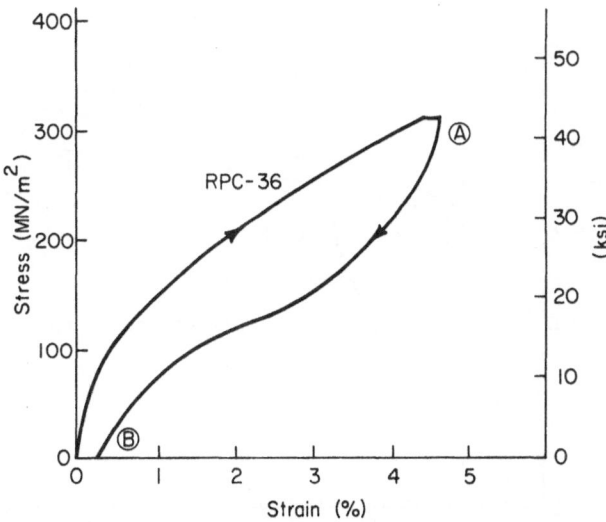

6. Stress-strain curve for a fine grained polycrystalline
    Cu-39.8 wt%Zn specimen deformed in tension below M_f and then
    maintained under stress while heating above A_f (12).

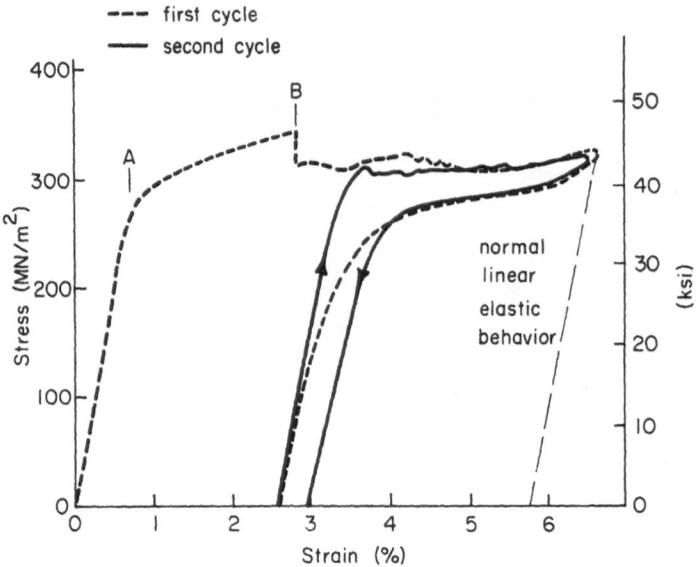

7.   Stress-strain curve for bamboo type Cu-39.8 wt%Zn specimen
     deformed above M_s (but below M_d) showing three general re-
     gions divided by points "A" and "B." The second loading cycle
     shows a reduction in plastic strain and a higher percentage
     of pseudoelastic recovery (12).

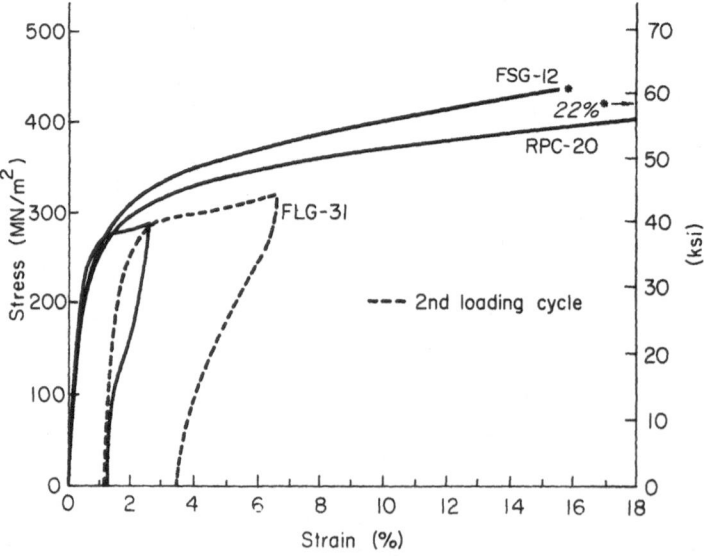

8.   Similar to Fig. 7, but for RPC, FLG and FSG (see text)
     specimens.

the higher flow stress, comparing the FLG specimen to one of the RB series, the stress-strain curve of the former shows no serration, a less distinguishable reverse yield effect upon unloading, and a substantial reduction in pseudoelastic strain due to the reversion of SIM. Fracture strains in FLG specimens were typically 10%. In further contrast, the FSG specimens (Fig. 8) showed no evidence of SIM formation and no reverse yield during unloading; in the case shown fracture occurred after 16% strain. With the finest grain size employed (RPC series) the stress-strain characteristics (Fig. 8) were similar to those of the FSG specimens, but strains to fracture were in excess of 20%. Incidentally, the strains to fracture for specimens tested at room temperature (above M_s) were all greater than for corresponding specimens tested below M_f (77°K).

The preceding results would appear to indicate that decreasing grain size suppresses serrated yielding, and hence the formation of SIM. This is further evidenced by the absence of the reverse yield effect in the fine grained specimens. However, metallographic examination showed that SIM was present in all specimens deformed at room temperature, and in fact the development of SIM during stressing was observed by means of a telemicroscope (12). In the RB specimens, the formation and reversion of SIM under a stress increase and decrease was followed, but in the finer grained specimens the SIM remained to a consdierably larger extent during unloading. The absence of serrations in the stress-strain curves of such specimens simply means that an averaging out occurs due to the increased polycrystallinity, but more importantly, the increased grain boundary area promotes irreversible deformation which acts to "pin" the SIM formed during deformation. This will be considered later.

In the large grained specimens, telescopic microscope observations show the formation of SIM and its subsequent disappearance to some extent upon release of the tensile load. Additional amounts of SIM could be made to disappear upon placing the specimen in compression.

The previous observations thus suggest certain parallels in behavior considering the formation of SIM above M_s and the deformation of athermal martensite below M_f. In fine grained specimens, localized irreversible deformation at grain boundaries inhibits the recovery of strain (by heating) of athermal martensite deformed below M_f and stress-induced martensite (by unloading) induced above M_s. In specimens loaded, for example, in tension, martensite strained below M_f will recover below M_f by application of a compressive stress; non-reversed stress-induced martensite formed above M_s will reverse when a compressive stress is applied.

C. Metallographic Analysis of Martensite
Deformation and SIM Formation

The previous results give insight to the macroscopic stress-
strain behavior of polycrystalline CuZn under a variety of conditions, but provide little information on the microscale. Thus, a
parallel study was carried out (13) using a specially constructed
tensile apparatus for use with an optical metallograph. This apparatus could be used over a temperature range of -185 to 400°C and
at strains up to 25%; heating cooling and strain rates over a substantial range of flexibility could be employed, and details are
provided elsewhere (13). Polycrystalline Cu-Zn alloys containing
38.5 to 40.0 wt%Zn of various grain sizes were studied, and supplemental studies on single crystal Cu-37.90wt%Zn-0.76wt%Sn and polycrystal Cu-Zn-Si specimens were also carried out. For the most part,
results are presented here for a Cu-38.9wt%Zn alloy where the M_s
(-50°C) and A_f (-40°C) temperatures are high enough to avoid problems associated with moisture condensation during cooling. In addition to results obtained from stressing specimens other results
from heating and cooling applicable to thermoelastic martensite
under no-stress conditions are given.

The thermoelastic martensite in Cu-38.9%Zn is typified by the
formation of clusters of plates. Within each cluster, a group of
four variants will form each of which is a variant of $\{2, 11, 12\}$ β
distributed about a common $\{110\}$β' pole. As many as six different
plate groups (i.e., all 24 habit planes) have been observed in
some grains. The groups per se play an important role in deformation as is described later. At M_s, all four variants in a cluster
can nucleate simultaneously. However, in a nearby region in the
same grain, only one variant may appear at first, which thickens on
further cooling. This latter growth mode is generally limited because a number of parallel variants usually join up (coalesce).
Fig. 9 is a sequence taken during cooling (spaced 2°C apart) which
shows that large plates appear to evolve from the coalescence of
parallel variants. Further, the plates which grew by joining up on
cooling transform back on heating by the formation of parallel segments, i.e., the inverse of the cooling process (Fig. 9f). This
coalescence of plates to form larger ones appears to be a rather
common characteristic of thermoelastic martensite. In general the
formation and reversal sequences were virtually identical in subsequent cycles provided the cooling and heating rates were constant.

D. Deformation of Martensite Below M_f

Fig. 10 shows a sequence displaying the deformation behavior of
a polycrystalline specimen deformed below M_f to 4.5% strain. The
regions designated A, B and C are entirely within an original
parent grain. Each of these consists of a self accommodating group

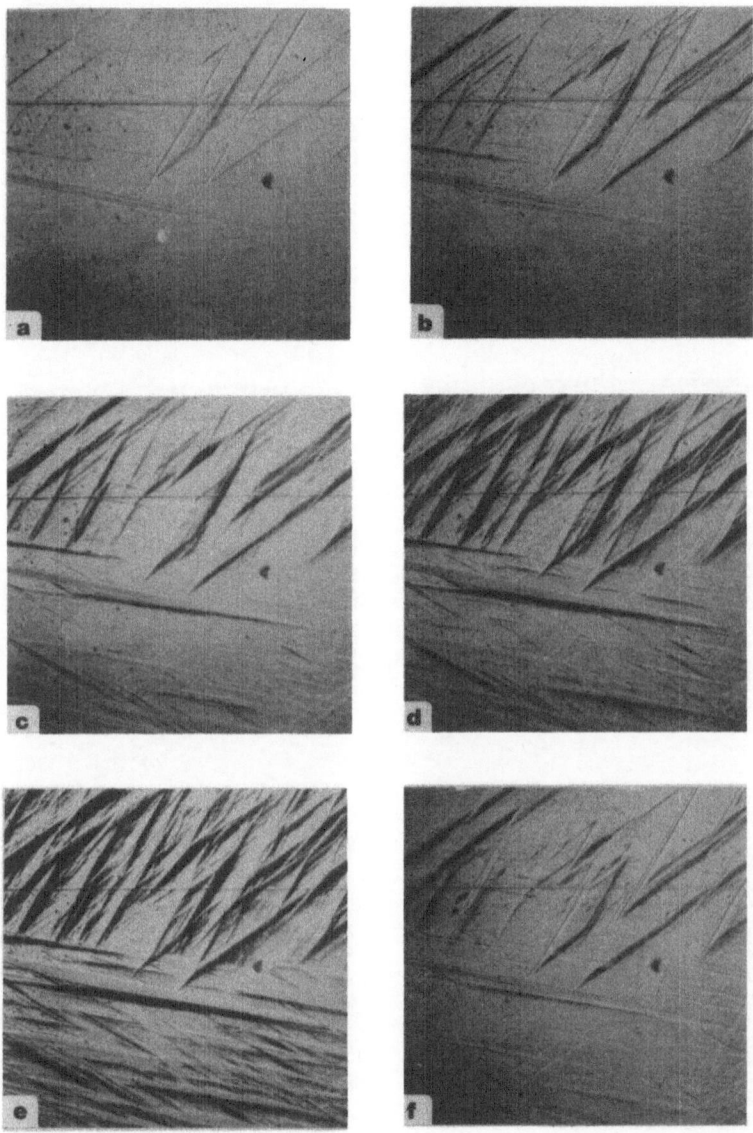

9. Surface relief micrographs (50X) obtained during cooling a
   Cu-39.8 wt%Zn alloy: (a) -52°C (2°C below M_s); (b) -54°C;
   (c) -56°C; (d) -58°C; and (e) -60°C. Note the coalescence of
   parallel plates to form thicker ones. (f) was taken at -44°C
   during heating (reverse transformation) and shows that the
   arrangement of plates is similar on heating and that thicker
   plates undergo segmentation into parallel thin plates (13).

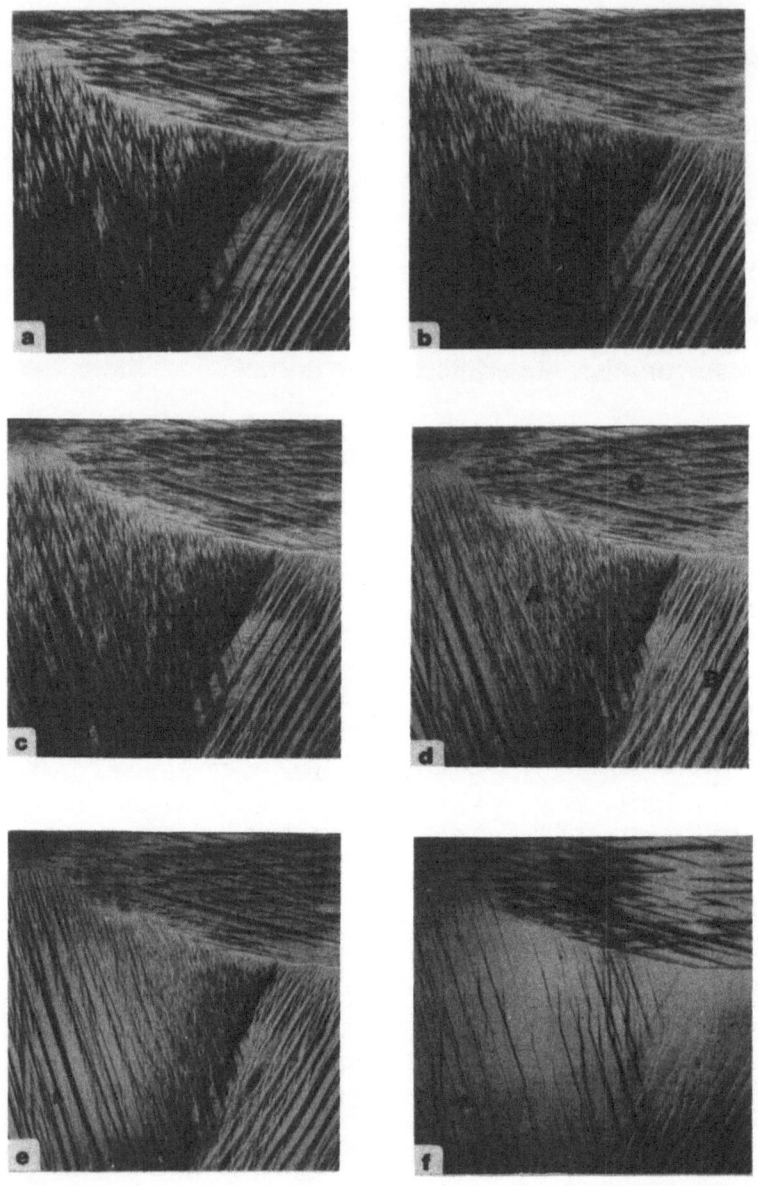

10. Surface relief micrographs (50X) from Cu-39.8 wt%Zn alloy
     showing a region within a grain containing three plate groups,
     A, B and C. The specimen was deformed in tension below M_f.
     (a) 0%; (b) 2%; (c) 3%; (d) 4%; and (e) 4.5% strain. (f) shows
     the same region at -35°C during the reverse transformation
     after removal of the stress below M_f. Note that the last
     plates to reverse are those which grew preferentially during
     straining (13).

of four variants. The variant in group A which is closest to 45° with the tensile axis grows with respect to the other variants with a less favorable Schmid factor. Similar behavior is noted for group B whereas virtually no change is seen in group C. In general, the variants observed to undergo preferential growth under stress were the last to disappear during the reverse transformation (Fig. 10f). For specimens deformed as above, i.e., 4-5% strain, all plates disappeared on heating to ~10°C above A_f, and no permanent damage in the β'matrix is observed. Different results are obtained at higher strains, as will be discussed later.

Another deformation mode observed below M_f consists of the formation of small deformation bands within a martensite plate. Another sequence shown in Fig. 11 (see arrow) indicates this behavior. As before, all deformed martensite reversed completely when heated slightly above the A_f temperature.

For strains in excess of those shown in Figs. 10 and 11, the variants which previously grew under stress continue to grow until a smooth surface typical of a single tilt (i.e., single crystal) is approached, Fig. 12 (the same region as shown in Fig. 9). With further deformation, the appearance of parallel deformation bands occurs, and further strain results in the widening and coalescence of these bands to result in one large band resembling a kink in deformed single crystals. In contrast to the yellow appearance of the deformed martensite, these bands are pink in color, indicative of a new phase. And whereas the adjacent deformed martensite reverts on heating to A_f or slightly above, the pink bands are stable, reversing only slightly on heating as high as 200°C (13).

In one case, a specimen was strained to 4.5% below M_f and heated under that stress. Complete reversion under stress did not occur until the specimen was heated to some 90°C above A_f.

Another more macroscopic type of deformation behavior was found in specimens deformed below M_f. Entire plate groups (each consisting of four variants) were observed to expand by consuming neighboring groups, (and thus, other groups would contract). This is shown in Fig. 13. In some cases the boundaries between groups were highly immobile. These boundaries served to concentrate the applied stress and resulted in localized irrecoverable deformation.

E. Formation of Martensite Under Stress Above M_s

Experiments were carried out at 10-15°C above the M_s temperature to observe the behavior of SIM under stress. Fig. 14 shows the formation of SIM in the same area shown previously in Fig. 10. Note that in region A the SIM variant which is preferentially formed is parallel to the athermal martensite variant which grew

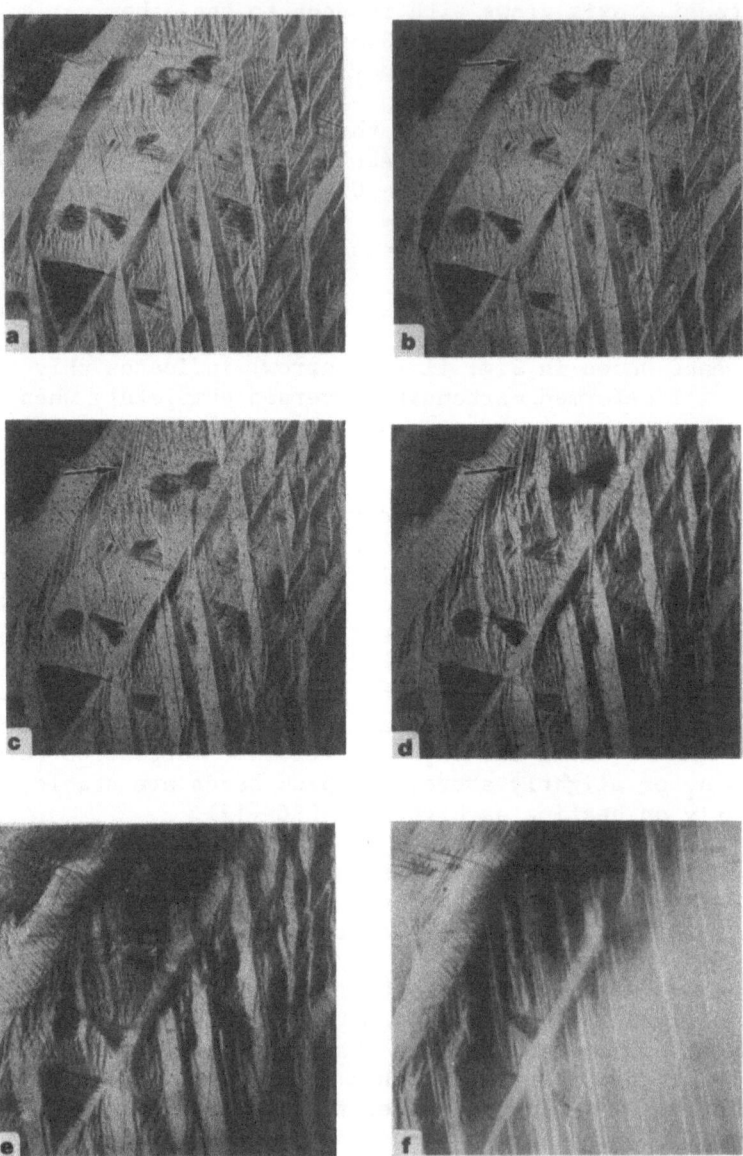

11. Deformation sequence below M_f illustrating the formation of small deformation bands (see arrow) within a martensite plate. (a) 0%; (b) 2%; (c) 3%; (d) 3.5%; and (e) 4% strain. (f) was taken at -35°C during the reverse transformation under no stress. Magnification 50X (13).

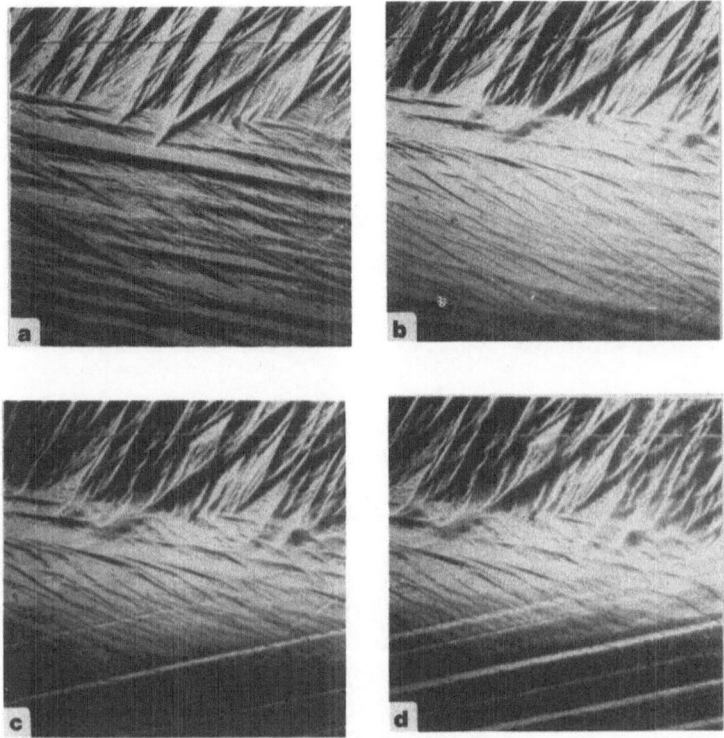

12.   Surface relief micrographs of same region shown in Figs. 9
      showing the progressive development of a near single crystal
      region and the formation of parallel bands within this region
      as the strain is increased below M_f. See text for discussion
      (13).

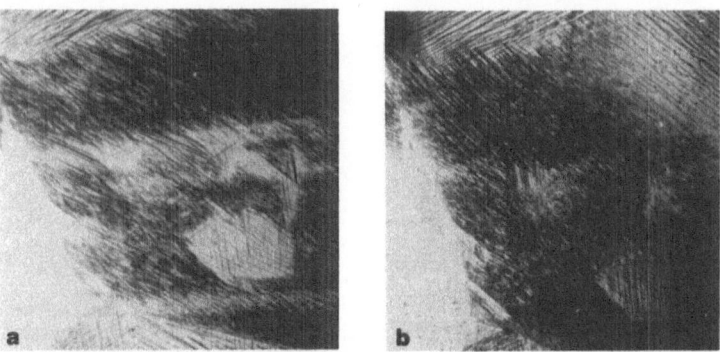

13.   Surface relief micrographs (50X) for Cu-39.8 wt%Zn alloy
      obtained below M_f for (a) 0% and (b) 6% strain showing the
      change in relative size of plate group clusters as a con-
      sequence of deformation (13).

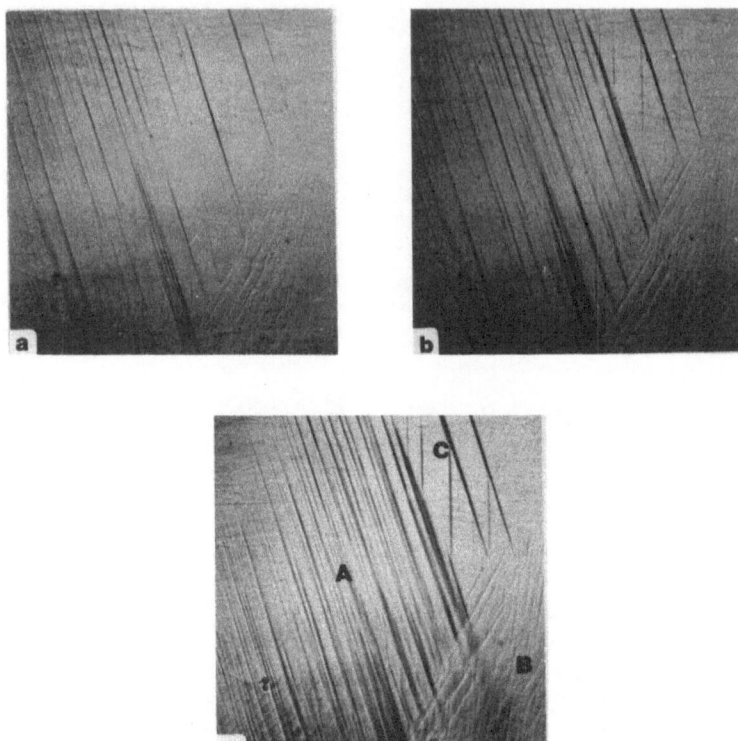

14. Surface relief micrographs showing the formation of SIM at -35°C in the same region as shown in Fig. 10. (a) 3%; (b) 3.5%; and (c) 4% strain. Regions designated A, B and C are the same as before (13).

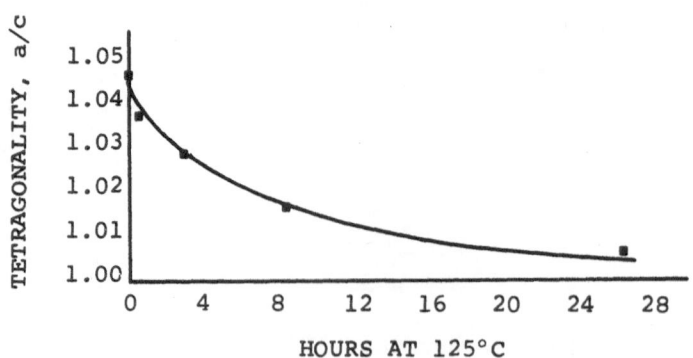

15. Effect of tempering time at 125°C on the tetragonality of martensite formed by cooling a Cu-39.1 %Zn alloy to -196°C (19).

preferentially under stress below M_f. The parallel SIM plates
widen and coalesce under increasing stress, and those which formed
first were the last to disappear when the stress was removed.
Higher stresses yet resulted in the formation of other SIM variants
which intersected the first formed, and the intersection boundaries
so formed tend to inhibit the reversion process when the stress is
released. Provided a critical stress was not reached, i.e., no
permanent deformation where the variants intersected, the multi-
variant SIM morphology could be reversed by heating.

As with the athermal martensite, continued stressing resulted
in the coalescence of parallel variants to form an apparent single
crystal region, and further stress results in permanent deformation
in the form of pink colored bands which subsequently coalesce and
form "kinks." This type of deformation was enhanced by a finer
grain size and occurred at a lower stress level. The kinks are
parallel to the plates initially formed under stress.

The above discussion serves to illustrate a number of similarities comparing the formation of stress-induced martensite to that
of the deformed athermal martensite. However, some basic differences were also observed, which are worthy of brief mention. The
stress-induced martensite plates formed above M_s do not grow on
cooling further below M_s and do not act as preferential sites for
the formation of athermal martensite. Actually the formation of
athermal martensite is uninfluenced by the prior presence of SIM,
and moreover the athermal martensite actually appears to grow
through the smaller SIM plates. (The same effect was observed
during the formation of athermal martensite in specimens containing
needle-like martensite which formed during the quench to room temperature.) On the other hand the athermal martensite formed under
no stress will not grow and parallel variants will not form under
stress. However, preferred variants of SIM will form amidst the
athermal martensite and in some cases appear to grow through the
athermal plates when the stress is increased.

A final demonstration of the difference between SIM and athermal martensite is provided by the following experiment. A specimen
was mildly deformed below M_f and warmed up under the same stress
until all martensite disappeared. The specimen was then cooled,
still under the same applied stress. Following this treatment, the
initial deformed athermal martensite (as in Fig. 10) is replaced by
plates of SIM, between which plates of athermal martensite formed.

F. Further Observations on Martensites in Cu-Zn Alloys

Although the athermal martensite in CuZn alloys has been termed
"burst" martensite (14), it has been observed that this product
forms uniformly between M_s and M_f and can be halted at any

temperature. Moreover, measurements of electrical resistivity between these temperature limits do not reveal any abrupt changes
over a small temperature interval, as would be expected for a burst
effect. The terminology athermal or thermoelastic is therefore
preferred. Careful analysis of the process of plate formation on
cooling showed that the growth of a particular variant occurs
mainly by the joining up of parallel variants, and that the large
plates decompose into the same parallel variants during the reverse
transformation. The apparent coalescence process during cooling
indicates that the highly mobile martensite-matrix boundaries are
coherent and can move without creating defects in both the parent
and product phases. It would appear (at least on the optical
microscope scale) that the adjoining boundaries annihilate each
other, like dislocations of opposite sign, leaving no residual defects. However this does not explain the segmentation of the larger
plates during the reverse transformation.

Two sources of irrecoverable deformation were observed. Pink
deformation bands, indicative of a martensite-to-martensite transformation, occur at high strain levels both during the formation of
martensite under stress and during the deformation of athermal martensite below M_f. The second type of irreversible deformation
occurs at higher temperatures where the flow stress of the matrix is
exceeded before SIM occurs. This type of deformation is enhanced
near grain boundaries and triple points and thus is more predominant
in fine grained specimens. (However, as discussed earlier, permanent deformation at grain boundaries also occurs below M_f at relatively high strains.) The recoverable deformation due to the reversal of SIM was observed to be markedly lower in fine grained
specimens. At lower strains, the less negative effect of grain
boundaries on the shape recovery of deformed athermal martensite
can be attributed to the mutual interaction of the plate group
clusters, i.e., strain can also be accommodated by the growth and
shrinkage of such domains at the expense of each other. These domains are not present in the case of SIM in which case one variant
is initially present which can extend from one grain boundary to the
other, resulting in grain boundary deformation as the applied stress
is increased.

Although the experiments described above concerning both the
macroscopic and microscopic aspects of athermal martensite deformation and stress-induced martensite formation raise a number of
questions, at the same time they are highly valuable in pointing out
the complexity and variety of the related reversible and irreversible
deformation processes. At least two additional processes have been
verified. In the reversible category it is clear that plates of
certain variants can grow at the expense of those having different
variants. Further, entire clusters of plates consisting of four
variants per cluster can mutually interact by growing and shrinking
under stress. In the irreversible classification, grain boundaries

and certain boundaries between plate group clusters serve to concentrate stress and produce localized plastic deformation. At
comparatively high strain levels, a new phase, apparently a stress-
induced martensite from an existing martensitic structure (either
athermal or stress induced) results which does not readily revert
on heating to temperatures substantially above A_f.

G.   Comments on the Shape Recovery in Cu-Zn and Cu-Zn-Si Alloys

A number of Cu-Zn and Cu-Zn-Si alloys were examined for shape
memory behavior (13). In all cases, specimens were approximately
1 X 1 X 90 mm in size, freshly quenched from the β region (840-850°C),
and deformed by bending below M_f or in liquid nitrogen into a "U"
shape. Following bending, the specimens were heated to effect the
shape recovery.

Cu-Zn alloys containing 38.5, 39.0, 39.5 and 39.8 wt%Zn were
treated as above, and in all cases, the shape recovery on heating to
room temperature was complete. These specimens all had M_s temperatures below room temperature.

Three Cu-Zn-Si alloys containing Cu-34.46Zn-0.97Si, Cu-34.44Zn-
0.93Si and Cu-34.42Zn-0.89Si (all in wt %) were investigated. The
first of these alloys had an M_s temperature at -10°C, the second
contained 40-60% martensite after quenching to room temperature,
and the third was completely martensitic after the room temperature
quench. As mentioned, all specimens were bent into a 180° "U"
shape below M_f. After heating to 30°C, the first alloy completely
regained its shape and the second partially recovered to an included
angle of 120°; the third alloy only recovered to an included angle
of 60° even after heating to 200°C. The latter two alloys exhibiting an imperfect shape recovery transformed only partially on
heating, even in the undeformed condition, and moreover, an incomplete shape recovery was found for the first alloy after two or
three cooling-bending-heating cycles. In view of the rather imperfect memory exhibited by the third alloy, further experiments
showed that a 90% shape recovery could be obtained, but only after
limiting the initial deformation at 77°K to 2%.

In view of the behavior mentioned above, some comments are now
given concerning the crystal structure of the martensite in the
various alloys. In the Cu-Zn alloys, it has been shown that the
martensite in Cu-38.5%Zn is internally twinned, and the structure is
probably a distorted version of the CuAu I structure derived from
the CsCl matrix by the Bain strain (15). This conclusion was obtained from transmission electron microscopy and diffraction. However, the situation is not entirely clear for this alloy because
subsequent analysis using slab type bulk specimens and a low temperature diffractormeter indicate the possibility of several different

martensites with different crystal structures (16). Further experimental work is in progress to resolve this difficulty. On the
other hand, similar diffractometer experiments using bulk type slab
specimens have been carried out for the Cu-39.0, 39.5 and 39.8%Zn
alloys (16), and in these cases the martensite appear to be untwinned, and possesses a "modified" 9R structure, i.e., a long
period stacking structure modified by shuffles. Since all of the
above mentioned Cu-Zn alloys exhibit complete shape recovery under
the conditions described, it may be that internally twinned and internally faulted martensites (and possibly mixtures of the two) are
capable of shape memory behavior.

On the other hand, the situation becomes less clear when the
Cu-Zn-Si alloys are considered (13) because the Cu-34.44Zn-0.93Si
and Cu-34.42Zn-0.89Si alloys with highly imperfect shape memories
appear to have essentially the untwinned 9R structure (and possibly
as well for the Cu-34.46Zn-0.97Si alloy which exhibited a complete
shape recovery) (17). In light of the above, it would appear that
some internally faulted martensites do not produce complete shape
memory behavior, while others (also internally faulted) do exhibit
a complete shape memory.

Although there are no known cases where internally twinned,
thermoelastic martensites do not exhibit a complete shape memory,
the question arises as to why some of the internally faulted martensites do and some do not. The answer certainly awaits further
work, but some indications point to a possible explanation. Besides
the examples noted, it has also been shown that internally faulted
martensites in the Cu-Al system exhibit an imperfect shape memory
(18). For the specific case of Cu-12 at%Al with an M_s temperature
~350°C, deformation below M_f at 20°C resulted in a consequent
shape recovery of ~50% after heating to above 500°C. It may be
significant that the internally faulted Cu-Al and Cu-Zn-Si alloys
which showed an incomplete shape recovery are characterized by M_s
temperatures somewhat above room temperature. Thus, on heating
through the reverse transformation interval (to effect the shape
recovery) it is possible that the martensite would undergo some
structural modification or "tempering" which would inhibit its
thermoelastic reversal. Some evidence for this is now discussed.

Some years ago, Kaminski (19) reported on an extensive study of
the martensite transformation in Cu-Zn alloys. He investigated
"impure" alloys made from electrolytic copper with zinc contents
("technical" Zinc) ranging from 37.3 to 38.4; typical impurity levels
were 0.5% Pb, 0.2% Fe and 0.5% Sn. He concluded that the martensite
transformation was basically a b.c.c. to f.c.c. transformation and
that the as-formed martensite was tetragonally distorted in the
manner of CuAu I. However, upon "tempering" the f.c.t. martensite
changed into the cubic α-lattice. For example, considering a
39.1%Zn alloy, in the as-quenched condition the axial ratio, a/c,

was 1.045; after tempering at 125°C for 0.5 hr, the ratio dropped
to 1.035, and after 3 and 26 hours at 125°C, the axial ratio decreased to 1.028 and 1.012 respectively. Fig. 15, from Kaminski's
paper, shows this effect quite clearly. It is implied from
Kaminski's results that the initially formed martensite, obtained
by cooling in liquid air, does not revert at the tempering temperature, and therefore the changes noted are intrinsic to the martensite. It such is the case, it is then expected that changes in the
martensite occur during heating which may indeed alter the characteristics of the reverse transformation, and thus restrict the
degree of the shape recovery.

It would appear that additional work dealing with aging or
tempering effects in relation to the shape memory process might
provide some valuable insights.

IV. DISCUSSION

A. Shape Memory and Related Mechanisms

In Au-Cd and In-Tl alloys there is a "rubberlike" memory associated with the martensitic state. Under an applied stress, twins
of orientation A grow at the expense of twins of orientation B by
the movement of a twin boundary normal to itself (20). A shape recovery occurs as the stress is released which is due to the "backwards" motion of the original twin boundaries which restores the
original twinned configuration. This effect is not typical of
most alloys which exhibit the SME because it is associated with the
martensitic state, and not the reverse transformation. However,
these same alloys also exhibit the SME effect provided that aging
effects below M_f do not occur (1).

In the general SME case, there are numerous plate variants of
martensite present below M_f. Under "pseudoplastic" deformation some
variants grow at the expense of others which shrink, at least in
later stages of the recoverable deformation process. Eventually, a
single orientation, or nearly so results, which reverts to give the
shape memory on heating, i.e., the modified martensite reproduces
the original orientation in the parent phase it came from. In fact,
it should be emphasized that in general only martensites which have
been modified by deformation are capable of memory behavior. We do
not expect the "consumed" variants to become restored because this
would be a martensite phenomenon. In effect, by deforming a multi-
variant specimen to result in a single variant, a situation is produced like that in a single interface transformation (21) which
"unkinks" or reproduces the original transformation during the reverse transformation. In the case of no strain below M_f, even
though the reverse transformation paths are followed, there is no

bending of the specimen as is likewise the case when random
variants of martensite form from the parent during cooling under
no stress. For small deformations short of those producing the
single crystal morphology, only those plates which grow under stress
will result in a volume fraction or preferred orientation which will
produce the shape memory. It is expected that "unconsumed" volumes
will produce no memory because of their random orientation resulting
from the initial transformation. Increased deformation results in a
near single crystal state (13) with a high volume fraction of martensite in a new orientation (i.e., the consumed plates) and a substantial shape recovery results when the fraction in new orientation
transforms back to the parent phase. Of course if a critical strain
is exceeded irreversible defects such as slip dislocations and the
formation of a new martensitic phase will inhibit the reversal
process and result in an incomplete shape memory.

Both "rubberlike" and shape memory effects are possible for the
case of stress-induced martensite. A pseudoelastic effect results
when stress-induced martensite plates spring back to their original
parent phase orientation when an applied stress is released. This
involves a reverse transformation and is not a phenomenon peculiar
to martensite as is the rubberlike effect in Au-Cd and In-Tl alloys.
For SIM, if random variants were introduced by the applied stress
one would expect no specimen bending since the situation would be
like that of forming random variants athermally on cooling. However if preferred variants of SIM are formed under stress the degree of preferred orientation would cause a specimen bending which
would recover when the applied stress is released. As discussed
earlier (13), preferred variants of SIM do indeed form under stress,
and provided the stress is sufficient the morphology resembles the
single crystal morphology obtained when athermal martensite is deformed below M_f. Thus the same type of reversion process and shape
recovery to the original austenite orientation should occur, only
in this instance driven by a decrease in stress rather than an increase in temperature. Thus SIM and SME are closely related and
depend on the reversion of martensite by the application of an external variable: stress (decrease) or temperature (increase). It
should be mentioned however that although the athermal and stress
induced martensites are both thermoelastic, they do not necessarily
have the same crystal structure.

The above mentioned processes all share the common characteristic that glissile boundaries can be displaced under stress. For
the rubberlike effect (twin boundary displacement) and SIM (mar-
tensite-parent boundary advancement), the situation is relatively
straightforward. It is less clear, in the case of the deformation of athermal martensite below M_f, how the apparently coherent
interplate boundaries move so that one variant grows at the expense
of the other. This is even true when parallel variants merge to
produce an apparent single crystal at the scale seen by the

optical microscope. There are certain difficulties here if the
internal twins or faults meet not "in phase" and this problem will
be treated in a later paper (22).

B. Reverse Transformation of Deformed Martensite

It has been emphasized that most memory alloys are ordered and
thermoelastic. The general matter of ordering and the thermoelastic
state needs considered. Work on Fe-Pt alloys shows that thermoelastic behavior results as a consequence of ordering, so it may
not be trivial to separate these two phenomena. It has been argued
(5) that ordering enhances thermoelastic behavior because it increases the stress required to deform the matrix (and martensite)
in an irreversible manner. Not only is there "order hardening,"
but also interphase boundaries take on an even more special significance when the martensite transformation occurs between two
ordered phases. In this case the habit plane must satisfy the
ordering requirements of both phases. It is thus a highly regular
boundary between two superlattices and as such becomes a rather in-
destructible "super" boundary (5).

As has been argued before, ordering places severe restrictions
on the reverse transformation if it is to occur martensitically
(1,23). Whereas a number of variants of the lattice correspondence
can produce the forward transformation, the martensite so produced
is usually a phase of lower symmetry which thus restricts the reverse correpondence (i.e., inverse Bain deformation). Of course,
when the transformation is thermoelastic, existing plates of martensite can simply shrink on heating, and the backwards movement of
the same interface automatically restores the initial condition.
On the other hand, the case can arise where, as a result of a single
interface transformation or deformation of an aggregate of martensite plates, a single crystal of martensite (apart from the internal
twins) is formed. In this case, there is no backwards movement of
an existing boundary, and the parent phase must be nucleated. However, even here because of the ordering and symmetry change, the
reverse correspondence is fixed, which in turn will reproduce the
same habit plane and parent orientation during the reverse transformation.

Finally, a few comments are made regarding the shape memory
effect and martensite crystal structure, i.e., internal twins vs.
faults. During early assessments of the SME (1,23), where sufficient information was available it seemed apparent that memory martensites were internally twinned. To examine this point in further
detail, the author retrieved some earlier data on Cu-Zn alloys and
came to the same conclusion. Cu-Zn alloys containing 38.5 - 40.0
wt%Zn showed a shape memory. Since it was well established at the
time that the change in shape (rubberlike effect) in Au-Cd alloys

was due to twin boundary movement (the boundaries moving normal to
themselves under stress (20)), a similar situation was envisaged to
apply to Cu-Zn alloys, and contrary to some earlier reports, it was
suggested that the memory behavior is indirect evidence for internal
twinning in the Cu-Zn martensite. It is now believed, as described
earlier, that some Cu-Zn alloys which exhibit the SME are untwinned.
While unlike the case with twins faults are not expected to move
normal to themselves, so it would seem that the transformation
partial dislocations by gliding in their own plane can produce a
deformation recoverable during the reverse transformation. Such a
situation may well apply during the _early_ stages of deformation of
memory martensites. For example, as Fig. 10 shows, there is no
obvious change in microstructure at low strain levels, even though
the specimen is recoverably deformed. At later stages, as has been
discussed, select variants grow at the expense of others, etc. It
is thus envisioned that internally twinned martensite plates can
undergo deformation by changing the twin ratio in addition to
thickening (or shrinking) whereas internally faulted martensites
when undergoing deformation result in the movement of the martensite interface per se.

ACKNOWLEDGEMENTS

    Various aspects of work reported here have been supported by
the Army Research Office (Durham), the Atomic Energy Commission, and
the National Science Foundation. Particular thanks are due
Dr. I. Cornelis and Mr. T. A. Schroeder for allowing me to present
results from our yet unpublished work. I am also indebted to
Professor K. Shimizu for cooperative work on the crystal structure
of Cu-Zn martensites (to be published) and to Professor K. Otsuka
for many helpful discussions.

REFERENCES

1. C. M. Wayman and K. Shimizu, Metal Sci. J., 6, 175 (1972).
2. L. Delaey, R. V. Krishnan, H. Tas and H. Warlimont, J. Mat. Sci., 9, 1521, 1536, 1545 (1974).
3. H. C. Tong and C. M. Wayman, Acta Met., 22, 887 (1974).
4. L. Kaufman and M. Cohen, J. Metals, 8, 1393 (1956); Prog. Met. Phys., 7, 165 (1958).
5. D. P. Dunne and C. M. Wayman, Met. Trans., 4, 137, 147 (1973).
6. H. C. Tong and C. M. Wayman, Acta Met., in press.
7. H. C. Tong and C. M. Wayman, Met. Trans., 6A, 29 (1975).
8. P. Hernandez, Lawrence Radiation Laboratory, U. of Cal., Priv. Comm., (1974).
9. M. Ahlers, Scripta Met., 9, 71 (1975).
10. C. M. Wayman and H. C. Tong, to be published.
11. C. M. Wayman, Scripta Met., 5, 489 (1971).
12. T. A. Schroeder, I. Cornelis and C. M. Wayman, to be published.
13. I. Cornelis and C. M. Wayman, to be published.
14. H. Pops and T. B. Massalski, Trans. AIME, 230, 1662 (1964).
15. I. Cornelis and C. M. Wayman, Acta Met., 22, 291, 301 (1974).
16. I. Cornelis, K. Otsuka, K. Shimizu, T. Tadaki, and C. M. Wayman, unpublished work.
17. I. Cornelis and C. M. Wayman, Mat. Res. Bull., 9, 1057 (1974).
18. K. Otsuka, Ph.D. Thesis, Univ. of Tokyo (1972).
19. E. Kaminskij, Tech. Phys. of USSR, V, 953 (1938); E. Kaminskij and G. V. Kurdjumov, Metallwirtschaft, 15, 905 (1939).
20. H. K. Birnbaum and T. A. Read, Trans. AIME, 218, 662 (1960).
21. L. C. Chang and T. A. Read, Trans. AIME, 189, 47 (1951).
22. K. Otsuka and C. M. Wayman, to be published.
23. K. Otsuka, Jap. J. Appl. Phys., 10, 571 (1971).
24. see for example, H. Warlimont, Iron and Steel Inst., Spl. Rep. 93, 58 (1965).

THE MECHANICAL PROPERTIES OF SME ALLOYS

C. Rodriguez* and L.C. Brown

Department of Metallurgy
University of B. C.
Vancouver, B. C., V6T 1W5, Canada

ABSTRACT

The mechanical properties of SME alloys are reviewed. The conventional pseudoelastic and strain-memory effects are discussed with some recent observations being presented on the effect of strain rate and grain size. A full description is given of the stress-induced transformations in the Cu Al Ni system at various temperatures above and below Ms. In this system, two types of stress-induced martensite, β_1', and γ' may form. It is found that the transformation from the β_1' phase to γ' is responsible for an extended range of metastability of the γ' phase above Af, and that the transformation from $\gamma' \to \beta_1'$ is responsible for a high degree of pseudoelasticity below Ms. The reversible strain-memory effect is reviewed. This is the reversible dimensional change in a specimen cycled though the martensite transition temperature range. The fatigue properties of pseudoelastic metals are discussed in some detail with some recent observations in single crystals and polycrystals being presented. Finally the damping capacity of these alloys are briefly mentioned.

---

*Now at Area de Investigaciones, Comision Nacional de Energia Atomica, Av. del Libertador 8250, Buenos Aires, Argentina.

INTRODUCTION

In the past few years there have been a large number of papers published on alloy systems which show stress-induced martensite (SIM) formation. The main reason for the interest in these systems is their unusual mechanical properties, especially the pseudoelastic effect and strain-memory effect (SME). The subject of SIM formation has been reviewed recently in an excellent series of three papers by Delaey et al [1-3]. Earlier review papers have been published by Wayman and Shimizu[4] and by Tas et al[5]. In the present work emphasis will be placed on recent observations of mechanical properties, to avoid conflict with these earlier papers. Mention will be made of the crystallographic and thermodynamic aspects of the transformation only when required to explain mechanical properties.

All alloys which show pseudoelasticity and the SME undergo a reversible martensitic transformation with relatively little temperature hysteresis. These alloys are generally β-phase intermetallic compounds in the Cu, Ag, Au and Ni - base systems. In most of these alloys both the matrix and martensite structures are ordered and the martensite has an internally twinned structure[4].

The unusual mechanical properties of β phase alloys were first observed in Au Cd by Chang and Read in 1951[6] and in In Tl by Burkart and Read in 1953[7]. Interest in the field waned until the work of Buehler and associates on Ni Ti in 1968[8]. This system shows some differences compared with other β phase alloys. For example, it shows some structural instabilities prior to the martensitic transformation. However it has been recently emphasised by Wasilewski[9] that the main phenomena in Ni Ti are the same as in any other β phase alloy. In the past five years there have been a large number of investigations carried out on SME alloys. Systems which have been investigated to date (giving only the most recent references) include the following β phase alloys: Ag Cd[10-12], Ag Zn[13], Au Cd[14], Cu Al[15], Cu Al Ni[16-17], Cu Zn [18-19], ternary Cu Zn alloys [20-21], Fe Be[22], Fe Pt[23], In Tl[24], Nb Ti[25], Ni Al[26-27] and Ni Ti[9,26,28]. Limited SME behaviour has also been observed in Fe Mn[29] and Fe Ni[30-31] steels and in the pure elements Co Ti and Zr[32].

In the first part of the paper a general review will be given of the pseudoelastic and strain memory effects. This discussion will follow conventional ideas with consideration being made of the formation of one type of martensite only. Some new results for Cu Al Ni will be presented in which transitions between two martensite structures were observed, giving rise to an extended range for the SIM at high temperatures and to low temperature pseudoelasticity. Finally some secondary mechanical properties will be discussed - the reversible strain-memory effect, fatigue properties and damping capacity.

PSEUDOELASTICITY

Pseudoelasticity occurs when a SIM transformation can be produced in a range of temperatures in which the SIM becomes unstable when unloaded. In this case the transformation strain obtained during loading is recovered on unloading when the SIM reverts back to the matrix phase.

A typical single crystal tensile curve illustrating pseudo-elasticity is shown in Fig. 1. It can be seen that recoverable strains of greater than 7 pct can be obtained, compared with less than 0.5 pct for the best conventional materials. The portion up to point A represents elastic deformation of the matrix. At A, SIM starts to form and from A to B the specimen progressively transforms to martensite. A specimen strained beyond B would show elastic deformation of the martensite followed either by fracture or by plastic deformation, depending on the alloy system. On unloading there is elastic recovery of the martensite up to point C followed by a progressive reversion of the martensite to the matrix phase between C and D and finally elastic recovery of the matrix phase. It is found that the sequence of martensite reversion is the exact opposite of martensite formation, with the first martensite plate to form being the last martensite plate to disappear.

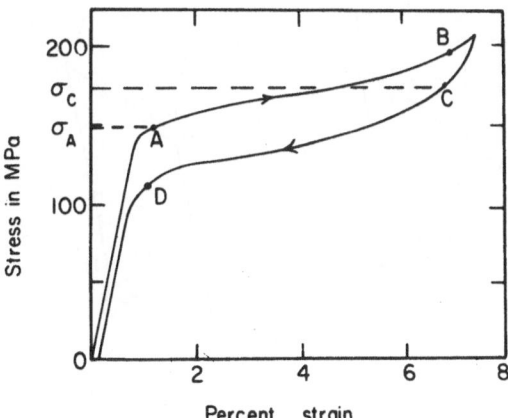

Fig. 1   Pseudoelastic stress-strain curve for a single crystal CuZnSn alloy at 24°C (76°C above Ms). The martensitic transformation occurs between A and B and the reversion to the matrix phase occurs between C and D[33].

   Most of the recoverable strain in Fig. 1 is associated with the
reversible martensite formation and is due to the macroscopic shear
strain caused by the transformation. The magnitude of the pseudoelastic strain in tensile, compressive or shear loading can be calculated from a knowledge of the habit plane of the martensite and
the magnitude of the shear strain for the transformation. It has
been found experimentally[12] that the martensite variant which forms
on stressing is the one for which the resolved shear stress of the
applied stress in the habit plane and in the shear direction is a
maximum. Since the habit plane of the martensite is irrational,
there is a choice of 24 possible planes and so there is always one
fairly closely oriented with respect to the plane of maximum shear.
Fig. 2(a) shows a single crystal of the matrix phase transformed to
martensite in the region ABCD. In the diagram shown the stress causes
an extension of the specimen. Owing to the constraints of the specimen grips, there will be transition regions with varying amounts of
martensite adjacent to the region which has completely transformed

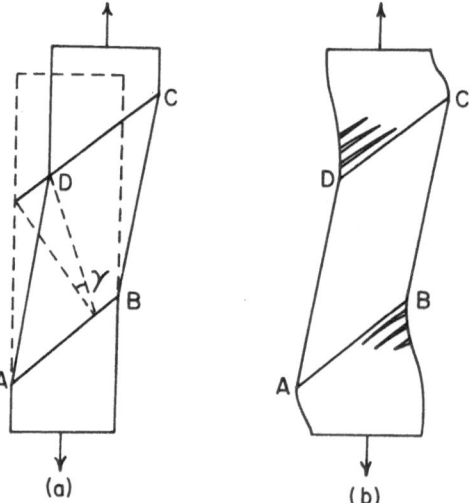

                (a)   (b)

Fig. 2   Schematic diagram showing partial transformation from the
          matrix phase to martensite. In (a) lateral displacement
          of the specimen is permitted, whilst in (b) the specimen
          is held in position by the specimen grips.

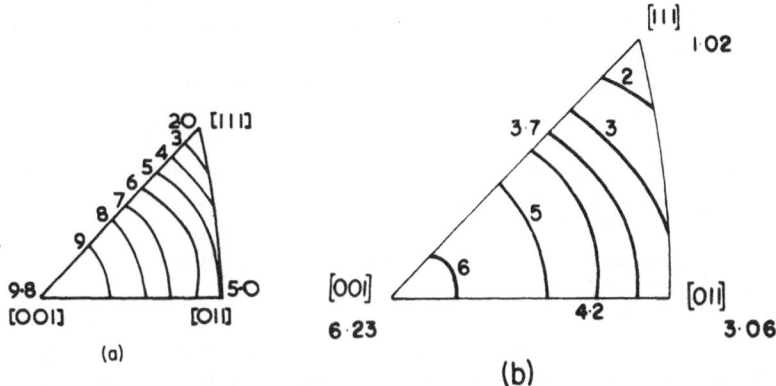

Fig. 3   Calculated values for the maximum strain associated with
the formation of SIM at different orientations of the
tensile axes in single crystals of (a) Cu Zn Sn[33] (b) Ag
Cd[12].

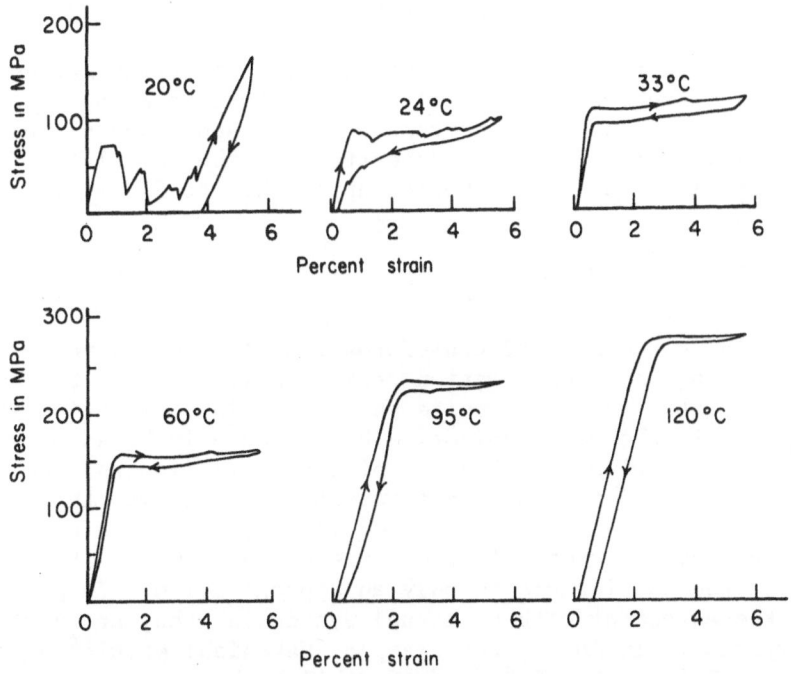

Fig. 4   Effect of temperature on the stress-strain curves in Cu Al
Ni single crystal specimens. In this alloy Mf = -10°C,
Ms = 11°C, As = 30°C, Af = 45°C[34].

(Fig. 2(b)). The shape of these regions will depend on the particular
experimental arrangement being used. Neglecting this, values of
tensile strain can be calculated for various orientations of the
tensile axis of the specimen assuming values for the macroscopic
shear calculated from the phenomenological theory. Results
are shown in Fig. 3 for Ag Cd and Cu Zn Sn. Agreement with experimentally determined strains for complete transformation to martensite
is fairly good. In Ag Cd, for example, strains of 6 pct were found
for specimens oriented close to [001], in good agreement with the
calculation in Fig. 3(b).

Fig. 4 shows the effect of temperature on the stress-strain
curves. Provided the temperature is above Af essentially complete
recovery of the strain will occur. With increasing temperature, a
progressively higher stress is required to nucleate the martensite
and the curves are displaced to higher stress levels with little
change in the magnitude of the pseudoelastic strain corresponding
to complete transformation to martensite. At sufficiently high
temperatures, plastic deformation in the matrix occurs prior to
martensite formation and some of the strain becomes non-recoverable.
This effect is greatest in the first cycle of loading and unloading.
In subsequent cycles the matrix has work hardened sufficiently to
allow martensite formation prior to plastic deformation and the
specimen shows a higher percent recovery.

Fig. 5 shows that the stress necessary to initiate martensite
(σ_A, Fig. 1) decreases linearly with temperature becoming zero at
Ms, the temperature at which martensite forms spontaneously on cooling. The variation of σ_A with temperature can be calculated from
the Clausius-Clapeyron Type equation applicable to SIM :-

$$\frac{d\sigma_A}{dT} = \Delta H / \Delta \epsilon \ T_0 \ V_m$$

where ΔH is the heat of transformation, $\Delta \epsilon$ the strain corresponding to complete transformation to martensite, T_0 the temperature
at which the matrix and martensite phases are in equilibrium at zero
stress and Vm is the molar volume. This formula is based on a balance between the volume free energy of transformation and the mechanical work involved in the transformation. It assumes that the
matrix and martensite are always in equilibrium and does not take
into account any hysteresis in the stress-strain curve. Hence for
any real system the formula is only an approximation. In practice,
however, the agreement with an actual system is good, at least for
single crystals. In Au Cd, for example, Nakanishi et al[14] have calculated $d\sigma_A/dT$ to be 0.265 Kg/mm^2 °C, whilst the measured value is
0.25 Kg/mm^2 °C.

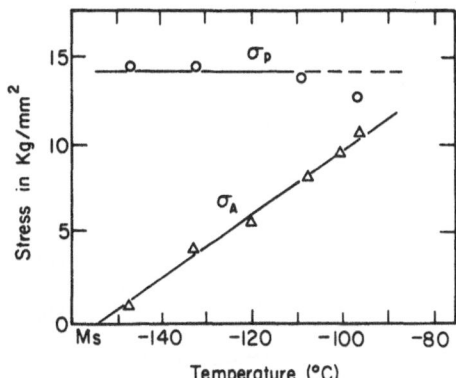

Fig. 5   Plot of stress for formation of SIM (σ_A) and stress for
          plastic deformation of the matrix phase (σ_p) as a function
          of temperature in Cu Zn single crystals[19].

Effect of Strain Rate

Some recent results[35] have shown that the shape of the pseudoelastic stress-strain curve depends on the rate of straining the
specimen. This is due to significant temperature changes which take
place in the specimen at high strain rates (Fig. 6). These temperature changes are associated with the heat of transformation (ΔH),
there being a temperature increase during the parent-martensite transformation and a temperature decrease during the reverse transformation. The maximum temperature change in the specimen is given by
$\delta T = (\Delta H)/c$ where c is the specific heat of the martensite for the
forward transformation or the specific heat of the matrix for the
reverse transformation. In Cu Al Ni this works out to ~12°C, in
good agreement with the measured temperature changes at high strain
rates. At lower strain rates the measured temperature increase is
smaller since there is more time available for heat to be conducted
away from the specimen.

Fig. 7(a) shows stress-strain curves in Cu Al Ni at strain
rates of .001 and 0.1/sec. In both cases there was a delay after
completion of loading before the specimen was unloaded in order to
allow it to return to ambient temperature. The slopes of the curves
in the transforming region are significantly higher at the high
strain rate compared with the low strain rate. This is because at

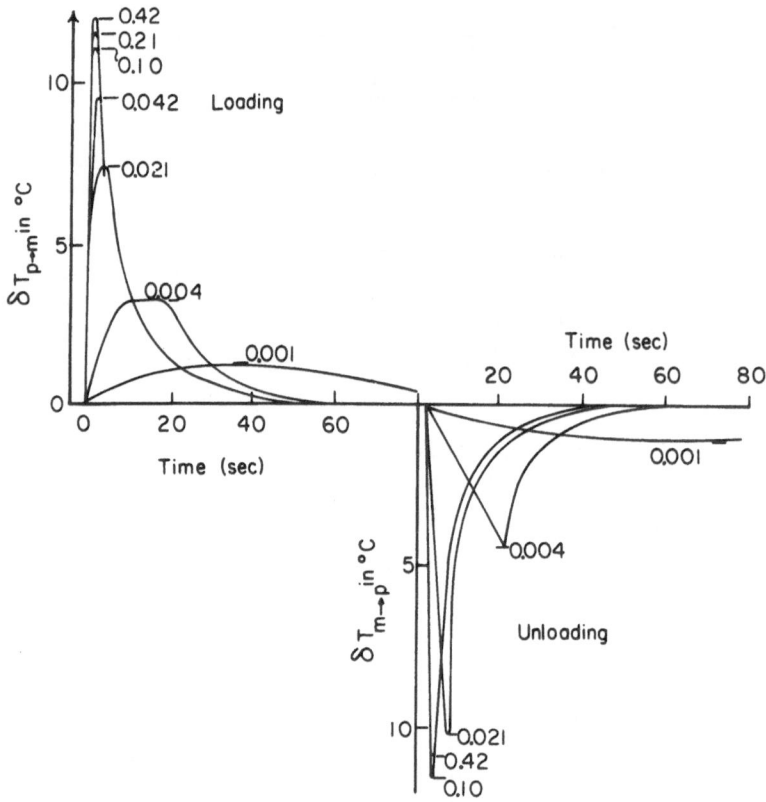

Fig. 6   Thermal effect on loading and unloading Cu Al Ni single
          crystal specimens at 24°C (23°C above Ms). The numbers on
          the plot give the strain rate in sec^{-1} [35].

the high strain rate the specimen is getting progressively warmer
as the transformation proceeds and so the stress necessary for continued growth of the martensite is progressively increasing. The
final difference in critical stress between the two strain rates
is due to the final temperature reached in each case. Fig. 6 shows
that the difference in δT for the two cross-head speeds used is
9.7°C for p → m and 10.1°C for m → p. Stress-strain curves taken
in the temperature range 20 to 70°C indicate that the critical stress
(σ_A, Fig. 1) increases by 0.19 Kg/mm^2 for a 1°C rise in temperature.
This enables the temperature changes of 9.7°C and 10.1°C to be con-
verted to stress changes, giving ($\sigma_{B'} - \sigma_B$) = 1.8 Kg/mm^2 and ($\sigma_D -
\sigma_D$) = 1.9 Kg/mm^2. It can be seen that these values are in good
agreement with those observed experimentally (~ 1.9 Kg/mm^2).

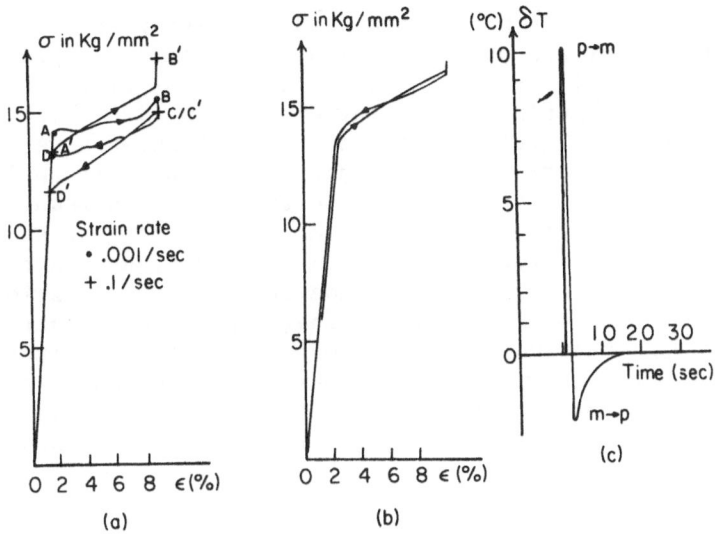

Fig. 7   Stress-strain curves for Cu Al Ni single crystals at 24°C
          (24°C above Ms) showing the effect of strain rate.   (a)
          Strain rates of 0.001 and 0.1/sec. with the specimen being
          allowed to cool down between loading and unloading.   (b)
          Strain rate of 0.1/sec with a minimum delay between loading
          and unloading.   (c) Thermal effect for curve (b)[35].

     Fig. 7(b) shows the stress-strain curve at a strain rate of 0.1/
sec with the loading and unloading being carried out immediately one
after the other, in order to minimize the conduction of heat to the
surroundings.   There is very little hysteresis in the stress-strain
curve although the slopes of the curves are quite steep in the transformation region.   Fig. 7(c) shows that the temperature changes in
the two directions cancel out one another almost completely so that
there is no significant temperature change in the specimen after
completion of the loading-unloading cycle.

                        Effect of Grain Size

     It has been established in a large number of investigations that
polycrystalline specimens have inferior pseudoelastic properties compared with single crystals.   This is to be expected since in single
crystals only one martensite variant need form whilst in polycrystals,
several variants (generally 5) must form in different regions of any
given grain to allow for the effects of constraint from adjacent
grains.   The stress concentrations due to the presence of the grain
boundaries produce non-reversible deformation martensite and prevent
full development of the pseudoelastic effect.

Dvorak and Hawbolt[20] have emphasised that the properties of the
polycrystalline alloy are not controlled specifically by grain size
but rather by grain size relative to the dimensions of the specimen.
In standard tensile specimens, the ratio of grain size/thickness
(gs/t) is a measure of grain constraint. Fig. 8 shows several stress-
strain curves for different gs/t ratios. The main characteristics
of decreasing gs/t ratio are an increase in the stress required for
nucleation of martensite, an increase in slope of the stress-strain
curve in the martensite formation region and a decrease in the amount
of recoverable strain. Fig. 9 shows the effect of gs/t ratio on the
stress for nucleation of martensite. There is a sudden increase in
this for gs/t less than unity. This corresponds to the situation
where there is more than one grain across the thickness of the specimen and so the problem of grain constraint becomes much more severe.
In specimens with gs/t less than 0.5, only very limited pseudoelastic
strains were observed.

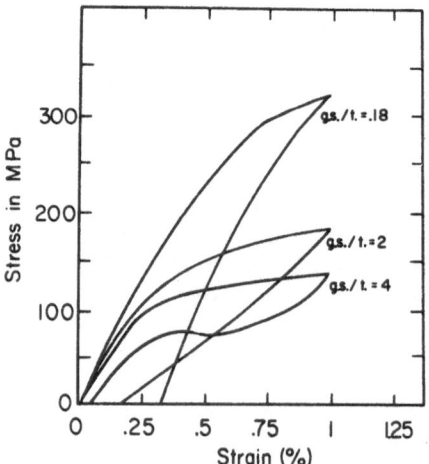

Fig. 8   Stress-strain curves in polycrystalline Cu Zn Sn alloys
          at 24°C (9°C above Af) showing the effect of grain size/
          thickness ratio[20].

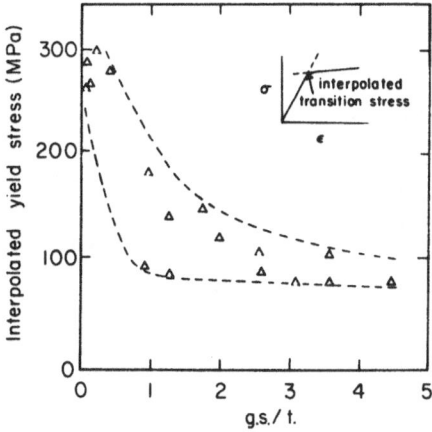

Fig. 9 Effect of grain size/thickness ratio on the stress for martensite formation in Cu Zn Sn alloys at 24°C (9°C above Af)20.

STRAIN MEMORY EFFECT

The strain-memory effect can most easily be discussed by considering a specimen having the structure of the matrix phase at a temperature close to Ms where the martensite once formed is thermodynamically stable or metastable. On stressing the matrix phase, SIM will nucleate, the particular variant or variants which form presumably being those for which the resolved shear stress in the habit plane and in the shear direction is a maximum. On release of the stress, the martensite will not revert back to the matrix phase since it is metastable and so there is little recovery in the dimensions of the specimen (Fig. 10). On heating, however, the martensite becomes unstable and between As and Af it progressively reverts back to the matrix phase, causing the specimen to recover its original dimensions. The amount of recovery which can be obtained corresponds to the strain for complete transformation of the matrix phase to martensite with the most favoured orientation. This equals the maximum strain for pseudoelastic transformation, Fig. 3, assuming the martensite structures are the same.

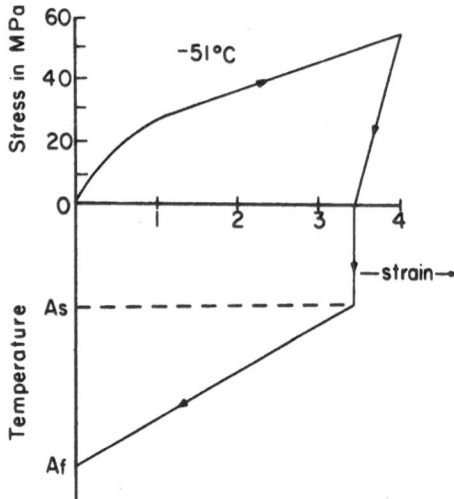

Fig. 10   Stress-strain curve for Cu Zn Sn alloy deformed at M_s
           $(-51°C)$[33], together with a schematic curve showing the
           strain-memory recovery of the alloy on heating.

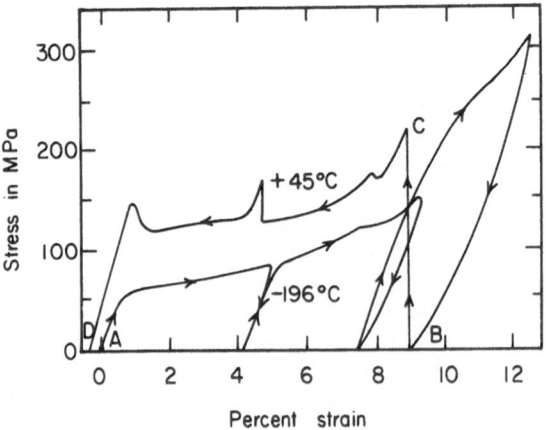

Fig. 11   Strain-memory recovery in a Cu Al Ni single crystal speci-
           men with A_s = 2°C and A_f = 30°C.   From A to B the specimen
           was cycled 3 times to 5 pct strain.   From B to C it was
           heated at constant strain to +45°C.   From C to D the speci-
           men was allowed to recover its strain[36].

    The strain memory effect at temperatures below Ms is of more
practical importance. In this case, the initial strucuture is thermal martensite. On stressing, the large number of variants associated
with the thermal martensite progressively transform to one martensite
variant with little strain recovery on release of the stress. On
heating, this variant transforms back to the matrix phase between
As and Af giving the strain recovery. Hence the process is basically
the same as the strain-memory effect at Ms except for the development of one variant from the large number of possible ones. The
details of this particular process are not fully understood. Wasilewski[9] has suggested that on stressing, the thermal martensite transforms to the parent phase which in turn transforms to SIM. However
there has been no direct evidence as to the existence of this transient matrix structure. Some experimental observations of this effect
in Cu Al Ni alloys will be presented in the next section.

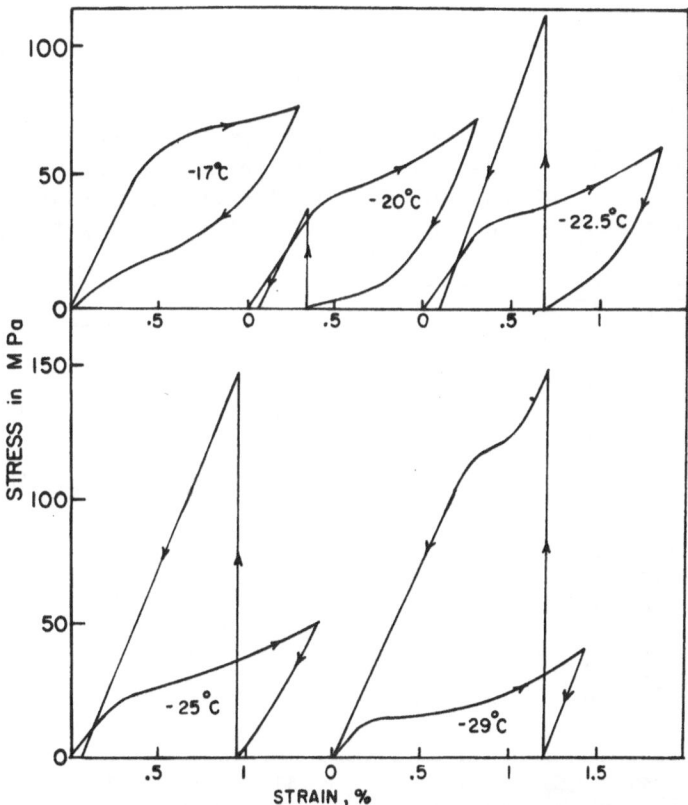

Fig. 12   Effect of temperature on the stress-strain curves in poly-
          crystalline Cu Zn Sn with As = -27°C and Af = -14°C. The
          effect of heating the specimen to room temperature at con-
          stant strain and then releasing the load is also shown[33].

The amount of recoverable strain associated with the stressing of thermal martensite is quite large as can be seen in Fig. 11. Here an alloy is cycled three times to 5 pct strain with relatively little recovery on unloading except in the last cycle. On heating above Af, 9.4 pct strain is recovered and the specimen essentially recovers its original dimensions.

The strain recovery seen in the last cycle in Fig. 11 has recently been shown to be due to a reversible transformation between two martensitic structures. This and other pseudoelastic effects observed in the martensitic state will be discussed in the section on recent observations in Cu Al Ni alloys.

Fig. 11 shows that when an alloy deformed below Ms is heated above Af it requires a large stress to prevent the transformation back to the matrix phase. This stress will be the same as the stress involved in the reversion of martensite to the matrix phase in pseudoelastic experiments (σc in Fig. 1). This stress will increase rapidly with temperature until above a certain temperature plastic deformation of the martensite will occur and recovery will not be complete.

There is a very close relationship between pseudoelasticity and the strain memory effect when only one martensite structure is formed. This is shown in Fig. 12 for polycrystalline Cu Zn Sn alloys deformed at various temperatures from As to Af. After straining the specimen

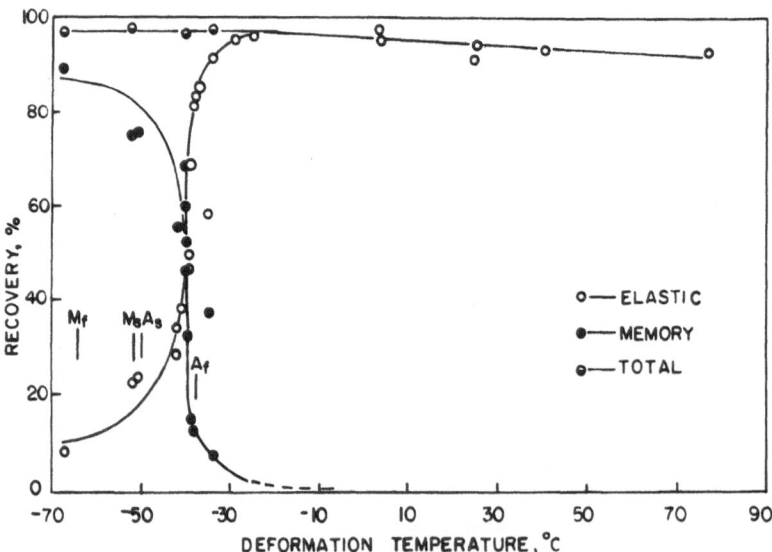

Fig. 13 Effect of deformation temperature on the pseudoelastic
        and strain-memory recovery in Cu Zn Sn single crystals
        showing almost complete recovery of strain[33].

1 pct and releasing the load, it is then heated above A_f and the strain-memory recovery is measured. It can be seen that when the pseudoelasticity is large the strain-memory recovery is small and similarly when the pseudoelasticity is small the strain-memory recovery is large. However the total recovery is always close to 100 pct and there is very little plastic deformation. This relationship is shown more clearly in Fig. 13 for single crystal Cu Zn Sn alloys.

RECENT OBSERVATIONS IN Cu Al Ni ALLOYS

The new results included in this section were carried out on single crystal specimens of Cu - 14 wt. pct Al - 3 wt. pct Ni with the tensile axis near the [001] direction of the parent β_1 phase. Table 1(a) gives the characteristic transformation temperatures for the thermal γ' martensite for the alloys studied. The microstructures were observed by optical microscopy whilst X-ray diffraction techniques were used to identify the crystal phases. The observations are summarised in the present report. More complete results will be presented in two papers to be published elsewhere[35,37].

Fig. 14 shows the stress-strain curves for sample A. These are typical for the range of temperatures in which the transition from SME to pseudoelasticity is observed. Just above Ms the stress-strain

Table 1(a)   Characteristic transformation temperatures in °C for thermal (γ') martensite.

Sample	$\beta_1 \rightarrow \gamma'$		$\gamma' \rightarrow \beta_1$	
	M_f	M_s	As	A_f
A	−31	−18	2	10
B	−38	−22	−16	−9
C	−4	10	20	27

Table 1(b)   Temperature range in °C for the occurrence of stress-induced martensite.

Sample	$\beta_1 \rightarrow \gamma'$	$\beta_1 \rightarrow \beta_1' \rightarrow \gamma'$		$\beta_1 \rightarrow \beta_1'$
		Loading	Unloading	
A	−15 to 11	11 to 16	11 to 23	23 to 60
B	−22 to −8	−8 to 4	−8 to 4	11 to 60
C	10 to 30	30 to 39	30 to 50	50 to 60

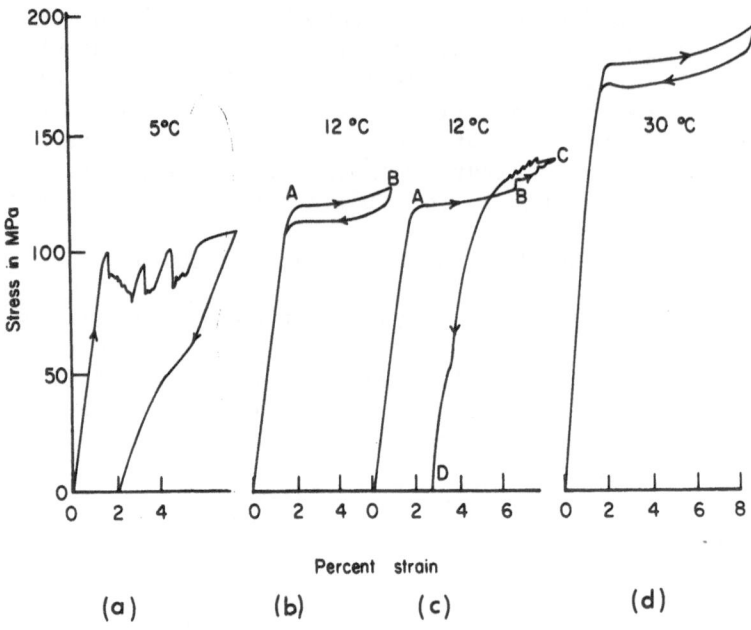

Fig. 14   Stress-strain curves in Cu Al Ni single crystal with As =
         2°C Af = 10°C, showing the formation of γ' martensite at
         5°C, β_1' martensite at 30°C and both β_1' and γ' at 12°C [37].

curve is jagged and corresponds to the stress-induced formation of
γ' from the β_1 matrix phase. This has a thick plate morphology and
a complex internal structure formed by several systems of parallel
plates, which are presumably twins.

    The deformation produced during the $\beta_1 \rightarrow \gamma'$ transformation is
non-recoverable near Ms, and becomes partially recoverable (as in
Fig. 14(a)) for higher temperatures. There is a narrow range of
temperature between the curves of Fig. 14(a) and (b) that corresponds to simultaneous formation of both the β_1' and γ' phases. In
this range the stress-strain curve presents a jagged shape and a
fairly large hysteresis, but the deformation due to both transformations is almost completely recovered.

    At temperatures well above Af (Fig. 14(d)), the stress-strain
curve shows the typical pseudoelastic behavior described earlier,
due to the stress-induced $\beta_1 \rightleftarrows \beta_1'$ transformation[16,38]. The β_1'
has a thin plate morphology and growth of a single crystal of β_1'
is accomplished by thickening and coalescence of the previously
nucleated thin plates.

Two kinds of stress-strain curves are obtained at temperatures just above A_f, depending on whether the stress level in the sample is above or below some critical value (point B, Fig. 14(b), (c)). For stress levels below point B the stress-strain curve shows typical pseudoelasticity. The structure and morphology of the SIM is observed to be the same as the β_1' formed at higher temperatures. At point B only a fraction of the β_1 single crystal is transformed to a β_1' single crystal. If the deformation is carried further to a stress level beyond point B, the nucleation of plates with a twinned thick plate morphology is observed inside the β_1' single crystal region. Simultaneously with this a sharp increase in load is observed in the stress-strain curve (point B, Fig. 14(c)). The structure of these plates was identified by X-ray diffraction as γ' martensite. The habit plane of the twins as determined by two - surface trace analysis are consistant with (101), a common twinning plane for this structure. Two {101} twin systems were found in some γ' plates.

The increase in load associated with the $\beta_1' \rightarrow \gamma'$ transformation is due to the smaller length change associated with the γ' compared to β_1' martensite, both relative to the original β_1 parent phase. For a sample oriented close to [001] the transformation strain was 7.5 pct for $\beta_1 \rightarrow \beta_1'$ and 5.6 pct for $\beta_1 \rightarrow \gamma'$. This implies a shortening of 1.9 pct for the $\beta_1' \rightarrow \gamma'$ transformation.

The $\beta_1' \rightarrow \gamma'$* transformation is partial during loading and occurs more extensively during unloading. The final amount of γ' forming increases with decreasing temperature. Complete transformation from β_1' to a twinned γ' single crystal can be obtained even above M_s. When this twinned γ' single crystal is loaded and unloaded again the disappearance and appearance of one of the twin variants is observed. This process produces a small pseudoelastic effect (<1 pct).

Table 1(b) shows the ranges of temperature in which the different stress-induced transformations are obtained in these alloys. It can be seen that the range of metastability of γ' formed from β_1' is extended around 20°C with respect to the A_f temperature.

The $\beta_1' \rightarrow \gamma'$ transformation can also be obtained when the sample is transformed from β_1 to β_1' above the metastability limit for γ', and then cooled to the range of temperature in which the transformation from $\beta_1' \rightarrow \gamma'$ is obtained. Fig. 15(a) shows a micrograph of alloy C transformed to stress-induced β_1' at 40°C and cooled to 25°C,

---

*See also the paper by K. Otsuka, H. Sakamoto and K. Shimzu presented at the present symposium.

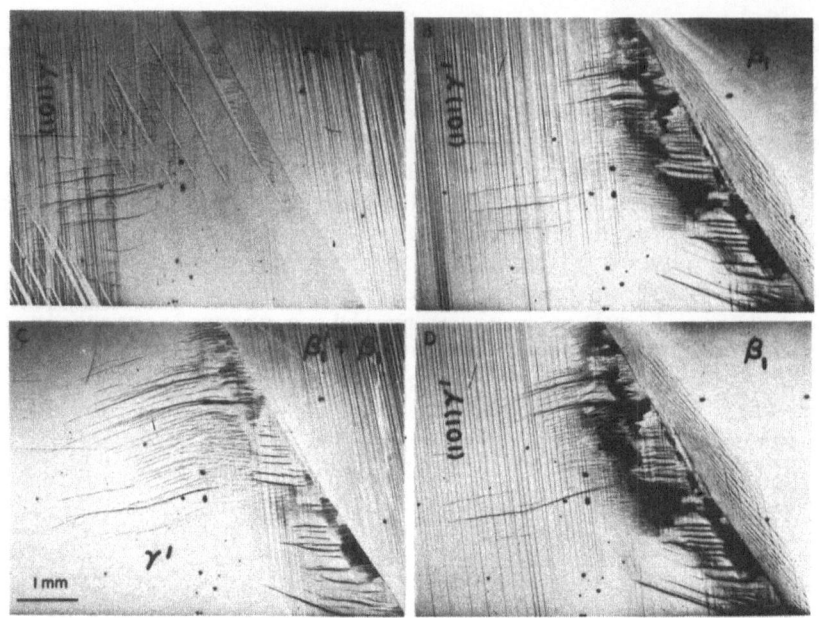

Fig. 15 Micrograph corresponding to a single crystal (alloy C, Table
         1) transformed to SIM β_1' at 40°C. (a) The sample was cooled
         to 25°C and shows partial transformation to a twinned γ'
         structure. (b) The sample is unloaded at 25°C and total
         transformation to twinned γ' is observed. (c) The sample
         is reloaded at 25°C and the disappearance of one twin var-
         iant is observed. (d) The sample is unloaded again at 25°C.
         The appearance of the twinned structure is observed[37].

maintaining the total length of the sample constant. In this condition some relaxation of load occurs due to the cooling of the sample.
It can be seen that partial transformation to a twinned γ' structure
occurs. If the load is then released, the transformation from β_1' to
γ' goes to completion (Fig. 15(b)). At the same time some of the
β_1' adjacent to the β_1/β_1' interface reverts back to β_1.

   In Fig. 15(c) and (d) the disappearance and appearance of one
of the twin variants is observed when the sample is loaded and unloaded. It can be seen that in the unloaded state (Fig. 15(b), (d))
there is a complex transition structure between γ' and the β_1.
Probably some β_1' plates are retained at this transition interface
even at zero stress.

   The extended metastability range for the stress-induced γ'
phase means that the resistive factors opposing the $\gamma' \to \beta_1$ transformation are greater for stress-induced martensite than for thermal

martensite. This is not surprising in view of the fact that the stress-induced γ' forms from β_1', rather than directly from β_1 as is the case for thermal martensite.

In recent experiments at temperatures below M_f[35], it was found by the present authors that a γ' single crystal transforms to a β_1' single crystal when loaded over a certain critical value. A typical series of stress-strain curves are shown in Fig. 16. During the first 4 cycles of loading to 1 pct strain and unloading, there is very little recovery of the deformation as γ' thermal martensite progressively transforms to a single variant. The way in which the reorientation of the thermal γ' martensite variants takes place was not observed in detail. Nevertheless macroscopic observation showed the gradual growth of flat regions over the complex tilted surface of the thermal martensite variants. Sometimes an audible click occurred simultaneously with the disappearence of a self-accommodating set of variants. In the middle of cycle 5, formation of the single crystal γ' variant is complete. At this point further stressing gives a sharp increase in load due to elastic deformation of the specimen. When the specimen is unloaded at this stage, some pseudoelastic recovery takes place as can be seen in cycles 6 and 8. This deformation appears

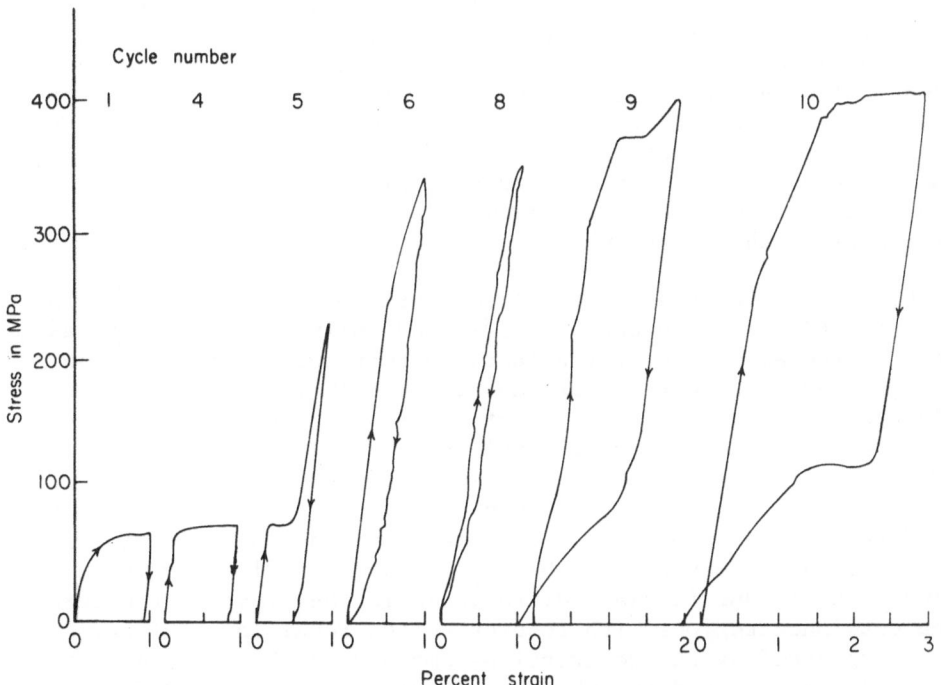

Fig. 16   Effect of cycling Cu Al Ni single crystal to 1 pct strain at -196°C[35].

to be due to the growth and disappearance of a (101) twin variant
when the sample is alternately unloaded and loaded. In cycle 9,
the specimen is taken to a higher stress and at this time a yield
point appears in the stress–strain curve. This is due to the formation of β_1', growing as thick bands within the γ' martensite. The
strain associated with complete transformation of γ' to β_1' is ~2 pct,
comparable to the value of 1.9 pct found earlier. On unloading, the
stress–strain curve shows appreciable pseudoelastic recovery with a
large hysteresis associated with the reversion of β_1' to γ'. The
γ' forms inside the β_1' with a twinned thick plate morphology similar
to the $\beta_1' \to \gamma'$ transformation at high temperatures. At a stress
level of 100 MPa the β_1' has completely reverted to a twinned γ'
single crystal. On further unloading the sample transforms to a
single variant of γ'. At a later stage of unloading the nucleation
of one or two new sets of parallel bands is observed. These produce
an extra shortening of the specimen and probably are the cause of
the crossover in the stress–strain curve. The habit plane of these
bands are found consistent with (121) γ' twins.

Further work is under way to get a closer picture of the complex sequence of SIM transformations observed in this alloy.

Two main types of pseudoelastic behavior have thus been observed
in Cu Al Ni at temperatures both above and below M_S. One type, the
more significant, is related with the SIM $\beta_1 \rightleftarrows \beta_1'$ above Ms and with
SIM $\gamma' \rightleftarrows \beta_1'$ below Ms. The other less important effect occurs when
a twinned γ' single crystal detwins to a single variant when an external stress is applied to the sample. The pseudoelastic effects
of the first type are both explained on the same basis. A SIM transformation is obtained by the effect of applied stress. The product
martensitic phase becomes unstable when the sample is unloaded and
reverts back to the parent phase.

There are references in the literature to pseudoelasticity in
the martensitic state in both Au Cd[39], and in In Tl[7,24]. This was
called the rubber-like or ferroelastic effect and was found to be
related to the disappearance of one of the twin variants when the
sample was loaded. The magnitude of the associated pseudoelastic
effect was small (less than 0.6 pct) as in Cu Al Ni. It is not
clear what is the source of the driving force or the mechanism to
produce back the original twin structure in single crystals when the
load is released and no other constraint acts on them (for example
the grip of the deformation machine).. Maybe the origin of this
driving force is due to small differences in the crystal structure
of the detwinned regions with respect to the remaining variant.
This is suggested by the restrictions imposed on the occurence of
mechanical twins in ordered phases. According to the analysis of
Laves[40], if the alloy is ordered the twinning shear leaves the atoms

in wrong places and it would be more accurate to regard the process
as a martensitic transformation. On the other hand, according to
a recent paper by Otsuka et al.[17], there are indications that no
structural change is introduced by deformation twinning in the case
of (121) and (101) γ' twins.

Cu Al Ni is the first SME alloy in which two stress-induced
martensitic structures have been recognized. In Au Cd[14], the stress-
strain curves show two separate plateaus and it is possible that
this is due to the formation of two different martensitic structures
at the two stress levels. In Ag Cd[12] there is a very marked change
in the slope of the initiation stress (σ_A, Fig. 1) - vs. - temperature plot at a temperature of 22°C above Ms. This has been explained as due to a change in the mode of nucleation of SIM[2]. However
it seems possible[16] that it may be due rather to the formation of
two different martensitic structures. Stress-strain curves with
two separate plateaus have also been observed in Cu Zn,[41] and it
seems very possible that more than one martensitic structure is
formed on stressing in a significant number of alloy systems.

The main features of the recent observed effects, such as the
competition between stress-induced γ' and β_1' and the transition
from β_1 to γ' above Ms, and the pseudoelastic $\gamma' \rightarrow \beta_1'$ transformation
below Ms can be understood qualitatively with a fairly simple analy-
sis[37]. This analysis takes into account the ease of nucleation of
the β_1' and γ' phases and requires that the minimum driving force
to produce γ' from β_1 be greater than that to produce the β_1'. It
also takes into account the effect of stress on the free energy - vs
- temperature curves and requires that the free energy of the β_1'
phase varies more rapidly with stress than the free energy of the
γ' phase, both relative to β_1.

REVERSIBLE STRAIN MEMORY EFFECT

In the reversible strain-memory effect, a specimen whose temperature is cycled between the martensite and parent phases shows a
reversed dimensional change on both heating and cooling. Normally
when a specimen is cooled through the martensite transition temperature, it shows no dimensional changes[34]. This is because a large
number of variants form in thermal martensite and any dimensional
change associated with one variant is cancelled out by opposite
changes from other variants. In the reversible strain-memory effect,
preferred growth of a favoured variant occurs and this causes the
dimensional change on cooling. Growth of a favoured variant can
occur either by cooling under stress or by prior plastic deformation.

The reversible strain memory effect under stress is illustrated
in Fig. 17. A specimen is stressed at a temperature above Af such

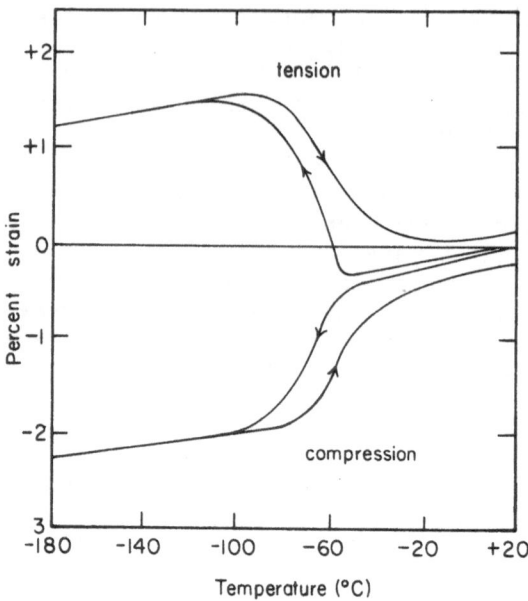

Fig. 17 Dimensional changes during thermal cycling in a Cu Zn
         single crystal under tensile and compressive stresses of
         $\pm$ 5 Kg/mm^2 [42].

that the stress level is insufficient for SIM to form. On cooling,
the stress required for martensite formation becomes smaller (Fig.5)
and at some particular temperature this becomes less than the applied
stress. The specimen therefore progressively transforms to a favoured variant of the martensite, giving a significant dimensional change.
Once transformation is complete no further dimensional changes take
place except for thermal contraction effects. On subsequent heating,
reversion of the martensite occurs and the specimen regains its original dimensions. This effect will be repeated on subsequent thermal cycles. Indeed, it has been suggested that the effect form the
basis for the operation of a heat engine[43-44]. Hornbogen and Wassermann[42] find that as the applied stress is increased, the amount
of strain associated with the transformation increases although at
the same time there is an increasing amount of non-recoverable strain-
the transformation plasticity.

    The second type of reversible strain memory effect is shown in
Fig. 18. A specimen is plastically deformed 1.4 pct at a temperature
above A$_f$. On cooling, ~0.2 pct strain is recovered due to preferred
orientation of the martensite that forms, and this strain is recoverable on subsequent heating. This effect will be repeated on temper-

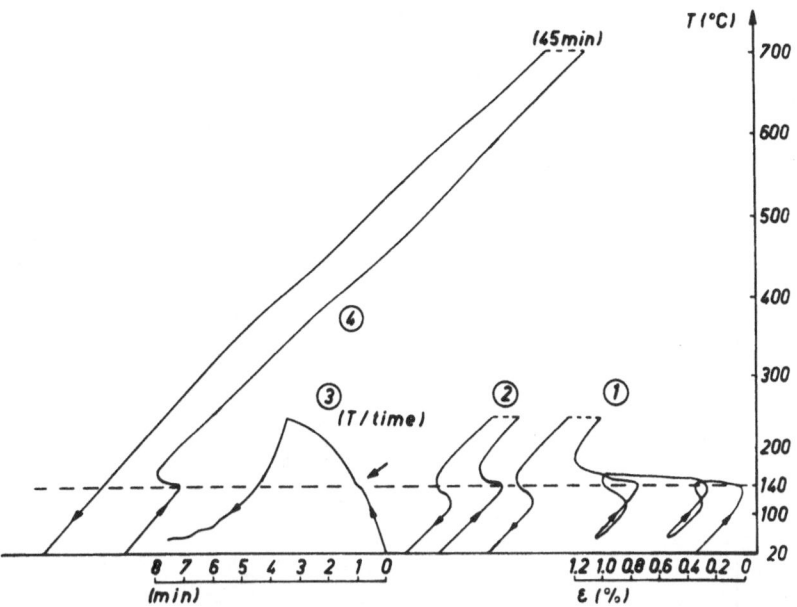

Fig. 18 Reversible strain-memory effect in Cu Zn Al Ni alloy.
1) 1.4 pct plastic strain interrupted by two cooling
cycles showing progressive development of the effect.
2) reversible strain-memory effect. 3) temperature-
time plot showing endothermic reaction at 140°C.
4) loss of the effect after heating to 700°C[5].

ature cycling provided the nuclei for the martensite are not lost as
they can be, for example, by holding the specimen at a high temperature for a short time. It should be noted that the force causing
the dimensional change on cooling will be small but the force on
heating will be large. It should also be emphasised that the recovery of the original deformation is very incomplete, being only ~14
pct for the case shown. Hence this effect, although interesting,
is of little practical importance.

FATIGUE PROPERTIES

No systematic study of the fatigue properties of pseudoelastic
metals have been published thus far, although the inference has been
that they are very good. Buehler[8], for example, found the fatigue
life of Ni Ti to be greater than 10^7 cycles at a stress level of
480 MPa.

A study has just been completed of the fatigue properties of
Cu Al Ni single crystals[45]. Specimens were cycled from zero stress

to a fixed strain, and the fatigue life was found as a function of strain magnitude, temperature (relative to M_S) and crystal orientation. The results are summarised in the stress -vs.- number of cycles plot in Fig. 19 and in Table 2. The fatigue life is suprisingly short, and this means that the use of these alloys in cyclic applications is severely restricted.

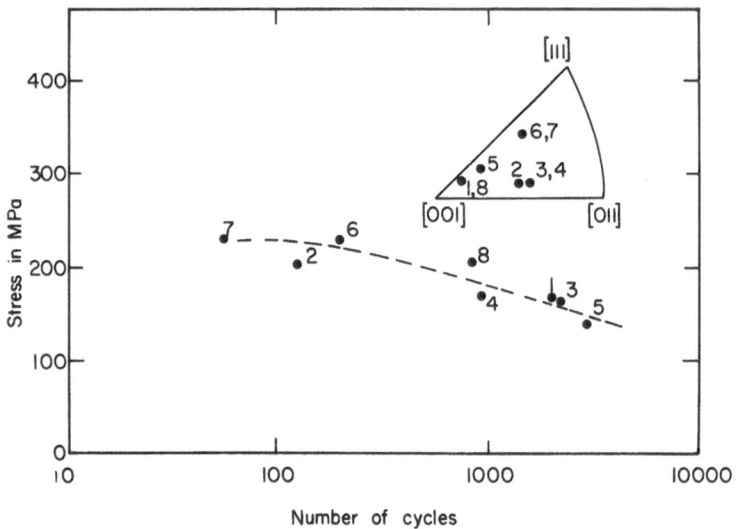

Fig. 19   Stress -vs.- number of cycles for the fatigue life of Cu Al Ni single crystals. Details of temperatures and strains are given in Table 2.

Table 2   Experimental Conditions for Fatigue Tests in Cu Al Ni Alloys

Sample Number	Temperature (°C)	Maximum Strain (pct)
1	24	3.7
2	55	5.0
3	24	2.1
4	24	7.0
5	24	3.6
6	40	5.0
7	55	5.0
8	55	3.7

There seems little doubt that it is stress rather than strain
which controls the fatigue life of pseudoelastic metals. For example,
specimens 1 and 8 were identical in terms of crystal orientation and
Ms temperatures, and were cycled to the same strain. When they were
fatigued at two different temperatures, 24°C and 55°C, they had significantly different fatigue lifes, 2000 cycles at 24°C and 830 cycles
at 55°C. The difference between the two tests was the magnitude of
the stress involved, being 167 MPa at 21°C and 208 MPa at 55°C. Similar specimens cycled to different strains at the same temperature
(specimen 3 - 2.1 pct, specimen 4-7.0 pct) showed significantly decreased fatigue life at the higher strain. This was attributed to
the higher stress reached at the end of the 7.0 pct cycle relative
to the 2.1 pct cycle. No effect of specimen orientation could be
detected, at least for the range of orientations studied.

Observations of the shape of the stress-strain curves during
cycling showed essentially no change in the shape of the curves,
right up to the cycle at which fracture took place (Fig. 20). This
implies that right until the end there was little fatigue crack propagation since a significant reduction in cross-sectional area would
reduce the stress for martensite initiation (σ_A in Fig. 1). Direct
observation of a fatiguing specimen also did not show any fatigue
cracks until final fracture took place. Hence the fatigue behaviour
of pseudoelastic metals is quite different from that in normal metals.

No detailed mechanism can be presented at this time for the fatigue failures. Unquestionably cycling introduces progressive changes
in the internal structure of the martensite. These changes are caused
mainly by the stress applied to the specimen. After a fairly reproducible number of cycles, the specimen has developed an internal structure which causes fracture on loading rather than the production of
SIM i.e. the stress for brittle fracture has decreased on cycling to
below that for SIM formation.

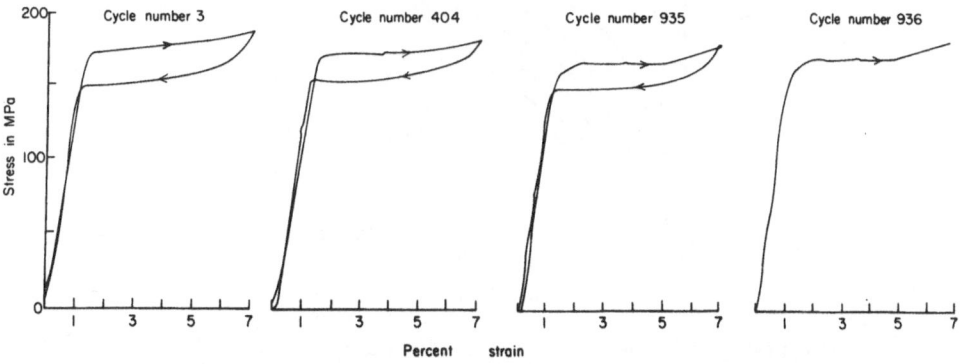

Fig. 20   Stress-strain curves at various stages in the life of
          stages Cu Al Ni crystal number 4 (see Table 2).

Dvorak and Hawbolt[20] have studied fatigue in polycrystalline Cu Zn Sn alloys. Emphasis was laid on the variation in shape of the stress-strain curves with cycling and no measurements were made of the number of cycles to failure. As discussed earlier, polycrystalline specimens show some non-recoverable strain after one cycle of loading and unloading due to the formation of deformation martensite as the grains try to accommodate one another. On subsequent stressing this deformation martensite gives rise to stress concentrations so that SIM forms at a lower applied stress. On continued deformation the characteristic stress for formation of SIM (σ_A in Fig. 1) becomes completely masked and the curve has essentially a constant slope from zero stress with little hysteresis (Fig. 21). At the same time the maximum amount of reversible strain becomes progressively smaller, again due to the deformation martensite interfering with the SIM (Fig. 22).

INTERNAL FRICTION

An important property of pseudoelastic alloys is their high damping capacity at temperatures below M_s. This makes them of potential value for sound damping and attenuation of vibrations. There is a sudden decrease in damping capacity as the temperature as raised above A_s (Fig. 23) and it is obvious that the high damping is due to the martensitic structure. This is easily demonstrated by comparing

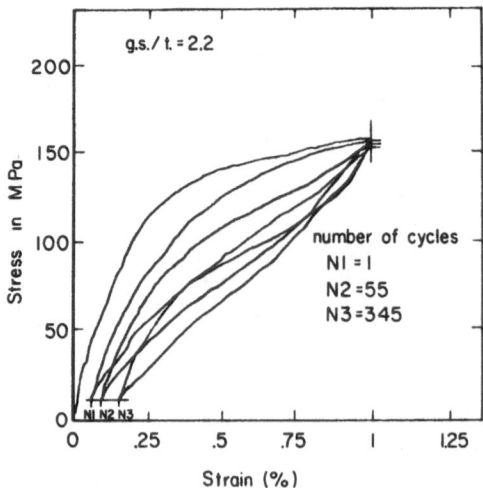

Fig. 21 Development of the stress-strain curve in a polycrystalline Cu Zn Sn alloy at 24°C (9°C above A_f) on extended cycling to 1 pct strain[46].

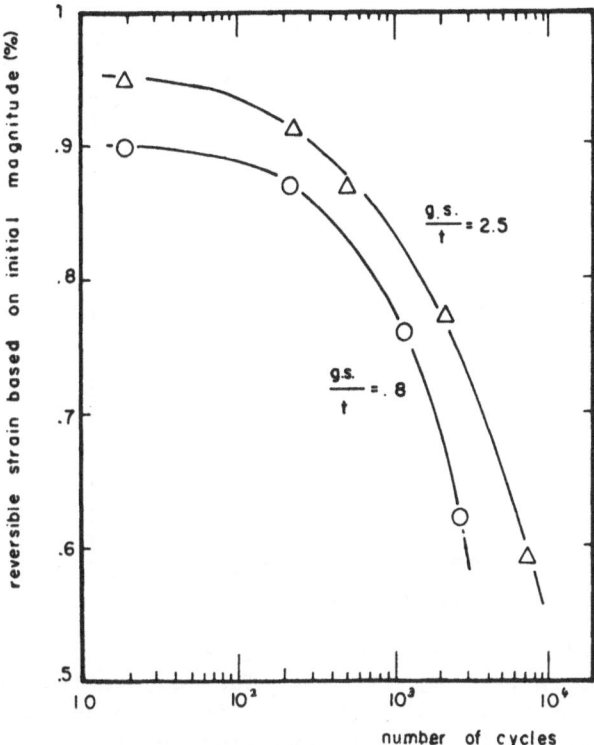

Fig. 22   Effect of extended cycling on the magnitude of the max-
          imum reversible strain in polycrystalline Cu Zn Sn alloy
          at 25°C (9°C above A_f)[20].

the resonant sound on striking a pseudoelastic metal at a temperature
above A_s with the dull sound obtained at lower temperatures.   These
alloys have good damping capacity over a wide temperature range
(~100°C) and as may be seen in Fig. 23 they are much better than grey
cast iron in this application.

The internal friction appears to have a maximum at temperatures
somewhat below M_f and hence is associated with the structure of martensite rather than with the transformation to martensite.   The main
energy loss appears to be due to friction associated with reversible
movement of the internal twins of the martensite and reversible movement of the martensite boundaries.   In Mn Cu[48], two peaks are observed in the internal friction-temperature curves- a main peak corresponding to movement of the twin boundaries and a secondary peak associated with movement of the martensite/matrix interface.   The main
peak occurred ~50°C below M_s whilst the secondary peak occurred mid-
way between Ms and M_f.

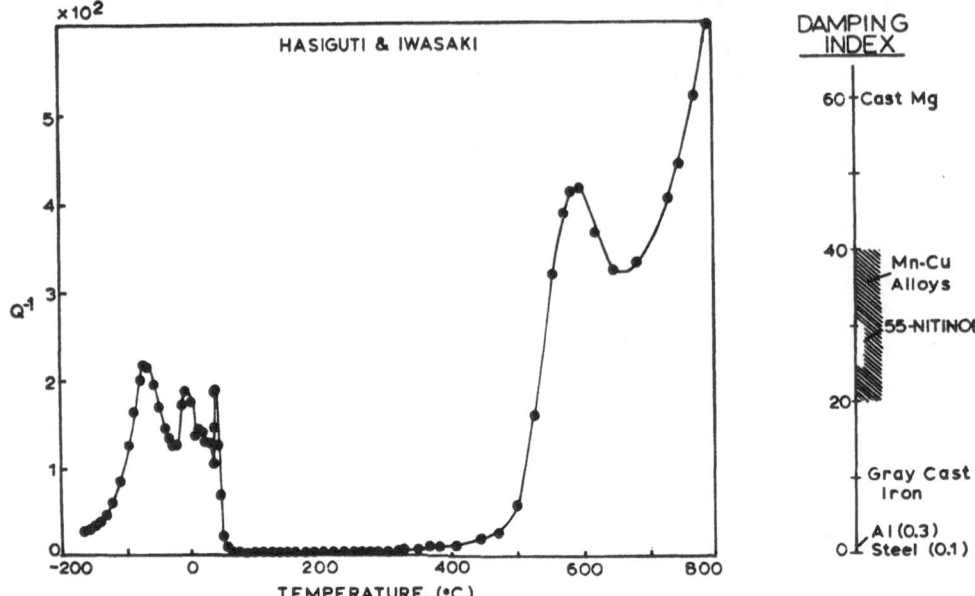

Fig. 23   Internal friction of polycrystalline Ni Ti as a function
          of temperature (A$_S$ = 60°C). The comparative damping
          index for Ni Ti and Cu Mn relative to other materials
          is given on the right of the figure[8,47].

ACKNOWLEDGEMENT

This work was financed by the National Research Council of
Canada under grant A 2549.

REFERENCES

1.   L. Delaey, R.V. Krishnan, H. Tas and H. Warlimont:
         J. Materials Science, 1974, vol. 9, p. 1521.
2.   R.V. Krishnan, L. Delaey, H. Tas and H. Warlimont:
         J. Materials Science, 1974, vol. 9, p. 1536.
3.   H. Warlimont, L. Delaey, R.V. Krishnan and H. Tas:
         J. Materials Science, 1974, vol. 9, p. 1545.
4.   C.M. Wayman and K. Shimizu:
         Metal Science J., 1972, vol. 6, p. 175.
5.   H. Tas, L. Delaey and A. Deruyterre:
         J. Less Common Metals, 1972, vol. 28, p. 141.
6.   L.C. Chang and T.A. Read:
         Trans. Met. Soc. AIME, 1951, vol. 191, p. 47.
7.   M.W. Burkart and T.A. Read:
         Trans. Met. Soc. AIME, 1953, vol. 197, p. 1516.

8. W.J. Buehler and F.E. Wang:
    Ocean Engineering, 1968, vol. 1, p. 105.
9. R.J. Wasilewski:
    Metallurgical Transactions, 1971, vol. 2, p. 2973
10. H.C. Tong and C.M. Wayman:
    Scripta Met., 1973, vol. 7, p. 215.
11. S. Muira, T. Mori and N. Nakanishi:
    Scripta Met., 1973, vol. 7, p. 697.
12. R.V. Krishnan and L.C. Brown:
    Metallurgical Transactions, 1973, vol. 4, p. 432.
13. I. Cornelis and C.M. Wayman:
    Scripta Met., 1974, vol. 8, p. 1321.
14. N. Nakanishi, T. Mori, S. Miura, Y. Murakami and S. Kachi:
    Phil. Mag., 1973, vol. 18, p. 277.
15. H. Tas, L. Delaey and D. Deruyterre:
    Zeit. fur Metallk., 1973, vol. 64, p. 855.
16. K. Otsuka, K. Nakai and K. Shimizu:
    Scripta Met., 1974, vol. 8, p. 913.
17. K. Otsuka, T. Nakamura and K. Shimizu:
    Trans. Japan. Inst. Met., 1974, vol. 15, pp. 103, 109,
    201, 211.
18. R. Pascual, M. Ahlers, R. Rapacioli and W. Arneodo:
    Scripta Met., 1975, vol. 9, p. 79.
19. W. Arneodo and M. Ahlers :
    Acta Met., 1974, vol. 22, p. 1475.
20. I. Dvorak and E.B. Hawbolt:
    Metallurgical Transactions, 1975, vol. 6A, p. 95.
21. D.V. Wield and E. Gillan:
    Scripta Met., 1972, vol. 6, p. 1157.
22. G.F. Bolling and R.H. Richman:
    Acta Met., 1965, vol. 13, pp. 709, 723, 745.
23. D.P. Dunne and C.M. Wayman:
    Metallurgical Transactions, 1973, vol. 4, pp. 137, 147.
24. Z.S. Basinski and J.W. Christian:
    Acta Met., vol. 2, p. 148.
25. C. Baker:
    Metal Science J., 1971, vol. 5, p. 92.
26. A. Nagasawa, K. Etiani, Y. Ishino, Y. Abe and S. Nenno:
    Scripta Met., 1974, vol. 8, p. 1055.
27. Y.K. Au and C.M. Wayman:
    Scripta Met., 1972, vol. 6, p. 1209.
28. C.M. Wayman, I. Cornelis and K. Shimizu:
    Scripta Met., 1972, vol. 6, p. 115.
29. H. Schumann:
    Kristall. und Technik, 1974, vol. 9, pp. 33, 281.
30. A. Nagasawa:
    J. Phys. Soc. Japan, 1971, vol. 30, p. 1505.

31. K. Enami, S. Nenno and Y. Minato:
        Scripta Met., 1971, vol. 5, p. 663.
32. A. Nagasawa:
        Phys. Stat. Solid., 1971, vol. (a)8, p. 531.
33. J.D. Eissenwasser and L.C. Brown:
        Metallurgical Transactions, 1972, vol. 3, p. 1359.
34. K. Oishi and L.C. Brown:
        Metallurgical Transactions, 1971, vol. 2, p. 1971.
35. C. Rodriguez and L.C. Brown, 1975:
        To be published.
36. L.C. Brown:
        Metallurgical Transactions:
        To be published in May 1975.
37. C. Rodriguez and L.C. Brown:
        Submitted to Metallurgical Transactions, May 1975.
38. K. Otsuka and K. Shimizu:
        Phil. Mag., 1971, vol. 24, p. 481.
39. H.K. Birnbaum and T.A. Read:
        Trans. Met. Soc. AIME, 1960, vol. 218, p. 662.
40. F. Laves:
        Naturwissenschaften, 1953, vol. 39, p. 546.
41. W. Arneodo and M. Ahlers:
        Scripta Met., 1973, vol. 7, p. 1287.
42. E. Hornbogen and G. Wassermann:
        Zeit fur Metallk., 1956, vol. 47, p. 427.
43. M. Ahlers:
        Scripta Met., 1975, vol. 9, p. 71.
44. H.C. Tong and C.M. Wayman:
        Metallurgical Transactions, 1975, vol. 6A, p. 29.
45. L.C. Brown:
        To be published.
46. I. Dvorak:
        M. A. Sc. Thesis, University of B.C., 1973.
47. R.R. Hasiguti and K. Iwasaki:
        J. Applied Physics, 1968, vol. 39, p. 2182.
48. K. Sugimoto, T. Mori and S. Shiode:
        Metal Science J., 1973, vol. 7, p. 103.

OPTICAL AND ELECTRON MICROSCOPE OBSERVATIONS OF TRANSFORMATION

AND DEFORMATION CHARACTERISTICS IN Cu-Al-Ni MARMEM ALLOYS[*]

K. Shimizu and K. Otsuka[**]
Institute of Scientific and Industrial Research
Osaka University
Yamadakami, Suita, Osaka 565, Japan

I. Introduction

The shape memory effect is now generally known as the phenomenon that a specimen apparently plastically deformed at some lower
temperature reverts to its undeformed original shape on heating to
a somewhat higher temperature in virtue of the reverse martensitic
transformation. This phenomenon is very peculiar compared to the
ordinary plastic deformation behaviour. Thus, it has been of
great interest for many workers of both academic and technological
fields, and is now found in a lot of alloy systems as recently
tabulated by Wayman and Shimizu (1) and Delaey et al. (2). A common property for all of the alloy system is that each alloy exhibits a martensitic transformation, and the shape memory effect
is commonly observed when the alloy is deformed at a partly or
wholly martensitic condition and then heated to the matrix phase
(3, 4, 5). These common property and observation indicate that
the shape memory effect is originated in the behaviour of materials upon martensitic transformation, deformation of the martensite
and reverse transformation to the matrix phase. Therefore, in
order to know the mechanism of the shape memory effect, the trans-

* This paper is based on a presentation made at an International
Symposium on "Shape Memory Effect and Applications" held at the
Spring Meeting in Toronto, CANADA, on May 19, 1975, under the spon-
sorship of the Physical Metallurgy Committee, TMS/AIME.
** Presently with the Department of Metallurgy and Mining Engine-
ering, University of Illinois, Urbana, Illinois, USA.

formations and deformation characteristics must be clarified for
alloy systems exhibiting the effect. In this paper, the term
"marmem" (1) will be used as descriptive of the shape memory behaviour, since the martensite phase exhibits a memory except for
the special case known as the two-way (2) or reversible (6) shape
memory effect.

It is to be noticed here that almost all of the marmem alloys
also exhibit a superelasticity effect (7, 8). This effect is defined as a phenomenon in which shape change is attained through a
stress-induced martensitic transformation, and it is completely
recoverable upon removing the stress by the reversible reverse
transformation*. Obviously this effect occurs in the temperature
range where the stress-induced martensites are unstable without
stress. Thus, the shape memory effect and the superelasticity
effect are quite similar in nature, both being associated with the
direct and reverse martensitic transformation. In short, strain
is attained by the deformation of martensites or stress-induced
transformation in the former, while it is essentially attained by
the stress-induced transformation in the latter. The driving
force for the reverse transformation is provided as thermal energy
in the former, while it is provided simply by reliesing the stress
in the latte . Therefore, both effects are treated in the present
paper, as has been done in several recent review papers (1, 2, 9,
15, 16).

Some marmem alloys such as AgCd (17) and CuZn (17, 18) exhibit
a mottled or complicated structure in their electron micrographs
taken from the matrix phase as-quenched to room temperature, while
Cu-Al-Ni alloys do not exhibit such a complicated structure in the
electron micrographs. It makes it simpler to analyse the electron
micrographs and diffraction patterns. Moreover, the Cu-Al-Ni
alloys exhibit red and yellowish colors in the states of the matrix and martensite phases, respectively. Therefore, the matrix
to martensite transformation and its reverse transformation associated with cooling and heating, or loading and unloading (or
shape change and its recovery) can be distinguished by the color
change even with unaided eyes (3). The crystallography and morphology of the Cu-Al-Ni martensites have been well established so
far, comparing with those of other marmem alloys, and the thermoelastic nature of the transformation, which is generally considered to closely be related to the marmem and superelasticity behaviours, has been well observed as will be described in more
detail later.

---

* Several other different terms are used for this phenomenon, as
tabulated (2, 9). Another pseudoelastic effect, which is similar
in phenomenon and different in nature, is also observed in the
perfectly martensitic state of AuCd (10) and In-Tl (11, 12) alloys,
and it has been known as the "rubber-like" (13) or "ferroelastic"
(14) behaviour. This behaviour will not be discussed in the present paper, since it is not realized in the present Cu-Al-Ni alloys.

Because of the above experimental easinesses and the well-established data, the present authors have carried out microscope and diffraction studies of the Cu-Al-Ni marmem alloys, and obtained usefull results on the marmem and superelasticity behaviour. In the present paper, those results will collectively be described, comparing with those of other marmem alloys. Some new aspects will also be presented, focusing on the transformation and deformation characteristics of the martensite, the stress-induced transformation in the matrix phase, along with the reverse transformation characteristics of the martensite.

II   Transformation characteristics of thermally and stress-induced martensites.

(a) Crystallography and morphology

<u>Thermally-induced martensite</u>: When the alloys with composition near Cu-14.2Al-4.3Ni (wt%) are cooled below the Ms temperature, the β_1 matrix phase transforms to the γ_1' martensite phase. The crystal structures of the β_1 matrix and γ_1' martensite phases were determined by Duggin and Rachinger (19) to be the DO_3 type and Cu_3Ti type (equivalently 2H structure in Ramsdell notation) ordered structures, respectively. The γ_1' martensite usually appears in a spear-like form consisting of two parts divided by a ridge (this was originally termed a mid-rib) at the center (21, 22), as shown in Fig. 1(a). The indices of habit planes between martensite and matrix, and those of plane of ridge were reported to be $\{331\}_{\beta_1}$ and $\{110\}_{\beta_1}$, respectively (21). In those previous works, however, the relation between the two halves of a

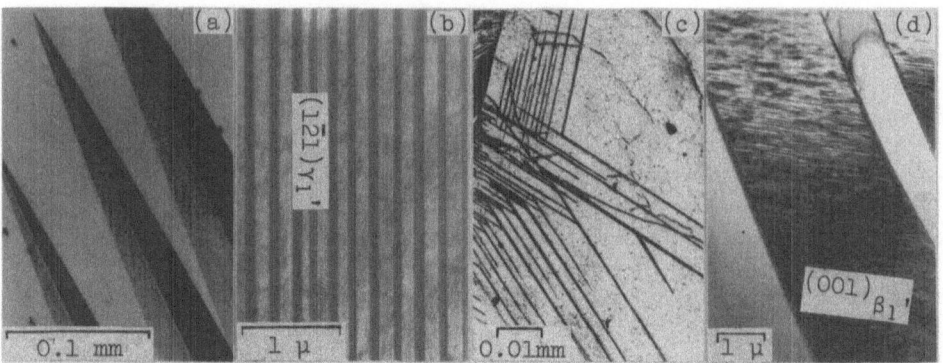

Fig. 1. Optical (a)(c) and electron (b)(d) micrographs of thermally formed γ_1' (a)(b) and stress-induced β_1' (c)(d) martensite. (b) and (d) show internal twins and stacking faults in γ_1' and β_1' martensites, respectively.

spear-like martensite, the nature of a ridge between the two halves
and the nature of striations in each half (Fig. 1(a)) were not established. In order to clarify them, the present authors carried
out an experiment by means of optical and electron microscopies
and selected area electron diffraction (23). As a result, the
plane of central ridge in a spear-like martensite was identified
as $(121)_{\gamma_1}$' twinning plane, and each half divided by the plane of
ridge was found to be internally twinned on $\{121\}_{\gamma_1}$' plane other
than the $(121)_{\gamma_1}$' ridge plane (Fig. 1(b)) and furthermore each
internal twin is faulted on the basal plane. A new orientation
relationship consistent with the morphology was proposed, according to which the two halves divided by the plane of ridge can be
regarded as variants of martensites as well as twins to each other.

Boundaries of internal twins in each half of a spear-like
martensite were found to be mobile under an applied stress, as
will be shown later, and this mobile nature seems to play an important role in the marmem behaviour of Cu-Al-Ni alloys. The reason
why two variants of martensites take the spear-like form can be
attributed to the self-accommodation effect so as to minimize the
strain energy set up by the formation of martensites (24, 25).
The self-accommodation effect of the strain energy is well explained based on the result of the phenomenological calculations (24).

The spear-like formation of two martensites, however, is not
always realized, but a thick band of a single variant of martensite is formed in single crystal alloys. In case of single crystals, the single interface type transformation is also realized
(26) by cooling slowly from one end of specimen as in the cases of
AuCd (13) and In-Tl (11, 12) alloys. In this way the morphology
of martensite is different from polycrystal alloys to single
crystal alloys, but the habit plane index and the twinning plane
index of internal twins are the same for both cases of poly and
single crystals.

Stress-induced martensite: When Cu-Al-Ni alloys are slightly
deformed by a punch, acicular martensites are stress-induced (27),
as seen in Fig. 1(c), which differs from the thermally formed γ_1'
martensite with a spear-like form. By studying with the method of
electron microscopy and selected area electron diffraction (28),
the crystal structure was identified to be the 18R type long period stacking order structure with AB'CB'CA'CA'BA'BC'BC'AC'AB'
stacking sequence, which was the same as that of the thermally
formed β_1' martensite in a Cu-Al binary alloy (29). Thus, the
stress-induced acicular martensite in the Cu-Al-Ni alloys was also
termed β_1'. Internal defects of the β_1' martensite were stacking
faults on the basal plane, as seen from Fig. 1(d), the habit plane
of the β_1' martensite was close to $\{155\}_{\beta_1}$, and the orientation
relationship between matrix and martensite was consistent with
that in the thermally formed β_1' Cu-Al martensite (30). In this
way, the stress-induced β_1' martensite is different from the thermally formed γ_1' martensite not only in morphology but also in all

crystallographic characteristics, even if the alloy composition is the same.

As will be described later, the stress-induced acicular β_1' martensite is stable under stress roughly above Af temperature, but between Af and Ms the γ_1' martensite is stress-induced (31, 32), which is the same crystal structure as the thermally formed one. The stress-induced γ_1' martensite usually appears in a spear-like form (thick band form in single crystals, as mentioned above), but in a rare occasion it was observed to appear in an acicular form similar to the β_1' martensites (33). This acicular γ_1' martensite is different from the spear-like or thick banded γ_1' martensite in lattice invariant strain, habit plane and orientation relationship as well as in shape, even though both have the same crystal structure and lattice parameters. That is, the lattice invariant strain was stacking faults on the basal plane, the habit plane was roughly located between $\{321\}_{\beta_1}$ and $\{332\}_{\beta_1}$, and the orientation relationship was deviated from that for the spear-like γ_1' martensite by about $3°$. The acicular β_1' and γ_1' martensites are easily distinguished even in bulk specimens by the difference in the habit planes.

(b) Thermoelastic nature of the γ_1' and β_1' martensites.

Thermally formed γ_1' martensite: The thermoelastic nature of the thermally formed γ_1' martensite in Cu-Al-Ni alloys was first observed clearly by Kurdjumov and Khandros (21). According to their experiment, a given martensite plate grows or shrinks as the temperature is lowered or raised, and the growth or shrink rate is governed solely by the rate of temperature change. The thermoelastic transformation has also been found in AuCd (13), In-Tl (11, 12) and CuZn (34) alloys and presently in many other marmem alloys, and has been explained (21) as being due to the equilibrium between a chemical driving energy of the phase change and a non-chemical elastic energy stored during the phase change. According to this explanation, the equilibrium may be broken and then a given martensite plate may grow or shrink not only by a temperature change but also by an elastic external force, that is, the martensite plate is also mechanically elastic (35). In fact, Arbuzov et al. (36) observed by an optical microscopy that the interface boundary between martensite and matrix was mobile under an applied stress.

The elastic movement of interface boundaries associated with a temperature change or applied stress may be realized only when a martensite crystal is coherently connected to the surrounding matrix crystal and the coherency is maintained during the whole process of the growth or shrinkage. In order to examine the interface structure, a scanning electron microscopy was recently carried out with γ_1' martensites subjected to the single interface transformation (37). Even though constraints from the surrounding matrix are absent in this type of transformation, $\{121\}_{\gamma_1'}$

internal twins were observable in the γ_1' martensite, manifesting
that the twins are necessary in order to make the interface an invariant plane, as is assumed in the phenomenological theory. The
width of the twins were different from one specimen to the other,
but they extended right up to the interface under an optical microscope. The density of twins are usually higher just behind the
interface than in the interior of the martensites, as seen from
Fig. 2(a) and as suggested for In-Tl alloy (12). This is because
the reduction of elastic energy by thinning of twins becomes more
important near the interface. Although the interface is usually

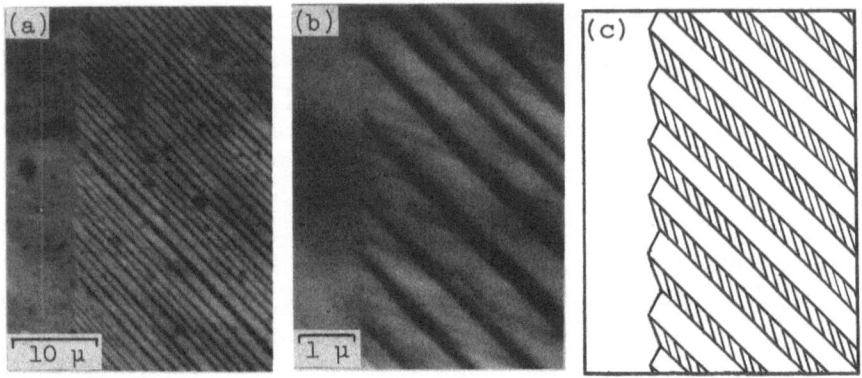

Fig. 2. Scanning electron micrographs of a thermoelastic in-
    terface between β_1 matrix and γ_1' martensite, and a schemat-
    ic interface expected from the phenomenological theory.

straight under an optical microscope, some deviations are observed
under the scanning electron microscope. The internal twins are
usually tapered toward the interface and pointed there, as seen
from Fig. 2(b). It was also sometimes observed that twins are
projected into the matrix as if they are independent variants of
martensite. These observations indicate that, although the elastic energy at the interface can macroscopically be minimized at
the interface by the introduction of twins, as is schematically
shown in Fig. 2(c), the local stress concentration or the stress
due to coherency are always present at the interface. This local
stress concentration probably causes the interface very mobile,
as is actually observed.

Stress-induced β_1' martensite: In order to know the growth
characteristics of the stress-induced acicular β_1' martensite,
point-indented alloys were observed under an optical microscope
using a cold stage. An example of the observations is shown in
Fig. 3 (28). With decreasing temperature, pre-existed acicular
β_1' martensites grow in length and other new acicular ones are
nucleated to grow in neighbouring regions, as seen from the series

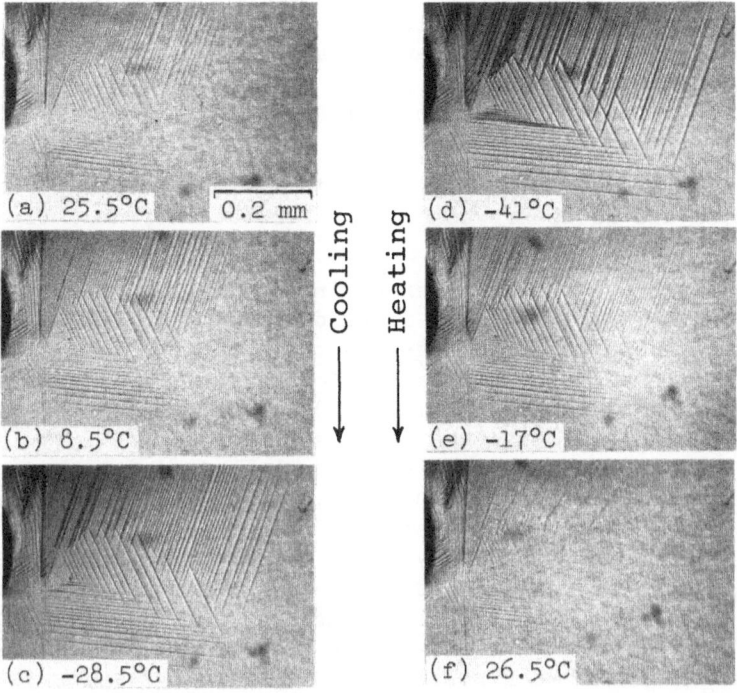

Fig. 3. A series of optical micrographs showing the growth and shrinkage of stress-induced β_1' martensites with changing temperatures.

of photographs (a), (b), (c) and (d). This tendency continues nearly up to the Ms temperature of the alloy in the undeformed state. When the temperature is raised, the martensites shrink and disappear without leaving any surface relief, as seen from the series of photographs (d), (e) and (f). Thus, the transformation is thermoelastic. As mentioned before, the thermoelastic transformation is also accompanied by mechanically elastic behaviour. Such an elastic behaviour for the stress-induced acicular martensites was already observed by Oishi and Brown (38). In Fig. 3, the β_1' martensites are seen to grow essentially in the lengthwise direction and not much in the lateral direction. This is probably because of the back stress of the surrounding matrix.

If the alloys are cooled beyond the Ms temperature, the growth of β_1' acicular martensites is replaced by the nucleation and growth of spear-like γ_1' martensites similar to those in the undeformed alloys, although thermoelastic β_1' acicular martensites are still present. On heating the alloys, spear-like γ_1' martensites shrink and disappear in the temperature range between As and Af,

and then acicular β_1' martensites start to disappear (28). These
observations show that acicular β_1' martensites are stable under a
stressed condition roughly above the Ms temperature, while that
spear-like γ_1' martensites are more stable below Ms temperature,
as naturally expected.

(c) Shape memory effect originated in the thermoelastic transfor-
     mation and necessary conditions for the thermoelastic trans-
     formation

Shape memory effect: After having found (3) the marmem effect
in Cu-Al-Ni alloy, like those in AuCd (13), In-Tl (12) and TiNi
(39), the present authors suggested that the necessary conditions
for the marmem alloys would be the following. (i) The martensitic
transformation is thermoelastic, (ii) the martensite and matrix
phases are ordered, (iii) the martensite is internally twinned,
and (iv) if the ordering is disregarded, the matrix and martensite
phases have a BCC and HCP structures, respectively (3, 40).
Wayman (41) came up with the similar idea a little later by finding it in CuZn and Fe_3Pt alloys, and he also emphasized that the
above (i), (ii) and (iii) conditions are necessary for many marmem
alloys, as reviewed in Ref. (1). Recently, however, lattice invariant shears in martensites of some other marmem alloys such as
CuZn and CuZnX have been reported to be stacking faults rather
than twins, and accordingly the above condition (iii) appears not
to hold universally, as will be discussed later. However, the
conditions (i) and (ii) are generally accepted and considered to
be essential for the marmem behaviour as well as superelasticity
(1, 2, 9, 15, 16), although the conditions (i) and (ii) are not
independent to each other, but the latter may be the origin of the
former.

Necessary conditions for the thermoelastic transformation:
Let us consider the necessary conditions for the thermoelastic
transformation in a little more detail. As mentioned earlier, a
thermoelastic martensite crystal must coherently be connected to
the surrounding matrix crystal, and the coherency must be maintained during the whole process of the growth or shrinkage. In
order to maintain coherency at an interface, the volume change associated with the transformation must be small in the first place
(42). In fact, the volume changes for many marmem alloys are
nearly 1/10 of those for ferrous alloys or steels (about 4 %) subjected to the burst type transformation, and thus the condition of
small volume change is satisfied, as seen from the 2nd column in
Table 1. In all thermoelastic transformations, the thermal hysteresis of the electrical resistivity vs. temperature curves is
small as seen from the 3rd column in Table 1. The small hysteresis means that the driving force necessary for the transformation
is small. Thus, the small driving force is also a controlling
factor, as previously suggested by Christian (43).

Table 1

Alloy Composition	Volume Change (%)	Thermal Hyster. (deg.)	Order	Lattice Softening	References
$(CuNi)_3Al$	-0.30	35	Ordered	c'	(4)(24)(23)(70)
AuCd	-0.41	15	"	c'	(12)(71)(72)
In-Tl	-0.2*	4	?	c'	(10)(73)(72)
CuAuZn	-0.25	6	Ordered	c'	(75)(55)(76)(75)
CuZn	-0.5	10	"		(77)(53)
TiNi	-0.34	30	"		(3)(78)
NiAl	-0.42	10	"		(79)(80)(83)
AgCd	-0.16	15	"		(81)(82)(84)
Fe_3Pt	(1.1)**	4	"		(44)(85)

* According to Ref.(73), the c/a ratio of the martensite phase changes with temperature. The present calculation was carried out for c/a = 1.02.
** This value corresponds to the volume change in disordered state. No accurate data are available for ordered state at present.

        According to Table 1, in which other properties are listed, ordered structure also seems to be necessary for the thermoelastic transformation, as the present authors (3, 40) and Wayman (41) have suggested for the marmem effect. Although how the ordering is related to the thermoelastic transformation is not clear at present, the ordering may probably be useful for the maintening coherency at the interface by raising the yield stress in the matrix phase (44). So far In-Tl is the only alloy, which is not confirmed to be ordered. However, there is a view that Tl atoms are arranged preferentially at sites along the c axis (43). If that is the case, all marmem alloys in Table 1 have the ordered structure.
        In the above, three factors of small volume change, small driving force and ordered structure have been suggested as the necessary ones for the thermoelastic transformation. However, the three factors may not be independent, and may mutually be corre lated to one another. In addition to the three, a smallness of transformation shears (43) and the lattice softening (45) have also been proposed as the controlling factors. However, the former is not satisfied in ordered Fe_3Pt alloy and others, and the latter is found in the burst type transformation of Fe-Ni alloy (46) and should rather be considered to be a common phenomenon for all martensitic transformations irrespective of the thermoelastic or burst types. Thus the above three factors seem to be necessary conditions for the thermoelastic transformation.

III   Deformation characteristics of the matrix phase

(a)   Two types of superelasticities.

As described previously, the superelasticity is due to the stress-induced martensitic transformation and its reversion upon unloading. This superelasticity in Cu-Al-Ni alloys was first found by Rachinger (7) and then investigated extensively by Oishi and Brown (38). The present authors also found that there were two distinct superelasticities depending upon the temperature tested, and that the two superelasticities were originated in the two distinct stress-induced transformations (31, 32). It will be shown in detail in the following how the two superelasticities are associated with the two transformations.

Fig. 4 (31) shows representative stree-strain curves tensile tested at various temperatures and two distinct stress-strain characteristics are observed depending upon the temperatures, re-lative to the characteristic transformation temperatures such as Ms, Mf, As and Af shown at the right upper corner. The curves at

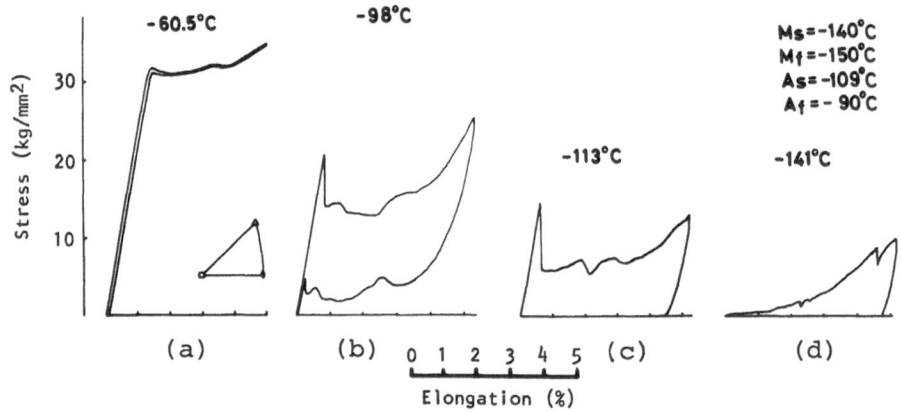

Fig. 4. Temperature dependence of stress-strain characteristics, along with the characteristic transformation temperatures. Strain rate: 2.5 x 10^{-3} /min.

temperatures roughly above Af ((a)) are characterized by extremely small hysteresis and no peak at the yield point, whereas those at temperatures roughly below Af (and above Ms) are characterized by a sharp peak at the yield point, and large hysteresis ((b)) or large residual strains ((c)) which are recoverable only by heating above Af temperature (this corresponds to the marmem effect). For reference, an example of the curves at temperatures below Ms is also shown in (d), which is characterized by the absence of linear elastic region.

It is to be noted that both curves (a) and (b) represent the

superelastic behaviour but their characteristics are extremely
different. Further there is a transition temperature range between (a) and (b) where the yield is initiated by the type of curve
in (a) and then followed by a large stress drop like those in (b)
and (c). The above two distinct stress-strain characteristics
seem to originate in two distinct transformations. This fact will
be macro- and micro-scopically verified in the next section.

(b)   Formation of two kinds of martensites associated by the
      two types of superelasticities.

Two distinct stress-strain sharacteristics have been observed
in the two temperature ranges above Af and between Af and Ms.
Thus, the process of martensitic transformations in the two temperature ranges were observed macroscopically and microscopically
during tensile tests (32).

Observations in the temperature range above Af: The typical
macroscopic change in the entire specimen in this temperature
range is shown in the series of Fig. 5 along with the stress-
strain curve corresponding to them. In the figure, (a) to (g)
correspond to the loading process, and (a') to (g') to the unloading process. Upon loading, starting from the single crystal state
of the β_1 matrix phase (a), no structural change occurs in the
linear elastic range until the point b is reached in the stress-
strain curve, where a few acicular martensites appear on the
upper right corner, (b). With increasing strains acicular
martensites are successively nucleated and some of them coalesce
to thicker plates since there is only one variant present under
the favorable Schmid factor. At point f the specimen essentially
becomes a single crystal, although a few acicular matrices are
still observable. Thus, the elongation between b and f ($\sim$7.2%)
corresponds to the one obtained by the shape strain of the stress-
induced transformation. The further increase of stress to point
g makes the specimen into a complete single crystal of the
martensite (at least in an optical microscopic scale). With further increase of stress no change is observable in the structure,
and the curve between g and h in the stress-strain curve is almost
linear. In some specimens, however, the nucleation of the
martensites during the process looked more like the Lüders band
propagation. Thus, the process of the transformation was slightly
different from specimen to specimen, but the important point to be
emphasized here is that the rate determining process for the
transformation is the nucleation of acicular martensites in both
cases, rather than the propagation of the pre-existing plates.
Upon unloading from point h, no structural change occurs until
point a' is reached, and at point b' some acicular matrices are
nucleated in the martensite. With decreasing strains more and
more matrices are nucleated, but the reverse transformation proceeds more or less like the Lüders band propagation in contrast
to the loading process as seen from the series of macrographs (a')

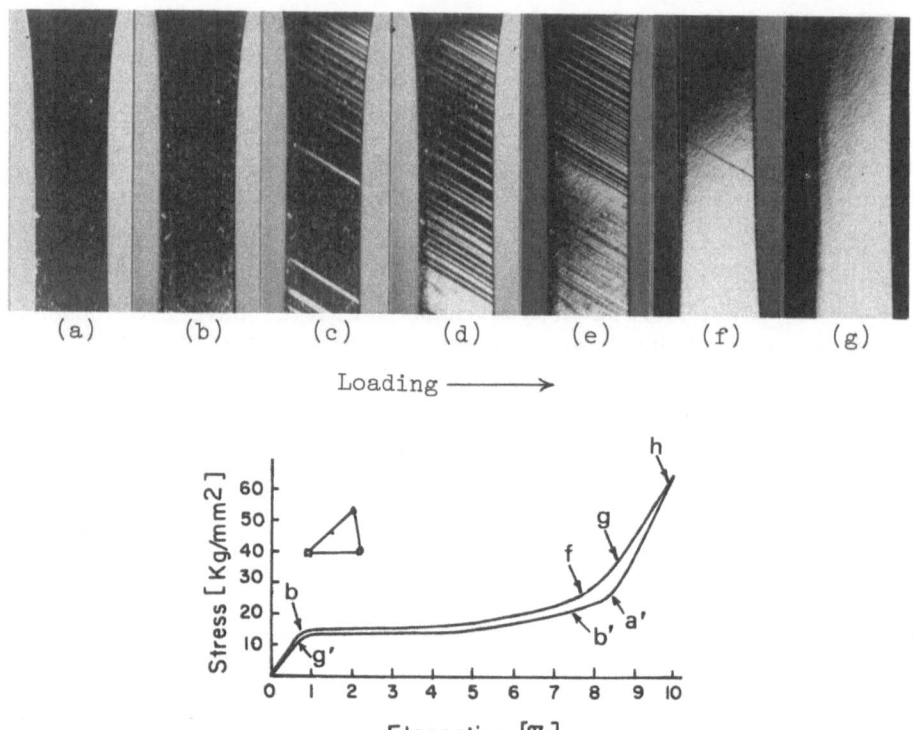

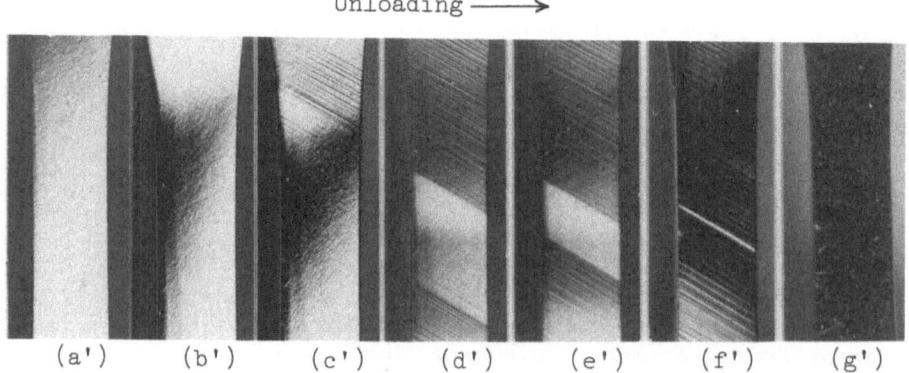

Fig. 5. Macroscopic change corresponding to the stress-
strain curve of the middle.

to (g'). At point g' the specimen is reverted to the original
matrix.
    Next the same process was observed on a more microscopic scale
using an optical microscope, and the result for the same specimen

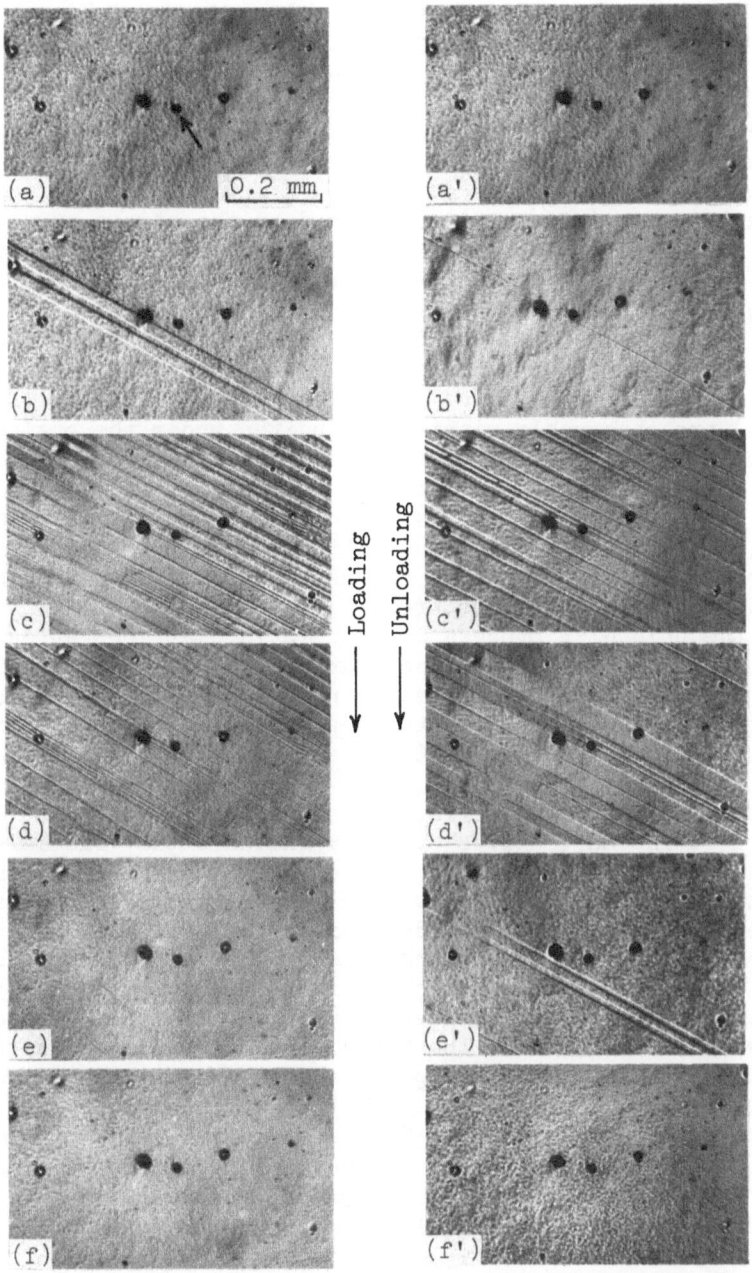

Fig. 6. Microstructural change corresponding to the stress-strain curve in Fig. 5. (a) to (f) correspond to loading and (a') to (f') correspond to unloading process. (e) corresponds to point f in the stress-strain curve of Fig. 5.

as for Fig. 5 is shown in a series of micrographs of Fig. 6. (a) to (f) correspond to the loading process and (a') to (f') correspond to the unloading process, all the micrographs being taken from the same area as seen from the existence of the same small hole (↑). (a) shows the initial β_1 matrix phase without stress. Upon loading no morphological change occurs in the linear elastic region of the stress-strain curve, and a few acicular martensites appear after the yield point is reached on the stress-strain curve, (b). With increasing strains, more and more martensites appear and after coalescence between the martensites the specimen eventually becomes almost a single crystal on a microscopic scale, (c) to (e), although a few traces of matrices are still observable in (e), which corresponds to point $\underline{f}$ in the stress-strain curve of Fig. 5. It is to be noted here that no internal defects (such as twins) are observable within martensites under the optical microscopy. With further increase of stress the specimen finally becomes a complete single crystal in (f). Although the outlook of (a) and (f) is quite similar, since both represent single crystal state, it is easy to tell by the difference in color, as stated in the introduction, that the two phases corresponding to (a) and (f) have different crystal structures.

On the unloading process a few acicular matrices appear after the point $\underline{a}$' in the stress-strain curve of Fig. 5 is reached (b'). With further decrease of stress the region of matrices increase, and finally the specimen reverts to the original β_1 phase in the same orientation as before without leaving any surface reliefs or upheavals in the specimen.

The following characteristics in the morphology of stress-induced acicular martensites strongly indicate that the martensite is of the 18R type β_1' structure described in the previous section : i) It has acicular morphology consistent with the β_1' martensite stress-induced by point indentation. ii) The lattice invariant strain is not observable under an optical microscope consistent with the fact that the stacking faults in long period stacking order structures are not observable under an optical microscope. iii) The trace analysis of the habit plane of the martensites in Fig. 5 was very close to $(\overline{1}\overline{4}4)_{\beta_1}$ consistent with that of the β_1' martensites shown in Fig. 1(c). Further, the direct evidence for the 18R structure was obtained by the back-reflection Laue method for another specimen subjected to the same morphological changes as Fig. 5 and 6. Thus, it is concluded that the $\beta_1 \rightleftarrows \beta_1$' martensitic transformation is responsible for the type of superelasticity as shown in Fig. 4(a) and Fig. 5.

Observation in the temperature range between Af and Ms: A typical macroscopic morphological change during the tensile test in the temperature range $Af > T > Ms$ is shown for the same specimen in Fig. 7, along with the corresponding stress-strain curve. With increasing stress, no morphological change is observed until the peak appears in the stress-strain curve. At the peak point a sin-

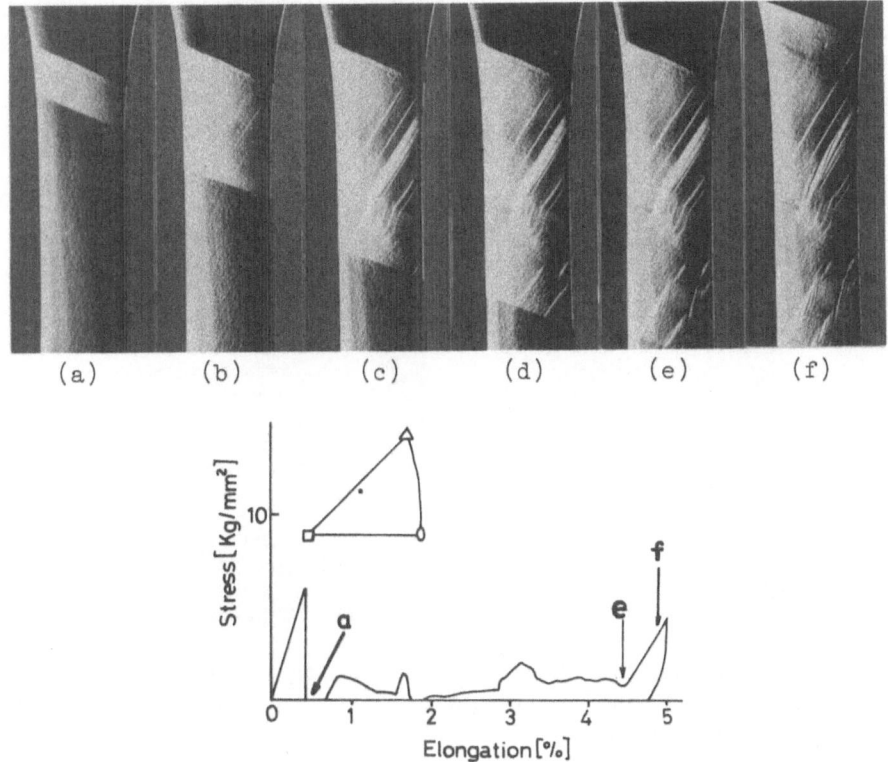

Fig. 7. Macroscopic change during a tensile test and the corresponding stress-strain curve. (a), (e) and (f) correspond
to a , e and f points of the curve.

gle variant of banded martensite appear and its interface between
matrix and martensite propagate with extremely high velocity at
this stress level. The rapid movement of the interface is associated with the rapid drop in stress as shown in the stress-strain
curve. With increasing strains thereafter the stress level does
not change greatly and the transformation proceeds by the propagation of both sides of interfaces of the pre-existing martensite
plate, no other martensites being nucleated in the untransformed
area. This is in sharp contrast to that described for the $\beta_1 \rightleftharpoons \beta_1'$
transformation. At point e the transformation within the guage
length is almost completed and the stress again goes up rather
rapidly. Thus, the strain between a and d (~4%) represents the
elongation associated with this stress-induced martensitic transformation. This value is quite different from that obtained for
$\beta_1 \rightleftharpoons \beta_1'$ transformation (~7.2%).

The same process as above was observed on a more microscopic
scale using an optical microscope, and the result is shown in Fig.
8. (a) shows the situation in which the interface of the martens-

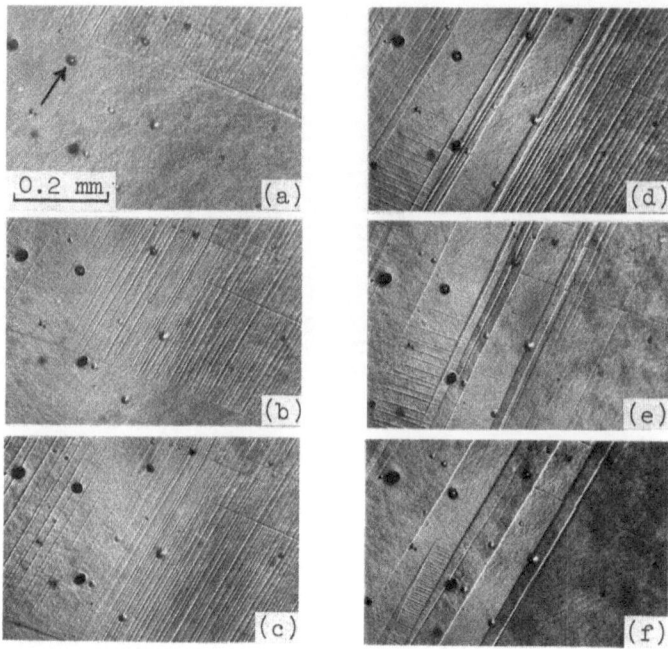

Fig. 8. Microscopic change corresponding to the stress-strain
curve in Fig. 7. (e) corresponds to the point e of the
stress-strain curve.

ite has just arrived in the view. It is clearly seen that many
fine striations are present within the martensite and they are
twins, just like those in the thermally formed $\gamma_1{}'$ martensite.
With increasing strains the interface moves gradually downwards,
leaving the twins behind. (e) represents the morphology corresponding to the point e in the stress-strain curve of Fig. 7, that
is, the one corresponding to the completion of the transformation
within the guage length. If the micrographs from (b) to (e) are
compared, it is clearly seen that the coalescence of twins has
started under stress long before the completion of the transformation. With further increase of stress, the twin with a favorable orientation is seen to grow at the expense of the other. However, it was not possible to eliminate the twins completely in
this particularly oriented specimen. Whether a specimen becomes
a single crystal martensite or not seems to depend upon the orientation of specimen, and in fact another specimen in different
orientation became a single crystal by a similar experiment. The
orientation dependence for becoming a single crystal can be explained by the Schmid factor analysis (32) for the twinning shear
in a given specimen, but the actual process in becoming a single
crystal seems more complicated.
     The following morphological characteristics described above

indicate that the stress-induced banded martensite is γ_1' phase
with the Cu_3Ti type ordered structure: i) Thick banded morphology
consistent with that observed in the single interface martensitic
transformation (26) in the same alloy. ii) Presence of twins inside the martensite which is characteristic of the γ_1' martensite.
The pole of the twin plane analyzed by the two surface analysis
was fairly close to $\{110\}_{\beta_1}$, which indicates that the twin plane
is $\{121\}_{\gamma_1'}$. iii) The habit plane of the martensite analyzed by
the two surface analysis was $(\bar{1}\bar{3}3)_{\beta_1}$, consistent with that of the
γ_1' martensite. Moreover, the back-reflection Laue pattern taken
from the single crystal martensite is consistently indexed by the
Cu_3Ti type γ_1' martensite structure. Therefore, it is safe to
conclude that the martensite stress-induced at temperature range
between Af and Ms and characterized by a sharp peak in the stress-
strain curve is the γ_1' martensite whose structure is the same as
that of thermally formed martensites in the same alloy.

(c) Occurrence of β_1' martensite to γ_1' martensite transformation

It was demonstrated in the above that the β_1' martensite is
stress-induced when the tensile test temperature is higher than
Af, and that the γ_1' martensite is stress-induced when the tensile
test temperature is between Af and Ms. Recently, the present
authors carried out an experiment (48) to know what will happen if
the temperature is lowered keeping the stress constant, after a
β_1' martensite single crystal has been stress-induced. As a result, a direct martensitic transformation from the β_1' martensite
to the γ_1' martensite was found to occur under certain conditions.
That is, a stress induced β_1' single crystal martensite was cooled
down to a temperature below Af keeping the stress constant, and
then unloaded at the temperature. At a certain stress level, a
banded γ_1' martensite appears in the β_1' single crystal martensite,
and the product martensite grow with decreasing stress and finally
becomes to the single crystal. The formation of the γ_1' martensite in the β_1' martensite is characterized by the existence of
habit plane and lattice invariant strain as in the ordinary transformation from matrix to martensite.
The details of the β_1' to γ_1' martensitic transformation will
be described in another paper of this symposium, and the similar
transformation is found independently and reported in this symposium by Rodriguez and Brown. Then, more description will not be
given here. But the important point to be emphasized here is that
such a martensite to martensite transformation may also be expected when a martensitic alloy is deformed in the marmem behaviour.

IV Deformation characteristics of the martensite phase.

In their earlier work of the marmem behaviour in Cu-Al-Ni
alloys (3, 40), the present authors observed the twin boundary
movement and the matrix-martensite interface movement under an ap-

plied stress, and interpreted the marmem effect on that basis. However, the later work stimulated by the investigation by other workers in other materials showed that the deformation mode in martensite phases are more versatile and complex. Thus, the deformation modes in Cu-Al-Ni alloys in partly or wholly martensitic state are discussed in this section, along with some other important deformation modes in other alloys. As such deformation modes, the following may be operative depending upon the condition given.

    i) Martensite-matrix interface movement,
   ii) Twin boundary movement of pre-existing transformation twin,
  iii) Deformation twinning with different twinning shear system,
   iv) Martensite-martensite interface movement,
    v) Formation of new variants of martensites in pre-existing
      martensites,
   vi) Transformation to another martensite, as described in the
      previous section.

Out of these, the deformation modes i)~iii) operate in a relatively low stress level, while those iv)~vi) operate in a rather high stress level. It is also important to notice that all the deformation modes listed above proceed by the movement of some kind of coherent or partially coherent boundaries, which capable of annihilating upon the reverse transformation. In fact, this characteristic in the deformation modes is responsible for the realization of the marmem effect, along with the reversible nature of the transformation.

    In the following, the deformation modes listed above are described more in detail in that order.

(a) Martensite-matrix interface movement.

    This is the most important deformation mode when specimens are partially transformed, and the example is shown in Fig. 9 (40), in which (a) represents the original state and (b) the state after bending. Obviously, the growth of martensites are observed by the interface movement. This mobile nature of the interface under

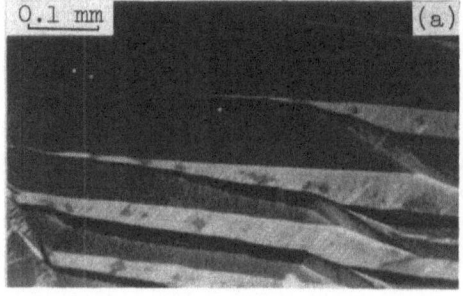

Fig. 9. Optical micrographs (taken in polarized light) showing the growth of martensites by external bending stress.

stress comes from the thermoelastic (or mechanically elastic) nature of the transformation. If the applied stress is favourable,
then the martensites grow as shown here, but if it is unfavourable,
then the martensites shrink on the contrary. If specimens are
heated thereafter, the reverse transformation proceeds by the re-
treat of these boundaries and they revert to the original orientation in the matrix phase, thus causing the marmem effect.

(b) Twin boundary movement of transformation twins.

   This type of deformation mode in martensites has been first
found in In-Tl (11, 12) and then in AuCd (49), and the "rubber-
like" behaviour in these alloys was explained on that basis.
Similar twin boundary motion was also observed in the γ_1' Cu-Al-Ni
martensite by straining specimens inside the electron microscope,
as shown in Fig. 10 (40), in which (a) represents the original
state and (b) represents the state after straining (Notice the reversal of contrast due to the change in diffraction condition).

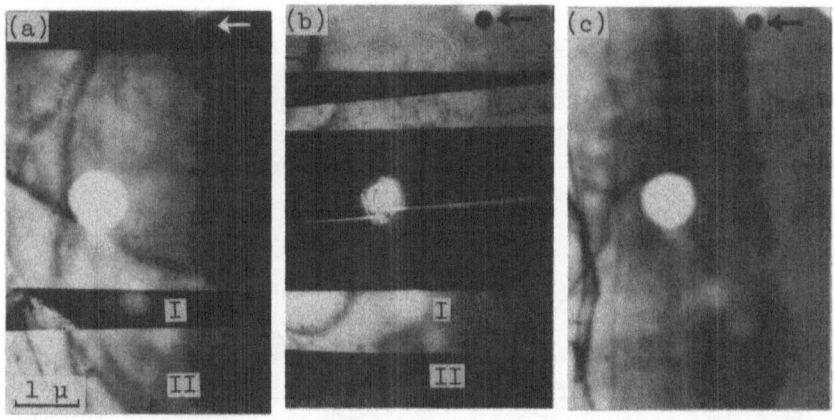

Fig. 10. Electron micrographs showing the twin boundary movement. ↑ is an indication of the same place. Twin I in (a)
   grows in (b) though the contrast is inversed (Cu-Al-Ni).

If the specimen is heated, these twins annihilate upon the reverse
transformation, as seen in (c), and the cause of the deformation
is removed completely, thus resulting in the marmem effect. Based
on these observations, the present authors (3, 40) and Wayman (41)
argued that one of the necessary conditions for realizing the marmem effect is that the lattice invariant strains in the martensite
are not slips by dislocation movement but twinnings. In fact, in
most of the marmem alloys internal twins are observed in the
martensites as listed in Ref. (1), and this argument is still
valid, since there is no reason why the slips occurred in the
martensite phase recover by the reverse transformation. However,

the case where the lattice invariant shear is stacking faults are
controvertial. In case of Cu-12.0(wt%)Al alloy, in which the
structure is a 18R type long period stacking order structure and
the lattice invariant shear is stacking faults on the basal plane,
the polycrystal specimens showed only a partial recovery in shape
upon reverse transformation (86). On the other hand, it is
recently reported in CuZn and CuZnX alloys (2, 9, 50) that they
exhibit perfect marmem effect even though they are not internally
twinned but faulted on the basal plane. Thus, is seems true that
even the faulted martensites potentially exhibit the perfect marmem effect, and the reason why Cu-12.0(wt%)Al does not show it may
be due to the eutectoid decomposition which may occur and inhibit
the perfect marmem effect during the reverse transformation, since
the Af temperature of this alloy is pretty high (about 400°C).
The reasons why faulted martensites can exhibit the marmem effect
are believed to lie partly in the fact that the slip is carried
out not by perfect dislocations but by partial dislocations which
bound the faults, and partly in the fact that there are other deformation modes described later in these martensites. The former
point will be described in detail elsewhere (87).

     There is another new fact concerning the crystal structures of
the above-mentioned CuZn and CuZnX martensites. The crystal
structures of these faulted martensites have been considered to be
of normal 9R or 18R type long period stacking order structure.
However, it was found recently (50, 51, 56) that the exact struc-

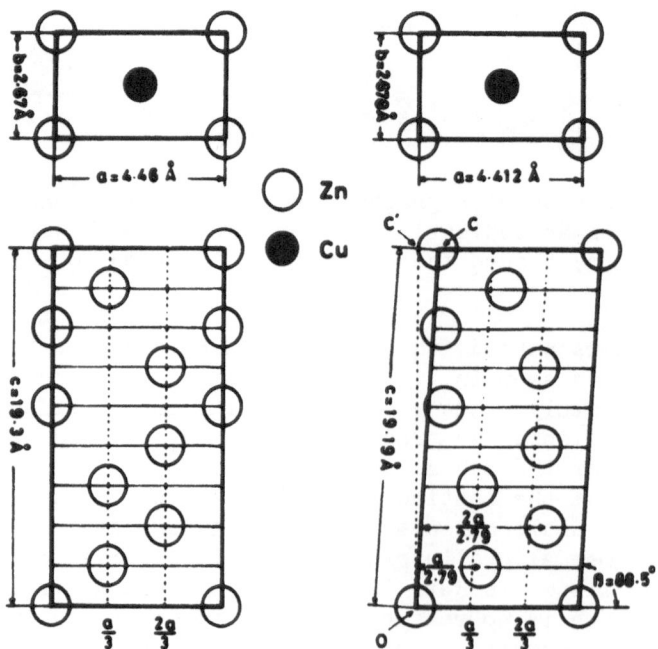

Fig. 11. The unit cell of the normal 9R structure (52)(a)
and the modified 9R structure (51).

ture of these martensites are slightly deviated from that of the
normal one, as shown in Fig. 11, and the new structure was termed
the modified 9R structure. The essential differences of the new
structure from the previous one are in that the stacking position
in each layer is deviated from the exact 1/3 or 2/3 position for
satisfying the close packing condition, and in that the unit cell
changes from an orthorhombic cell to a monoclinic cell because of
that. These modified structures are also found in CuZnGa (53),
CuZnAl (56) and AuCuZn$_2$ (54), and the deviation in the famous
Ölander structure in Au-47.5(at%)Cd marmem alloy is analogous to
those in the above modified structure. Therefore, the deviation
from the normal structure may be related to the marmem behaviour
for some reason.

Wasilewski (88) argued that the usual twinning in ordered
alloys is hardly possible, and thus proposed an alternative mechanism in which the boundary motion is made possible only by the
successive transformations from a martensite to matrix to another
martensite in the twinned orientation. It is true that the twinning in ordered alloys is difficult, as Laves predicted (57).
However, the twinning in ordered alloys is not always impossible.
In fact, in case of γ_1' Cu-Al-Ni martensite it has been shown by
using the Bilby-Crocker theory (58) of deformation twinning that
both {121}$_{\gamma_1}$' and {101}$_{\gamma_1}$' twinning are possible without destroy-
ing the original structure (55).

(c) Deformation twinning with different twinning shear system.

This type of deformation mode is shown in Fig. 12 (40). (a)
represent the original martensitic state. Fine striations on both
sides represent {121}$_{\gamma_1}$' transformation twins, and a vertical line
at the center is the interface between the two variants of
martensite. (b) represents the morphological change after slight
bending. Obviously new twins are introduced in the right-hand

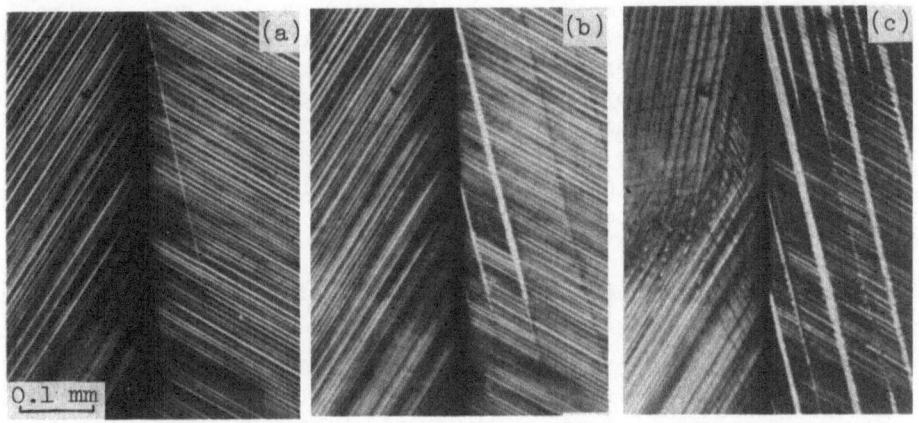

Fig. 12. Optical micrographs showing the occurrence of defor-
    mation twinning associated with a bending in Cu-Al-Ni alloy.

variant. These twins are believed to be $\{101\}_{\gamma_1'}$ twins because of
the geometry. With further bending (c), the same twins in the
right-hand variant increase both in size and number, and similar
new twins are introduced in the left-hand variant also, while the
original transformation twins contract in the upper left corner.
These twins disappear by the reverse transformation without leaving any traces, irrespective whether they are introduced by the
transformation or deformation.

(d) Martensite-martensite interface movement.

     This type of deformation mode is shown in the electron micro-
graphs of Fig. 13 for the β_1' martensite in a Cu-Al-Ni alloy with
low Al content. (a) represents the original state, and (b) represents the state after straining. As is seen with respect to a
fiducial marker (↑), one of the interfaces is obviously displaced.

Fig. 13. Electron micrographs showing the martensite-martens-
     ite interface movement for the β_1' martensite in Cu-Al-Ni.

Thus, the interface is mobile under the stress, but some contrast
is observable in the original position of the interface in Fig. 13
(b). This type of interface movement is also observed in Cu-Zn-Al
alloys (56, 59) by an optical microscope, and it is termed the reorientation of a martensite. Similar interface movement in the
γ_1' Cu-Al-Ni martensite as shown in Fig. 13 is also possible, but
it occurs usually at a high stress level. Thus, the twin boundary
movement is more important in this case at a usual low stress
level.

(e) Formation of another variant of martensites in pre-existing
     martensites.

     During the tensile tests of Cu-Al-Ni alloys in the complete
γ_1' martensitic state, the formation of new variants in pre-existing martensites were sometimes observed at higher stress level,
associated with the sharp stress drop in the stress-strain curve.
However, it is often hard to make distinction between new variants

and deformation twins described in Sec. (c) only by the optical microscope observations.

(f) Transformation to another martensite.

It was described in an earlier section that the martensitic transformation from the β_1' martensite to the γ_1' martensite occurred in a Cu-Al-Ni alloy by successive cooling and unloading. Thus, if the above-mentioned deformation modes are all used up, even the martensitic transformation from one structure to another may occur by deformation. In fact, such a structure change by deformation is observed in a few alloys. The thermally-formed β_1' martensite (18R) in Cu-Al alloys is reported to transform either to fcc or hcp structure by deformation, depending upon the Al content (60, 61, 62, 63). The β'' martensite (trigonal) in Au-49.5 % (at)Cd is reported to transform to the β' martensite (2H) under stress (64), and the modified 9R structure in CuZn is also reported to transform to an unknown structure, which is possibly a fcc (89). However, it is not clear whether or not fcc or hcp such as the γ_1' phase in Cu-Al-Ni alloys also transforms to another structure by deformation, because they are considered to be the stable phases. This point will be discussed more in detail in a separate paper presented in the present symposium. As is described in the same paper, it is confirmed in the Cu-Al-Ni alloy that specimen returns to the original dimension by the reverse transformation upon heating, even though they are subjected two successive martensitic transformations.

V. Reverse transformation characteristics of martensites.

The reverse transformation directly corresponds to the shape recovering process of the marmem behaviour (3). So the investigation of this process is important to know the mechanism of the marmem behaviour. However, it has not been studied systematically yet like those of the direct transformation and the deformation process. Thus, a few comments and some experimental results are presented in the following by focussing attention on the marmem behaviour.

As will be easily understood from the descriptions in the previous section and those of earlier works (3, 40), the complete shape recovery can be realized only when the following two conditions are satisfied,

i) A specimen reverts to the original orientation of the matrix phase by the reverse transformation (i.e. crystallographic reversibility).

ii) All lattice defects (including several boundaries described in the previous section), which cause the deformation in a wholly or partially martensitic state, annihilate by the reverse transformation.

The crystallographic reversibility is usually understood by the thermoelastic nature of the transformation. In this type of

transformation, the reverse transformation usually proceeds by the
reverse movement of the interface. So, if the matrix phase remains before the reverse transformation starts, a specimen regains
the original orientation by this process. Now, what will happen
if a specimen is completely transformed before the reverse transformation ? In this situation, the nucleation process must proceed prior to the reverse movement of the interface. Many experiments showed that only the variant in the original orientation is
chosen upon reversion, even though the nucleation process is involved.

   A typical example of this situation is shown in Fig. 14 (65).
(a) represents a γ_1' Cu-Al-Ni single crystal martensite without
any interface produced by the stress-induced transformation (32).
(b) clearly shows that a variant of the β_1 matrix phase is nucleated inside the γ_1' martensite by the reverse transformation.

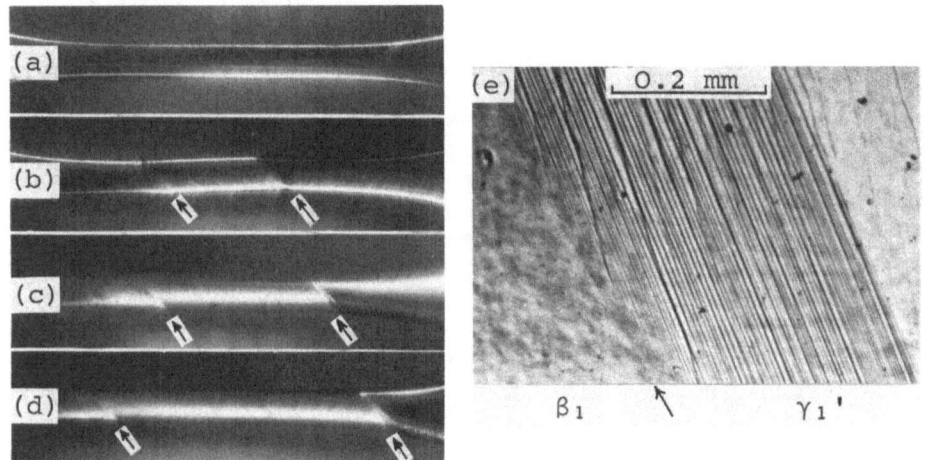

Fig. 14.   Macro- and micro-graphs showing the γ_1' to β_1 reverse
      transformation upon heating.

With further progress of the reverse transformation, the variant
grows ((b)$\sim$(d)) and the specimen reverts to the β_1 matrix phase
in the original orientation. (e) represents an enlarged micrograph of the interface during the process. It is interesting to
note that twins appear in the region near the interface inside
the γ_1' martensite phase, even though they are not present in the
initial state (a). The width of the twinned region is about a
few microns, and the width is maintained during the growth of the
β_1 matrix phase. The introduction of the twins may be ascribed
to the invariant plane condition at the interface similar to the
case of the direct martensitic transformation, but the interesting

point is that they are introduced not in the product phase (matrix) but in the parent phase (martensite). As a matter of fact, the phase, in which the lattice invariant shear is introduced, will be determined by the easiness of its availability in the two phases.

The crystallographic reversibility in the nucleation process is not physically obvious, and the reason why it is always realized in the thermoelastic transformation must be explained. Two sources are conceivable. One is sought in the ordered structure of the matrix and martensite (1, 66), which is usually observed in the thermoelastic materials as described in Sec. II, and the other is sought in the fact that the martensite structure usually has a lower symmetry than the matrix structure (90). That is, if the structure is ordered, the number of possible variants upon reverse transformation is limited to only a few and possibly one (i.e. only one in case of γ_1' Cu-Al-Ni martensite), since the choice of the variants other than the one in the original orientation usually destroys the matrix structure, if ordering is taken into account. In fact, it is confirmed in a disordered Fe-Ni alloy that several variants of matrices are nucleated by the reverse transformation, and that the original matrix single crystal become polycrystals after the successive martensitic and reverse transformations (67). Thus, these materials show only a partial memory effect (68, 69). Likewise, the lower symmetry of the martensite structure limits the possible number of matrix variants upon reversion to a few and perhaps even only one when the symmetry is very low. The energetic condition, such that if only one variant in the original orientation is nucleated, the total strain energy due to the presence of the interfaces be reduced, also favours the crystallographic reversibility, but that alone does not guarantee the crystallographic reversibility.

The necessity of the annihilation of internal defects, which were described in Sec. IV, upon reversion may be explained in the following way, by noticing that all these boundaries are coherent or semi-coherent (40). Let us start with the case of the twin boundaries. In case of Cu-Al-Ni alloys, two types of twin boundaries are present, such as $\{121\}_{\gamma_1'}$ and $\{101\}_{\gamma_1'}$. According to the lattice correspondence, these planes transform into the $\{110\}_{\beta_1}$ and $\{100\}_{\beta_1}$ planes, respectively. Since both are mirror planes, they can not be twin planes in the matrix phase. Thus, they must annihilate upon the reverse transformation. The annihilation of the stacking faults in 18R and 9R type martensites may be explained in a similar way. Again according to the lattice correspondence, the fault plane $(001)_{\beta_1'}$ in these structures transforme into $\{110\}_{\beta_1}$ planes in the matrix phase, on which stacking faults can not exist. Thus, they annihilate upon the reverse transformation. Now, how about the matrix-martensite and martensite-martensite interfaces ? We note that these interfaces are introduced because both sides of the interface have a different crystal structure or different orientation, and that they are not incoherent. Thus, if both sides revert to the matrix phase in the same orientation, it is a natural consequence that they annihilate.

Now, what will happen if the internal defects were perfect dislocations ? There is no reason why they must annihilate upon the reverse transformation. Rather it is expected that they inherit upon both direct and reverse transformations. In fact, if a shape memory experiment starts with a specimen, which has been deformed in the matrix phase, the specimen reverts to the pre-deformed configuration after the experiment. This simple experiment simply indicates that there must be some thing which inherit during both direct and reverse transformation in order to restore the original pre-deformed configuration. Those will be the perfect dislocations. The behaviour of partial dislocations, which are always associated with some kind of faults, upon transformation, will be different, and it will be discussed elsewhere (87).

Acknowledgements

The authors would like to thank Prof. C. M. Wayman, University of Illinois, for providing informations on his research and for many valuable comments on their research. Many thanks are also due to Messers. H. Sakamoto, K. Nakai, T. Nakamura and M. Takahashi for their collaboration on the present Cu-Al-Ni works. Finacial support over the course of the Cu-Al-Ni works has been provided as a series of the Grant-in-Aid for Fundamental Scientific Research (Ippan B, C, and Tokutei, 1972-73) by the Ministry of Education of Japan.

References

(1) C. M. Wayman and K. Shimizu, Met. Sci. J. $\underline{6}$ (1972), 175.
(2) L. Delaey, R. V. Krishnan. H. Tas and H. Warlimont, J. Mat. Sci., $\underline{9}$ (1974), 1521.
(3) K. Otsuka and K. Shimizu, Scripta Met., $\underline{4}$ (1970), 469.
(4) C. M. Wayman, I. Cornelis and K. Shimizu, Scripta Met., $\underline{6}$ (1972), 115.
(5) K. Otsuka, K. Shimizu, I. Cornelis and C. M. Wayman, Scripta Met., $\underline{6}$ (1972), 377.
(6) A. Nagasawa, K. Enami, Y. Ishino, Y. Abe and S. Nenno, Scripta Met., $\underline{8}$ (1974), 1055.
(7) W. A. Rachinger, Brit. J. Appl. Phys., $\underline{9}$ (1958), 250.
(8) H. Pops, Met. Trans., $\underline{1}$ (1970), 251.
(9) J. Perkins, Met. Trans., $\underline{4}$ (1973), 2709.
(10) A. Ölander, Z. Krist., $\underline{83A}$ (1932), 145.
(11) M. W. Burkart and T. A. Read, Trans. AIME, $\underline{197}$ (1953), 1516.
(12) Z. S. Basinski and J. W. Christian, Acta Met., $\underline{2}$ (1954), 101.
(13) L. C. Chang and T. A. Read, Trans. AIME, $\underline{191}$ (1951), 47.
(14) D. S. Lieberman, "Phase Transformations", (1971), 1, Metal Park, Ohio (Amer. Soc. Metals).

(15) H. Tas, L. Delaey and A. Deruyttere, J. Less-Common Met., 28 (1972), 141.
(16) H. Warlimont and L. Delaey, Prog. Mat. Sci., 19 (1974).
(17) C. M. Wayman, private communication.
(18) K. Takezawa and S. Sato, J. Jap. Inst. Met., 37 (1973), 793.
(19) M. J. Duggin and W. A. Rachinger, Acta Met., 12 (1964), 529.
(20) M. J. Duggin, Acta Met., 14 (1966), 123.
(21) G. V. Kurdjumov and L. G. Khandros, Dokl. Acad. Nauk USSR, 66 (1949), 211.
(22) D. Hull and R. D. Garwood, J. Inst. Metals, 86 (1957-58), 485.
(23) K. Otsuka and K. Shimizu, Jap. J. Appl. Phys., 8 (1969), 1196.
(24) K. Otsuka and K. Shimizu, Trans. Jap. Inst. Met., 15 (1974), 103.
(25) H. Tas, R. V. Krishnan and L. Delaey, Scripta Met., 7 (1973), 2833.
(26) K. Otsuka, M. Takahashi and K. Shimizu, Met. Trans., 4 (1973), 2003.
(27) K. Otsuka and K. Shimizu, Phil. Mag., 24 (1971), 481.
(28) K. Otsuka, T, Nakamura and K. Shimizu, Trans. Jap. Inst. Met., 15 (1974), 200.
(29) Z. Nishiyama and S. Kajiwara, Jap. J. Appl. Phys., 2 (1963), 478.
(30) S. Kajiwara and Z. Nishiyama, ibid, 3 (1964), 749.
(31) K. Otsuka, K. Nakai and K. Shimizu, Scripta Met., 8 (1974), 913.
(32) K. Otsuka, K. Nakai, H. Sakamoto, K. Shimizu and C. M. Wayman to be published.
(33) K. Otsuka, T. Nakamura and K. Shimizu, Trans. Jap. Inst. Met., 15 (1974), 211.
(34) A. B. Greninger and V. G. Mooradian, Trans. AIME, 128 (1938), 337.
(35) C. S. Barrett and T. B. Massalski, "Structure of Metals" 3rd Ed., McGraw-Hill Book Comp., New York, (1966), 529.
(36) I. A. Arbuzova, G. V. Kurdjumov and L. G. Khandros, Fiz. Met. i Metall., 11 (1961), 272.
(37) K. Otsuka, M. Takahashi and K. Shimizu, Intern. Cong. on Cryst. Growth, Tokyo, Japan, March (1974), Coll. Abstracts, 191.
(38) K. Oishi and L. C. Brown, Met. Trans., 2 (1971). 1971.
(39) W. J. Buehler, J. V. Gilfrich and R. C. Wiley, J. Appl. Phys., 34 (1963), 1475.
(40) K. Otsuka, Jap. J. Appl. Phys., 10 (1971), 571.
(41) C. M. Wayman, Scripta Met., 5 (1971), 489.
(42) R. W. Cahn, Il Nuovo Cimento (Suppl), 10 (1953), 350.
(43) J. W. Christian, "Theory of Transformation in Metals and Alloys", Pergamon Press, Oxford, (1965).
(44) D. P. Dunne and C. M. Wayman, Met. Trans., 4 (1973), 147.
(45) N. Nakanishi, Bull. Jap. Inst. Met., 11 (1972), 435.
(46) G. Hausch and H. Warlimont, Acta Met., 21 (1973), 401.

(47) R. V. Krishnan and L. C. Brown, Met. Trans., 4 (1973), 432.
(48) K. Otsuka, H. Sakamoto and K. Shimizu, to be published in
     Scripta Met., 9 (1975), No. 5.
(49) H. K. Birnbaum and T. A. Read, Trans. AIME, 218 (1960), 662.
(50) T. Tadaki, M. Tokoro and K. Shimizu, to be published in
     Trans. Jap. Inst. Met., 16 (1975), No.5.
(51) T. Tadaki, M. Tokoro and K. Shimizu, Scripta Met., 8 (1974),
     1077.
(52) S. Sato and K. Takezawa, Trans. Jap. Inst. Met., 9 (1968)
     Suppl. 925.
(53) T. Saburi, S. Nenno and S. Kato, Annual Autumn Meet. of Jap.
     Inst. Metals, Osaka, Japan, Oct. (1974).
(54) H. Kubo, Y. Yoshida and K. Shimizu, to be published in Trans.
     Jap. Inst. Met., 16 (1975).
(55) K. Otsuka and K. Shimizu, Trans. Jap. Inst. Met., 15 (1974),
     109.
(56) K. Takezawa and S. Sato, Proc. Symp. on Atomic Interaction in
     Sol. Solut., held at Shirahama, Japan, Jan. (1975).
(57) F. Laves, Naturwissenschaften, 39 (1952), 546.
(58) B. A. Bilby and A. G. Crocker, Proc. Roy. Soc. (London), Ser.
     A, 288 (1965), 240.
(59) H. Tas, L. Delaey and A. Deruyttere, Z. Metallk., 64 (1973),
     855.
(60) A. B. Greninger, Trans. AIME, 133 (1939), 204.
(61) I. Isaitschev, E. Kaminsky and G. V. Kurdumov, Trans. AIME,
     128 (1938), 361.
(62) S. Kajiwara, Trans. Jap. Inst. Met., 9 (1968) Suppl., 543.
(63) H. Tas, L. Delaey and A. Deruyttere, Scripta Met., 5 (1971),
     1117.
(64) N. Nakanishi, Proc. Symp. on Thermoel. Mart. Transf. and
     Shape Mem. Effect, held at Tokyo, Japan, June (1974).
(65) H. Sakamoto, K. Shimizu and K. Otsuka, unpublished work.
(66) K. Otsuka, Dr. Eng. Thesis, Univ. of Tokyo, Japan, (1972).
(67) H. Kessler and W. Pitsch, Acta Met., 15 (1967), 401.
(68) A. Nagasawa, J. Phys. Soc. Jap., 30 (1971), 1505.
(69) K. Enami, S. Nenno and S. Minato, Scripta Met., 5 (1971), 663.
(70) M. Suezawa and K. Sumino, Annual Spring Meet. of Phys. Soc.
     Japan, (1973).
(71) D. S. Lieberman, M. S. Wechsler and T. A. Read, J. Appl.
     Phys., 26 (1955), 473.
(72) S. Zirinsky, Acta Met., 4 (1956), 164.
(73) L. Guttman, Trans. AIME, 188 (1950), 1472.
(74) D. B. Novotny and J. F. Smith, Acta Met., 13 (1965), 881.
(75) N. Nakanishi, Y. Murakami and S. Kachi, Scripta Met., 5 (1971),
     433 .
(76) Y. Murakami, N. Nakanishi and S. Kachi, Jap. J. Appl. Phys.,
     11 (1972), 1591.
(77) A. L. Titchener and M. B. Bever, Trans. AIME, 198 (1954), 303.
(78) K. Otsuka, T. Sawamura and K. Shimizu, phys. stat. sol., 5

(1971), 457.

(79) Y. K. Au and C. M. Wayman, Scripta Met., 6 (1972), 1209.

(80) S. Chakravorty, Ph. D. Thesis, Univ. of Illinois, (1975).

(81) H. C. Tong and C. M. Wayman, Scripta Met., 7 (1973), 215.

(82) D. M. Masson and C. S. Barrett, Trans. AIME, 212 (1958), 260.

(83) K. Enami, S. Nenno and K. Shimizu, Trans. Jap. Inst. Met., 14 (1973, 161.

(84) A. Nagasawa, J. Phys. Soc. Japan, 35 (1973), 489.

(85) T. Tadaki and K. Shimizu, Trans. Jap. Inst. Met., 11 (1970), 44.

(86) K. Shimizu, H. Okamoto and K. Otsuka, unpublished work.

(87) K. Otsuka and C. M. Wayman, to be published.

(88) R. J. Wasilewski, Scripta Met., 5 (1971), 127.

(89) C. M. Wayman and I. Cornelis, to be published in Scripta Met., 9 (1975), No. 4.

(90) R. J. Wasilewski, Scripta Met., 5 (1971), 131.

CRYSTALLOGRAPHY AND THERMODYNAMICS OF SME-MARTENSITES

L. Delaey ° and H. Warlimont[x]

° Dept. Metaalkunde, K.U.L. Leuven (Belgium)

[x] Alusuisse, Forschung und Entwicklung, Neuhausen (Switzerland)

Introduction

SME-martensites are characterized by stress-induced macroscopic shape changes and reversal of these shape changes by simple heating. The shape changes result from stress-induced martensite formation or reorientation. The essential parameters in which the martensitic phase transition will be described are structural, crystallographic and thermodynamic. Three conditions are essential for the occurrence of the shape memory effect : 1. a low energy of nucleation (if the SME is associated with transformation) such that the martensite plates form at moderate rates of growth and, thus, with a high degree of structural perfection and reversibility; 2. a medium degree of frictional stress during growth or reorientation, respectively, such that structural reversibility is maintained while the impediment of interface motion is sufficient to prevent isothermal reversion; 3. a high capacity for elastic energy storage permitting large shape changes to be realized without the production of irreversible defects. The purpose of the present contribution is to show the close interrelationship, which exists for the SME martensites between these three parameters.

The structural transitions will be analysed in the first part taking into account the pre-martensitic properties and transitions of the parent phase. Among other points it is demonstrated that the β'(3R) type martensite with an ABCBCACAB stacking sequence of the close packed planes offers no difficulties in understanding its thermoelastic and SME-behaviour.

The microstructural changes are discussed in the second part, in particular with regard to the question as to how a macroscopic shape change can be achieved. The mechanical coupling which exists between the self-accommodating or strain-accommodating martensite plates and the significance of the orientation of a martensite plate with respect to the stress-tensor will form the centre of discussion.

Due to the interchangeability of temperature and stress as variables affecting the transformation and thus the nucleation and growth of a martensite plate, energetic considerations and estimates concerning the SME martensites will be given in a third part.

The discussions will be confined to the bcc β-phase alloys, such as the Cu-, Ag- and Au-based alloys (1), the NiTi and NiAl β-phase alloys and Ti- and Zr-based alloys, for which many crystallographic as well as thermodynamic data are available. The transformation in these systems is also characterized by a high degree of crystallographic perfection; moreover, all of these systems have a common feature which is the positive temperature coefficient and low absolute value of the shear elastic constant $C' = (C_{11} - C_{12})/2$ in the vicinity of the transformation temperature. The In-Tl alloys, being also characterized by a positive temperature coefficient of C', and the other alloy systems showing SME, such as the Fe-Pt, Fe-Be, Fe-Ni, Fe-Cr-Ni, Fe-Mn-C and cobalt alloys, will only be discussed where qualitatively different.

A. Structural Transitions

1. The β-phase and premartensitic lattice transitions

It has often been argued that the premartensitic structural changes and the positive sign of dC'/dT are strongly linked with the nucleation mechanism of martensite. Although no detailed nucleation mechanism can yet be established, some important properties of the β-phase and the local structural changes occurring prior to the martensite formation will be discussed in more detail in order to give a short summary of the factors which should be taken into account in constructing a nucleation model. In addition some observations of a structural transition, which could be one of the nucleation steps, will be shown. Furthermore, a model will be proposed based on zonal dislocations, shear strain and simple atomic shuffling, which allows to derive the 3R structure directly from the bcc β-phase.

One of the important properties of the β-phase alloy systems is their high quenched-in vacancy concentration, which may be as high as 1 %. It can be shown (2) that even a small degree of vacancy clustering is sufficient to give rise to local lattice instabilities in the β bcc-lattice. Energetic calculations have also shown that lo-

cal atom rearrangements take place leading to sheared and shuffled
zones with a structure corresponding to the ω-structure, to collaps-
ed regions with a hexagonal structure and to dislocation loops.
Electron microscopy of the β-phases has shown anomalous mottling
and diffraction effects (3). The mottling is due to localized
structural singularities and to small dislocation loops (9).
The diffraction effects can be explained in terms of ω-type and
shear-type displacements (3,10). Atomic displacements (transversal
waves associated with a 1/3 <110> wave vector and a <1$\bar{1}$0> polarization vector, and a 1/2 <110> wave vector and a <1$\bar{1}$0> polarization
vector) were found which correspond to those observed in NiTi alloys
(12). These pre-martensitic diffraction phenomena have been interpreted in terms of softening of certain vibrational modes in the
high temperature structure as the temperature is decreased; this
new concept has been referred to as phonon nucleation (13). Fig. la
shows that ω-type diffraction spots not only occur in CsCl-ordered

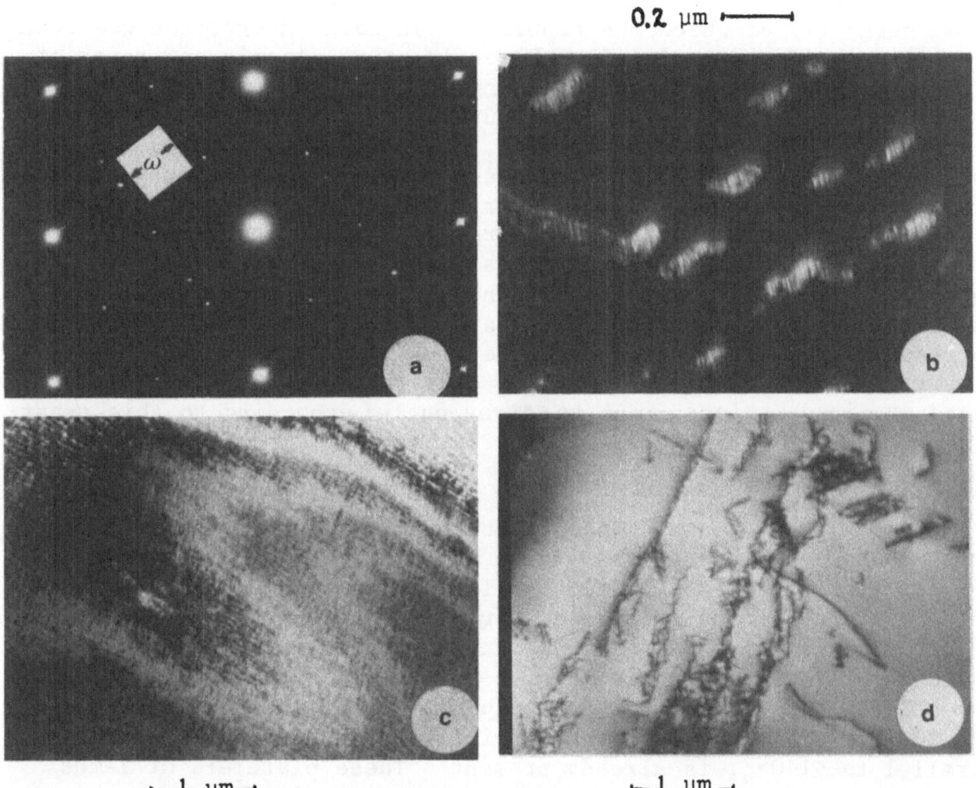

0.2 μm ⊢────┤

Fig. 1 : Electron diffraction pattern and micrographs corresponding
         to the β to ω transition in β-Cu-Zn-Al.
         1a : ω-type diffraction spots in a DO_3-ordered β-phase;
         1b : ω-type domains in quenched β-phase;
         1c : ω-domains in a previously deformed and annealed β-phase
         1d : dislocation structure after flash-heating

β-phases but also in DO_3-ordered β-phases. The ω-type domains (figure 1b), obtained in alloys which were previously deformed and annealed, can be large in size if compared with the mottled areas usually observed in quenched β-phases (figure 1c). Figure 1d, taken from a β-phase after cycling through the M_s-temperature and a subsequent flash-heating and quenching treatment, shows only dislocations. No ω-type diffraction spots and mottling could be detected, although martensite could be induced by simple cooling. From this experiment it can be derived, that the appearance of ω-type or shear type displacements at temperatures above but not close to M_s is not a prerequisite for martensite formation.

The most important common property is the temperature dependence of the C' elastic constant. C' shows a positive temperature coefficient up to 200 K and more above M_s and, moreover, C' has a low absolute value on approaching M_s (11) indicating an instability with respect to {110} <1$\bar{1}$0> shear. This anomalous elastic behaviour has been treated in terms of the different contributions to the lattice energy (4) of which the repulsive ion core interaction was found to play an essential role (5). Recently, the elastic constants of CuZn have been redetermined including the third and fourth order elastic constants (6). The contribution of the next nearest neighbours to the third order constant C_{111} was found to be greater than the nearest neighbour contribution and to stabilize the structure with respect to the <1$\bar{1}$0>{110} shear. Taking higher order elastic constants into account the free energy versus lattice strain relation was calculated, which indicates the existence of a "strain spinodal" (7). Thus, it can be found, at least for β-CuZn, that the lattice can be decomposed into areas with higher and lower strain corresponding to the initial and the final structure. The local distorted areas discussed above are sites which could act as localized strained regions prone to act as nuclei according to this treatment.

Recent electron microscopic observations show that, at least in the observed samples, transition structures different from β but very similar to martensite occur. In a splat cooled Au-47.5 at % Cd alloy internal small platelet formation was observed and the electron diffraction pattern was interpreted as being due to a 3R-structure, <100> zone axis (fig. 2 of ref. 14). However, a careful analysis shows that the spots do not lie on the exact 3R reciprocal lattice position, although a tripling of the repeat distance, parallel to $<110>_\beta$, is already present. These platelets or laths behave thermoelastically. In order to confirm similar observations in other alloy systems, a β-Cu-Zn-Al alloy with an M_s-temperature close to but below room temperature was analysed by electron microscopy (8). Figures 2a and b show an early resp. progressed state of martensite lath formation, micrographs which are remarkably similar to those observed in Au-47.5 at % Cd. The electron diffraction pat-

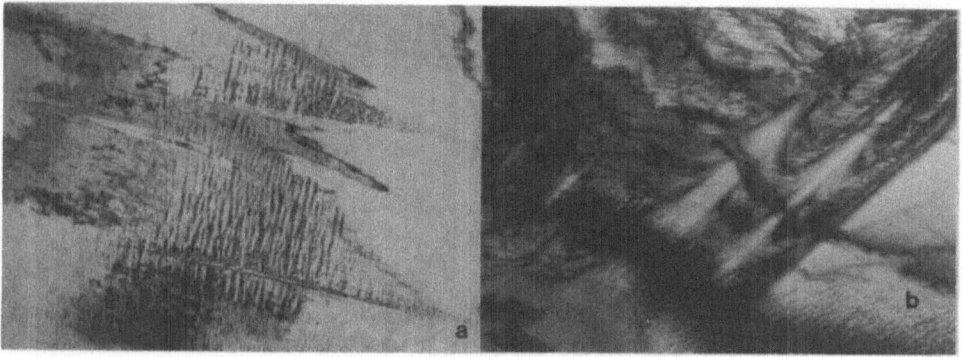

├─1 µm─┤   ├─1 µm─┤

<u>Fig. 2</u> : Electron transmission micrographs of the intermediate mar-
         tensite lath formation in β-Cu-Zn-Al.

terns confirm the observation in Au-Cd. The electron diffraction
pattern was a superposition of the $<100>_\beta$ -pattern and the pattern
corresponding to the new phase which is characterized by a tripling
of the repeat distance.

     Similar observations were made in AgCd-alloys, but here, an
intermediate phase characterized by a doubling of the repeat distance is formed (15).

     Some attempts have been made already to explain this two-step
transformation from β to an intermediate structure in lath-shaped
regions and, finally, to the 3R or 2H martensite. Using the group
theory approach to phase transitions, it has been argued that the
a crystal structure intermediate between that of β and the resp.
martensite is required (17). The point group to which the product
phase belongs should be a subgroup of the point group of the parent
phase (16,17).

2. The β to 3R- and 2H-martensite transition

     With the atomic displacements discussed above and the inter-
mediate martensite structure of the initial stage of transition in
mind, a geometric model is constructed and proposed which allows the
direct derivation of the 3R-stacking sequence (ABCBCACAB) from the
β-phase, without introducing slip or stacking faults (9). This geometric model allows, also, to explain the diffraction patterns of
the intermediate phase and the other observations.

     The theory of zonal dislocations and of lattice matching, ap-
plicable to twinning was recently proposed to describe the martensite
transition (19). It was applied to the β → 3R and β → 2H transi-

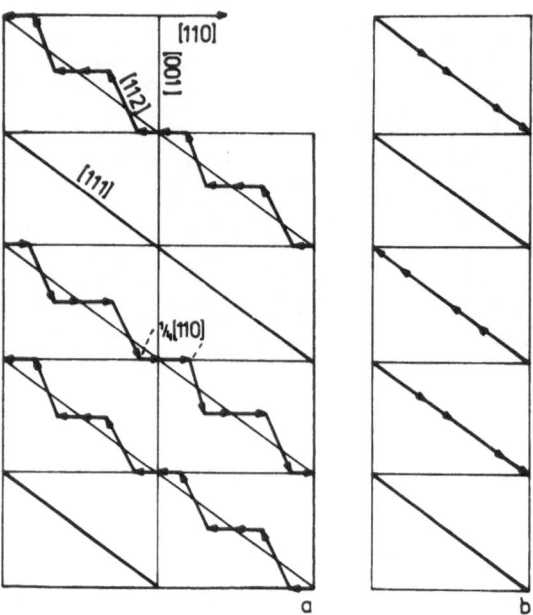

Fig. 3 : Distribution of the dissociated dislocations in the (1T̄0)
plane of the bcc β-lattice (ref. 9).

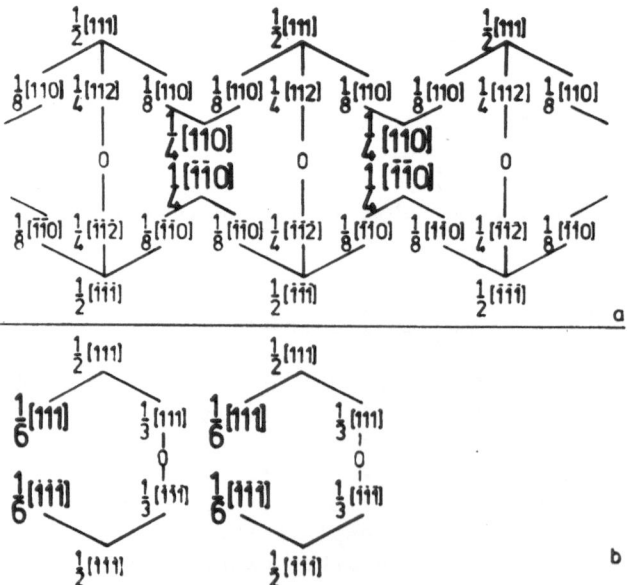

Fig. 4 : Dislocation dissociation and reactions used in the transformation model (ref. 9).

tion in β-brass type alloys (9). In the present paper, the theory will not be explained in detail since it will be discussed by S. Mendelson at the same symposium (19). The basic idea is the introduction of a shuffle mechanism which consists of the dissociation of dislocations combined with an additional small shear. In order to apply this method the presence of a sufficient number of dislocations in the β-phase is required; these dislocations are thought to be dissociated in the habit plane. The dislocations may be formed during the premartensitic lattice stabilisation and/or during the growth of the martensite plates. In addition, an acceptable lattice matching along the normal to the plane of shear, the normal to the habit plane and along the direction of shear is required between the parent and the martensite phases. In order to fulfil these requirements a second shear is introduced.

Among the various dissociation mechanisms discussed in the literature the following two were selected

$$1/2 \ [1 1 \bar{1}] \qquad 1/3 \ [1 1 \bar{1}] \ + \ 1/6 \ [1 1 1] \qquad \cdots \ [1]$$

and $\quad 1/2 \ [1 1 \bar{1}] \qquad 1/8 \ [1 1 0] \ + \ 1/4 \ [1 1 2] \ + \ 1/8 \ [\bar{1} 1 0] \qquad \cdots \ [2]$

The dissociated dislocations were distributed in the β -lattice as represented by fig. 3. The dissociation and dislocation reactions which are allowed to occur are given schematically in figure 4. The two reactions, [1] and [2] , together with their spatial arrangement lead to two atomic movements, a 1/6 <111> movement and a 1/4 [1 1 0̄] movement. Thus, the β-lattice consists of rows of atoms which are displaced either in the <111> - or in the <110> -direction. Due to the low C' elastic stiffness the <110> -movement will be predominant.

The resulting cooperative, alternating atom displacements have been transposed into the three-dimensional β-lattice, a front view of which is given in figure 5 (projection on a (0̄1̄0)-plane). The structure shown in figure 5a is characterized by a tripling the repeat distance [1̄0̄1̄] and the structure shown in 5b by a doubling of the repeat distance [1̄0̄1̄]. The structures represented by figure 5 can now explain some of the diffraction patterns.

The atomic positions as shown in figure 5a can be transferred into the exact 3R-(orthorhombic)-positions by introducing a second shear (1̄01̄) [1̄0̄1̄] with γ = 6°. With a 9° shear the atoms can almost be transfered into a 2H position, but small adjustments are required.

A good agreement exists for 2H if the second shear is applied to the structure represented by figure 5b. However, according to Mendelson's theory the lattice matching should be controlled for 3R and 2H, i.e. the amount of extra shear η_{12} should be calculated. The calculations show a shear angle γ = 3 to 6° which means

L. DELAEY AND H. WARLIMONT

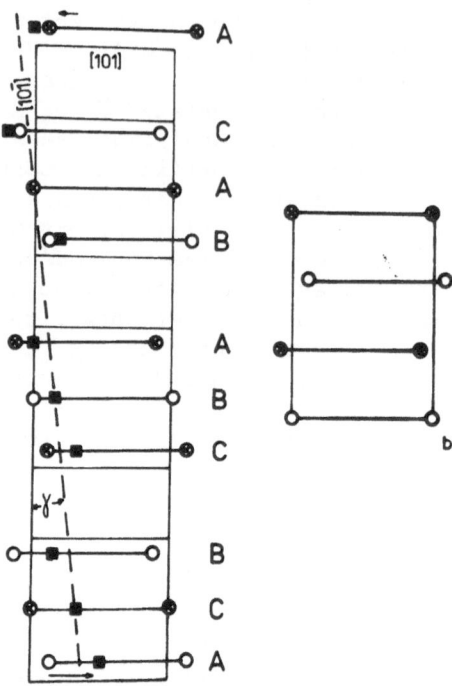

Fig. 5 : Atomic displacements in the β-lattice to form the 3R-struc-
         ture (a) and the 2H-structure (b). Projection onto the
         $(0\bar{1}0)_\beta$-plane. o, ⊗ displacements due to the shuffling, ■
         final displacements into the 3R-atom positions (ref. 9).

that 3R should be monoclinic. Indeed, the β_2(3R)-Cu-Zn martensite
has been proved to be monoclinic(20). In order to obtain a 2H-
structure a shear angle of 9° was calculated (9).

By calculating the rotation of the habit plane $\bar{K}_{11}$ assumed in
the theory, it could be shown that the rotation is 2 to 3° for 3R at
39 at % Zn and 8 to 9° at 42 at % Zn. The rotation for 2H is 30° at
39 at%Zn but only 1 to 2° at 42 at % Zn. It could also be shown
that the matching for an f.c.c. martensite is best at lower zinc-
concentrations. Figure 6 is a schematic representation of the rotation of the habit-plane versus concentration, showing the concentration dependence of the different martensite structures.

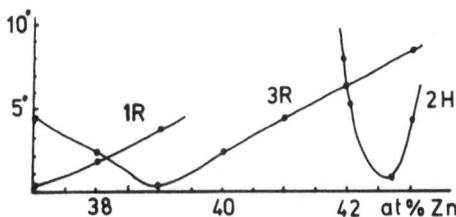

<u>Fig. 6</u> :   Rotation of the habit with composition and crystal struc-
            ture (ref. 9).

Thus, a geometrical model exists for obtaining the 3R-structure from
the bcc β-phase by a combination of shuffling and shear, a mechanism
which was already proposed earlier (10). By adopting this mechanism
many of the apparent anomalous structural features of the β-phase
mentioned above can be explained.

3. Conclusion

    In view of the elastic properties, lattice displacements,
electron microscopic observations and geometric calculations, any
nucleation model and transition mechanism (18,15,14,13) will have
to take these features into account.

    Due to local lattice instabilities at internal and free sur-
faces and lattice defects, local strained regions are created in the
β-phase; the elastic constant C', which becomes very small on approaching the M_s temperature favours static lattice displacements
with respect to a $(110)_\beta [1\bar{1}0]_\beta$ shear; taking the higher order elastic constants into account, a strain spinodal decomposition in the
vicinity of free surfaces and lattice defects can be assumed to occur; it has been proved that in a copper-, a silver- and a gold-
based alloy, a quasi-martensitic product intermediate between the

parent and the final 3R or 2H martensite exists; a geometrically acceptable model can be constructed for the structural transition
intermediate martensite → final martensite.

Before these findings can be fully reconciled to form an energetically acceptable model for nucleation, more experiments and calculations are needed to confirm these findings and to elaborate the
interrelationships. Furthermore, the "critical event" of nucleation
has not yet been identified by structural observations or a theoretical model for the bcc to close-packed structures martensitic transformations.

To understand the very nature of stress-induced martensite
formation, the influence of an externally applied stress on the various aspects of nucleation should be studied and measured carefully.

B. The Microstructural Analysis of Martensite Induced Shape Deformation

The microstructural analysis is concerned with three problems :
1. The selectivity of the stress-induced martensite plate variant
with changing the parent crystal orientation, taking into account
the macroscopic shear;   2. The amount and direction of the recoverable shape strain, taking into account the mechanical coupling
between the martensite plates;   3. The mechanical metallurgical
treatment of the influence of a complex stress distribution on a
polycrystalline microstructure.

The most straightforward situation is that of a single crystal in which a single martensite variant has been induced, a case
which is depicted in figure 7a. A single martensite plate variant
M is formed by the application of a tensile load on a parent single
crystal. This martensite plate is characterized by a habit plane
normal $\bar{p}_1'$ and the macroscopic martensite shear direction $\bar{d}_1'$.

If a 3R-type of martensite is taken as an example, $\bar{p}_2$ represents the normal to the close packed basal plane and $\bar{d}_2$ direction
of secondary shear. Arneodo and Ahlers (22,23) studied in detail
the deformation behaviour of a β-phase single crystal and a 3R-
martensite single crystal. Their results can be discussed in terms
of the highest resolved shear stress component. After having been
stressed to a stress-level above σ_M a β-phase single crystal shows
surface traces due to the formation of plates belonging to a single
martensite variant. The habit plane normal $\bar{p}_1$ and the shear direction $\bar{d}_1'$ were measured and plotted in such a way that the orientation
of the specimen axis was located within the unit triangle shown in
figure 8, which is the triangle with the highest shear stress com-

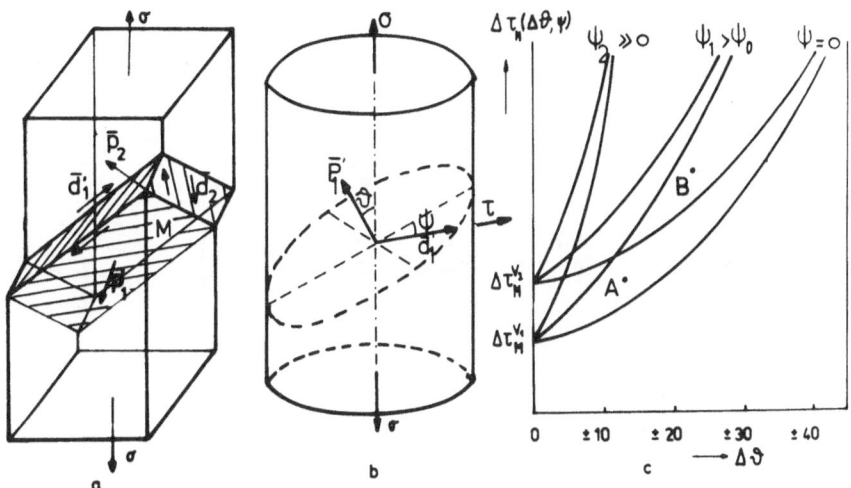

Fig. 7 : Schematic representation of the stress-induced formation of a single martensite variant in a single β-phase crystal.
a. Geometry with the position of habit plane and basal plane.
b. Relation between uniaxial stress and the shear system resolved on the macroscopic shear system $(\overline{p}_1, \overline{d}_1)$ of a martensite plate.
c. Orientation dependence of the shear stress for martensite growth. The stress acting on an orientation variant V_1 may first be characterized by point A. The rotation during tensile deformation may take the path A-B such that the orientation variant V_2 is favoured. It should be noted that a rotation of the crystal axes is actually equivalent to relative displacement of the two sets curves along the ordinate in addition to the rotation A-B shown. This displacement is not shown for clarity.

ponent on the $\overline{p}_1'$, $\overline{d}_1'$ system. In the same figure the present authors plotted the habit plane normals (●), the basal plane normals (▲), together with the shear directions as determined elsewhere (25) for the β-to-3R trnasformation. The latter calculations were performed for Cu-Al alloys but it has been shown (26) that the same relations hold for the β-to-3R transformation in Cu-Zn alloys. As can be seen in figure 8, the measured data (22) coincide with the calculated habit plane normal and basal plane normal of the 3R-martensite plate variant 1 of the self-accommodating group of martensite plate variants 1,2,3 and 4.

If the crystal axis is close to [001], it can be seen that both the habit plane normal and the secondary shear plane normal are favourably oriented with respect to the tensile axis, whereas

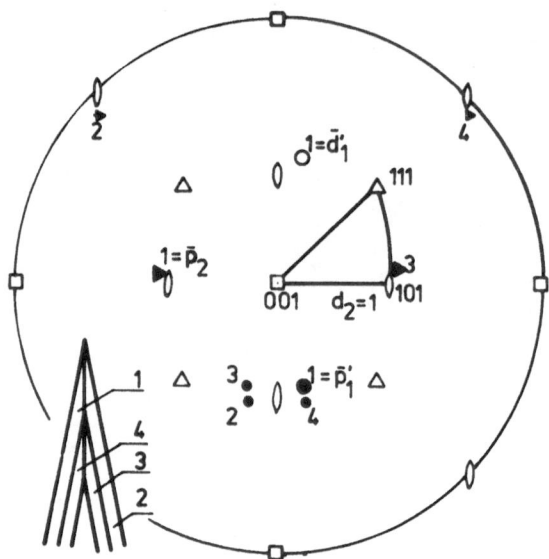

<u>Fig. 8</u> :   Stereographic projection of the orientation of the mar-
tensite variants with respect to the specimen axis.

the situation for the three other variants is less favourable.

The single crystal of martensite obtained was stressed further
and, after the disappearance of the first set of surface traces, a
second set of surface traces appeared as soon as the stress level
exceeded a stress σ_B (23). The measured normal of the second set of
planes coincided with $\bar{p}_2$, which is the normal to the close packed
planes of variant 1. This means that a shear occurred along the
close packed plane, $\bar{p}_2$, which are now favourably oriented with respect to the stress axis after the sample has been completely transformed into martensite. In other words, a process of stress-induced
transformation succeeded by reorientation is accomplished.

On stressing β-Cu-Zn-Al single crystals, it was possible to obtain a 3R-single crystal by an equivalent succession of shear processes (27). Upon subsequently compressing the 3R-single crystal
at a temperature below A_s, a new single variant was obtained showing
the rubber-like behaviour. In order to explain the pseudoelastic
behaviour in Cu-Zn (22,23) and the rubber-like behaviour in Cu-Zn-Al
(27) the following crystallographic considerations should be analysed (33).

It has been shown earlier (1) that two modifications of the

β'' type martensite occur, consisting of a mixture of lamellae of either 3R + IR or 3R + 2H type stacking, depending on the composition. This lamellar mixture, which is dictated by the invariant plane strain condition (28), may be removed by continued straining as soon as the β-phase is completely transformed to martensite (a habit plane does no longer exist). Upon releasing the stress or upon heating to a temperature above A_s, the habit plane reappears invoking the invariant plane strain condition again and the lamellae will also reform. Thus, part of the pseudoelastic behaviour or part of the memory effect find their origin in the invariant plane strain condition.

A second aspect which should be considered is the fact that the close packed 3R-variant has three shear directions located in the close packed plane. But due to the long range order which is present in these martensites (1), only one shear direction will not violate the next-nearest neighbour relationships, whereas the other two will change the degree of long range order. Similar arguments can be put forward for the stress-induced twinning which has been observed to occur in these martensites (29). Consequently, shear along a "wrong" direction does not correspond to simple reorientation or twinning but a phase transformation takes place. This situation is similar to that found in Fe-Be (30), giving rise to the pseudoelastic $(\sigma.\varepsilon)$ behaviour, depending on the orientation of the shear system, i.e., whether the first or one of the two latter systems will be activated. Thus, the long range order which may be violated during the stress induced reorientation will be responsible in part for the observed pseudoelastic and memory effects.

The above examples have shown that more complete crystallographic analysesof the stress-induced transformation and reorientation are needed, in order to understand the crystallography of pseudo-elasticity and the memory effects.

Once the crystallography is completely understood, it becomes straightforward to calculate the amount of recoverable shear. In dealing with the memory effects which occur in polycrystalline material the mechanical coupling which exists between the various martensite plate variants (25), even across grain boundaries (31), should be brought into the mechanical metallurgical calculations as an important mechanism.

C. Thermodynamic and Kinetic Considerations

Applied stress and the stress generated by the martensite formed are dominating the shape memory behaviour. Therefore, only a detailed analysis of the effects of stress in combination with those of temperature on phase equilibria, nucleation and growth will

permit one to understand the transformation and reorientation phenomena involved in the SME completely.

A disposition of these aspects has been given in a recent review (32). In the present paper we are elaborating on some of them
with regard to further experimental observations.

1. Effects of stress on the equilibrium between parent and martensite phases.

Whereas the purely internal energy changes of martensitic transformations are fairly well understood, the effects of stresses and
strains enter in several different ways which have not yet been fully assessed; some of these are discussed in the present section.

The stresses applied and generated during a martensitic transformation act as a state variable affecting the phase equilibria.
Since the volume changes accompanying the martensitic transformations of the present interest are comparatively small, only the
shear stresses and strains will be considered. In terms of the
Clausius-Clapeyron relation the effect of stress σ on the equilibrium temperature T_o between parent phase P and martensite phase M
may be written

$$\frac{dT_o^\sigma}{d\sigma} = \frac{\varepsilon^{P-M} T_o}{\rho \Delta H^{P-M}} = \frac{\varepsilon^{P-M}}{\rho \Delta S^{P-M}} \qquad [3]$$

where ε^{P-M} is the macroscopic transformation strain, T_o is the equilibrium temperature in the absence of stress, ρ is the density, ΔH^{P-M}
is the latent heat, and ΔS^{P-M} the entropy of the transformation
(at T_o).

In numerous systems the M_s (σ_a) dependence has been determined and was found to be linear in the stress range accessible to
measurements, i.e., $o < \sigma < \sigma_y$ (yield stress). Assuming that the
driving force for nucleation is practically independent of temperature and applied stress, which appears to be a reasonable assumption
in the light of existing experimental evidence, the relation

$$\frac{dT_o^\sigma}{d\sigma_a} \simeq \frac{dM_s^\sigma}{d\sigma_a} \qquad [4]$$

may be used to compute the transformation entropy from equation 3.
Characteristic values in Cu, Ag and Au based alloys are $0.8 \lesssim \Delta S^{P-M}$
$\lesssim 3.1$ J mol^{-1} K^{-1} (0.2 ... 0.75 cal mol^{-1} K^{-1}) (32).

If the applied stress is increased further after the transformation is completed, reorientation will usually set in. The critical stress σ_B for this process is not associated with an entropy
change (unless shear on a "wrong" shear system leads to a change in
crystal structure as mentioned above). Therefore, its temperature
dependence is considerably smaller than that of σ_M (= σ_a at M_s). In
Cu-Zn σ_B was actually found to be independent of temperature within
the experimental accuracy, fig. 9; on Cu-Zn single crystals both σ_M
and σ_B were determined (23).

In several alloy systems two types of martensite phases occur
at different temperatures at one composition. This has been known
for some time for the $\gamma \rightarrow a'$ and $\gamma \rightarrow \epsilon'$ transformations in Fe-Mn alloys
and stainless steels and was found, e.g., for the $\beta_1 \rightarrow \beta_1'$ and $\beta_1 \rightarrow$
γ_1' transformation in Cu-Sn (34). It is to be expected that the two
martensite phases form with different transformation entropies
$\Delta S^{P-M1} \neq \Delta S^{P-M2}$, such that the critical stress of formation shows
a different temperature dependence : $\sigma_{M1}(T) \neq \sigma_{M2}(T)$. Since $\Delta S =$
$d\Delta G^{P-M}/dT$ the situation may be depicted as shown in fig. 10

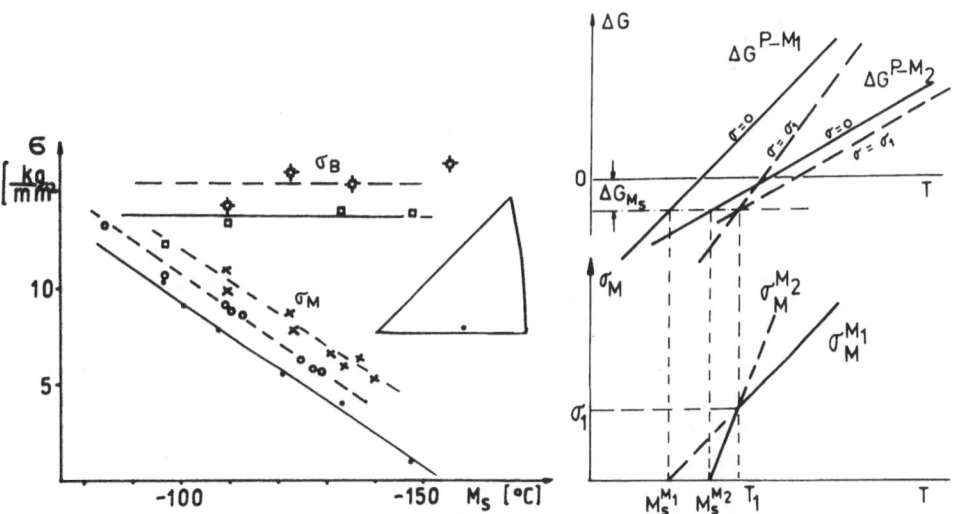

Fig. 9 : Critical stress of martensite formation, σ_M, and reorienta-
          tion, σ_B, of a β–Cu-Zn single crystal (23); σ_M – cycled
          at $\sigma < \sigma_B$; $\square \sigma_B$ – cycled to $\sigma \gg \sigma_B$ at least once; $O \sigma_M$ –
          during subsequent cycles; $\times \sigma_M$, $\Phi \sigma_B$ after an interruption of
          the experiment for 30 h at room temperature.
Fig. 10 : Schematic representation of the free energy changes ΔG and
          the critical stress to form martensite σ_M as a function of
          temperature if two different martensite phases M1 and M2
          are formed.

The $\Delta G(T)$ and corresponding $\sigma_M(T)$ functions are shown with the simpli-
fying assumptions that all are linear and that the driving force for
nucleation $\Delta G_{M_s}^{\beta \rightarrow M}$ is independent of structure, temperature and stress
level. The principal relations and conclusions are not altered if
these restrictions are removed. It appears that several complex
interrelations between temperature, stress and structures found,
e.g. in Cu- 14 % Al - (3...4) % Ni, which were reported in recent
papers and are presented at this symposium can be understood, in
terms of fig. 10. If we set Ml = β_1' and M2 = γ_1' we see that γ_1'
forms first upon quenching because $M_s \left[\gamma_1' \right] > M_s \left[\beta_1' \right]$ but that stress-
induced formation of β_1' can occur above $M_s \left[\gamma_1' \right]$; this is in agreement with the observations by Otsuka, Nakai and Shimizu (35). It is
obvious that the pseudoelastic behaviour of the two martensites differs not only because of their different boundary structures but also
because of their different thermodynamic properties. Both Otsuka,
Sakamoto and Shimizu (36) and Rodriguez and Brown (37) report the
occurrence of a stress-induced martensitic transformation $\beta_1' \rightarrow \gamma_1'$
in Cu-Al-Ni which had previously been observed in binary Cu-Al, too
(β_1' (3R) $\rightarrow \beta_1'$ (1R), γ_1')(38,39). With reference to fig. 11 this is
to be expected at $T' < T_1$ where $\sigma_M \left[\gamma_1' \right] < \sigma_M \left[\beta_1' \right]$; strictly speaking,
the stresses to transform one martensite variant into another are not
represented by fig. 10; but the relative magnitude of $\Delta\sigma_M$ will indicate the correct sequence. (While this paper was under review, K.
Otsuka, H. Sakamoto and K. Shimizu (52) have drawn the same conclusions based on their former and additional experiments). A $\sigma_M(T)$
curve of the type shown in fig. 10 was found experimentally in an
Ag - 45 at % Cd alloy (40). Although a transformation to a single
martensite phase, namely γ_1' only, was assumed in the original paper
the same author showed in a later publication (41) that both β_1' (1R)
and γ_1' are formed at different compositions and degrees of plastic
deformation. Therefore, a situation very similar to that in Cu-Al-Ni
alloys can be surmised to exist, with the $\sigma_M(T)$ dependence indicating
the formation of two different martensite phases.

In considering the stresses required and the associated free
energy changes in stress-induced transformation and reorientation of
SME treatments, two particular structural changes mentioned in section B yield additional effects : (i) the removal of the 1R and 2H
structures from β''(3R/1R) and β''(3R/2H) type martensites upon continued stressing after complete transformation; (ii) the shear of
3R and 2H type martensites in directions which are non-conservative
with respect to the superlattice structure. In both cases a change
in crystal structure is induced, implying that an additional change
in free energy occurs. As this change will usually be an increase,
these secondary stress-induced structural transitions of martensite
are contributing, in turn, to the driving force for the shape recovery during the heating cycle of the SME treatment. Thus, two-stage

transformations of this type will produce higher stresses during shape recovery which may be of interest for potential applications.

A detailed analysis of the effects of stress has to take the anisotropy of the transformation shear into account. An orientation dependence of $d\sigma_M/dT$ was observed, e.g., in In-Tl (42) and AgCd (40) single crystals. Several factors contribute to this anisotropy : (i) the orientation variant on whose macroscopic shear system the maximum resolved shear stress is acting will be growing preferential-ly; (ii) near certain symmetric orientations (37) individual plates and self-accommodating groups (being associated with different transformation strains and stresses) will compete for the maximum resolved shear stress; (iii) the shear stress acting on the effective (macroscopic) transformation shear system (habit plane $\bar{p}_1'$, macros-copic shear $\bar{d}_1'$) of the martensite plate most prone to transforma-tion will, still, be a function of orientation. This latter problem has been treated by Patel and Cohen (43). However, they based their calculation, designed to compute $dM_s/d\sigma$ (i.e., the reciprocal of $d\sigma_M/dT$) for uniaxial tension and compression and for hydrostatic pressure, on the Mohr's circle contruction. This takes into account the angle θ between the stress axis and the normal to the habit plane $\bar{p}_1'$ but neglects the possible angular deviation ψ of the shear direc-tion $\bar{d}_1'$ from the direction of maximum shear stress τ_{max} in the ha-bit plane. In other words, a formulation in terms of the Schmid law is more appropriate (the latter is usually written for constant vo-lume, which is sufficient here, but it could be modified according to the treatment by Patel and Cohen to account for volume changes as well). Fig. 7b shows the situation graphically which leads to the resolved shear stress

$$\tau = \frac{\sigma}{2} \sin 2\theta \cos \psi$$

It is obvious that $\tau(\theta, \psi)$ will enter into the relations $\Delta G^{P-M}(T)$ and $\tau_M(T)$ in such a way as to raise τ_M as $\theta \gtrless 45°$ and $\psi > 0°$. This is an essential modification of the thermodynamic relations which shows that an experimental $\sigma_M(T)$ determination based on single crys-tal specimens would have to take the orientation into account. Clear indications of this orientation dependence are given by Burkart and Read for In-Tl (42) and by the results of Krishnan and Brown for single crystals of AgCd with 100, 111 and 133 tensile axes (40). In In-Tl where the transformation is cubic $\rightarrow$ tetragonal the sign of the uni-axial stress (tension or compression) leads to considerable addi-tional differences in $\tau_M(T)$ as is to be expected.

Orientation dependence $\tau_M = f(\theta + \Delta\theta, \psi + \Delta\psi)$ may be repre-sented in a schematic and simplified manner as shown in fig. 7c. This diagram was drawn to emphasize that (i) a field of critical

stress vs. orientation relations exists for each orientation variant;
(ii) further orientation variants may be formed as the stress axis is
rotating, e.g., due to the transformation strain; (iii) an orientation variant that had formed under the initial conditions may even
undergo reversion and a different one may be formed due to changes
in macroscopic and/or local stress conditions. The same general
relation applies not only to orientation variants but also to structural variants. In other words the local stress conditions determine the growth of a given martensite plate or self-accommodating
group even to the point of shrinkage with increasing stress and decreasing temperature.

Recently, Tong and Wayman (44,45) have considered the non-
chemical energy $\Delta G_{nc}^{P \to M}$ stored during thermo-elastic growth as an additional state variable. Although this stored energy is influenced
by the microstructural configuration and surface effects the repro-
ducibility of the hysteresis curve (M_s-M_f-A_s-A_f) appears to support
this approximation for sufficiently large, polycrystalline samples.
The upper temperature limit of stability of a fully martensitic spe-
cimen is designated T'_o; it is given by $(G^{P \to M} + G_{nc}^{P \to M})_{VM=1} = D$ and can
be determined from the approximate relation $T'_o = (A_s - M_f)/2$. It
will be discussed below that the non-chemical free energy terms can
vary in wide limits depending on the structural state of the specimen and will, thus, affect T'_o. Therefore, their treatment in the
energy balance of growth appears to be more generally applicable than
their inclusion in the thermodynamic treatment of the phase transition.

Recently, G.B. Olson and M. Cohen have pointed out that the
elastic strain energy stored in the case of thermoelastic martensite
formation cannot be separated from the chemical free energy change
(53). This means that calorimetric determinations of total enthalpy
changes and applications of the Clausius-Clapeyron relation to thermoelastic martensite formation do not represent measurements of the
chemical enthalpy change of the transformation. The elastic strain
energy, being distributed quite inhomogeneously, enters the macroscopic thermodynamic property changes in a complicated fashion and is,
thus, difficult to assess and to subtract. Non-thermoelastic martensitic transformations which are associated with a comparatively
small fraction of stored elastic strain energy are, therefore, best
suited for determinations of the chemical free energy changes.

2. Particular Effects on Nucleation

As pointed out in the introduction the SME behaviour is favoured
by a low free energy of nucleation because this leads to a small (or

zero) excess driving force at the onset of growth which, in turn, ensures a high degree of crystallographic perfection of the growth process. We shall investigate the particular properties of   phases which contribute to a low free energy of nucleation by writing the free energy of a nuleus

$$\Delta g^{P-M} = a\Delta G^{P-M} V_n + bC_{ij}(\varepsilon_M - \varepsilon_o)^2 V_n + b(\gamma_M - \gamma_o)A_n, \qquad [6]$$

where a,b,c are constants (mainly geometric), C_{ij} is the elastic constant pertaining to the nucleation (shear) strain, ε_M is this strain (tensor), ε_o is the strain component of a heterogeneous nucleation site providing part of the nucleation strain, γ_M is the parent-to-martensite interfacial energy, γ_o is a fraction of this energy provided by the structure of the nucleation site, V_n and A_n are the volume and the interfacial area of the nucleus.

In the β-phases C_{ij} may be set proportional to $C' = (C_{11} - C_{12})2$. This she ar constant has an extremely low value in non-ferous β-phase alloys which transform martensitically. Experimentally $C' = (0.1 ... 0.35)x10^{10}$ Nm^{-2} $(0.1...0.35)x 10^{11}$ dynes cm^{-2}) immediately above M_s. Thus, a first contribution to the low value of Δg^{P-M} is the anomalous elastic behaviour of the β-phase alloys. It should be noted that both elastic shear constants, C' and C_{44}, of Fe_3Pt are lowered, mainly by the magnetic exchange energy contribution (54). Therefore, similar conditions as those in β-phase alloys exist and SME behaviour is, indeed, observed (55). Thus, it appears that a low value of C_{ij} in eqn. [6] is an essential contribution to the occurrence of SME martensite irrespective of the origin of the low elastic stiffness (56).

The magnitudes of ε_o and γ_o can be considerable as far as the premartensitic structural transitions discussed in section A yield displacements with components of the martensitic shear and structures similar to the martensitic structures. This is clearly the case f or the structure of the lath shaped regions shown in fig. 2. But the ω-type and shear type displacements are contributing to ε_o, too. Since some of the premartensitic structural transitions depend sensitively on the mechanical and thermal history of the parent phase it is to be expected that the nucleation kinetics is affected accordingly.

Quantitatively treatments of nucleation will only be possible a after the structural nature of the subcritical nucleus and the path of its structural changes at the onset of growth have been determined. The effects of stress reside mainly in the stress dependence of ΔG^{P-M} discussed in section Cl.

3. Factors affecting growth and reversion

Apart from the low driving force for nucleation it is the kinetics of growth, reorientation and reversion of the martensite that determines the shape memory behaviour. It was shown explicitly in (32) that the energy terms for the growth and reversion of an individual martensite plate in the parent phase may be formulated in close analogy to the treatment of elastic twinning (30,48) :

$$\Delta G_T^{P-M} \, dV + \sigma_a \epsilon_M dV \gtreqless \sigma_1 \cdot \epsilon_M dV + \gamma_M dA + \xi \epsilon_M dV \qquad [7]$$

where ΔG_T^{P-M} is the "pure" chemical driving force, dV is an incremental volume change of the martensite plate and dA the corresponding change in interfacial area, $\sigma_a \epsilon_M$ is the stress-induced fraction of the driving force, σ_1 is the internal stress produced by the martensite plate, γ_M is the interfacial energy of the parent-martensite interface and ξ is a frictional stress arising from all processes which give rise to energy dissipation during growth and reversion of the martensite plate. This schematic equation would have to be expanded by geometric factors for quantitative applications.

The condition for the first stage of the SME treatment, i.e. the stress-induced martensite formation, is (after dividing by dV, thus referring to a unit volume) :

$$\Delta G_T^{P-M} + \sigma_a \epsilon_M > \sigma_i \epsilon_M + \gamma_M (dA/dV) + \xi \epsilon_M \qquad [8a]$$

where the left-hand side comprises the (negative) driving forces and the right-hand side the (positive) resistive forces; if the first stage occurs by reorientation only : $\Delta G_T = 0$, but eqn. [8a] must still hold. During the second stage of the SME treatment, i.e. the release of the applied stress, perfect SME behaviour requires

$$\Delta G_T^{P-M} + \sigma_i \epsilon_M + \gamma_M (dA/dV) < \xi \epsilon_M \qquad [8b]$$

which means that the frictional force must be sufficiently high and /or the magnitude of ΔG_T^{P-M} (< 0) must suffice (i.e. $T \ll T_o$) to prevent isothermal reversion of the martensite. During the third stage of the SME treatment, i.e. reversion of the martensite to the parent phase by heating, the required condition is

$$\Delta G_T^{P-M} + \sigma_i \epsilon_M + \gamma_M (dA/dV) > \xi \epsilon M \qquad [8c]$$

In what follows we shall discuss some of the particular properties of eqns. [7] and [8.]

The effect of σ_a will change as the shape deformation progresses during a stress-induced transformation, because the accompanying lattice rotation results in a change of the shear stress τ (θ, ψ) acting on the macroscopic shear system of the martensite plate as shown schematically in fig. 7c. This means that the increase in driving force is not a linear function of the applied stress. The effect of σ_a is modified further by the stress field of martensite plates growing concurrently. This interrelation is treated in the context of the next paragraph in which the internal stresses are discussed.

The strain energy produced by a growing martensite plate and opposing its growth is proportional to $\sigma_i \cdot \varepsilon_M = C_{ij} \cdot \varepsilon_M^2 \approx C' \varepsilon_M^2$; the latter approximate equality may be used because the habit plane and macroscopic shear direction of martensite phases in β-phase alloys are often sufficiently close to $\{110\}$ $<1\bar{1}0>$ (1). Therefore, the low magnitude of C' in β-phases gives rise to a low amount of stored energy per unit volume of martensite formed as compared to the strain energy associated with ferrous martensites. This ensures, furthermore, that σ_i remains below the yield stress of both the parent and the matrix phase in a considerable temperature range such that the rate of production of lattice defects and associated degeneracies of the transformation process (and the concomitant increase in ξ) is low. In this context it should be noted that the positive temperature coefficient of C' leads to an increase in strain energy per unit strain with increasing temperature. For the energy balance of growth in a SME cycle this means that the reversion of the martensite and the accompanying shape recovery upon heating is not only due to the change in sign of ΔG_T^{P-M} at T_o but also to the increase in stored energy (at a constant volume fraction of martensite). The comparatively low strain energy in β-phase alloy martensites is further due to the formation of self-accommodating groups of plates. The lattice strain per unit volume of martensite formed as a single plate (1), group of two (2) and group of four (4) is

$$\varepsilon_M^{(1)} > \varepsilon_M^{(2)} > \varepsilon_M^{(4)} \qquad\qquad [9]$$

such that the structural configuration has an essential bearing on the growth kinetics as would have been expected. An additional effect of internal stress minimization arises by the formation and reversion of individual plates as the crystal orientation is rotating with increasing transformation strain. Fig. 11 gives a simple and qualitative example. A plate I grows preferentially under the initial stress conditions. As the specimen axis rotates due to the increasing shear strain, growth of a plate II is initiated. Since the rotation of the specimen persists plate II continues to grow, partly at the expense of plate I. This is a typical situation observed in studies of microstructural changes during stress-induced martensite formation and reorientation (49,50).

The interfacial energy term $\gamma_M(dA/dV)$ is difficult to assess quantitatively. However, it will generally play a minor role in the energy balance because of its small magnitude. A simple esti- mate assuming $\gamma_M = 0.2$ J m^{-2} (200 ergs cm^{-2}) and a plate thickness of 1 µm yields 0.2 J mol^{-1} (0.05 cal mol). In cases where energies as small as this determine the local configuration the different inter- facial structure of β' and γ' martensites may play a role. Whereas β' type martensites have essentially a "smooth" habit plane the internal twin lamellae of γ' martensite give rise to a corrugated interface with a strain field associated with each twin lamella. Therefore, the strain energy component of γ_M and, thus, its total magnitude is expected to be higher for γ' than for β' type martensite. The important contribution of strain energy to the interfacial energy of γ' martensite is, also, borne out by the microstructural observa- tions of a single interface transformation, where the internal twins were found to be tapered toward the interface and pointed there, as reported by Otsuka, Shimizu and Takahashi (51). Since the interfa- cial energy is recovered during the reversion of the martensite when heating during the SME cycle it is part of the driving force of the process of shrinkage.

Implicitly, eqns. [7] and [8] contain a term accounting for the interfacial energy of the faults (in β' martensite) or interfaces (in β'' and γ' martensite) of the internal structure. Since the den- sity of faults or interfaces is proportional to the volume, we have

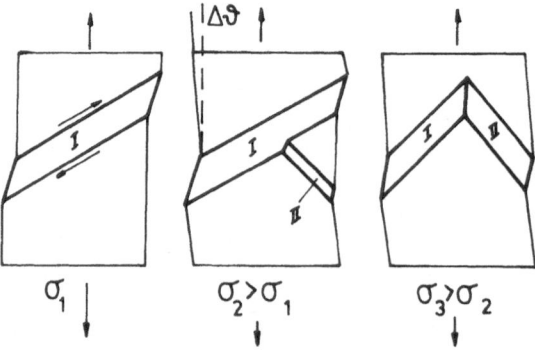

Fig. 9 : Schematic representation of the growth sequence of two mar- tensite plates during stress-induced growth when the orien- tation changes by $\Delta\theta$ such that initial variant I is partial- ly reverted and replaced by the growth of variant II.

included their energy contribution in σ_i in our previous treatment
(32). But it is more appropriate to take account of this contribution in a separate term, because the internal structure of the martensite plates is differently dependent on the effects of neighbouring plates than the internal stress σ_i. The term could be written

$$\gamma_M^i \delta \qquad\qquad\qquad [10]$$

where γ_M^i is the interfacial energy of the internal defects and δ the
internal interfacial area per unit volume of martensite. Assuming
$\gamma_M^i = 2 \times 10^{-2}$ J m^{-2} (20 ergs cm^{-2}) and a lamellar thickness of 10 nm
($\delta = 10^{-8}$ m^{-1}) we obtain $\gamma_M^i \delta \simeq 15$ J mol$^{-1} \simeq 4$ cal mol^{-1}. This
is of the same order of magnitude as the driving force for nucleation
$\Delta G_{M_s}^{P-M}$ indicating that the internal defect structure constitutes an essential part of the energy balance and, thus, of the driving force
for reversion, during an SME cycle. It is probably this term which
causes the removal of 1R and 2H lamellae, respectively, upon stress-
induced reorientation of β'' type martensite (see section B) and of
detwinning of γ' martensite upon stress-induced reorientation as
well as further behind the interface in a single interface transformation (51).

The energy stored during reorientation can be composed of
two terms : (i) an elastic energy which may be written formally
$\sigma_i \varepsilon_R$, where ε_R describes the strain attained beyond ε_M, and (ii)
a chemical free energy change $\Delta G^{M-M'}$, where M' is a structural
variant produced by shearing the martensite on the "wrong" shear
system as discussed above. The high rate of energy absorption observed at $\sigma > \sigma_B$ (i.e. in the range of reorientation) indicates that
both processes give rise to considerable additional energy storage.
This work does not seem to have been evaluated in any of the investigations of SME alloys thus far.

The last term of equs. 7 and 8 plays a decisive role for the
SME behaviour by determining whether (thermo- or pseudo-) elastic
equilibrium prevails in the growth process or the stoppage of moving
interfaces is caused by the frictional forces. It is implicit in
many experimental results of the shape memory effect and has been
pointed out perspicuously by Krishnan and Brown (40) that there is
a transition from pseudo-elasticity to the shape memory behaviour
as the temperature of stress-induced transformation is lowered.
Near T_o the stress-induced transformation and reversion progress
rather perfectly as far as atomic movements accompanying the displacement of interfaces is concerned. Therefore, ξ is small and
elastic equilibrium is maintained between the driving forces (i.e
all except $\xi \varepsilon_M$ in equs. 7 and 8). With an increasing degree of undercooling below T_o, i.e. parallel to an increase of the thermal component of the driving force, the rate of defect production during martensite growth is enhanced such that, finally, condition 8b is at
tained.

Thus, a detailed analysis of the quantities involved in the growth, reorientation and reversion processes during an SME cycle is the key to its understanding and to a quantitative prediction of the potential of SME behaviour for its use in heat engines and other devices where one may envisage to utilize the transformation energy for mechanical work.

Acknowledgements

The authors would like to thank ir. J. Van Paemel, lic. R. Rapacioli, Dr. M. Chandrasekaran for valuable discussions and for agreeing to the incorporation in this paper of some micrographs and results of current work. The assistance of Miss A. Stesmans, Mrs. R. Verhoeven-Cieters, Miss N. Neefs and M. Degreef is much appreciated for bringing the present paper and also some of the other papers presented at the same symposium into ready manuscripts.

The present work is supported in part by the "Fonds Derde Cyclus" of the K.U.Leuven.

References

1. H. Warlimont and L. Delaey, Progress in Mat. Science, 1975, vol.18
2. J. Van Paemel et al., Zeitschrift f. Metallk., vol.11, Aug. 1975 and J. Appl. Cryst., 1975, vol. 8,p. 181.
3. L. Delaey, A.J. Perkins and T.B. Massalski, J. of Mat. Science, 1972, vol. 7, p. 1197.
4. N. Nakanishi et al., Trans. Jap. Met. 1965, vol. 6, p. 222 and Scripta Met. 1968, vol. 2, p. 673.
5. N. Rusovič and H. Warlimont, Proc. of Int. Symp. on SME, this volume.
6. K.D. Swartz, W. Bensch and A.V. Granato, Univ. of Ill. 1975.
7. P.C. Clapp, Phys. stat. sol. (b) 1975, vol. 57, p. 561.
8. M. Chandrasekaran et al., Zeitschrift f. Metallkde, Sept. 1975, vol. 11.
9. J. Van Paemel and L. Delaey, ibid. Okt. 1975, vol. 11.
10. L. Delaey, J. Van Paemel and T. Struyve, Scripta Met., 1972, vol. 6, p. 507.
11. N.Nakanishi, Y.Murakami and S. Kachi, Scripta Met.,1971,vol.5,p433
12. R.F. Hehemann and G.D. Sandrock, Scripta Met. 1971, vol. 5, p. 801 and Met. Trans. 1971, vol. 2, p. 2769.
13. A.J. Perkins, Scripta Met. 1974, vol. 8, p. 31 and 439.
14. P.L. Ferraglio and K. Mukherjee, Acta Met., 1974, vol. 22, p. 835.
15. A. Nagasawa, J. of Phys. Soc. of Japan, 1973, vol. 35, p. 489.
16. S.C. Abrahams, Mat. Res. Bull. 1971, vol. 6, p. 881.
17. A. Nagasawa, Phys. Letters, 1973, vol. 45A, p. 265.
18. M. Ahlers, Zeitschrift für Metallkunde, 1974, vol. 10, p. 636.

19. S. Mendelson, Phase Transitions, ed. L.E. Cross, Pergamon Press N.Y., 1973, p. 287, and Proc. of Int. Symp. on SME, this volume.
20. T. Tadaki, M. Tokoro and K. Shimizu, Scripta Met., 1974, vol. 8, p. 1077.
21. R. Rapacioli and M. Ahlers, Scripta Met. 1973, vol. 7, p. 977.
22. W. Arneodo and M. Ahlers, Scripta Met. 1973, vol. 22, p. 1287.
23. W. Arneodo and M. Ahlers, Acta Met. 1973, vol. 22, p. 1475.
24. S. Miura, S. Maeda and N. Nakanishi, Phil. Mag. 1974, vol. 30, p. 565.
25. H. Tas, L. Delaey and A. Deruyttere, Met. Trans. 1973, vol. 4, p. 2833.
26. Y. Cornelis, Ph.D. Thesis, Univ. of Ill., Urbana, 1973.
27. R. Rapacioli et al., Proc. of Int. Symp. on SME, this volume.
28. L. Delaey and Y. Cornelis, Acta Met. 1970, vol. 18, p. 1061.
29. H. Tas et al., Zeitschrift für Met., 1973, vol. 64, p. 855 and 866.
30. G. F. Bolling and R.H. Richman, Acta Met. 1965, vol. 13, p. 709, 723 and 745 and Met. Trans. 1971, vol. 2, p. 2451.
31. L. Delaey, E. Vandevoorde and R.V. Krishnan, Proc. of Int. Symp. on SME, this volume.
32. H. Warlimont et al., Journ. of Mat. Science, 1974, vol. 9, p. 1545.
33. L. Delaey, M. Ahlers, R. Rapacioli and M. Chandrasekaran, to be published.
34. H. Warlimont and D. Härter, Proc. 6 Int. Conf. Electr. Micros. Maruzen, Tokyo, 1966, vol. I, p. 453.
35. K. Otsuka et al. Scripta Met. 1974, vol. 8, p. 913.
36. K. Otsuka et al. Proc. of Int. Symp. on SME, this volume.
37. C. Rodriguez and L.C. Brown, ibid.
38. G.V. Kurdyumov and G. Bull. Acad. Sci. USSR, Chem. Scr. 1936, vol. 2, p. 271.
39. H. Tas et al. Scripta Met,, 1971, vol. 5, p. 1117.
40. R.V . Krishnan and L.C. Brown, Metal. Trans.1973, vol. 4, p. 423.
41. R.V. Krishnan and L.C. Brown, Met. Trans. 1973, vol. 4, p. 1017.
42. M.W. Burkart and T.A. Read, Trans. AIME, 1953, vol. 197, p. 1516.
43. J.R. Patel and M. Cohen, Acta Met., 1953, vol. 1, p. 531.
44. H.C. Tong and C.M. Wayman, Acta Met., 1974, vol. 22, p. 887.
45. Ibid, Scripta Met. 1974, vol. 8, p. 93.
46. C.L. Magee, in "Phase Transformations", ASM, Metals Park 1970, p. 115.
47. T. Nishizawa and K. Ishida, to be published.
48. A.M. Kosevich and V.S. Boĭko, Sov. Phys. Nsp. 1971, vol. 14, p. 286.
49. L. Delaey and J. Thienel, this symposium.
50. L. Delaey, F. Vandevoorde, R.V. Krishnan, this symposium.
51. K. Otsuka et al. Proc. 4th Int. Conf. Crystal Growth, Tokyo 1974, p. 191.
52. K. Otsuka, H. Sakamoto and K. Shimizu, Scripta Met., 1975, vol. 9, p. 491.
53. G.B. Olson and M. Cohen, private communication, to be published.

54. G. Hausch, J. Phys. Soc. Japan 1974, vol. 37, p. 819 and 824.
55. M. Joos, C. Frantz and M. Gantois, this symposium.
56. H. Warlimont, G. Hausch, to be published.

MARTENSITIC TRANSFORMATIONS IN β PHASE ALLOYS

S. Vatanayon and R. F. Hehemann

Case Western Reserve University

Cleveland, Ohio 44106

CsCl type ordered phases occur frequently in equiatomic alloys. Two broad classes of these phases are recognized (1). The first, with by far the largest number of members, involves transition metals in which one component lies to the left and the other to the right of the chromium group. The equiatomic Ti-Ni alloy is a member of this category, which, for the present purposes, will also be considered to include uranium and its alloys. In the latter systems, however, the high temperature phase generally exhibits a disordered BCC structure. The β phases of the noble metals constitutes the second class. These exhibit the CsCl structure when alloyed with group IIb elements and other ordering arrangements when alloyed with higher valance solutes.

Many of these systems undergo martensitic transformations at lower temperatures and these generally exhibit shape-memory effects (2). The transformations in the noble metal alloys have been reviewed critically (3) and the structures have been determined for many of the transition metal systems (4).

Frequently, a sequence of structures is formed martensitically as a function of composition and/or temperature. This is revealed most clearly in uranium-base alloys (5) as is shown by the M_s temperatures shown schematically in Fig. 1. It is apparent that transformation during cooling follows the sequence:

BCC (γ) $\rightarrow$ Tetragonal (γ_o) $\rightarrow$ monoclinic (α'') $\rightarrow$ orthorhombic (α').

Au-Cd alloys undergo a similar sequence as a function of composition (6). As Cd content increases the structure at room temperature exhibits the sequence:

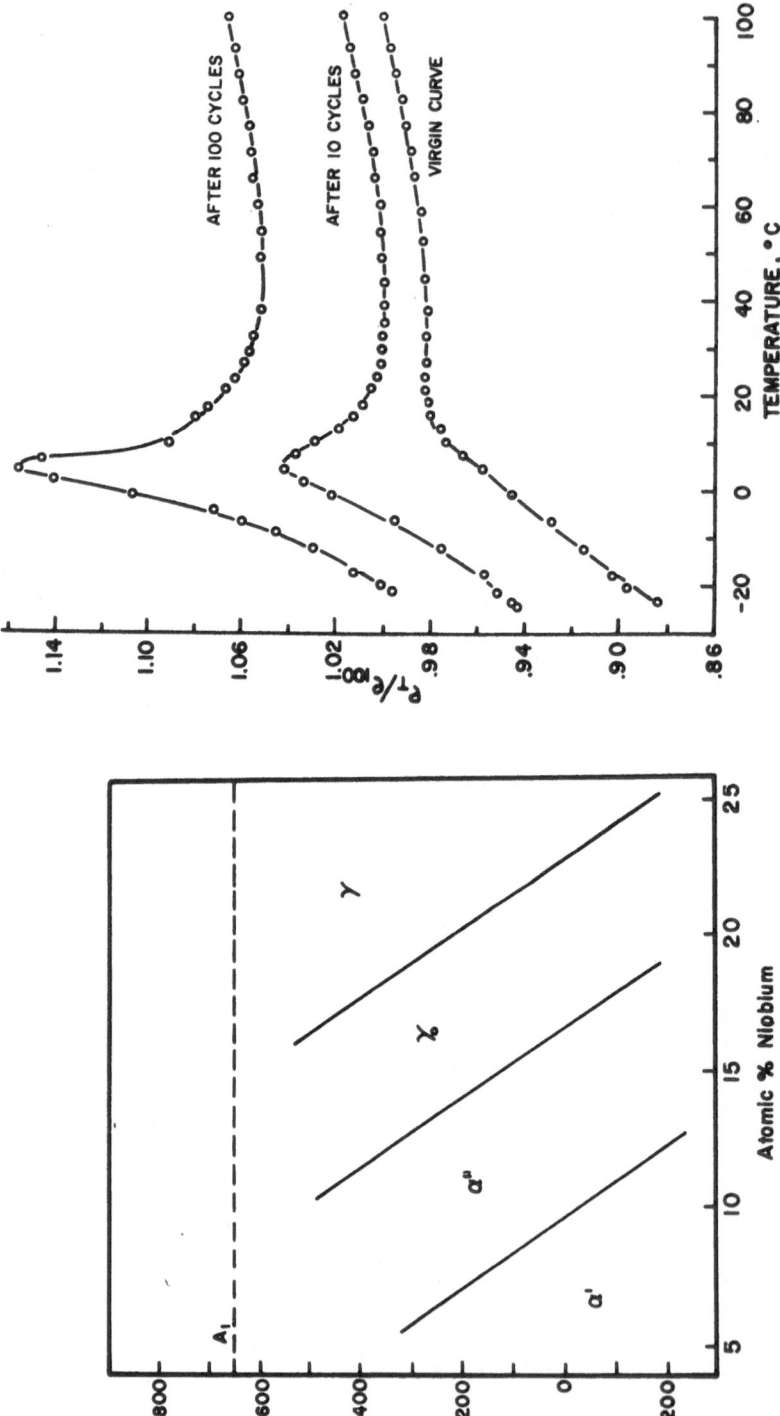

Fig. 2: Effect of cycling between 0° and 40°C on the resistivity, temperature cooling curve of as-arc-cast Ti-50.5 Ni. Curves normalized to 100°C on virgin cycle.

Fig. 1: The effect of alloy content on M_s temperatures of uranium base alloys.

$$B2 \rightarrow \alpha'(1R)* \rightarrow \beta'(3R) \rightarrow \gamma'(B19) \rightarrow \zeta'(Hex).$$

The sequential formation of these structures as a function of temperature has not been observed although lamellar mixtures of β' and γ' have been reported (6). Such sequential reactions occur also in many non-metallic systems of which $BaTiO_3$ is perhaps the most widely recognized (8).

Zener (9) first suggested that martensitic reactions in the noble metal alloys may result from an instability of the B2 structure to [110][1$\bar{1}$0] shear produced by softening of the 1/2 ($C_{11}-C_{12}$) elastic constant as the temperature is lowered. The lattice dynamical aspects of this problem were formulated in 1960 (10) by Cochran specifically for ferroelectric materials which lead to the concept that these transitions result from a "soft" phonon mode. In recent years, this concept has been studied extensively in a wide range of ferroelectric materials and existence of the soft mode demonstrated experimentally in some, but not all, systems. As summarized by Scott (11), these concepts have now been applied with varying degrees of success to a wide range of systems including the A15 superconductors such as Nb_3Sn.

A variety of anomalous effects, which are frequently called premartensitic phenomena, occur in the parent phase at temperatures above M_s. These vary widely from one system to another and include, among others, softening of elastic constants (12-17), resistivity anomalies (18-21), diffuse diffraction effects (3,19,22-24) and calorimetric effects (25, 26). The premartensitic phenomena and structural sequences appear to be different manifestations of closely related phenomena. Operation of specific vibrational modes can be invoked to rationalize these phenomena and, formally at least, the polarization vectors of these modes converts the parent to the product structure. The relationship between the premartensitic and martensitic reactions in Ti-Ni and Au-Cd will be employed to illustrate these concepts.

EXPERIMENTAL PROCEDURE

The Au-Cd alloys employed in this study were prepared from ultrahigh purity gold and cadmium. Melting was conducted in evacuated quartz capsules and charged weights were used as the chemical analysis. The single crystals for x-ray diffraction studies were grown by a modified Bridgman technique.

The experimental techniques employed in this study have been described previously (24). The Buerger precession camera was

---

* This structure may form by a massive rather than a martensitic transformation (7).

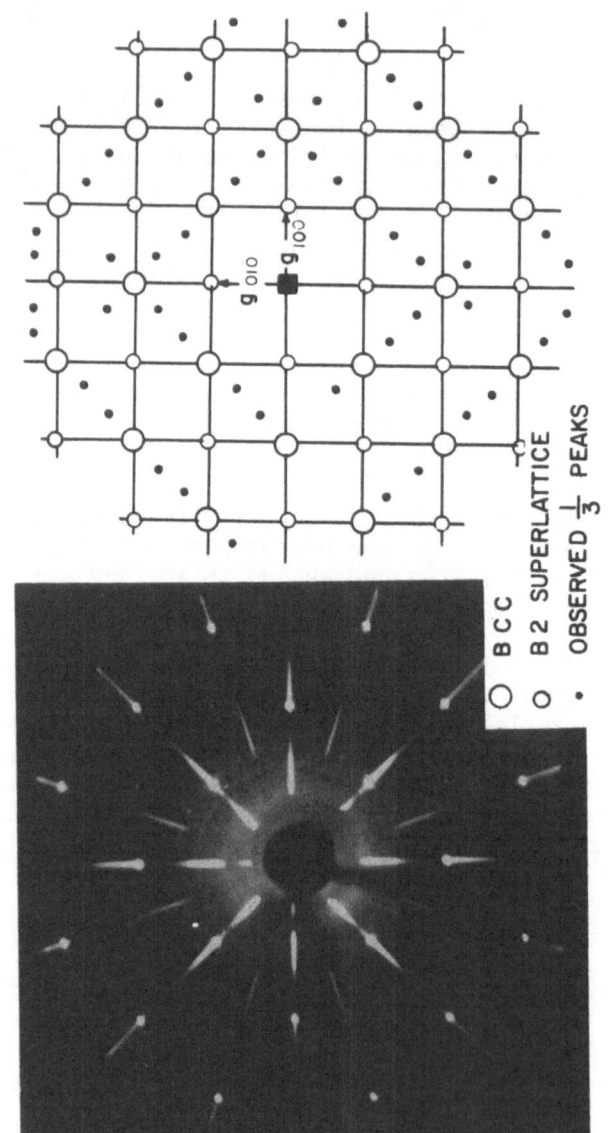

(a)   (001)B2

Figure 3:   Precession patterns taken below room temperature, showing locations of 1/3 reflections. Zr-filtered Mo radiation, 36 kV – 16 mA, 12 hour exposure, 10–15°C.

employed to determine the structure of the ζ'-phase Au–50% Cd alloy. A series of precession patterns were taken and the density method was used to obtain relative intensity data.

Electron microscopy was generally conducted on a Hitachi HU–650-B electron microscope equipped with a heating stage. Thin foil specimens were prepared for electron microscopy by electropolishing in a cyanide solution. All thinning was conducted at 80°C in the high temperature β phase.

RESULTS AND DISCUSSION

Resistivity and Diffuse Diffraction in Ti–Ni

The Ti–Ni system provides particularly clear indications of the premartensitic phenomena. These have been studied previously (24) and will be reviewed only briefly here for comparison with the behavior of Au–Cd alloys.

The shape change produced by martensitic reactions provides the most definitive means for delineating the premartensitic and martensitic temperature ranges for evaluation of the relevant resistivity and diffraction effects. Typical resistivity curves are shown in Fig. 2 which illustrates the pronounced change in behavior produced by cycling as reported originally by Wang (18) and confirmed by many others. Surface relief first appears at the maximum in the resistivity curve. Thus, cycling has lowered M_s from about 22°C in the virgin sample to about 0°C after 100 cycles and a clear premartensitic region with a negative temperature coefficient of resistivity is observed. The resistivity above M_s is reversible without hysteresis; however, the martensitic reaction exhibits a hysteresis of the order of 30°C. The ability of cycling to lower M_s, thereby revealing the premartensitic reaction more clearly, has been attributed, at least in part, to a stabilization effect associated with dislocation debris generated by the formation and reversal of the martensitic phase.

Single crystal x-ray diffraction as well as electron diffraction patterns exhibit diffuse diffraction effects in the premartensitic temperature range (19,22-24). These diffraction effects are more pronounced in electron diffraction than in x-ray diffraction patterns and consist of diffuse streaks as well as extra reflections. In (110) patterns, the streaks are in <112> directions indicating that they result from the intersection of (111) planes of diffuse intensity with the (110) reciprocal lattice plane. The extra reflections occur at 1/3 positions of the B2 reciprocal lattice as shown for the (001) relplane in Fig. 3. In this relplane, the reflections occur at 1/3 <110> positions; whereas, they occur at 1/3 <112> positions

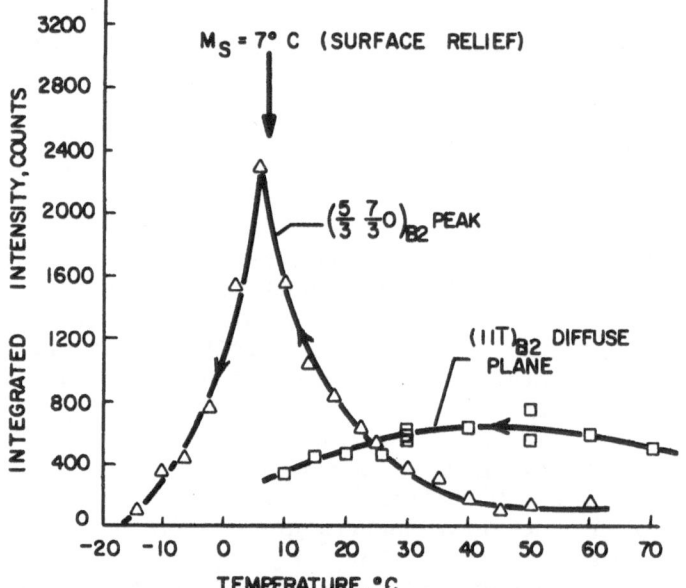

Figure 4: Variation in intensity of 1/3 reflection and {111} dif-
           fuse plane with temperature. Zr-filtered Mo radiation,
           36 kV, 1/3 peak at 16 mA, {111} diffuse plane at 25 mA.

of the (110) relplane. The reflections at 1/3 positions of the B2 reciprocal lattice constitute a tripling of the repeat distance and the reflections in the (110) relplane can result from simple <111> displacements analogous to those of the omega reaction in Ti- or Zr- base alloys. The omega displacement wave, however, produces no extra reflections in the (001) relplane so that additional displacements occur in these Ti-Ni alloys. As will be shown, these require 2/3 <110><110> displacement waves combined with the omega wave. In common with the resistivity results, the intensity of the diffuse streaks and extra reflections changes reversibly with temperature exhibiting essentially no hysteresis. Typical data are presented in Fig. 4. The (111) planes of diffuse intensity persist well above room temperature and have been noted at temperatures up to 300°C in electron diffraction patterns. At about 30°-40°C, the intensity of the (111) diffuse plane reaches a maximum and then decreases in intensity as the 1/3 reflections form with further reduction of temperature. The maximum intensity of the 1/3 reflections occurs at the M_s temperature; however, this results simply from the reduction in diffracting volume as the matrix is converted to the martensitic phase.

The intensity of the 1/3 reflections in these Ti-Ni alloys is too weak to permit an unambiguous interpretation of the diffraction effects; however, as will be shown, they are essentially the same as those of the hexagonal ζ' phase that forms in Au-50% Cd alloys.

The structure of the martensitic phase in Ti-Ni has been the subject of considerable controversy (22-24,27-29) and, as with other systems, more than one structure may exist. The martensitic structure in 50.5 At.% nickel alloy appears to be a monoclinic distortion of the B19 structure that forms in Au-47.5% Cd alloys (24,29) although there is not yet agreement on the positional parameters (shuffles). It thus appears that the Ti-50.5% Ni alloy exhibits, as a function of temperature, characteristics of both Au-47.5% Cd and Au-50% Cd alloys.

Transformations in Au-Cd

The martensitic transformations in Au-Cd have been studied extensively and the relevant literature has been reviewed critically (30,31). Attention here will be restricted to the Au-47.5% Cd and the Au-50% Cd compositions. It has long been known (32) that martensite in the 47.5% Cd alloy exhibits the B19 structure; however, it is only recently that the martensitic structure in the 50% Cd alloy has been identified as hexagonal or trigonal with 18 atoms in the unit cell (6,31). X-ray diffraction studies have been conducted in this investigation to obtain the positional

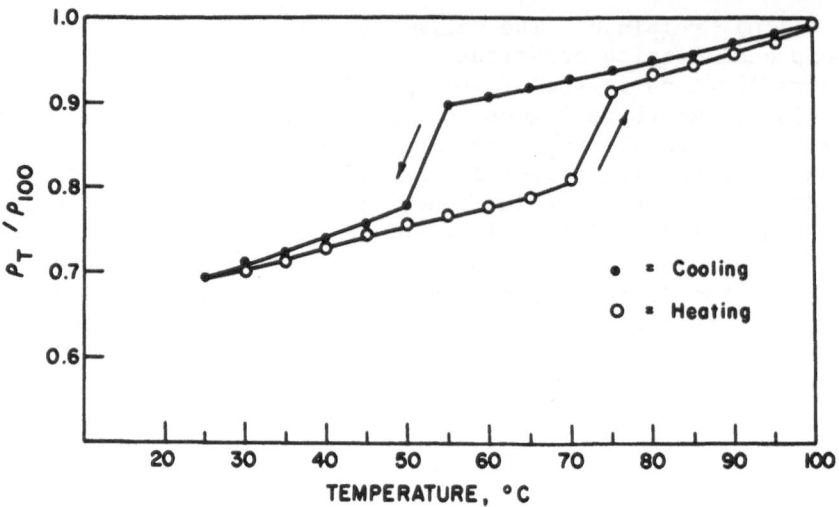

Fig. 5: The resistivity behavior of cycled Au–47.5 at. pct. Cd.

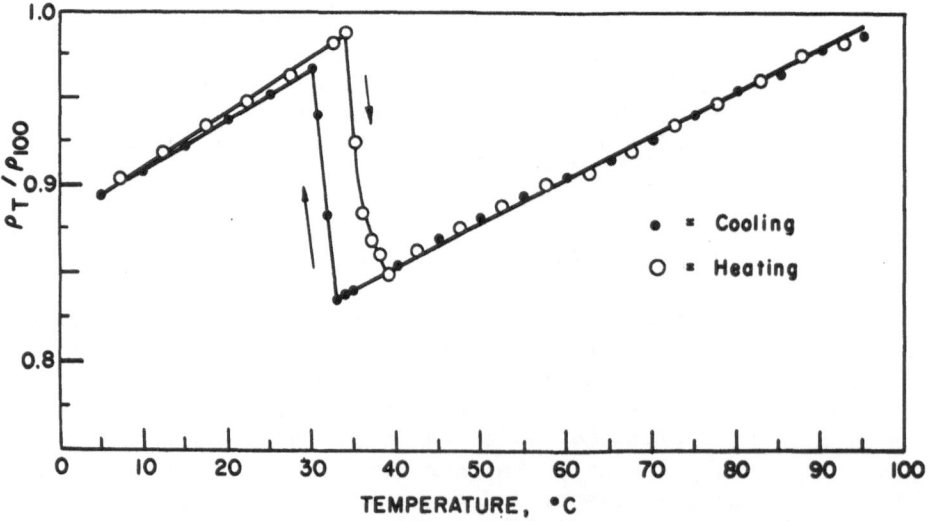

Fig. 6: The resistivity behavior of cycled Au–50.0 at. pct. Cd.

coordinates that are required for an analysis of the shuffles and their relationship to vibrational modes.

Premartensitic Diffraction Effects in Au–Cd. Electrical resistivity curves for the Au–47.5% Cd and Au–50% Cd alloys have been reported by many investigators (30) and are shown in Figs. 5 and 6 for comparison with the behavior of the Ti–Ni alloy (Fig. 2). It is apparent that the 47.5 Cd alloy exhibits the reduction in resistivity shown by the martensitic reaction in Ti–Ni; whereas, the resistivity increase associated with the martensite reaction in the 50% Cd alloy is similar to that of the premartensitic reaction in Ti–Ni. However, neither of the Au–Cd alloys exhibit any clearly defined resistivity anomalies above M_s. Such anomalies have been reported for the 47.5% Cd alloy when resistivity measurements were made in a magnetic field (21).

Both of the Au–Cd alloys do, however, exhibit some of the diffuse diffraction effects that were observed in Ti–Ni. Specifically, as shown for the 47.5% Cd alloy in Fig. 7, diffuse streaks appear on heating at the A_s temperature and virtually identical results have been observed for the 50% Cd alloy. The diffuse streaks persist up to about 300°C and disappear simultaneously with disappearance of the superlattice reflections (Fig. 7D).* The diffuse streaks reappear upon cooling (Fig. 7E); however, there were no indications of extra reflections at temperatures above M_s although the B19 martensitic structure was recovered at temperatures below M_s (∿50°C) (Fig. 7F).

Structure of Martensite in Au–50% Cd. X-ray diffraction patterns using the precession technique were employed to determine the martensitic structure in the Au–50% Cd alloy. Precession patterns at room temperature and at 60°C are presented in Figs. 8 and 9. The (111) plane of diffuse intensity is clearly evident, particularly in Fig. 8A. Thus, it is apparent that these diffuse diffraction effects do not require the relaxed restraints of thin foils but occur as well in bulk samples.

The reflections characteristic of the martensitic phase appear at the M_s temperature and, as with electron diffraction, there were no extra reflections at temperatures above M_s. This also is evident in the temperature dependence of the intensity of the extra reflections (Fig. 10). Comparison of Fig. 10 with Fig. 6 indicates that these reflections appear over the same narrow temperature range in which the martensite is formed and that they exhibit the temperature hysteresis characteristic of the martensitic

---

* Evaporation of Cd becomes significant at this temperature which may be responsible for the disordering as well as the slight distortion from the cubic structure evident in Fig. 7D.

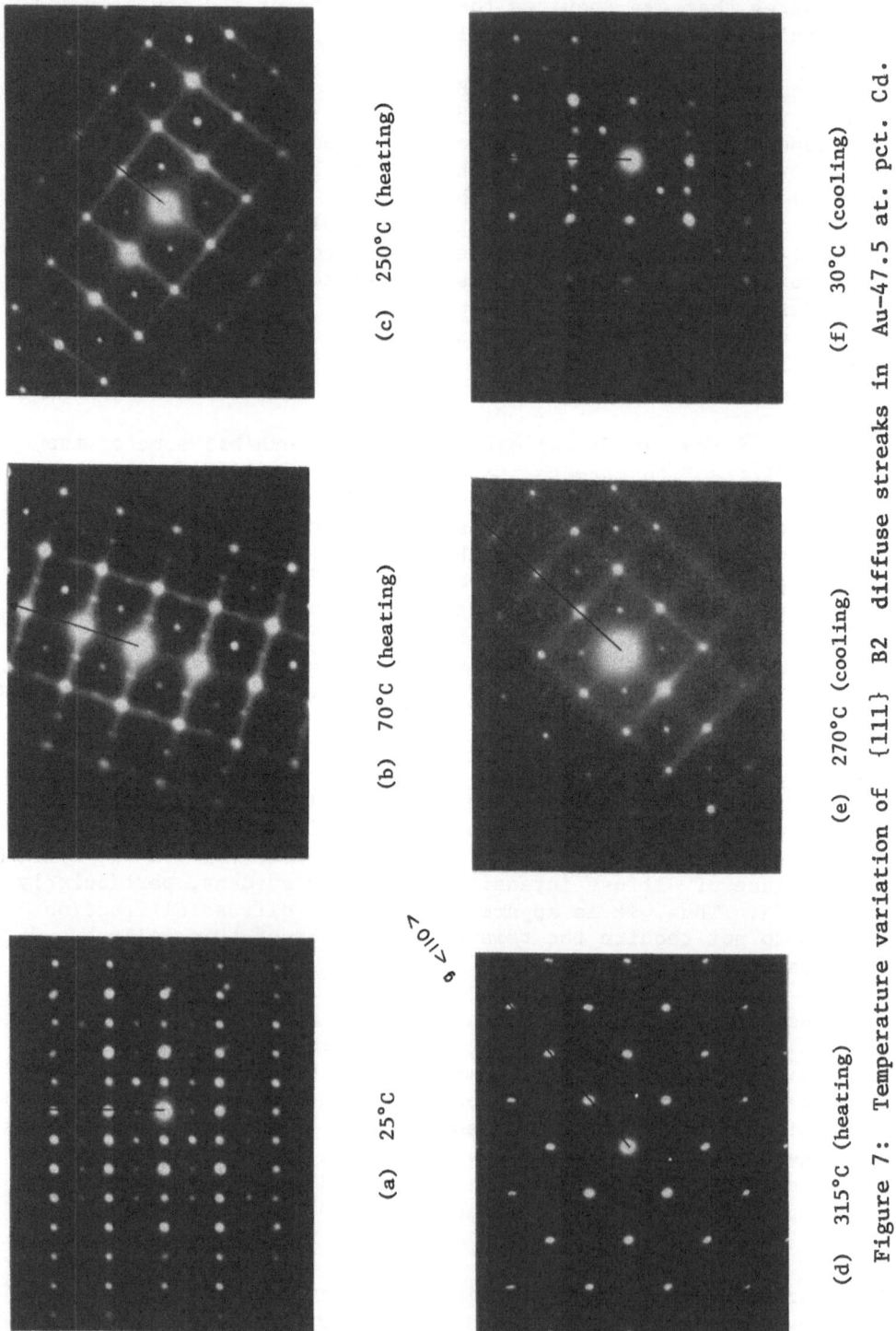

(a) 25°C

(b) 70°C (heating)

(c) 250°C (heating)

(d) 315°C (heating)

(e) 270°C (cooling)

(f) 30°C (cooling)

Figure 7: Temperature variation of {111} B2 diffuse streaks in Au-47.5 at. pct. Cd.

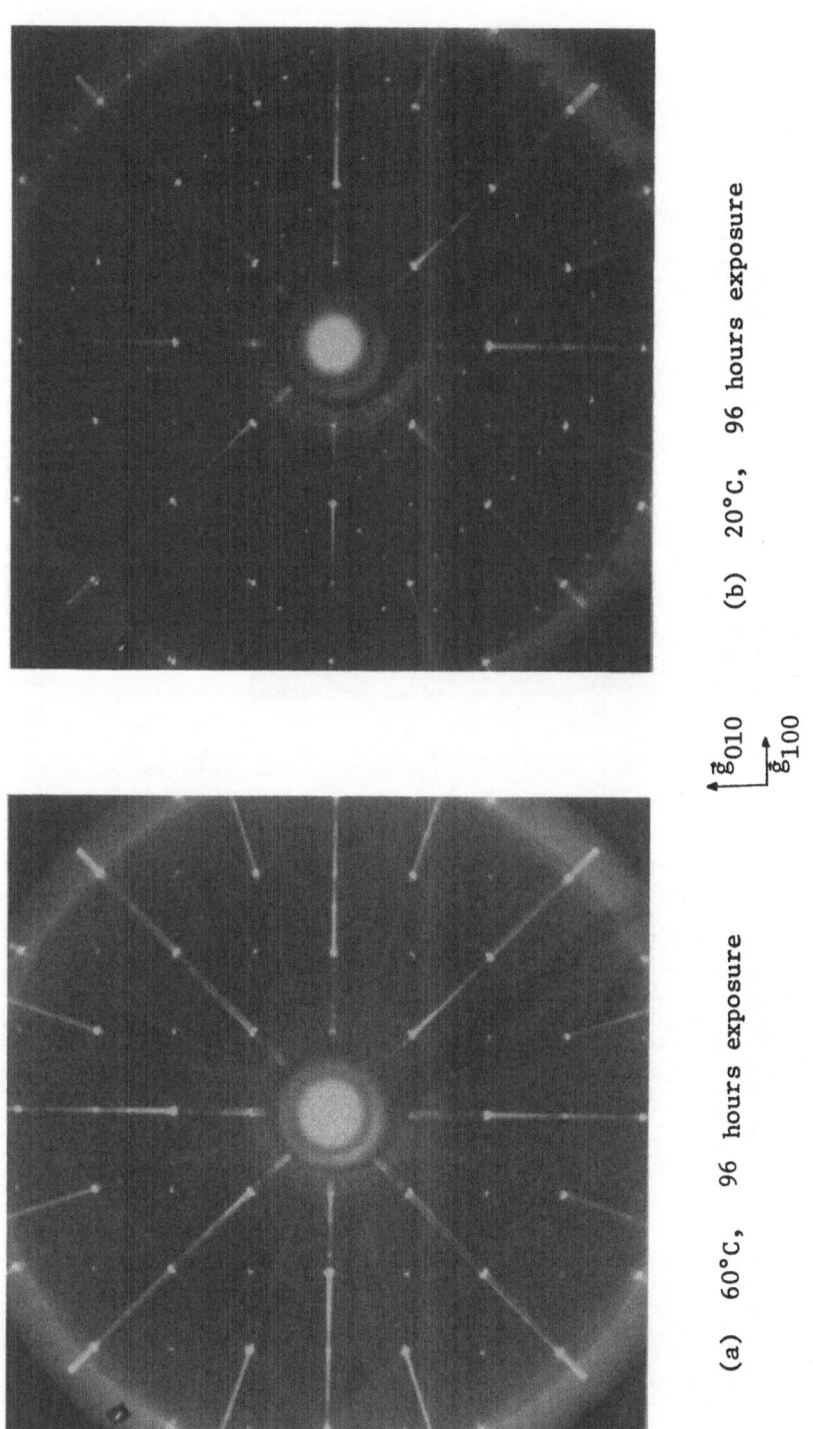

(a) 60°C, 96 hours exposure   (b) 20°C, 96 hours exposure

g_{010}
g_{100}

Figure 8: Precession patterns on the [001] B2 Au–50.0 at. pct. Cd, Zr-filtered MoK$_\alpha$ radiation.

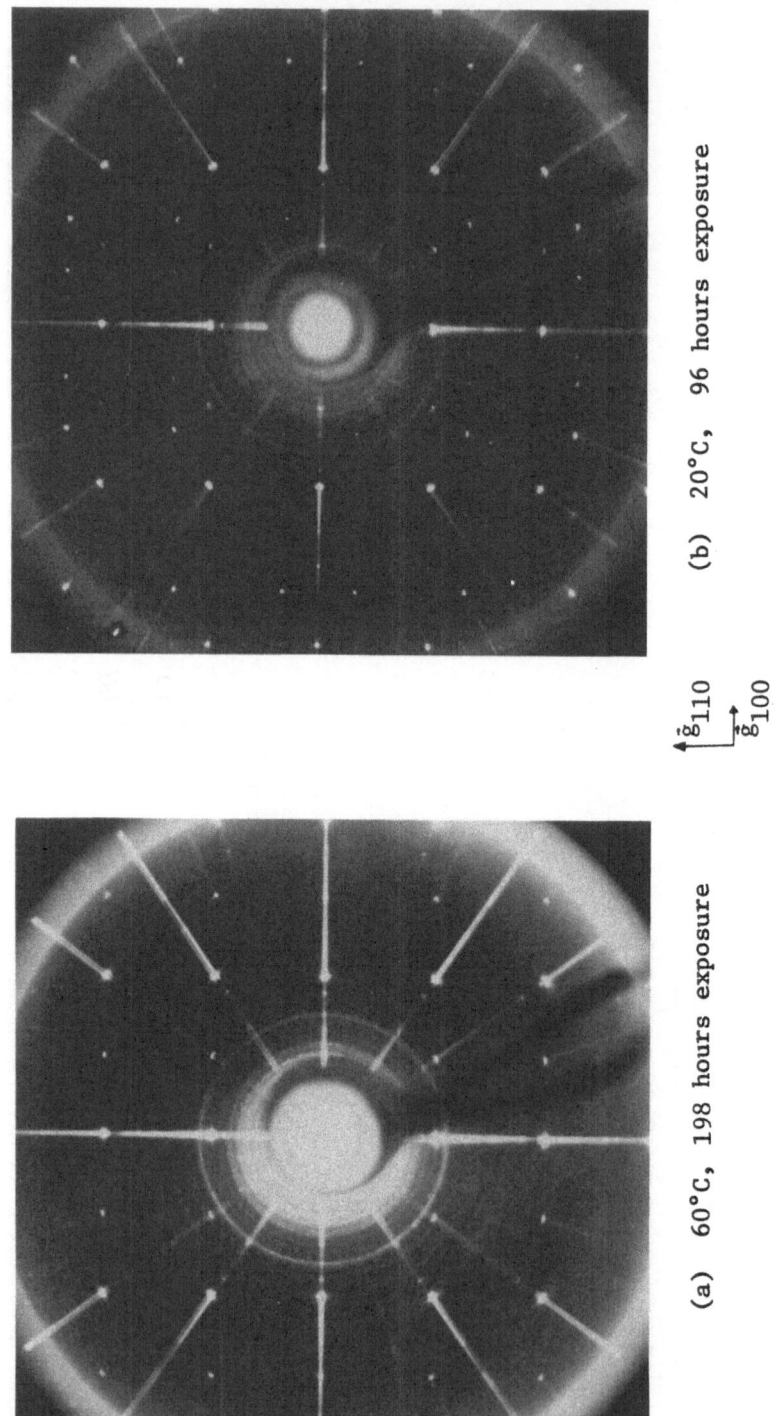

(a) 60°C, 198 hours exposure

(b) 20°C, 96 hours exposure

Figure 9: Precession patterns on the [011] B2 Au–50.0 at. pct. Cd, Zr-filtered MoK_α radiation.

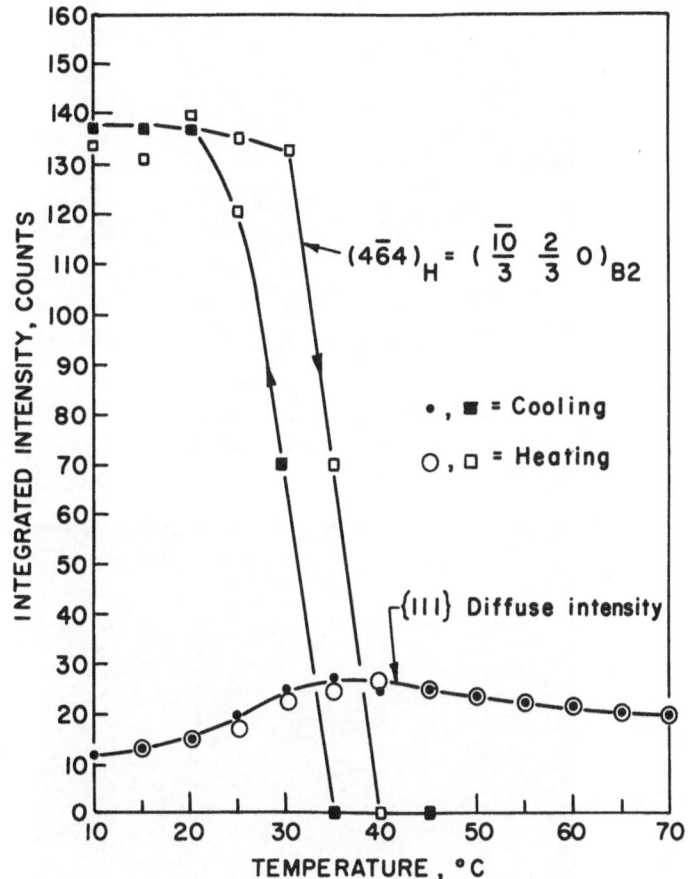

Figure 10: Variation in intensity of 1/3 reflection and {111} diffuse plane with temperature of Au–50.0 at. pct. Cd. MoK_α radiation.

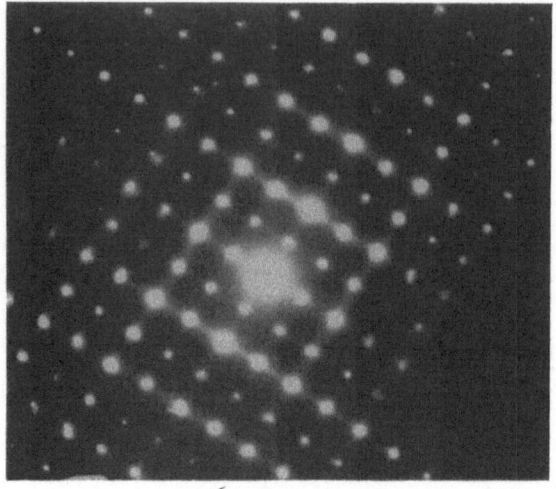

(a)   Au-47.5 at. pct. Cd
      (100) B2, 25°C

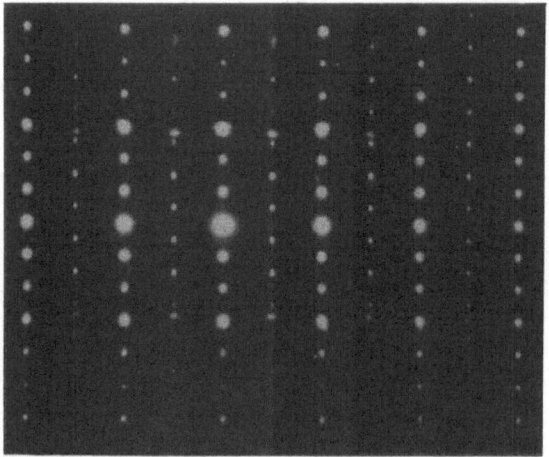

(b)   Au-50 at. pct. Cd
      (100) B2, 25°C

Figure 11: Diffuse intensity in the low temperature phase
            of Au-Cd alloys.

transformation. The {111} diffuse plane, on the other hand changes intensity in an essentially reversible fashion. The increase in intensity of the (111) diffuse plane on heating of the martensitic phase at temperatures below A_s is particularly significant and suggests that softening of the vibrational modes occurs as the transformation is approached in either direction. This also is evident in electron diffraction patterns which frequently exhibit diffuse streaks in the low temperature as well as the high temperature phase (Fig. 11).

Comparison of Figs. 3 and 8 indicates that the structure of the ζ' phase in Au-Cd is similar to, if not identical with, that of the transition phase produced by the premartensitic reaction in Ti-Ni. The (001), (110) and (111) precession patterns can be indexed in terms of the hexagonal or trigonal unit cell identified previously (6,31). The intensities of (hkl) and (hkl̄) reflections are not the same, however, which verifies the trigonal symmetry of the ζ' phase.

The correspondence between the β and ζ' phases is shown in Fig. 12a and is specified by:

$$(0001)_{\zeta'} \quad \| \quad (111)_{\beta}$$

$$[11\bar{2}0]_{\zeta'} \quad \| \quad [11\bar{2}]_{\beta}$$

Thus,

$$a_{\zeta'} \stackrel{\sim}{=} \sqrt{6}\, a_{\beta}$$

$$c_{\zeta'} \stackrel{\sim}{=} \sqrt{3}\, a_{\beta}$$

The Bain strains in this transformation are extremely small so that the lattice parameters of the ζ' phase deviate only slightly from those deduced from the correspondence. The precession patterns are consistent with the coexistence of four variants of this phase with "$c_{\zeta'}$" axis derived from the four <111> directions of the parent.

In order to determine the structure of the ζ' phase, intensities were determined by comparison of precession patterns exposed for various times with standard blacking films and a least squares technique was employed for the structural refinement. This study, which will not be reported in detail here, demonstrated that the ζ' phase exhibits the trigonal symmetry of space group 162-(P3̄1m) with atomic coordinates:

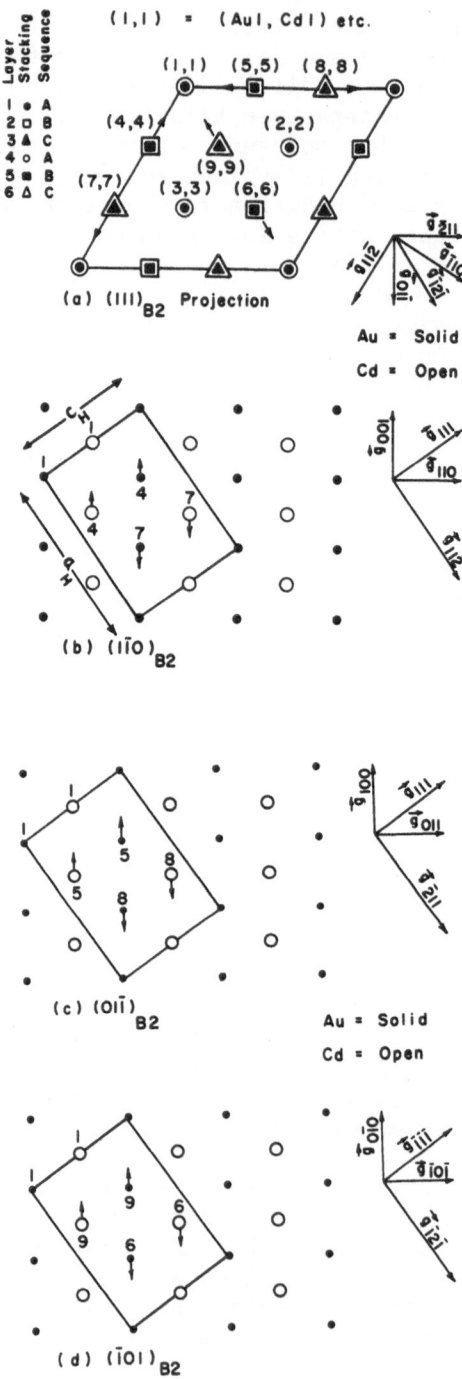

Figure 12: The trigonal unit cell as derived from B2 structure
            in Au-50.0 at. pct. Cd.

1   a   Au   0, 0, 0

2   c   Au   1/3, 2/3, 0; 2/3, 1/3, 0

6   k   Au   X, 0, Z; 0, X, Z; $\bar{X}$, $\bar{X}$, Z; $\bar{X}$ 0 $\bar{Z}$; 0 $\bar{X}$ $\bar{Z}$; X, X, $\bar{Z}$

        X = 0.308   Z = 0.692

1   b   Cd   0, 0, 1/2

2   d   Cd   1/3, 2/3, 1/2; 2/3, 1/3, 1/2

6   k   Cd   X, 0, Z; 0, X, Z; $\bar{X}$, $\bar{X}$, Z; $\bar{X}$ 0 $\bar{Z}$; 0, $\bar{X}$, $\bar{Z}$; X, X, $\bar{Z}$

where   X = 0.308   Z = 0.192

The shuffles involved in this transformation are associated with the 6 Au and 6 Cd atoms in the k positions. The remaining three atoms of each type undergo no displacements other than those required by the Bain strains. The atomic coordinates in the parent and product phases are presented in Table I along with the displacements (shuffles) deduced from these coordinates.

The shuffles in the 50% Cd alloy are more complicated than those in most other martensitic transformations. These shuffles occur in <$\bar{1}$01> directions of the ζ' phase, which, as shown in Figs. 12b to d, are derived from the three <100> directions of the parent phase.

Vibrational Modes and Structure of Martensite

The correspondence between the parent and product phases contains the essential information for examining the relationship between the lattice modes and the product structure as demonstrated by Cochran (33,34). It is convenient to consider separately the situations where the transformation produces no change in the number of atoms in the primitive cell and those where there is a change.

When the primitive cell of the parent and product structures contain one atom as in the FCC → BCC transition in steels, the Bain strains convert all of the atoms from their positions in the parent to those in the product and there are no shuffles. Thus, the potential role of vibrational modes in these systems is not revealed clearly in the structural change. Nevertheless, long wavelength acoustic modes can produce the dilations or shears required by the Bain strains. Direct experimental evidence for these dynamical effects at temperatures above M_s and their interpretation have been reported by Wayman (35).

Table I: The Atom Positions of ζ'-AuCd Based on Isotropic Temperature Correction

Atoms	Initial Positions			Displacements			Final Positions		
Au1	0	0	0	0	0	0	0	0	0
Au2	1/3	2/3	0	0	0	0	1/3	2/3	0
Au3	2/3	1/3	0	0	0	0	2/3	1/3	0
Au4	1/3	0	2/3	-0.6/24	0	0.6/24	7.4/24	0	16.6/24
Au5	0	1/3	2/3	0	-0.6/24	0.6/24	0	7.4/24	16.6/24
Au6	2/3	2/3	2/3	0.6/24	0.6/24	0.6/24	16.6/24	16.6/24	16.6/24
Au7	2/3	0	1/3	0.6/24	0	-0.6/24	16.6/24	0	7.4/24
Au8	0	2/3	1/3	0	0.6/24	-0.6/24	0	16.6/24	7.4/24
Au9	1/3	1/3	1/3	-0.6/24	-0.6/24	-0.6/24	7.6/24	7.6/24	7.6/24
Cd1	0	0	1/2	0	0	0	0	0	1/2
Cd2	1/3	2/3	1/2	0	0	0	1/3	2/3	1/2
Cd3	2/3	1/3	1/2	0	0	0	2/3	1/3	1/2
Cd4	1/3	0	1/6	-0.6/24	0	0.6/24	7.4/24	0	4.6/24
Cd5	0	1/3	1/6	0	-0.6/24	0.6/24	0	7.4/24	4.6/24
Cd6	2/3	2/3	1/6	0.6/24	0.6/24	0.6/24	16.6/24	16.6/24	4.6/24
Cd7	2/3	0	5/6	0.6/24	0	-0.6/24	16.6/24	0	19.4/24
Cd8	0	2/3	5/6	0	0.6/24	-0.6/24	0	16.6/24	19.4/24
Cd9	1/3	1/3	5/6	-0.6/24	-0.6/24	-0.6/24	7.4/24	7.4/24	19.4/24

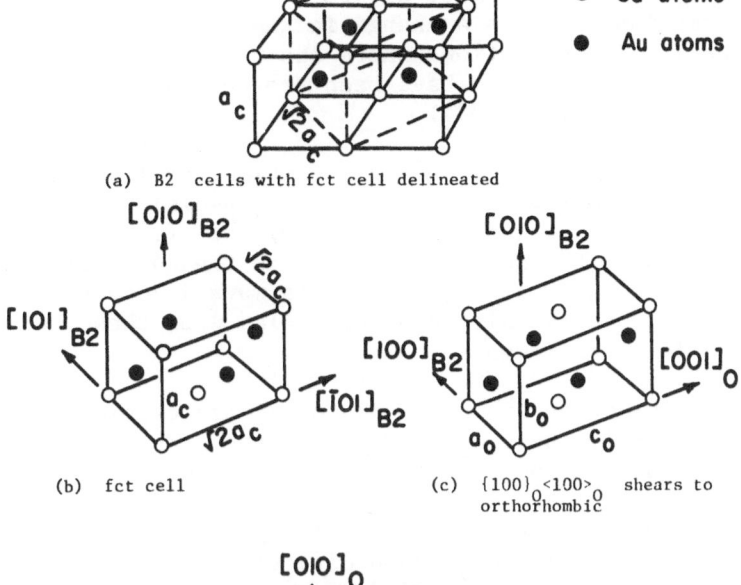

(a) B2 cells with fct cell delineated

(b) fct cell

(c) $\{100\}_0<100>_0$ shears to orthorhombic

(d) $(100)_0[001]_0$ planar shuffle to B19 structure

Figure 13: Crystallographic steps for B2 → B19 transformation
in Au–47.5 at. pct. Cd.

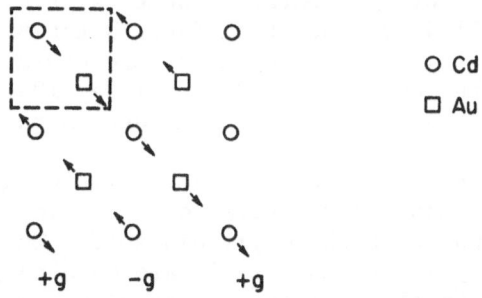

○ Cd
□ Au

+g   −g   +g

Figure 14: (010) cubic projection showing the π/a<110>
acoustic mode.

The potential importance of vibrational modes in martensitic transformations is perhaps most apparent in the situation where the transformation produces a change in the number of atoms in the primitive cell. The situations where the reaction doubles the cell are easiest to visualize and have been treated in detail for ferro-electrics such as $SrTiO_3$ (33).

In terms of the lattice wave description, cell doubling transitions occur when the wave vector occurs at special points on the Brillouin zone boundary such as the (110) or (111) points. For these wave vectors, the wavelength is twice the unit cell dimension and corresponding atoms in adjacent cells move in opposite directions. Thus, the repeat distance is doubled when these modes condense as static displacements.

The transformation in Au-47.5% Cd alloys provides a particularly simple example (36). In this case, the high-temperature phase is B2 and the low-temperature phase is orthorhombic. The usual lattice correspondence is shown in Fig. 13 and for comparison with data from Pearson, the crystal setting will be taken such that the b axis is derived from a cube direction of B2 and the a and c axes are derived from $<110>_{B2}$.

An (010) projection of the B2 structure is shown in Fig. 14 and the origin has been displaced off an atom to identify clearly the two atoms in the unit cell. The arrows represent the polarization vectors for a π/a [101] zone boundary acoustic mode and it is apparent that this mode displaces alternate ($\bar{1}$01) planes in opposite directions by an amount $\pm g$. These eigenvectors correspond to the directions required to produce the shuffles observed in the B19 structure as shown in Fig. 15. (The Bain strains again are not included.) The atomic positions in the B19 phase (37) are listed on the right and all atoms are converted from their positions in the parent to those in the product by this displacement wave. It is apparent that combination of this $q = \pi/a$ [101] zone boundary mode with $q = 0$ acoustic modes required for the Bain strains can convert the B2 to the B19 structure. Somewhat similar considerations have been applied by Moss et al (38) to the BCC → HCP transition as in Ti or Zr. In this case, a zone center $<112><11\bar{1}>$ mode is combined with the zone boundary $<110><1\bar{1}0>$ mode which doubles the cell as required for the hexagonal structure.

The B19 structure is a particularly simple example in which the shuffles can be described in terms of a single mode represented by the π/a [101] point of the Brillouin zone. The ζ' structure and the premartensitic state in Ti-Ni are more complicated and, as indicated by the tripling of the repeat distance, require $2/3\ \pi/a<hkl>$ displacement waves. The diffraction patterns for the premartensitic state in Ti-Ni and the ζ' phase in Au-50% Cd are virtually identical and require both $<111>$ and $<110>$ displacement waves.

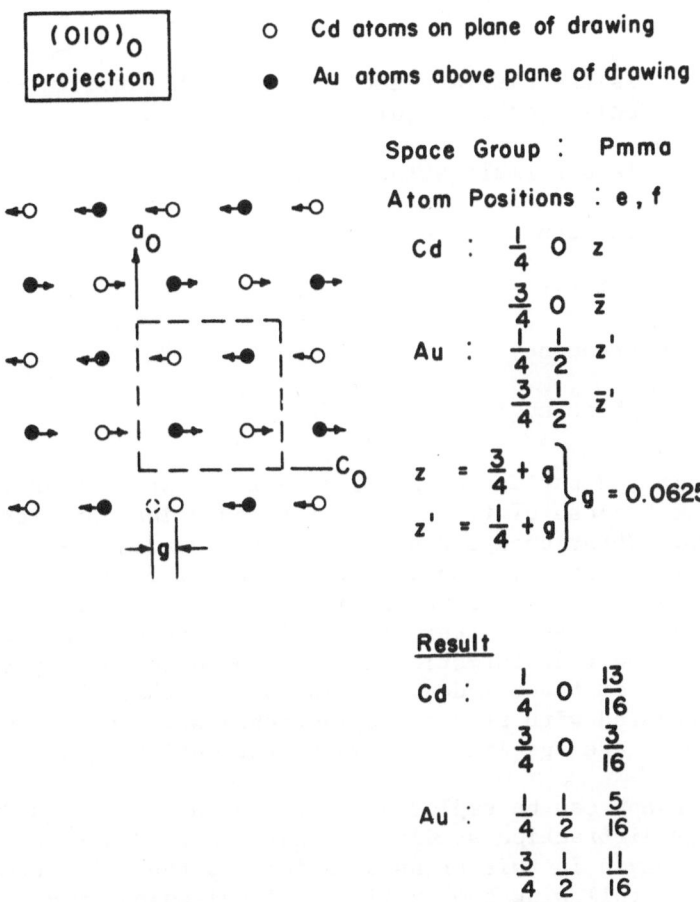

Figure 15: Comparison of B19 structure based on the presence
of a zone boundary acoustic mode with Pearson data (37).

The displacement of an atom by the phonon spectrum is given by a summation over the modes and can be represented by the expression (34):

$$\mu_{\ell k} = (Nm_k)^{-1/2} \sum_{q \atop j} B_j(q)e_{jk}(q) \cos [qr_{\ell k} - \omega_j(q)t + \theta_j(q)] \quad (1)$$

where

$\mu_{\ell k}$ is the displacement of the k'th atom in the ℓ'th cell from its equilibrium position

$(Nm_k)^{-1/2}$ is a normalization factor

$B_j(q)$ = amplitude of the jth mode

$e_{jk}(q)$ = polarization of jth mode

$\omega_j(q)$ = frequency

$\theta_j(q)$ = phase angle of jth mode

In general, a particular displacement has at best an infinitesimal lifetime because of the different frequencies associated with the different vibrational modes. These frequencies are, however, a function of temperature. Thus, as the dispersion curves change with temperature, specific modes may achieve the same frequency permitting their displacement pattern to persist. The displacements associated with the transformation can then be described by the polarization vectors of these modes. It is assumed that the transformation is associated with static displacements whose directions, but not magnitudes, are specified by the vibrational modes.

This concept can be employed to obtain the ζ' structure from β. The x-ray diffraction studies indicate that 2/3 <110><110> modes are involved in this transformation and the ζ' structure results from the following combination of longitudinal modes:

2/3 <111><111> + 2/3 <1$\bar{1}$0><1$\bar{1}$0>

+ 2/3 <01$\bar{1}$><01$\bar{1}$> + 2/3 <10$\bar{1}$><10$\bar{1}$>.

Since only the displacement directions are significant, arbitrary amplitudes will be employed in equation 1 and the combination of these modes then yields:

$$\mu_{\ell k} = <111> \cos\left[\frac{4\pi}{3a}(X + Y + Z)a - \frac{\pi}{2}\right]$$

$$+ <1\bar{1}0> \cos\left[\frac{2\pi}{3a}(X - Y)a + \frac{\pi}{2}\right]$$

$$+ <01\bar{1}> \cos\left[\frac{2\pi}{3a}(Y - Z)a + \frac{\pi}{2}\right]$$

$$+ <10\bar{1}> \cos\left[\frac{2\pi}{3a}(X - Z)a + \frac{\pi}{2}\right] \qquad (2)$$

where a is the lattice parameter of the cubic phase and X, Y, and Z are the coordinates of the atoms in the cubic phase.

Table II presents the results of applying equation (2) to the ζ' transformation. Atomic positions in the cubic phase are listed both in trigonal and cubic coordinates and the latter are used with equation (2) to obtain the displacements listed in the table. It is apparent that three atoms of each type are undisplaced by this combination of modes and that the remaining six are displaced in <100> directions as required by the structure of the ζ' phase (Fig. 12). Thus, formally at least, the shuffles in the β → ζ' transformation can be described by this combination of <111> and <110> displacement waves. The Bain strains, of course, have not been considered and require the operation of appropriate zone center modes.

It now becomes of interest to compare the behavior of the Ti-Ni and Au-Cd systems at temperatures above M_s. Both systems exhibit {111} planes of diffuse intensity which suggests the operation of a linear scatterer in crystal space. As has been suggested previously (24) the close packed [111] rows are believed to vibrate in a rigid and more or less uncoordinated manner thus acting as essentially linear scatterers. The observation that the temperature dependence of the diffuse intensity is opposite that for normal thermal diffuse scattering is consistent with a reduction in frequency and increase in amplitude of this mode as the temperature is lowered.

The Ti-Ni and Au-Cd systems differ with regard to the appearance of the extra reflections that characterize the ζ' phase. In the Au-Cd system these reflections appear at M_s, (which has been confirmed by observation of surface relief) and exhibit the hysteresis of the martensitic transformation. Thus, in this system, coordination of the <111> and <110> vibrational modes required for the shuffles along with the zone center modes that constitute the Bain strains apparently occurs simultaneously. In Ti-Ni, on the other hand, there are no indications of surface relief in the

Table II: Directions of Atom Displacements during the Martensite Reactions in β-Phase Au–50.0 At. Pct. Cd.

Atom	Hexagonal			Cubic			U_{1k}		
	X	Y	Z	X	Y	Z	X	Y	Z
Au1	0	0	0	0	0	0	0	0	0
Au2	1/3	2/3	0	-1	1	0	0	0	0
Au3	2/3	1/3	0	0	1	-1	0	0	0
Au4	1/3	0	2/3	1	1	0	0	0	1
Au5	0	1/3	2/3	0	1	1	0	1	0
Au6	2/3	2/3	2/3	0	2	0	1	0	0
Au7	2/3	0	1/3	1	1	-1	0	0	-1
Au8	0	2/3	1/3	1	1	1	0	-1	0
Au9	1/3	1/3	1/3	0	1	0	-1	0	0
Cd1	0	0	1/2	1/2	1/2	1/2	0	0	0
Cd2	1/3	2/3	1/2	-1/2	3/2	1/2	0	0	0
Cd3	2/3	1/3	1/2	1/2	3/2	-1/2	0	0	0
Cd4	1/3	0	1/6	1/2	1/2	-1/2	0	0	1
Cd5	0	1/3	1/6	-1/2	1/2	1/2	0	1	0
Cd6	2/3	0	1/6	-1/2	3/2	-1/2	1	0	0
Cd7	0	2/3	5/6	3/2	3/2	-1/2	0	0	-1
Cd8	1/3	1/3	5/6	-1/2	3/2	3/2	0	-1	0
Cd9	1/3	1/3	5/6	1/2	3/1	1/2	-1	0	0

premartensitic range and the 1/3 reflections change intensity reversibly within this range. Thus, it appears that the zone center instability required for the Bain strains is separated from the short wavelength modes that constitute the shuffles and this may result from differences in the coupling parameters between the two systems.

Although surface relief is not observed in the premartensitic range in Ti-Ni, striated contrast in extinction contours does occur (19, 24). Thus, it seems most likely that the extra reflections in Ti-Ni result from long-lived dynamic fluctuations between the CsCl and a ζ' transition structure in a manner similar to that which has been suggested for the athermal ω reaction in Ti or Zr alloys (39). Whether the anomalous resistivity increase in Ti-Ni results from scattering by these fluctuations or from an electronic transition cannot be answered at present. However, the absence of a resistivity anomaly in spite of the diffuse intensity in the Au-Cd system appears to favor an electronic origin.

The martensitic structure in Ti-Ni can also be considered in these terms. There is reasonable agreement that the structure is a monoclinic distortion of the B19 structure but the shuffles are still uncertain. As in the Au-47.5% Cd alloy (Fig. 13), the transformation involves contraction of the a_o and c_o axes and expansion of the b_o axis.* The monoclinic distortion can be described formally as a shear of $(100)_M$ in the $[010]_M$ and a planar shuffle occurs on a (110) plane of the parent phase. Both an $(010) [001]_M$ (29) and an $(001) [010]_M$ (24) shuffle have been proposed. Neither of these shuffles describes the diffraction pattern completely so that intensity calculations have been made for a variety of shuffles of the type $(001) [hk0]_M$.

These calculations indicate that shuffle of the (001) plane in the <120> direction by 1/16 provides reasonably good agreement with the diffraction pattern as shown for the low angle lines in Table III. In terms of displacement waves, this structure can be produced by combining a $q = o$ shear mode of the type $<112><11\bar{1}>$ with a $q = 2/3 \pi/a<111><111>$ and a $q = \pi/a<110><1\bar{1}0>$ displacement wave.

The intermediate structure produced by the $<112><11\bar{1}>$ shear and the $2/3 \pi/a<111><111>$ displacement wave is shown in Fig. 16. This produces the correct monoclinic angle and approximate lattice parameters (40) but shuffles alternate $(001)_M$ planes in $[110]_M$ rather than $[120]_M$ directions. However, the $\pi/a<1\bar{1}0><110>_\beta$ mode corresponds to displacing these same planes in an $[010]_M^\beta$

---

* Here the crystal setting is such that the a_o axis is derived from $[001]_\beta$.

Table III: Comparison of Observed Intensities and Structure
            Factors for Martensite in Ti-Ni

hkl	I_{obs}	$|F|$
010	W	6.3
001	0	0
011	10	8.6
100	10	9.9
1$\bar{1}$0	W	21.3
101	10	2.9
110 } 020	30	54.1 } 26.8
1$\bar{1}$1	70	65.5
002	70	67.0
111 } 021	100	41.7 } 60.8
1$\bar{2}$0	–	5.9
012	VVW	5.3
1$\bar{2}$1	10	7.1

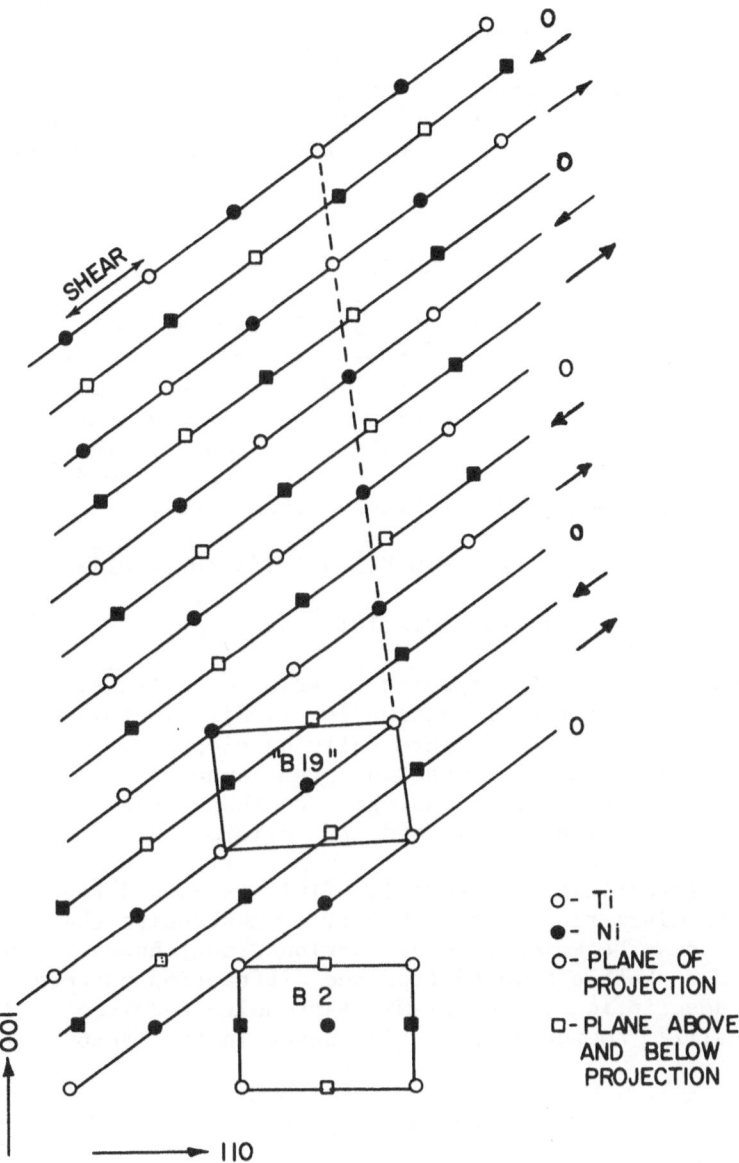

Figure 16: Intermediate structure of Ti-Ni martensite produced by <112><11$\bar{1}$> shear and 2/3 <111><111> displacement wave.

direction so that the resulting overall shuffle can occur in the $<120>_M$ direction. It also is evident in Fig. 16 that this combination of displacements results in a more complex ordering arrangement. The martensitic structure in Ti-Ni is not known in sufficient detail to pursue this question further although it is worth noting that a variety of long period superlattices as well as complex stacking arrangements of "close packed" planes are observed in martensitic structures formed in the β phases including Ti-Ni (41,43).

CONCLUSIONS

The trigonal structure of the ζ' martensite in Au-50% Cd has been confirmed and shown to exhibit P$\bar{3}$1m symmetry (space group 162). This structure has 18 atoms in the unit cell and a formal description of the transformation involves shuffling of six symmetry related atoms of each type in $<100>_\beta$ directions of the parent phase. This pattern of displacements can be described by the following combination of displacement waves:

$$2/3 <111><111> + 2/3 <1\bar{1}0><1\bar{1}0> + 2/3 <01\bar{1}><01\bar{1}>$$

$$+ 2/3 <10\bar{1}><10\bar{1}>.$$

This same structure appears to arise as a transition phase in the premartensitic transformation in Ti-Ni. In this case, however, evolution of the structure into fully developed martensitic plates is not observed and the transition phase appears to exist as long-lived dynamic fluctuations analogous to those that characterize the omega transformation in Ti and Zr alloys.

The martensitic structure in TiNi is considered to be a monoclinic distortion of the B19 (γ') martensite that forms in Au-47.5% Cd. The monoclinic distortion alone, however, does not produce the shuffle deduced from the diffraction pattern. In terms of wave descriptions, the Ti-Ni martensite requires a 2/3 π/a<111><111> component which is absent in the Au-Cd system.

ACKNOWLEDGEMENTS

The authors gratefully acknowledge support of this work by the National Science Foundation. Discussions with Profs. Gordon, Lando and Schuele of Case Western Reserve University as well as Prof. Wayman of The University of Illinois and Prof. de Fontaine of The University of California have been most helpful.

REFERENCES

1. M. V. Nevitt, Electronic Structure and Alloy Chemistry of the Transition Elements, edited by P. A. Beck, John Wiley, 1963, p. 101.

2. C. M. Wayman and K. Shimizu, Metal Sci. Journ.,6, 175 (1972).

3. H. Warlimont and L. Delaey, Progress in Materials Science, Pergamon Press, 18, 1974.

4. A. E. Dwight, R. A. Conner, Jr., and J. W. Downey, Acta Cryst., 18, 835 (1965).

5. R. J. Jackson and W. L. Larsen, J. Nuc. Mat. 21, 263 (1967).

6. R. S. Toth and H. Sato, Acta Met., 16, 413 (1968).

7. Reference 3, page 38.

8. P. W. Forsbergh, Phys. Rev., 76, 1187 (1949).

9. C. Zener, Phys. Rev., 71, 846 (1947).

10. W. Cochran, Advan. Phys., 9, 387 (1960); 10, 401 (1961).

11. J. F. Scott, Rev. Mod. Phys., 46, 83 (1974).

12. F. E. Wang, B. F. De Savage, W. J. Buehler and W. R. Hosler, J. Appl. Phys., 39, 2166 (1968).

13. S. Zirinsky, Acta Met., 4, 164 (1956).

14. R. R .Hasiguti and K. Iwasaki, J. Appl. Phys., 39, 2182 (1968).

15. N. G. Pace and G. A. Saunders, Phil. Mag., 22, 73 (1970).

16. R. J. Wasilewski, Trans. TMS-AIME, 233, 1691, (1965).

17. N. Nakanishi, Memoirs of the Konan University, Science Series, No. 15, Article 77, (1972).

18. W. J. Buehler and F. E. Wang, Ocean Eng., 1, 105, (1968).

19. D. P. Dautovich and G. R. Purdy, Canadian Met. Quart., 4, 129 (1965).

20. J. E. Hanlon, S. R. Butler and R. J. Wasilewski, Trans. TMS-AIME, 239, 1323, (1967).

21. H. Livingston and K. Mukherjee, J. Appl. Phys., $\underline{43}$, 4944 (1972).

22. F. E. Wang, W. J. Buehler and S. J. Pickart, J. Appl. Phys., $\underline{36}$, 3232 (1965).

23. K. Chandra and G. R. Purdy, J. Appl. Phys., $\underline{39}$, 2176 (1968).

24. G. D. Sandrock, A. J. Perkins and R. F. Hehemann, Met. Trans., $\underline{2}$, 2769 (1971).

25. R. J. Wasilewski, S. R. Butler and J. E. Hanlon, Met. Sci. J., $\underline{1}$, 104 (1967).

26. D. P. Dautovich, Z. Melkvi, G. R. Purdy, and C. V. Stager, J. Appl. Phys., $\underline{37}$, 2513, (1966).

27. F. E. Wang, S. J. Pickart, and H. A. Alperin, J. Appl. Phys., $\underline{43}$, 97, (1972).

28. M. J. Marcinkowski, A. S. Sastri, and D. Koskimaki, Phil. Mag., $\underline{18}$, 945, (1968).

29. K. Otsuka, T. Sawamura, and K. Shimizu, Phys. Stat. Sol., (a) $\underline{5}$, 457, (1971).

30. D. S. Lieberman, Phase Transformations, ASM, 1970, p. 1.

31. H. M. Ledbetter and C. M. Wayman, Met. Trans., $\underline{3}$, 2349, (1972).

32. A. Ölander, J. Amer. Chem. Soc., $\underline{54}$, 3819, (1932).

33. W. Cochran and A. Zia, Phys. Stat. Sol., $\underline{25}$, 273 (1968).

34. W. Cochran, The Dynamics of Atoms in Crystals, Crane and Russak, New York, 1973.

35. I. Cornelis, R. Oshima, H. C. Tong and C. M. Wayman, Scripta Met., $\underline{8}$, 133 (1974).

36. L. Delaey, J. Van Paemel and T. Struyve, Scripta Met., $\underline{6}$, 507 (1972).

37. W. B. Pearson, A Handbook of Lattice Spacings and Structures of Metals and Alloys, Pergamon Press, 1958.

38. S. C. Moss, D. T. Keating and J. D. Axe, BNL Report 18098, 1973.

39. S. C. Moss, D. T. Keating and J. D. Axe, BNL Report 18110, 1973.

40. R. F. Hehemann and G. D. Sandrock, Scripta Met., <u>5</u>, 801, (1971).

41. A. Nagasawa, T. Maki and J. Kakinoki, J. Phys. Soc., Japan,
     <u>26</u>, 1560 (1969).

42. Y. Murakami, N. Nakanishi and S. Kachi, Japanese J. Appl. Phys.,
     <u>11</u>, 1591 (1972).

43. M. Oka, Y. Tanaka and K. Shimizu, Japanese J. Appl. Phys., <u>11</u>,
     1073 (1972).

LATTICE SOFTENING AND THE ORIGIN OF SME

N. Nakanishi

Department of Chemistry, Konan University

Motoyama, Kobe, Japan

ABSTRACT

"Lattice softening" or "Soft shear mode" has recently been found in connection with the displacive or martensitic phase transition in thermoelastic alloys and compounds. It is here described that in the β_1 (ordered bcc) alloys such as $CuAuZn_2$ and $AuCd$, the lattice softening appears as a premonitory phenomenon in thermoelastic martensitic transformation and it has an intimate relation to the essential characters (shear mechanism and nucleation) of the thermoelastic martensite. Moreover, the lattice softening is found to be fundamental for understanding the mechanism of the so-called "pseudoelasticity" (superelastic and ferroelastic behaviour), which has been observed to be closely related to SME. The present author suggests that the origin of SME might be contained in the superelastic behaviour, which is a unique stress-strain feature observed upon deformation of both the β_1 and martensite phases. In order to make clear the origin of SME, therefore, it is most important to account for the role of lattice softening upon the mechanism of this superelastic behaviour, which basic nature may be due to the existence of coherent boundaries, such as mobile twin or domain boundaries in the martensite crystal and also interfaces between the stress-induced martensite and the β_1 (matrix) crystal.

1. LATTICE SOFTENING

The β-Tungsten compounds, such as V_3Si and Nb_3Sn, have very high critical transition temperatures (T_c) for superconducting state and their martensitic transition temperatures (M_s) exist a few degrees above T_c: For example, $T_c = 16.9$ K and 18.1 K, and $M_s = 22$ K

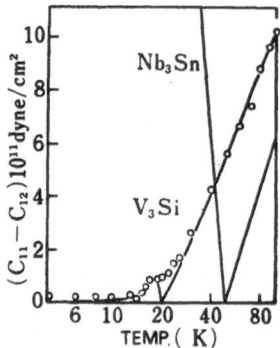

Fig.1: Temperature dependence of $(C_{11}-C_{12})$ in the vicinity of the transition temperature in Nb_3Sn and V_3Si (after Pytte).

and 43 K in V_3Si and Nb_3Sn, respectively. The shear elastic constant of both the phases, $C'=(C_{11}-C_{12})/2$, becomes soft above and below M_s as shown in Fig.1(1). This suggests that the resistance for the $\{110\}\langle110\rangle$ shear decreases foward zero in the vicinity of the M_s temperature.

The mechanism for the martensitic cubic⇌tetragonal transformation has been explained by the double shear operating on the $\{110\}\langle110\rangle$ system. The volume difference associated with the transition is very small, $\simeq 10^{-4}$, the tetragonality (c/a) in the martensite crystal is about 1.0024 (V_3Si), and the hysteresis of the transition temperatures is observed to be extremely small. Moreover, almost completely reversible transformation occurs upon heating and cooling, and thus the character of this transition seems nearly second order.

Recently this lattice softening of the β−W compounds has been theoretically explained by Pytte(2) as a "band-Jahn-Teller effect", in which an interaction between soft phonons and 3d−electrons on a linear chain was taken into account. According to this theory, the softening of acoustic waves with zero wave vector, q=0, caused by the electronic ordering of orbital state leads to the decrease in the resistance for the $(110)[1\bar{1}0]$ uniform shear, and the cubic crystal is transformed into tetragonal. In more general expression, the lattice softening can be understood as a premonitory phenomenon of the phase transition from a higher symmetry into a slightly

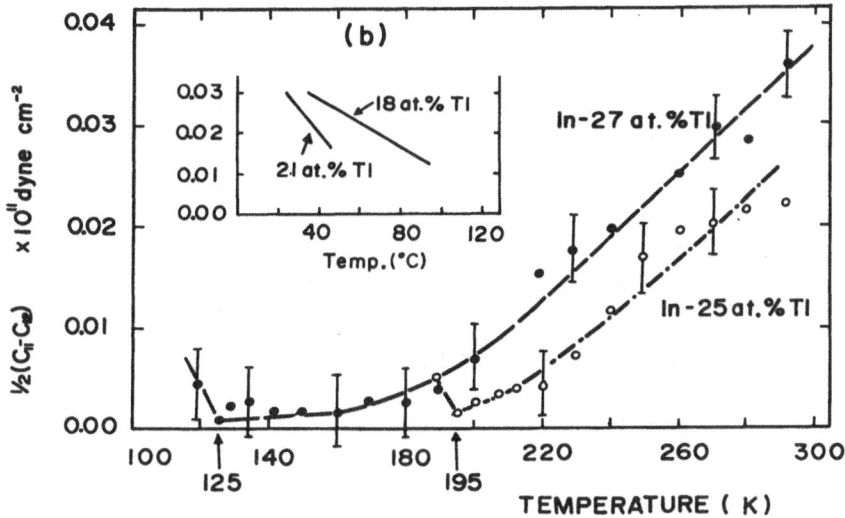

Fig.2: Temperature dependence of $(C_{11}-C_{12})/2$ above and below the
    transition temperature in In-Tl alloys (Gunton and Saunders).

distorted structure, and means that in the higher symmetrical
structure, the amplitude of phonon modes that coincide with the
lower symmetry structure can be remarkably enhanced immediately
above the transition temperature.

In the case of In-Tl alloys, the martensitic transformation
due to the lattice softening has been also observed and the shear
elastic constant C' of both the phases, fcc and fct, decreased
toward zero in the vicinity of the M_s temperature(3) (Fig.2).
In a 20.7at.% Tl alloy the temperature hysteresis between M_s and
A_s is found to be around 2-4 ^{o}C and the transition heat is about
0.4 cal/mole. Since the double shear is operated on the {110}
⟨110⟩ system, the habit plane is observed nearly the (110) plane,
this making us possible to regard the transition as nearly second
order. However, the origin of the lattice softening is not so far
obvious in this alloy, though some theoretical considerations have
been reported(4). In the case of Mn-Cu alloys, the M_s temperature
corresponding to the transition temperature of antiferromagnetic
γ-Mn, decreases with increasing Cu content and the double shear
transition accompanied with the lattice softening has been observed
by Sugimoto et al(5).

In Figs 3 and 4, a phase diagram of the pseudobinary β-Cu-Au-
Zn alloy and the temperature change in the shear elastic constants,
C_{44} and C', measured at temperatures above the transition temperature, $T_o (=(M_s+A_s)/2)$, are shown, respectively(6). It is found in

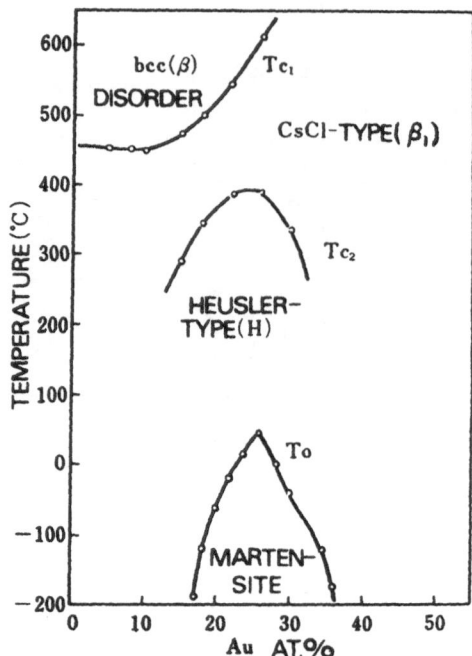

Fig.3: Phase diagram of the $Au_xCu_{55-x}Zn_{45}$ alloy.

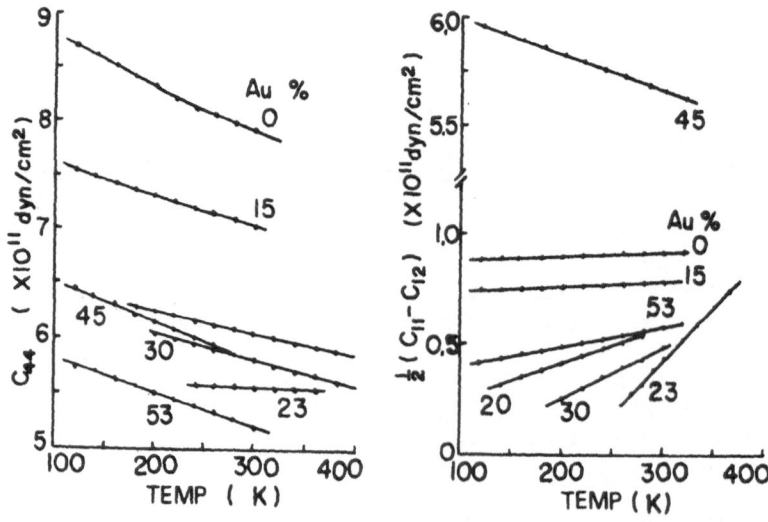

Fig.4: Temperature dependence of the elastic constants $(C_{11}-C_{12})/2$
and C_{44} of the $Au_xCu_{33-x}Zn_{47}$ alloy.

the diagram that the CsCl typed ordered β_1 lattice can be transformed into the Heusler typed ordered (H) lattice, its maximum
transition temperature existing at the stoichiometric composition,
$CuAuZn_2$, and upon subsequent lowering the temperature the martensite (M) phase appears. In Fig. 4 the positive temperature coefficient of C' is found to become steeper at the Au 23at.% corresponding to the Heusler composition. The crystal structure of the
martensite determined by means of X-ray diffraction and transmission
electron microscopy was explained as a mixed structure of 18R
(long period stacking structure)+2H(hcp) with ordered atomic con-
figuration(7). The lattice orientation between the Heusler and
the martensite phase was determined as $(110)_H //(001)_M : [1\bar{1}0]_H //$
$[010]_M$, with the twinning plane $(121)_M$, which is transformed from
a $(10\bar{1})_H$, and with stacking faults existing on a $(001)_M$, which
corresponds to a $(110)_H$. Here, the orthorhombic unit lattice is
used instead of 18R and 2H representation. The elastic anisotropy,
$2C_{44}/(C_{11}-C_{12})$, measured at a constant temperature just above the A_s
temperature is plotted as a function of Au content and compared
with a relation of the transition temperature vs. Au content as
shown in Fig. 5. The curves obtained are very much similar to
each other; that is, the maximum of both A_s and the anisotropy is
observed near Au 25at.% and there is no anisotropy at Au 45at.%
where the martensite cannot be observed. This large elastic anisotropy is caused by the decrease in the shear constant $(C_{11}-C_{12})/2$
near the transition temperature, while the C_{44} increases gradually
or is constant.

The lattice softening or the positive value of dc'/dT suggests
an entropy elasticity, just as same as in the case of rubbers, of
the β_1 phase alloys, and therefore the entropy term associated
with lattice vibration makes an important role on the stability of
the β_1 phase. The Debye temperatures θ_D in the β_1 phase were calculated using De Launey's Table(8) and the present data of elastic
constants with being extrapolated to 0 K. The results are shown
in Fig. 6. The Debye temperature changes rapidly with Au content
and shows a minimum at Au 25at.% near the Heusler composition and
a maximum near Au 45at.%. This signifies that the force constant
in the Heusler ordered lattice becomes so weak that the lattice is
prone to be unstable into phase transition. The shape of the θ_D vs.
composition curve is reciprocally very similar to the A_s temperature or the elastic anisotropy vs. composition curve (Fig. 5),
suggesting a close relation between them.

In order to clarify the thermodynamical properties of this
transition due to the lattice softening, the free energy difference
ΔF^{H-M} between the Heusler and the martensite phases was calculated.
The difference can be expressed generally as follows:

$$\Delta F^{H-M} = \Delta F_{ele}^{H-M} + \Delta F_{mag}^{H-M} + \Delta F_{latt}^{H-M}. \qquad (1)$$

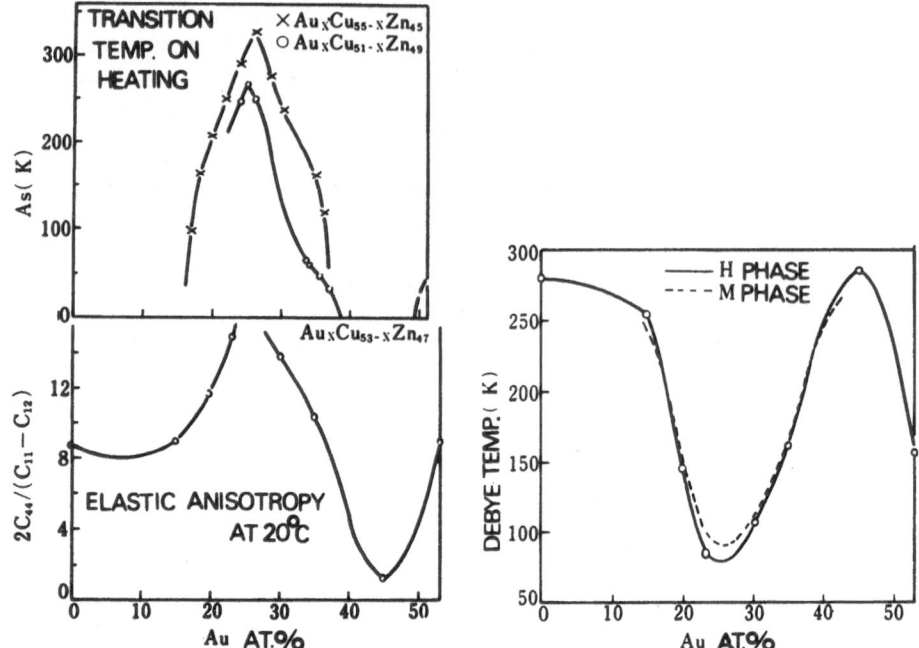

Fig.5 (left): Compositional dependence of the A_s temperature and
              the elastic anisotropy.
Fig.6(right): Debye temperature of the Heusler and the martensite
              phases of the $Au_x Cu_{53-x} Zn_{47}$ alloy.

Where ΔF_{ele}^{H-M}, ΔF_{mag}^{H-M} and ΔF_{latt}^{H-M} are respectively the difference of
electronic, magnetic and lattice free energies. Since the compo-
sition and the degree of order are generally inherited from the
Heusler into the martensite phase, the difference of configuration
entropy can be neglected. Also in this β–Cu–Au–Zn alloy the term
ΔF_{mag}^{H-M} may be neglected. It may be allowed to assume that the contributions from the temperature change in ΔF_{ele}^{H-M} can be involved in
the term ΔF_{latt}^{H-M} because it seems difficult to express precisely the
electron-phonon coupling interation. As a result, ΔF^{H-M} is simply
expressed, using Einstein model, as an approximation as follows:

$$\Delta F^{H-M} = \Delta E^{H-M} + \Delta F_{latt}^{H-M} = \Delta E^{H-M} + 3kT \ln \frac{1-\exp(-\theta_D^H/T)}{1-\exp(-\theta_D^M/T)}. \qquad (2)$$

Where θ_D^H and θ_D^M mean respectively the Debye temperature of the
Heusler and martensite phases.

        Seeing that ΔF^{H-M} is zero at the transition temperature T_0,
and that the latent heat of the transformation ΔH^{H-M} is given by
$\partial(\Delta F^{H-M})/\partial T = \Delta H^{H-M}/T_0$, one can estimate the values of θ_D^M and ΔF^{H-M}
as a function of temperature by introducing the experimental values

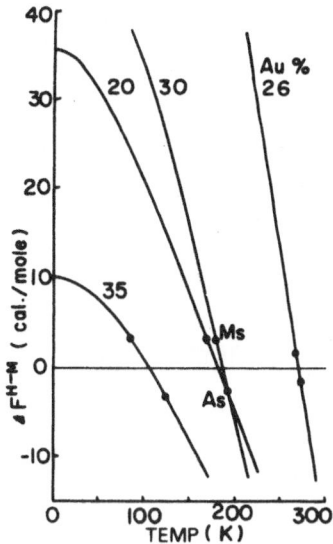

Fig.7: Free energy difference between the Heusler and the martensite phases of the $Au_xCu_{53-x}Zn_{47}$ alloy.

of ΔH^{H-M}, T_0 and θ_D^M into the above equation. The results are
shown in Figs 6 and 7. It is to be noticed that the Debye temperature curves in both phases of Fig. 6 have a minimum consistently
near the Au 25at.%. This may suggest that the martensite lattice
also has a large elastic anisotropy and becomes soft near the
transition temperature. From Fig. 7 and the measured transformation temperatures, we can estimate the value of driving force required to start the transformation and reversely to convert the
martensite to the Heusler phase. Both the values obtained are
consistently 3~5 cal/mole. These values are smaller than those
(about 300 cal/mole) in the ordinary martensitic transformation of
iron alloys. This is one of the characteristics of the thermoelastic martensitic transformation. On the other hand, the re-
straining force, $G=C_H'\Delta\mathcal{E}^2/2$, of the transformation (where $\Delta\mathcal{E}$ corresponds to the transformation shear strain $\frac{1}{6}[1\overline{1}0]$ $(110)_{bcc}$) is also
small being about 5~10 cal/mole and fairly consistent with that
of driving force. Of course, for the reverse transformation the
restraining force $C_M'\Delta\mathcal{E}^2/2$ is similarly small. A schematic graph
which represents the above-obtained shear constants vs. temperature relation around the transition temperature is presented in
Fig. 8. The shear elastic constants C_H' and C_M' become very small
near T_0, their temperature coefficient being positive and negative

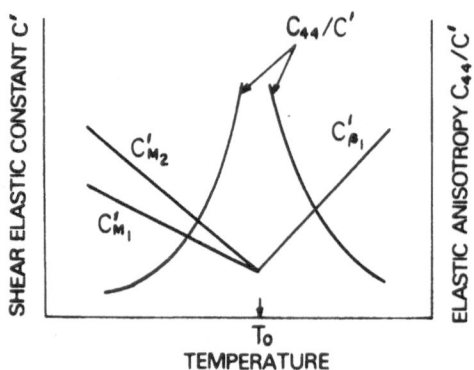

Fig.8: A schematic relation between the shear elastic constant C'
and elastic anisotropy in the vicinity of T_o.

above and below T_o, respectively, while the elastic anisotropy is
very large near T_o. It might be reasonable to assume that the
elastic constant C'_H in the Heusler phase, being a resistance against
the $(110)[1\bar{1}0]$ shear, is resolved respectively into C'_{M1} (a resistance against the $(001)[010]_M$ shear) and C'_{M2} (a resistance against
the twinning shear on the $(1\bar{2}1)[1\bar{1}\bar{1}]_M$) in the lower symmetry martensite phase.

This model shown in Fig. 8 suggests that there exists a small
and flat variation of potential energy for the atomic shear displacement near the transition temperature. In other words, this
may suggest an extremely small Peierls force for the motion of
transformation or twinning dislocations. Thus it is concluded that
the lattice softening is an essentially important property of the
β–Cu–Au–Zn alloy which can explain the thermoelastic behaviour of
the alloy; reversible nature of the transformation and very small
driving force required for the transformation (this leads to the
small hysteresis of the transformation temperature). As will be
shown later, the lattice softening is found to be fundamental for
understanding the mechanical (pseudoelastic) properties of the
thermoelastic alloys.

2. PREMONITORY PHENOMENA DUE TO LATTICE SOFTENING

The phonon dispersion relation obtained by neutron inelastic
scattering in an $AuCuZn_2$ alloy has been recently reported by
Yamada et al. (Fig.9)(9). Their important results are as follows:
(1) The TA_1 branch $((C_{11}-C_{12})/2)$ holds very low energy state compared

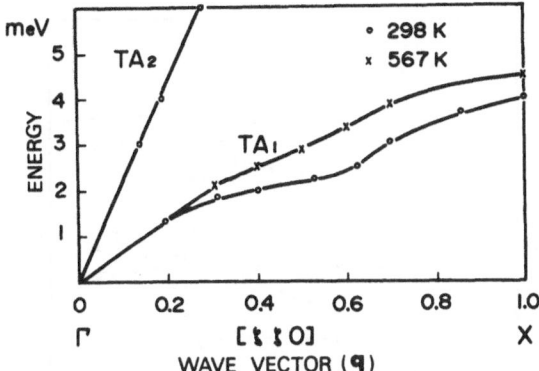

Fig.9: Phonon dispersion relation, "Energy vs. q", obtained by
      neutron inelastic scattering (Yamada et al.)

with that of the TA_2 branch (C_{44}). (2) Data obtained at 298 K$(>T_o)$
are in good agreement with the present ultrasonic results. (3)
However, the energy of TA_1 phonon with zero wave vector q=0 does
not so change against temperature as is expected from ultrasonic
velocity. (4) Instead, the TA_1 phonon with wave vector q=2/3 becomes soft toward the transition temperature. This phonon with
wave vector q=2/3 has a basic relation to the 18R(or 9R) structure
of the martensite, because this martensite structure corresponds
to a quenched state of the acoustic transverse modes of q=0 and
q=2/3.

      According to recent experimental data(10), as shown in Fig. 10,
the softening of the shear elastic constant C' has been widely
observed in many Cu-, Ag-, and Au-based β phase alloys, though
their different slopes are due to the content of alloying elements
(or the electron concentration e/a). Generally speaking, the
tendency toward softening is observed remarkably in the Au-based
alloys, while it is not so notable in the Ag-based alloys. However,
in the Cu-Sn alloys which martensite has not behaved as it was
thermoelastic, the temperature coefficient of C' becomes positive
very slightly. Therefore, the lattice softening as a premonitory
feature might be considered to be not a peculiar but a common
phenomenon in the β-brass alloys.

      Delaey et al.(11) have suggested that the 9R structure can
be produced by linear combination of the transverse wave with the
$\frac{1}{3}\langle110\rangle$ wave vector and the $\langle1\bar{1}0\rangle$ polarization vector and that with
the $\frac{1}{3}\langle112\rangle$ wave vector and the $\langle11\bar{1}\rangle$ polarization vector (this wave

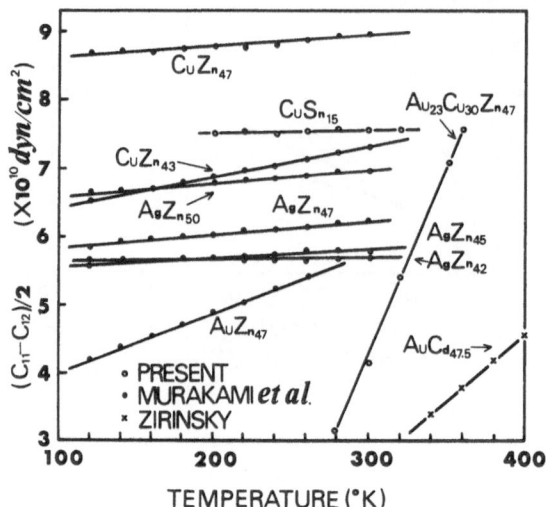

Fig.10: A premonitory lattice softening of $(C_{11}-C_{12})/2$ in the noble
        metal based β-phase alloys.

has been observed at the β→ω transformation in Ti, Zr alloys), and
pointed out that in thermoelastic martensitic transitions from not
only Heusler-type but also CsCl-type to 18R(or 9R) structure these

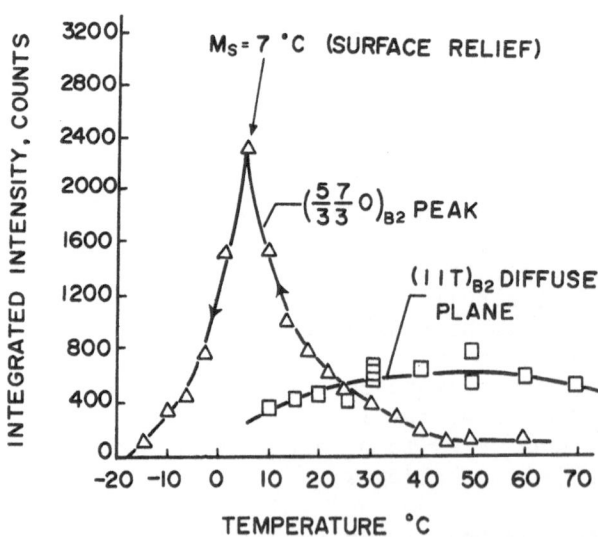

Fig.11: Intensity change of the 1/3 reflection and diffuse {111}
        planes in TiNi(after Sandrock et al.)

two transverse phonons become soft during the premartensitic process. A similar phenomenon has been observed in TiNi martensite
and its special feature which appeared in the diffraction patterns
is shown in Fig. 11. Sandrock et al.(12) (in Fig.11) have found
experimentally that a fairly strong intensity of diffuse $(11\bar{1})_{B2}$
plane exists at temperatures considerably higher than M_s and extra
maxima newly appear at about 1/3 positions between the normal Bragg
positions in the vicinity of the M_s temperature. The 1/3 reflection
can grow at the expense of diffuse $(11\bar{1})$ during cooling, its intensity reaching to maximum at M_s, and finally fades again below
M_s. They have pointed out that there exists a premonitory lattice
softening associated with the $(110)[1\bar{1}0]$ shear in the B2 structure
and simultaneously suggested the existence of soft phonons coupled
with the $\beta-\omega$ transformation (diffuse scattering of the $\{111\}_{B2}$
planes).

An interesting propose is found in "localized soft mode theory"
by Clapp(13), where he pointed out that an unusually large amplitude of lattice vibration exists at the special places such as
free surfaces, incoherent grain boundaries and local defects in
crystals. According to Clapp, several potential barriers, which
height is depending upon the kinds of defects, can exist in a
crystal, where one can expect nucleation sites for martensite. On
the other hand, as an effect of dynamical soft phonons, Perkins
has presented an idea of "phonon nucleation"(14). According to
his view, since the martensite embryos have not been observed experimentally so far, it is reasonable to consider that necessary
atomic displacements for starting the transition can be induced

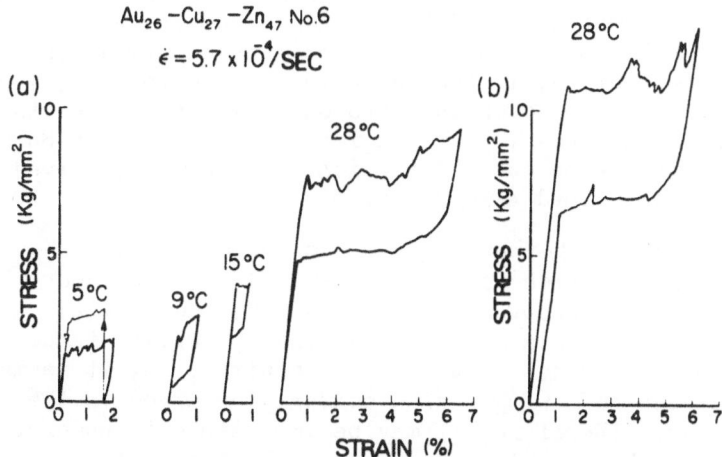

Fig.12: Effect of creep on the stress-strain curves in $Au_{26}Cu_{27}Zn_{47}$
alloys.

when the amplitude of soft phonons reaches to a certain critical
magnitude. Therefore, the martensite formed by "phonon nucleation"
needs no embryos.

In our recent study on a $\beta-Au_{26}-Cu_{27}-Zn_{47}$ alloy(15), when the
specimen is creeped for 21hrs at 27.5 °C (just above the M_s temperature) under a small tensile stress,$\sigma = 1.25Kg/mm^2$ ($\ll \sigma_M$), it was
found that the critical stress σ_M for SIM (stress induced martensite) after the creep increased up to the value of 1.5 times as
much as that for SIM without creep(Fig.12), and the plateau length
on stress-strain curve,$\Delta\epsilon$, decreased after the creep. Although
this reason might be explained by assuming that there exist martensite embryos even above M_s, which can be activated into martensite in sequence, another explanation will be as follows: During
the creep just above M_s, the condensation of strains has occurred
locally in the matrix phase even under the very small tensile
stress,$\sigma \ll \sigma_M$,and resulted in tiny martensite plates. Since these
tiny martensites have increased the enforced elastic stress of
matrix regions surrounding them, the critical stress for SIM could
accordingly be increased. Thus, after the creep the amount of SIM
was also reduced.

Nagasawa(16) has recently found experimentally that the martensitic transformation occurs in two-step process in AgCd alloys,
and proposed from a consideration of group theory that, in general,
the martensitic transition of B2(CsCl)$\rightleftarrows$9R(Close Packed Layer
structure) cannot be directly proceeded. In the case of AgCd
alloys, he reported that the first transition β' (B2)$\longrightarrow\beta''$(base-
centered orthorhombic) occurred at about -10 °C by the shear (110)
$[\bar{1}11]/2$ on alternate layers in the β' phase, and the second $\beta''\longrightarrow$9R
occurred at -175 °C, accompanying the soft B_{1u} mode. According to
his proposal, not only the case in an AgCd alloy but also the CsCl
$\rightarrow$CPL type martensitic transition found in such as AuCd, CuZn and
other β phase alloys must proceed through an intermediate step into
the final CPL structure, if the mechanism is due to the $(110)[1\bar{1}0]$-
type soft phonon. Based on this proposal, it is of importance to
re-investigate the two-step transition reported in a Cu-Sn marten-
site(17) and the anomalous diffraction effects(18) observed above
M_s in the β-brass typed alloys.

So far, in many β-phase alloys such as CuZn(10), AuCd(19) and
Cu-Al-Ni(20) in addition to the Cu-Au-Zn alloy, the shear $(C_{11}-C_{12})/2$
softening has been observed, and in Mn-Cu(5) and TiNi(21) alloys,
the softening of Young's modulus near the transition temperature
has been also observed. In view of this situation, it seems clear
that the lattice softening can be commonly observed in the thermoelastic alloys. Therefore, it may be impossible to understand the
transformation characteristics of thermoelastic martensite with the
exception of the knowledge of lattice softening. Since it seems
inappropriate to deal with soft modes as they are soft "phonons"

Table 1: Comparison between the thermoelastic martensite and typical ferrous martensite in their transformation strains.

Martensitic Transformation				
Second Order (or First Order?)		First Order, Nearly Second Order		First Order
$SrTiO_3$?	V_3Si	In-Tl	TiNi, AuCd	Fe-Ni
$BaTiO_3$	Nb_3Sn	Mn-Cu	AgCd, Cu_3Al	Fe-C
KTN			$CuAuZn_2$	Fe-Mn
GMO			Cu-Al-Ni	
Ferroelectric and Optical Soft Mode		Acoustic Soft Mode Entropy Elasticity		Ferromagnetic
Transformation Strains: $\simeq 10^{-5}$	$\simeq 10^{-4}$	$\simeq 10^{-3}$	$\simeq 10^{-2}$	$\simeq 10^{-1}$

in the vicinity of the transition temperature, more complicated mechanisms might be considered as to lattice instability near the transition temperature. An interesting premartensitic phenomenon, "streaming" named by Tong and Wayman(22) may present a significant example for solving these mechanisms.

In Table 1, is shown a classification of martensitic transformations according to the amount of their transformation strains. We can see a typical second order transition with very small strains, $\simeq 10^{-5}$, in the so-called displacive ferroelectric crystal $BaTiO_3$(23), in which the optical soft mode plays an important role in the transition mechanism. On the other hand, in the case of large transformation strains which appeared in ferrous martensites some plastic deformation produced by the transformation makes the temperature hysteresis and the driving force quite large, and thus the transition has to be first order type. In the thermoelastic martensitic alloys which occupy an intermediate state between them, the transition becomes first order, but rather having the nature of second order transition and is mainly characterized by the existence of acoustic soft modes, and the so-called "entropy elastic" state is realized.

3. PSEUDOELASTICITY AND LATTICE SOFTENING

As already has been described, because of the soft shear modes $\{110\}\langle 110\rangle$ (including both the $q=0$ and $q\neq 0$ modes) in the vicinity of the transition temperature, it is supposed that the motion of dislocations for the transformation and the lattice invariant shear takes place certainly in this shear system. In other words, since the Peierls force for moving the transformation dislocations

should be much small just above T_o, the {110}⟨110⟩ displacement can
be induced by a small applied stress, and consequently the martensite can be easily formed. However, when the applied stress is
released, this SIM can be returned to the original phase, which is
thermodynamically stable, because the slip deformation cannot occur
but the elastic restoring force due to the large elastic anisotropy
may act at interfaces between the SIM and the parent crystal. As
an example, stress-strain curves of Au-47.5at.% Cd single crystal
obtained above T_o are shown in Fig.13. A jerky yield (at an M
point) represents the SIM making a double hysteresis loop during
tension and compression. This is a phenomenon named "superelasticity"(24) and the critical stress at the M point increases with
increasing temperature. The reason why the shape of loops is
different for tension and compression, may be due to the different
martensite variants formed. On the other hand, when the applied
force is loaded in the martensite crystal (below M_s), an interesting stress-strain loop was obtained, as shown in Fig.14. The loops
are resulted from the five-cycles operation of tension—compression
The followings are some special features; a wider loop was obtained
in the first cycle, while in the range of second to fifth cycle the
stress-strain curves made a quite identical loop, and no cyclic
work hardening. This loop has been termed "ferroelastic loop"(25).
However, this ferroelastic loop is a little different from that
named by Lieberman(26), and the reason will be described later.

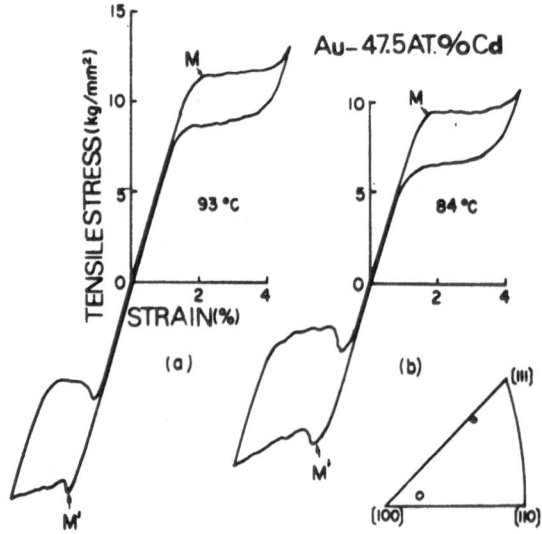

Fig.13: Double hysteresis curves obtained by tension—compression
test in the tempreature range T_o<T<M_d (superelastic loops).

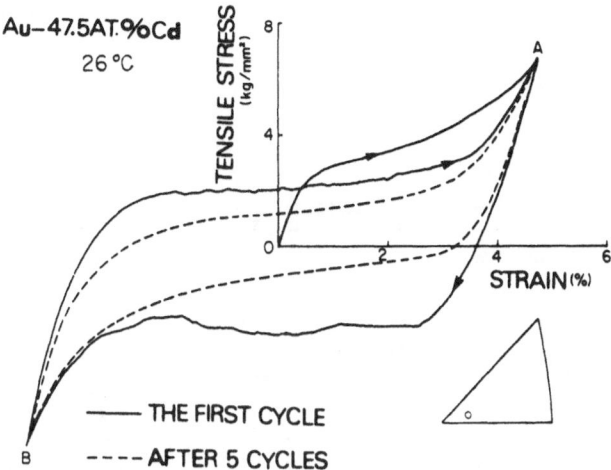

Fig.14: Tension-compression hysteresis loops below M$_s$ (ferroelastic loops).

When an applied stress is loaded on the martensite crystals, the movement of the twin boundary $(121)_M$, the expansion or contraction of stacking faults on the $(001)_M$ plane, and/or the formation of a new twinning deformation might be possible to occur. As the no work hardening cycle is preserved, however, these movements must be operated reversibly crystallographically. Being unlike the superelastic loop, the twin boundary which has been retained at the release of the first applied stress (tension) can move in the opposite direction by the application of a reverse stress (compression) and again stops at unloading. Then, upon successive application of the second tension, the twin boundary begins again to move in the direction of tension and moves back to the original position, thus a ferroelastic loop with a narrow width along the stress axis being made. If this mobile twin boundary is stopped by the locking interaction which is caused by some diffusional relaxation processes, such a small restoring force may not be able to overcome this locking force. Therefore, it is considered that the narrow ferroelastic loop is produced only when the elastic energy stored in the martensite crystal is not so large as it can move the twin boundary back to the initial position. As will be shown later, after ageing the quenched martensite at room temperature, this elastic energy which has increased during ageing can move the twin boundary reversely, and consequently the superelastic (not ferroelastic) loop in the stress-strain curve is formed.

Similar psuedoelastic (superelastic and ferroelastic) loops have been also found in the thermoelastic β-Cu-Au-Zn alloys. In

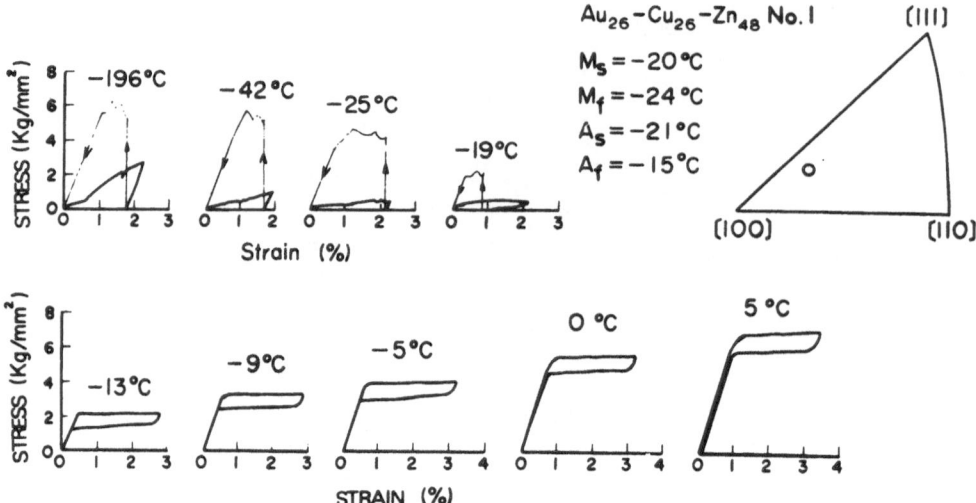

Fig.15: Typical stress-strain curves obtained above and below T_o in $Au_{26}Cu_{26}Zn_{48}$ alloys.

Fig. 15 a stress-strain curve obtained in the temperature range from -196° to 5° C in an Au_{26}-Cu_{26}-Zn_{48} single crystal (M_s=-20° C)(15) is shown. Below M_s the ferroelastic loops with fairly small critical stresses appeared and the martensite exhibited the shape memory upon heating, while above M_s the superelastic loops were obtained and their critical stresses increased with increasing temperature.

The pseudoelastic stress-strain behaviour obtained below the transition temperature in an Au-Cd single crystal is shown in Fig. 16. The crystal structure of martensites was found to be changeable according to the Cd content; the former (47.5at.% Cd) was named the β' martensite (orthorhombic) and the latter (49.5at.% Cd) the β" martensite (hexagonal or trigonal), respectively(27). The following characteristics obtained in the stress-strain curve are as follows: (1) Yielding occurred in two stages with serration in 47.5at.% Cd alloys. The first yielding leaves strain behind it (ferroelastic), but in subsequent yieldings a reversible loop is produced (superelastic) upon removal of the applied stress. (2) The critical stresses for the first plateau were extremely small and almost constant in the wide temperature range below T_o in 47.5at.% Cd alloys, but those for the second stage increased with decreasing temperature in 47.5at.% Cd alloys. (3) Particularly in the 49.5at.% Cd alloys, there existed only one-stage yielding with superelastic loop and also a higher critical stress even when close to T_o. The first yielding is thought to be due to the

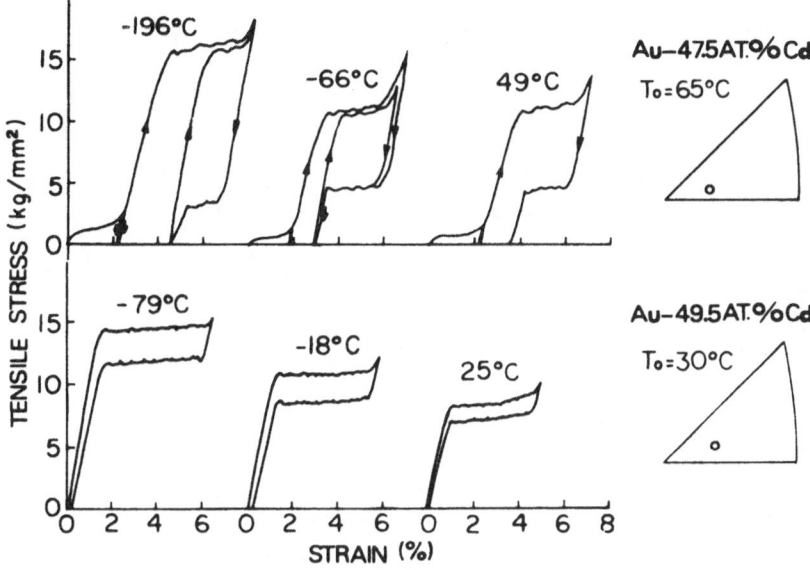

Fig.16: Stress-strain curves obtained below T_o in Au-47.5 and
        49.5at.% Cd alloys.

movement of the twin boundaries or stacking faults already induced
as lattice invariant deformations. However, according to Otsuka's
observation in Cu-Al-Ni martensite(28), a new twinning deformation
was found in this first yielding. Although it is unresolved what
mechanism is operated in the second yielding, the superelastic
loops obtained in 49.5at.% Cd alloys are at least considered to be
due to the formation of a new martensite upon tensile deformation
of the thermally formed β" martensite. Anyway, a future observation on the second loops in 47.5at.% Cd alloys is expected. From
the fact that the first yielding was induced by small critical
stresses even at -196° C, as can be seen in Fig. 16, some diffusional processes, which may lock these mobile interfaces as seen
in the case of strain ageing, seem to be improper to elucidate the
reason why the strains are retained after the first yielding.
Therefore, the ferroelastic strains would be retained in the first
yielding stage whenever the elastic recovery force stored in the
crystal is smaller than the critical force for moving the twin
boundaries and/or interfaces between martensite domains in the
reverse direction. While, in the second yielding stage the superelastic recovery would be probable because the increased recovery
force can overcome this critical stress. Since, as mentioned
earlier, the large elastic anisotropy may exist in the martensite
phase near the transition temperature, the larger the anisotropy
the larger the elastic recovery force stored should be. Thus, this
critical stress was termed "ferroelastic limit".

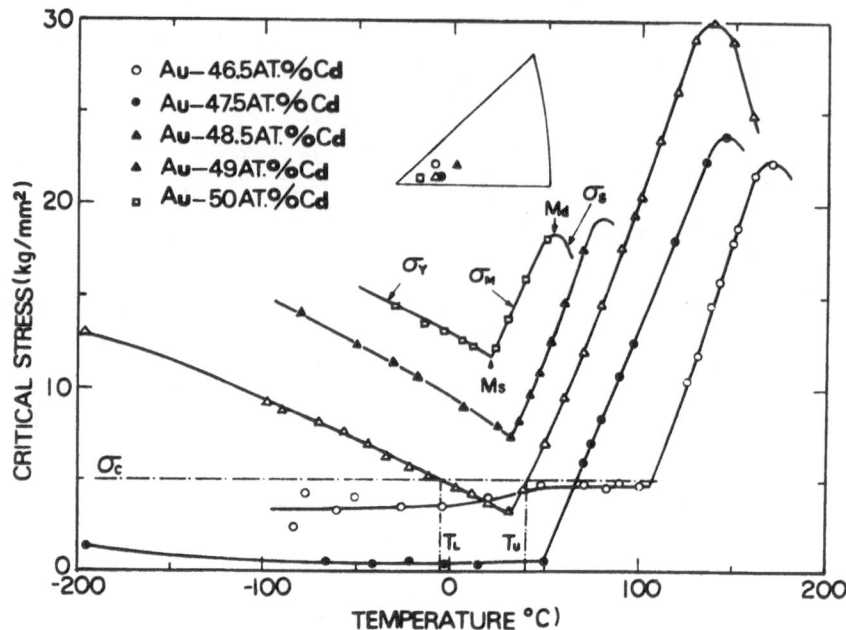

Fig.17: Temperature variation of the critical stresses, σ_M and σ_Y, above and below T_o, and σ_s for slip.

The temperature dependence of the critical stresses obtained below and above the transition temperature was shown in Fig. 17. Here, σ_M and σ_Y are respectively the critical stresses for SIM and the yielding in the martensite, and σ_s is that for slipping. Both the curves made the minimum, but not zero, at about the M_s temperature, and the $d\sigma_M/dT$ represented almost constant in all the Cd composition range. The σ_Y maintained certainly low values with decreasing temperature below M_s in 46.5 and 47.5at.% Cd, while in the 48.5at.% Cd or more σ_Y increased with decreasing temperature. As it is shown in the figure, in the temperature range above M_s in Au–Cd alloys the ferroelastic limit may correspond to the critical stress for the reverse transformation of SIM.

A similar pseudoelastic behaviour was also reported in In–Tl alloys(29), and the results obtained are as follows; (1) super-elastic (or rubber–like) behaviour was obtained just above M_s, and at the temperatures below $0°$ C, and (2) a perfectly plastic be-haviour was obtained in the temperature range from $0°$ C to the M_s temperature, and this phenomenon was considered to be associated with some relaxation process. These results and mechanisms are thought to be similar to those obtained in Au–Cd alloys.

In our recent studies on Cu-15.3at.% Sn alloys, some notable phenomena concerning the formation of SIM have been observed. In Fig. 18 are shown the stress-strain curves with different strain-rate at -41°C (about M_s) in a single crystal. The critical stress for SIM increased with increasing strain-rate and the superelastic recovery was attained, while when the strain-rate was decreased gradually the critical stress also proportionally decreased and some strains began to retain (i.e., a ferroelastic loop) upon removing the stress. These retained strains were removed completely when the alloy was heated up to 90°C, suggesting that the so-called shape memory effect can be observed in related to this SIM.

It is interesting to note that the shape of stress-strain curves markedly depended upon the strain-rate and the interfaces between the SIM and the matrix phase might be locked by some relaxation process due to Sn diffusion in the lower rate of strain(30). Therefore, when this locking interaction is released during heating, the specimen can recover its original shape, because coherent phase boundaries between the SIM and the matrix phase might be formed. When the deformation was interrupted for five minutes in the middle of the stress-strain curve, this locking circumstance was easily built up and the recoverable property of the interfaces was lost. Although the shape memory has not been so far discovered in the thermally formed martensite in Cu-Sn alloys, it is possible to actualize the shape memory associated with SIM by using this interface locking effect.

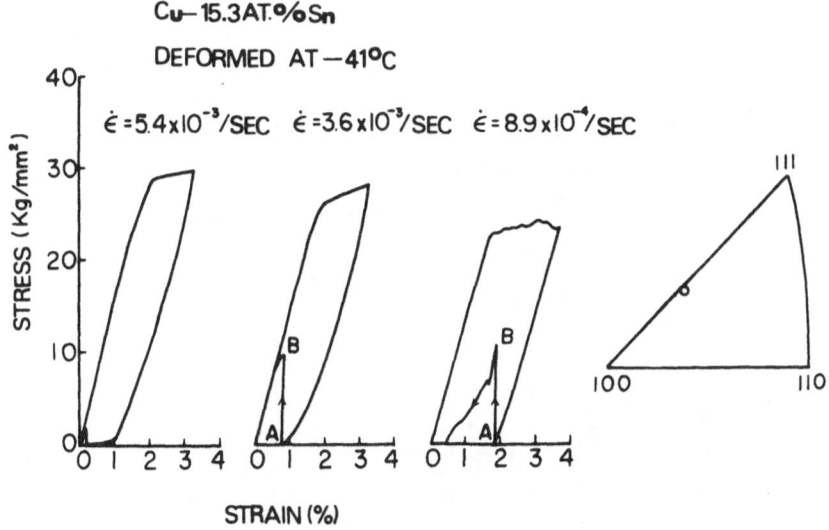

Fig.18: Variation of the tensile behaviour with strain rate in Cu-15.3at.% Sn single crystals.

In Au–Cd alloys, an interesting fact concerning ageing effect of martensite has been experimentally observed(25). In Fig. 19 are shown the effect of ageing at 22°C on stress–strain curves in the 47.5at.% Cd martensite obtained by single interface transformation. Shortly after the transformation (1 min.), the stress–strain curve was similar to that shown in Fig. 16. However, the yield stress of the first stage increased when the ageing time was increased, whereas the yield stress of the second stage did not. It is to be noted that after ageing at 22°C for 14hrs, the specimen exhibited a complete superelastic loop even in the first stage. A marked increase in Young's modulus and decrease in the internal friction appeared with increase of the ageing time up to about 4000 mins. It may be natural to suppose that some relaxation process which can stabilize the twin boundaries operated during ageing, though Lieberman pointed out(26) that atomic shuffling can occur during the ageing. This shuffling means that the transformation can be completed after the ageing in Au–Cd martensite. As the Young's modulus gradually increased during ageing, the critical stress for the first yielding must have become consequently higher. It was observed that upon tensile loading after the ageing one orientation of the twin relation grew at the expense of another and finally single oriented martensite was obtained, and that upon unloading these mobile twin boundaries, which had already been stabilized during the ageing, moved back to the original positions, and the superelastic loop was obtained. This reason may be due to the fact that the elastic restoring force, which increased with increasing yielding stress, can exceed the critical stress for moving these stabilized twin boundaries. Similar ageing effect has been observed in β–Cu–Au–Zn alloys(15). In our opinion,

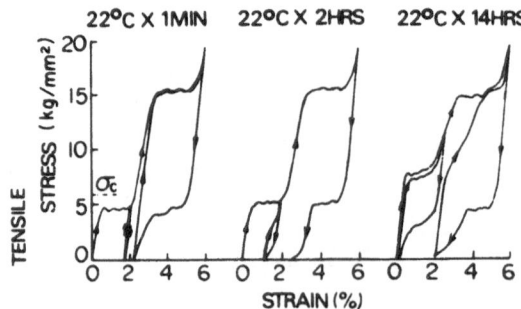

Fig.19: Effect of ageing time on the stress–strain curves in Au–47.5at.% Cd martensite.

the ferroelastic loop named by Lieberman appears to correspond to
the stress–strain loop obtained after the ageing in the Au–47.5at.%
Cd martensite, and it has to be called "superelastic loop."

Now, direct observations of moving interfaces which produce
the superelastic stress–strain loops by optical microscopy and X–
ray diffraction, are presented in β–Au–Cd–Cu alloys. In Fig. 20,
a single crystal of Au–47.5at.% Cd–0.7at.% Cu alloy was deformed
by tension at 25°C (T_o=6°C). When the critical stress for
yielding was reached an interface appeared at an edge of the specimen and moved into another. It is worthy of note that a lamellar
structure composed of alternate thin layers of bcc and martensite
appeared just behind this moving interface, disappearing further
from this interface and thus a single oriented martensite resulted.
It was confirmed from an X–ray Laue photograph that the SIM has the
same crystal structure as that of thermal β' martensite. Of
course, on removal of the stress, the interface moved back in the
opposite direction and a single crystal of the $β_1$ phase (bcc) was
recovered. This experiment suggests that the interface between the
SIM and $β_1$ phase is extremely mobile. In Fig. 21 is shown a case
of single crystal of Au–49.5at.% Cd alloy. As the M_s temperature
exists at 30°C, the stress–strain curve obtained at 18°C signifies
the fact that the thermally formed martensite (β") was deformed by

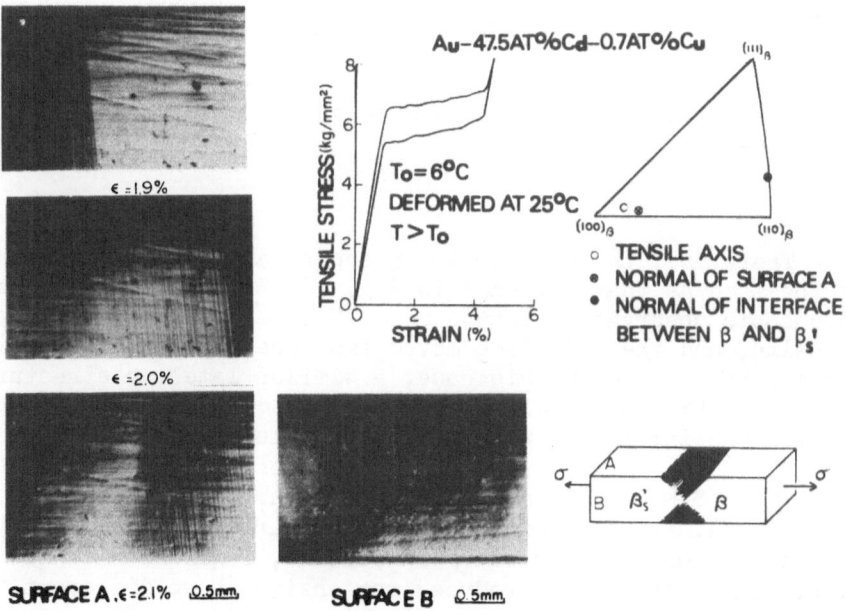

Fig.20: Superelasticity and direct observation of SIM formed in
    Au–47.5at.% Cd–0.7at.% Cu alloys.

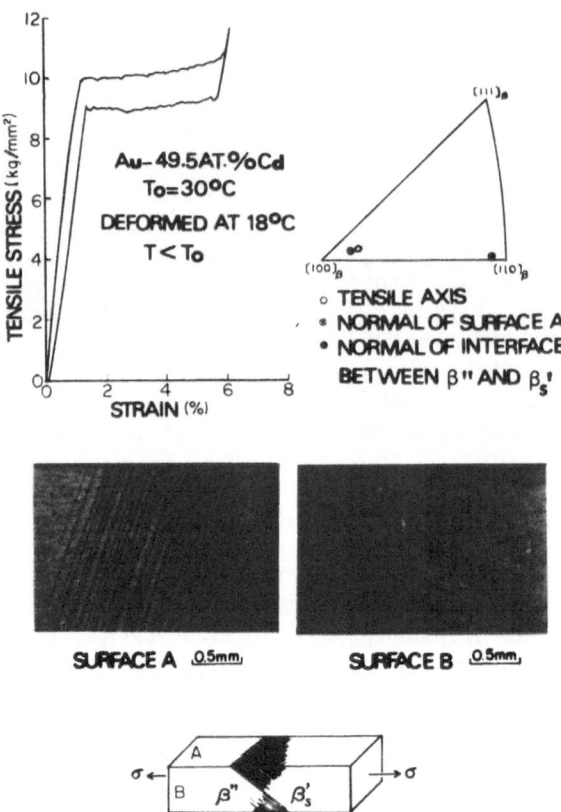

Fig.21: Superelasticity and surface observation of SIM formed
        in Au–49.5at.% Cd alloys.

tension. Therefore, the stress–strain superelastic loop obtained
is just the same as that obtained in Fig. 16. It was evident from
the microscopic observation that a new SIM appeared from the thermal
β″ martensite, and the interface moved from one edge of the specimen to the other. Also in this case, a similar lamellar structure
appeared just behind the moving interface, gradually disappearing
further from the interface and finally a single crystal of SIM was
obtained. Upon unloading the reverse motion of the interface was
observed. This SIM was, from the Laue pattern, confirmed as the
β′ structure. It is of importance that the martensite induced by
tensile deformation has only the β′ structure, regardless of the
Cd content, while thermally produced martensites have different
crystal structures, β′ and β″ according to the Cd content.

4. SHAPE MEMORY EFFECT

4.1. Definition of SME

The so-called "SME phenomena may involve the following three cases:

(1) A thermally produced martensite is first deformed, and upon subsequent heating its original shape of the higher temperature phase can be recovered in the temperature range from A_s to A_f, (and examples are found in the β-phase and In-Tl alloys).

(2) A martensite is induced by deformation in the temperature range from M_s to M_d, which deformation strains are retained upon removing the applied stress, and upon subsequent heating its shape can return into the original shape in the temperature range A_s-A_f or higher temperatures (Cu-Sn alloys (30) and others).

(3) A thermally produced martensite is first deformed, and upon subsequent heating the initial shape of the thermal martensite can be recovered at certain temperature below A_s (Ni-Al alloys by Nenno et al. (31)).

Here, first we intend to discuss the above case (1), and subsequently the cases (2) and (3) are discussed.

4.2. Mechanisms Already Proposed

Proposals associated with the origin or mechanism of the SME may be summarized as follows:

(1) A condition proposed by Wayman and Shimizu (32) that the martensitic transformation should be thermoelastic seems most important.

As mentioned earlier, from the themodynamical consideration in β-Cu-Au-Zn alloys, the lattice softening deeply concerns to the characteristics of thermoelastic martensite in which the first order transition, but neary second order character appears to occur; that is, complete reversibility and small driving force for the transformation. Although the existence of lattice softening may not be indispensable to form the thermoelastic martensite, it seems at least necessary to make the shape memory that a large elastic anisotropy exists in the vicinity of the transition temperature.

(2) The thermoelastic martensite has twin structures of stacking faults and no slip deformation as a lattice invariant deformation (33).

According to Otsuka (28), a new deformation twin was found to be induced by bending the thermal martensite in Cu-Al-Ni alloys, and this twin also disappeared at the A_s temperature upon heating. Delaey et al. (33) have suggested that in thermoelastic martensites each martensite is formed with self-accommodating manner to minimize three dimensional strains, and upon tensile or compressive loading re-orientation of each martensite would be carried out in favor of the applied orientation. Therefore, "SME" means that the re-oriented martensite transforms reversibly to the higher phase at A_s-A_f upon heating.

A similar proposal has been made by Brown et al. (34); that is, the thermally produced martensite having a lattice invariant shear of $(011)[0\bar{1}1]$ can be re-oriented by deformation into SIM having the $(110)[1\bar{1}0]$ lattice invariant shear and this SIM may transform to the higher phase in the temperature range A_s-A_f upon heating.

It has been frequently observed that the martensite has a twin structure as a lattice invariant shear. However, it is of importance that this twin boundary can easily move by a small applied force. Even in a martensite having no twin but stacking faults, such as Cu_3Al and Cu-Zn martensites (35), the existence of similar mobile planes or interfaces would be necessary for requiring the thermoelastic properties. Also, the interfaces between martensite domains should be mobile when the necessary re-orientation, due to deformation, takes place among these martensite domains which are accommodating in conformity with three dimensional strain circumstances.

In other words, it is requested that a stress-strain curve made by deformation has to form the "ferroelastic loop" which exhibits no cyclic work-hardening.

(3) The thermoelastic martensite has an ordered structure (33).

Since the β (cubic) phase has higher symmetrical structures than the martensite, a definite path may be selected in the reverse transformation. In the case of Fe-Pt martensite (36), the ordering can limit the direction of atomic displacements in the reverse transformation (to keep the ordered configuration). While, in the case of Fe-Ni martensite, bcc$\rightleftharpoons$fcc type, the reverse martensite arises from newly nucleated sites.

As pointed out by Delaey, since the martensite phase has usually lower-symmetrical structures than the matrix phase, the atomic displacement upon transformation from ordered martensite to ordered β might occur on the fixed planes and directions.

As already shown in Fig. 4, it is of interest that the lattice softening was most remarkably observed at the Heusler composition $CuAuZn_2$. However, whether the ordering configuration is necessary or not for obtaining the SME seems to be not conclusive. Considering that there exists the elastic anisotropy in the vicinity of the transition temperature, atomic displacements associated with both the martensitic shear transformations during heating and cooling might be forced to occur along the only path (or the soft path) due to the elastic recovery force, even if there is no ordered configuration in the specimen. This problem has to be confirmed in the In-Tl or Mn-Cu martensites having no ordered structure and also it may offer a good material to discuss whether "SME" can be observed or not in pure Na or Li martensite.

As already mentioned, it seems natural to consider that the softening of bonding force along the shear planes and directions in the cubic lattice is inherited in the martensite lattice. For example, the degenerate soft shear system {110}⟨110⟩ in the cubic lattice may be transformed to the main shear plane and direction and to the twinning shear plane and direction in the martensite lattice because of the depression of symmetry in the martensite crystal. Since the habit plane has been observed near (110) in In-Tl alloys, the soft character in the shear system {110}⟨110⟩$_{fcc}$ in the high temperature phase must result in the mobile property of the habit plane or twin boundary in the martensite phase. As it has been reported by Brown (37), this (110)[1̄1̄0]$β_1$ system became the lattice invariant shear in Ag-Cd and Cu-Zn-Sn martensite.

The following characters were obtained as a response for applied stress in the vicinity of the transition temperature; (i) the critical stresses σ_s and σ_Y are extremely small and cannot exceed the elastic limit of the specimen, and (ii) the twin boundaries, stacking faults and interfaces between martensite domains in the martensite are very much mobile.

4.3. Pseudoelastic Behaviour and SME

Now, in order to consider the origin of "SME", it may be available to pay attention to the superelastic property, because of its close relation to SME. In Fig. 22 a schematic relation between the transformation temperatures (M_s, A_s and M_d) and critical stresses (for slip, and yieldings of thermal martensite and of SIM) is shown in the typical thermoelastic alloys.

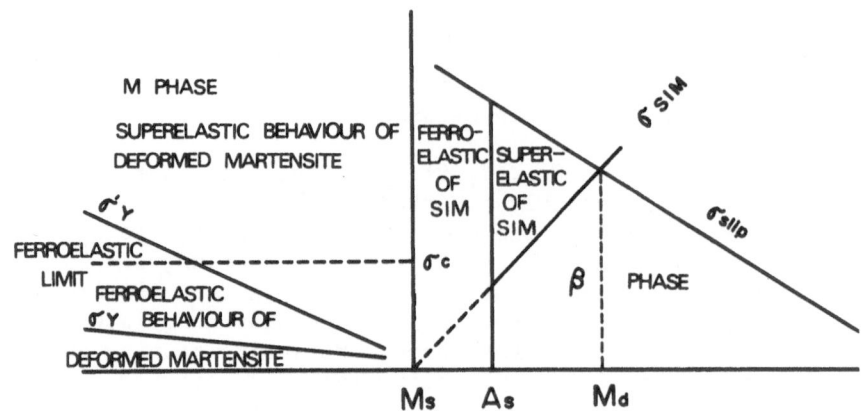

Fig.22: A schematic relationship among pseudoelasticity, critical
        stresses and transition temperatures.

        The discussion will be developed, referring to the figure, as
follows:

        (1) In the temperature range from A_s to M_d; "superelastic
behaviour of SIM (elastic shape memory(38))" is observed.

        The reasons are described as follows: (i) In this temperature
range the high temperature (β) phase is thermodynamically stable.
Therefore, it seems sure that the superelastic stress-strain curve
is observed upon SIM→β transformation. (ii) According to detailed
considerations, the interfacial strain resulted from the formation
of SIM is very small, and the interfaces between SIM and β hold a
good coherency. Thus the plastic deformation is not required at
the interfaces. Of course, a large elastic restoring force is
operating in these interfaces.

        It is interesting to notice that, as already shown in Figs 13
and 20, the amount of macroscopic strains resulted from the formation of SIM covered from 4% to 25% and this large strain was recovered superelastically, and that the critical stress for obtaining SIM increased lineally with increasing temperature and made a
maximum at M_d. When the interfaces are locked by some diffusional
process, the superelastic recovery cannot be expected, even if the
specimen is held in the temperature range from A_s to M_d. In this
case, the recovery may occur at some higher temperatures where
this locking is released. The case was observed in Cu-Sn alloys,
as mentioned earlier.

(2) In the temperature range from M_s to A_s ; "ferroelastic behaviour of SIM (plastic shape memory[38])" is observed.

The reasons are described as follows : (i) In this temperature range the high temperature (β) phase is still stable during cooling. However, below A_s of SIM (here, the A_s temperature of SIM is assumed to be equal of that of thermal martensite) the driving force for the reverse transformation is not enough to cause the transition, at which the summation of the interfacial and elastic energies may correspond to the driving force. (ii) Since the coherency is well kept at the interfaces, any plastic deformation cannot occur.

When the SIM is heated beyond the A_s temperature, the specimen can recover his original shape, because the specimen is now held in the temperature range from A_s to M_d. This phenomenon has to be called "SME of SIM".

(3-1) Below M_s ; "ferroelastic behaviour" is observed upon applied deformation when the critical stress σ_Y is lower than the ferroelastic limit σ_c.

The following reasons have to be considered : (i) In this temperature range the martensite phase is of course stable thermodynamically. (ii) The structure or configuration of thermally formed martensite is gradually changed into an analogous or favorable one to that of SIM with increasing the applied stress during tensile, compressive or bending loading. (iii) Since the ferroelastic strains amount to a few percents as well as those for obtaining SIM, the structure of thermally formed martensite can be reoriented into that of SIM upon a few percents tensile loading. (For example, in Au-47.5at.% Cd after twin boundaries had moved away a single oriented martensite crystal was produced.) (iv) When the critical stress for yielding the thermal martensite is not larger than the σ_c, the ferroelastic loop is formed. In this case, the elastic energy stored at the twin or martensite domain boundaries cannot exceed their critical stresses (i.e., ferroelastic limit σ_c). (v) When it is heating beyond A_s, the reoriented martensite which has a favorable orientation to applied stress can transform to the high temperature phase because the elastic recovery force can overcome the driving force for the reverse transformation, and simultaneously the ferroelastic loop disappears and the original shape can be recovered. Therefore, in an extreme case, the recovery can be carried out by the same mechanism as that which operated in the "SME of SIM". In general, this is called the shape memory effect.

(3-2) Below M_s ; "superelastic behaviour" is observed upon applied deformation when the critical stress σ_Y is higher than the ferroelastic limit σ_c.

We can see two cases in the superelastic behaviour, where the
first (i) is that the superelastic loop can be obtained in the martensite phase, and the second (ii) is that it concerns with the
formation of SIM from the thermal martensite. Each example is
shown below : (i) The stress-strain loop obtained after the ageing
of quenched martensite in Au—47.5at.% Cd, that obtained in the
second yielding stage in Au—46.5 and —47.5at.% Cd martensites, and
also that obtained at lower temperatures in In—Tl martensite are
typical. It is expected, in these cases, that the elastic recovery
force operating at the twin or domain boundaries can exceed the
critical ferroelastic limit. Therefore, the re-oriented or single
oriented martensite, being in favor of the applied deformation,
can recover the initial multi-domain orientation upon removal of
the load, thus the superelastic stress-strain curve could be obtained. (ii) The typical example was seen in Fig. 21, where the
SIM (β') was produced from the thermal martensite (β'') and the
associated stress-strain curve was superelastic.

It is believable that the ferroelastic strain induced by
deformation can be released during heating even below A_s, namely
the specimen can recover the initial shape of martensite, if the
elastic restoring energy was increased beyond σ_c by some reasons,
this being the case observed in the Ni—Al martensite(31).

REFERENCES

1. E.Pytte: Phys. Rev. Letters, 1970, vol.25, p.1176.
2. E.Pytte: Phys. Rev., 1971, vol.B4, p.1094.
3. D.J.Gunton and G.A.Saunders: Solid State Commun., 1974, vol.14,
   p.865.
4. D.J.Gunton and G.A.Saunders: Solid State Commun., 1973, vol.12,
   p.569.
5. K.Sugimoto, T.Mori, and S.Shiode: Metal Sci.J., 1973, vol.7,
   p.103.
6. N.Nakanishi, Y.Murakami, and S.Kachi: Scr. Met., 1971, vol. 5,
   p.433.
7. Y.Murakami, N.Nakanishi, and S.Kachi: Jap.J.Appl.Phys., 1972,
   vol.11, p.1591.
8. G.A.Alers: Physical Acoustics, Academic Press, 1965, vol.3B,
   p.5.
9. Y.Yamada, M,Mori, J.D.Axe. and G.Shirane: Report on the Autumn
   Meeting of Phys. Soc. Japan. 1973.
10. Y.Murakami and S.Kachi: Jap.J.Appl.Phys., 1974, vol.13, p.1728.
11. L.Delaey, J.Van Paemel, and T.Struyve: Scr. Met., 1972, vol.6,
    p.507.
12. G.D.Sandrock, A.J.Perkins, and R.F.Heheman: Met.Trans., 1971,
    vol.2, p.2769.
13. P.C.Clapp: Phys. Stat. Sol., 1973, vol.(b)57, p.561.
14. J.Perkins: Scr. Met., 1974, vol.8, p.31.

15. S.Miura, S.Maeda, and N.Nakanishi: Phil. Mag., 1974, vol.30, p.565.
16. A.Nagasawa: J.Phys. Soc. Japan, 1973, vol.35, p.1654.
17. Z.Nishiyama, K.Shimizu, and H.Morikawa: Trans. JIM, 1967, vol.8, p.145.
18. L.Delaey, A.J.Perkins, and T.B.Massalski: J.Mat.Sci., 1972, vol.7, p.1197.
19. S.Zirinsky: Acta. Met., 1956, vol.4, p.164.
20. M.Suezawa and K.Sumino: "Symposium on the Thermoelastic Martensite and Shape Memory Effect" Jap. Inst. of Metals. 1974, p.9.
21. R.R.Hashiguchi and K.Iwasaki: J.Appl.Phys., 1968, vol.39, p.2182.
22. H.C.Tong and C.M.Wayman: Phys. Rev. Letters, 1974, vol.32, p.1185.
23. D.S.Lieberman: Phase Transformations, ASM, 1970, p.1.
24. H.Pops: Met. Trans., 1970, vol.1, p.251.
25. N.Nakanishi, Y.Murakami, S.Kachi, T.Mori, and S.Miura: Phil. Mag., 1973, vol.28, p.277.
26. D.S.Lieberman: Dept. of Metallurgy, Univ. of Illinois, 1973, private communication.
27. H.M.Ledbetter and C.M.Wayman: Met. Trans., 1972, vol.3, p.2349.
28. K.Otsuka and K.Shimizu: Scr. Met., 1970, vol.4, p.469.
29. Z.S.Basinski and J.W.Christian: Acta Met., 1954, vol. 2, p.101.
30. S.Miura, S.Maeda, and N.Nakanishi: to be published.
31. K.Enami and S.Nenno: Met. Trans., 1971, vol.2, p.1487.
32. C.M.Wayman and K.Shimizu: Metal Sci.J., 1972, vol.6, p.175.
33. H.Tas, L.Delaey, and A.Deruyttere: Met. Trans., 1973, vol.4, p.2833.
34. R.V.Krishnan and L.C.Brown: Met. Trans., 1973, vol.4, p.423.
35. K.Takezawa and S.Sato: "Symposium on the Thermoelastic Martensite and Shape Memory Effect" Jap. Inst. of Metals, 1974, p.23.
36. D.P.Dunne and C.M.Wayman: Met. Trans., 1973, vol.4, p.137.
37. J.D.Eisenwasser and L.C.Brown: Met. Trans., 1972, vol.3, p.1359.
38. H.Tas, L.Delaey, and A.Deruyttere: J.Less-Comm. Metals, 1972, vol.28, p.141.

PREMARTENSITIC-MARTENSITE TRANSITIONS RELATED TO SHAPE MEMORY EFFECT

K. Mukherjee, M. Chandrasekaran and F. Milillo

Polytechnic Institute of New York

Brooklyn, New York, U.S.A.

The literature pertaining to premartensitic-martensitic transition is reviewed with special emphasis on the role of premartensitic lattice instability in shape-memory mechanism. A large elastic anisotropy and a low value of the elastic modulus $\frac{1}{2}$ $(C_{11}-C_{12})$ as necessary criteria for the thermoelastic type martensitic transition are scrutinized in light of published and new experimental results. Progressive lattice softening and instability of B_2 structure, with respect to the $\{110\}\langle 110 \rangle$ shear, approaching the transition temperature are analyzed in terms of martensitic nucleation. Some new experimental results on the premartensitic instabilities in the Ti-Ni alloy, the effect of stress on the x-ray diffraction peaks in the Ti-Ni alloys and some new results of frozen-in lattice instability and omega type premartensitic distortion in a number of splat quenched B_2 alloys are presented. The presence of omega type reflection, profuse streaking in electron diffraction patterns and anomalous broadening of (110) x-ray diffraction line indicate a pretransformation instability of the parent phase. These results also suggest a progressive cooperative shear and lattice distortion which produces the final martensitic structure.

I. INTRODUCTION

In recent years, a great deal of interest has been generated to study a special class of alloys which exhibit the so-called "shape memory effect" (SME). In these alloys, a change of dimensions or the shape, induced by an external stress, is completely recovered upon heating above a transition temperature range (TTR). Although this shape recovery phenomenon, in Au-Cd[1] and In-Tl[2] alloys was reported by Read and his associates [1,2] over twenty years ago, the term SME was first proposed by Wang et al. [3] for the similar recovery mechanism observed in Ti-Ni alloys. Ever since the SME has been studied in many binary and ternary alloys and in some pure metals such as: Au-Cd[1,4], In-Tl[2,5], Ti-Ni[3,6,7], Cu-Zn[8,9], Fe-Pt[9], Ag-Cd[10,12], Cu-Al[13,14], Cu-Al-Ni[15-17], Au-Cu-Zn [18,19], Ni-Al[20,21], Fe-Ni[22], Co-Ni[5], 304 stainless steel[23] Co[23], Ti[5] and Zr[5].

It is quite apparent that the genesis of the SME is intimately related to the diffusionless martensitic type of phase transformation and the instability of the parent lattice with respect to an applied stress and/or change of temperature in the vicinity of the TTR. There remain, however, certain ambiguities and uncertainties about the subtle nature of a specific type of displacive transformation that is directly responsible for the SME. Various terms such as "thermoplastic", "pseudoplastic", "superelastic", "ferroelastic" and "thermoelastic", etc., have been used [19,24,25] to characterize the necessary martensitic transformation modes associated with a SME alloy. It has also been proposed, for example, that a necessary criterion for the SME is that the lattice invariant deformation is twinning rather than slip[26]. However, examples are found in the literature where the internal structure of the SME martensite is not twinning, but consists of long period stacking order modulations such as 2H, 3R, 4H, 4R, etc.[25,27-31]. Interestingly, the most pronounced SME is observed in alloys which possess a high-temperature ordered Cs-Cl type structure and many of the martensitic products in these alloys are produced by periodic stacking faults. The ordered high-temperature phase in these alloys also manifest premartensitic lattice instability near the TTR. In this paper, therefore, we will first review the nature of premartensitic instabilities in B_2 lattices and present new experimental results which might shed some light on the mechanism of premartensitic and martensitic transitions related to SME. We will also present some results of splat quenching of martensitic ordered alloys where the martensitic transformation is suppressed and the premartensitic instabilities are frozen in.

II. INSTABILITY OF B_2 LATTICE

Born[32] first studied the stability of crystal lattices in terms of elastic constants. Zener[33] applied this concept to explain the anomalously low value of $C_{11}-C_{12}$ in β-brass and the possible instability of the b.c.c. lattice with respect to a (110) [$\bar{1}$10] shear. It appears now that many of the B_2 lattices, which transform martensitically, exhibit an anomalously low value of $\frac{1}{2}(C_{11} - C_{12})$ near the Ms temperature, corresponding to a

(110) $\langle$110$\rangle$ shear of the β-phase. Similar lattice softening
and associated instability of the β-phase resulting in a diffusionless transformation have also been reported for Au-Cd[34],
In-Tl[35], Cu-Zn[36], Fe-Ni[37] and Au-Cu-Zn[18]. Zirinsky[34] has
further shown that, in the Au-Cd system, as the alloy composition
changes, the value of $\frac{1}{2}$ $(C_{11} - C_{12})$ changes and there is an associated change of the Ms temperature. Robertson[38] has shown that
addition of Co in Fe-Ni alloy lowers the value of $\frac{1}{2}(C_{11} - C_{12})$

and there is a corresponding increase in the Ms. A similar correlation between the Ms termperature and the value of $\frac{1}{2}(C_{11} - C_{12})$
in Au-Cd alloys with Mg, Zn and Ga additions has also been reported
recently[39]. The mechanical instability of the β-phase with respect to a (110) [$\bar{1}$10] shear can be viewed as a long wave length
lattice displacement wave[40]. It is of interest to note that the
(110) [$\bar{1}$10] shear is a common lattice invariant deformation mode,
either twinning or slip, in martensitic reactions involving B_2 or
b.c.c. parent phase. Besides the low value of $\frac{1}{2}(C_{11} - C_{12})$ near
the Ms temperature, various other premartensitic anomalies are observed in β-phase alloys near the TTR. In the next section we
will review such anomalies as they relate to the premartensitic and
martensitic transformations and the SME.

III. PREMARTENSITIC TRANSITIONS

The pre-existing embryo theory of martensitic transformation[41,42]
assumes that large embryos of the order of 1000Å are formed at high
temperature and are triggered at or below the Ms temperature. However, such relatively large embryos have not been observed directly.
The concept that pre-existing embryos or strain centers much
smaller than the calculated ones exist, and atom shuffling and fluctuations around these singularities become more pronounced at temperatures very near the Ms temperature, is an attractive one.
Recently, atom shuffling and a metastable transition phase formation have been invoked[40,43,44] to explain {111} walls of diffuse
intensity in electron diffraction patterns of the Ti-Ni β-phase
above the Ms temperature. Similarly diffuse intensity streaking
along $\langle$110$\rangle$ and $\langle$112$\rangle$ directions in the β-phase of Au-Cd alloy
has also been observed[45,46] above the Ms temperature. Using a
novel experimental technique, the premartensitic instability in the
Au-47.5 at .% Cd and Fe-29.7 at .% Ni alloys has been studied recently[46]. It is known that structural singularities, such as
dislocations, which produce anisotropic strain field, display an
anomalous magnetoresistance effect[47,48]. An anomalous magnetoresistance peak was observed[46] for both of these alloys prior to the
martensitic transformation[46]. In the Au-Cd alloy, this peak was
2° - 3° C above the Ms and for the Fe-Ni alloy it was 5° - 6° C
above the Ms. Furthermore, when the Ms was lowered in the Au-Cd
alloy by a stabilization anneal above the Ms temperature, the peak

moved in conjunction with the new Ms temperature. This indicates that the observed peak is truly associated with the premartensitic instability of the parent phase. It can be speculated that the amplitude of premartensitic fluctuation grows and atom shuffling occurs around pre-existing structural singularities at a temperature, which is only a few degrees above the Ms temperature. Thus the frozen-in embryos could be much smaller in size than predicted[41,42] and therefore defy direct observation. Alternatively, the anomalous magnetoresistance peak could be related to the short wave length displacement waves and periodic shuffling, as a precursor to martensitic transformation, which introduce additional Brillouin zone boundaries and thus alter the energy of the conduction electrons. Such a change in the electronic energy states will give rise to second order galvanomagnetic effects such as magnetoresistance.

The premartensitic transition, which manifests itself through anomalous behavior of some physical properties above the Ms temperature, is quite pronounced in the near equi-atomic Ti-Ni alloys. Anomalous behavior of x-ray and electron diffractions[43, 44], electrical resistivity[40,49,50], internal friction and elastic modulus[51-53] and hardness values[7] have been observed to precede the actual crystal structure change. These anomalies have been variously interpreted to indicate a premartensitic instability of the parent phase and to suggest a progressive shear and distortion which carry the parent phase through a succession of metastable states to the final martensitic structure[44]. Therefore the transformation is not strictly "martensitic" in the classical sense where the parent phase is converted to martensite through a sharp displacive transformation characterized by an extremely rapid reaction rate[54,55]. Recently we have carried out some further detailed study of the premartensitic and martensitic transformations in Ti-Ni alloys and these results will be presented next.

IV. PREMARTENSITIC AND MARTENSITIC TRANSFORMATIONS IN Ti-Ni: SOME FURTHER STUDIES

The crystal structure of the parent high-temperature phase of the equi-atomic Ti-Ni alloy is commonly believed to be B_2 type[49,56,57]. However, evidence has been provided to indicate that the high temperature phase consists of a mixture of B_2 and $P\bar{3}m1$ structures[44,58]. The low temperature phase(s) has been reported to consist of stacking modulated structures[27] such as $2H$, $3R$, $4H$, $9 \overset{\circ}{A}$ cubic phase[59-61] and monoclinic martensites[28,56].

The purpose of the present study was to investigate the structural changes preceding the martensitic transformation and to relate these with the observed electrical resistivity anomaly and the instability of the parent phase (with respect to an applied stress)

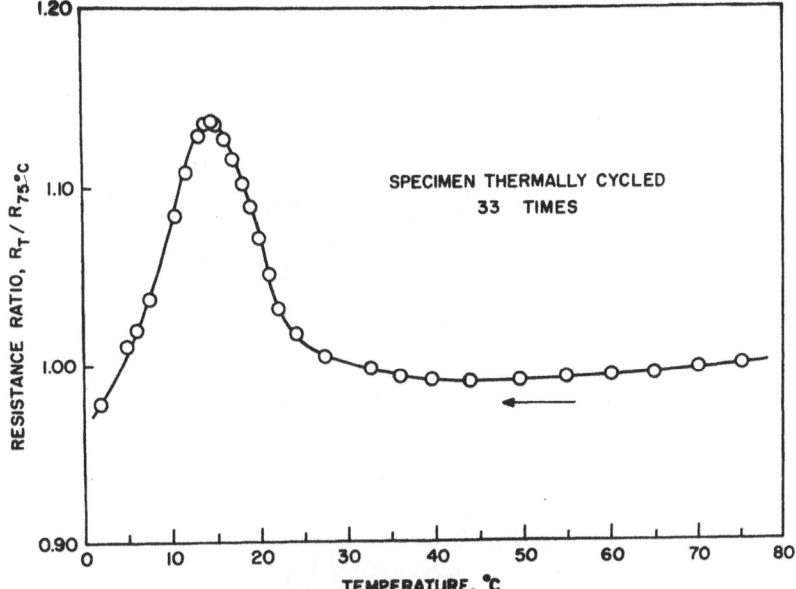

Fig. 1. Normalized resistivity v.s. temperature for the equi-atomic Ti-Ni alloy after 33 transformation cycles. The resistivity peak temperature corresponds to the Ms temperature.

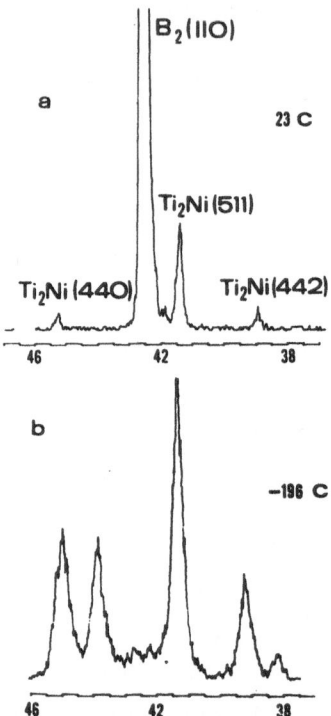

Fig. 2. Diffractometer trace of the important low angle peaks
        of the Ti-Ni SME alloy. (a) Diffraction peaks from
        the high temperature phase showing 110 and some
        Ti₂Ni peaks. (b) Diffraction pattern at -196° showing
        only martensitic peaks.

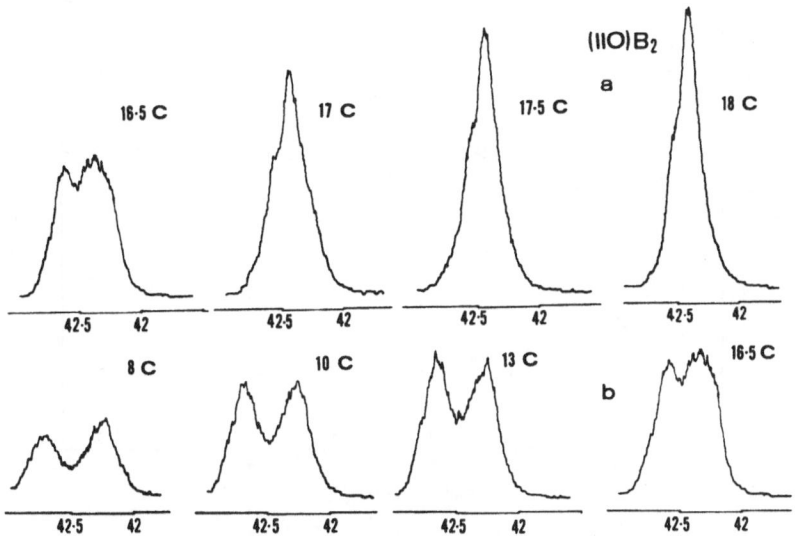

Fig. 3.   (a) and (b) Progressive broadening and splitting of $(110)_\beta$ peak in the TRR region.

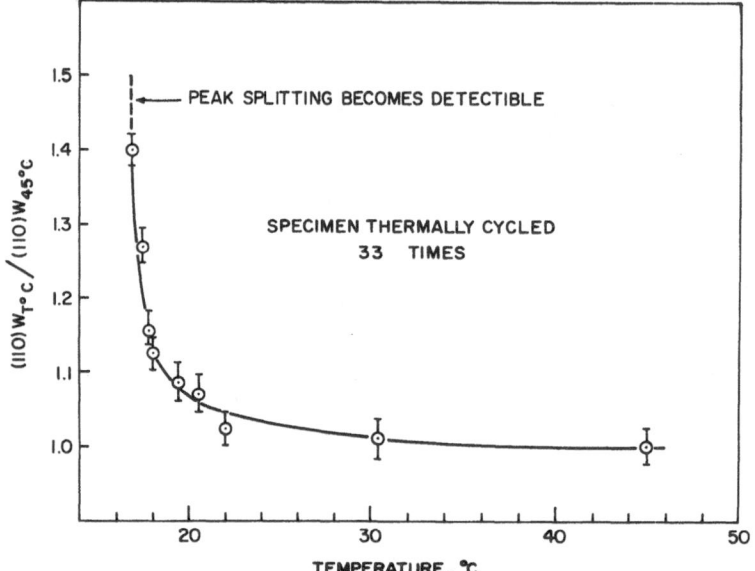

Fig. 4. Half width of 110 peak in the Ti-Ni alloy
         plotted as a function of temperature. The
         peak width is normalized with respect to the
         width at 45°C. Note that at or above 45°C
         no change in the 110 peak is observed.

in the vicinity of the TTR. Electrical resistivity, x-ray diffraction and stress induced transformations were studies. For the
sake of brevity detailed experimental procedure will not be discussed. Polycrystalline Ti-51.3 at .% Ni alloy was used for these
experiments.

V. RESULTS AND DISCUSSION

Figure 1 shows the resistivity peak after 33 complete cycles. The
peak temperature is relatively stable after about 30 cycles and
therefore this figure will serve as a reference for comparing other
results.

Figure 2 (a) shows a portion of the diffraction pattern around the
(110) reflection at 23°C after 33 cycles of transformation.
Figure 2 (b) shows the corresponding diffraction pattern at -196°C
in the fully martensitic condition. Note that the strongest martensite peak very nearly coincides with the (511) reflection of the
Ti_2Ni phase which might have produced some confusion in the past.

The next series of diffraction patterns of the (110) peak was obtained by using a specially designed cooling stage such that the
sample could be cooled or heated at temperature intervals of 0.5°C
with a temperature stability of $\pm 0.01^{\circ}$C at each temperature.
Figures 3(a) and 3(b) are photographic reproductions of the actual
diffractometer traces obtained from a sample after 33 cycles of
transformation. It is observed that in the transition temperature
range, the (110) peak first broadens and then splits. Comparing
with Fig. 1, it is noted that the peak broadening occurs at temperatures above the Ms (resistivity peak). Only a slight peak
splitting occurs at 16.5°C, which corresponds to the top of the
resistivity peak. A symmetric shift and separation of the split
peaks become pronounced at about 13°C. This temperature corresponds to the temperatures on the left hand side of the resistivity peak, i.e. lower than the Ms temperature.

The systematic broadening of the (110) peak is shown in Fig. 4.
The half width of the (110) peak, in this plot, is normalized
with respect to the corresponding half width of the (110)β peak
at 45°C (where no peak broadening is observed). An interesting aspect of this peak broadening is that even up to 13°C (in Fig. 3(b))
the integrated intensity under the peak is constant within the
range of experimental error. This is summarized in Table I.

Table I. Variation of Integrated Intensity of
          (110)   after 33 Cycles as a Function
          of Temperature

Temp. °C	18°	17.5°	17.0°	16.5°	13°	10°	8°
Integrated intensity of (110) (arbitrary scale)	723 +25	743 +25	715 +25	724 +25	722 +25	597 +25	346 +25

It must be mentioned that a somewhat similar experiment was reported by Wang et al.[44]. However, they used powder samples and diffraction patterns of the entire spectrum was scanned at 5°C intervals and therefore fine details of the (110) broadening and
splitting were not obvious.

The foregoing new results clearly indicate that a gradual structural distortion begins in the premartensitic range as indicated
by the (110) peak broadening and a negligible loss (if any) of
diffracted energy. The appearance of new martensitic peaks
coincide with the stage where a marked decrease in the integrated
intensity of the split peak is apparent. It is important to note
that the observed broadening is highly reversible (no hysteresis)
up to the temperature at which no detectable loss of integrated
intensity is observed.

VI. STABILITY OF THE   -PHASE WITH RESPECT TO AN APPLIED STRESS
IN THE PREMARTENSITIC RANGE

Next a series of experiments were performed to study the effects
of an applied stress in the premartensitic range. Figure 5(a) -
5(c) show the typical nature of the structural changes as obtained
by an uniaxial tensile stress at 23°C (in the TTR). It is clear
that even a small strain, such as 0.8% near TTR produces martensite
at the expense of the[7]-phase. This result then is complementary
to our hardness study[7] as a function of thermal cycling of the
Ti-Ni alloy. In a next series of experiments, samples were cooled
down to 0°C and strained various amounts and then diffraction
patterns were taken at 0°C. It is seen in Figure 6(b) that a 1.2%
strain produces an amount of   -phase as indicated by the appearance of the (110) reflection. If the sample is now cooled to
-196°C, the (110) intensity almost vanishes. Therefore it can be
said that the Ad temperature is at least as low as 0°C and the Mf
temperature is below 0°C. We are presently conducting more detailed studies to accurately determine Ad and Mf temperatures in
this alloy.

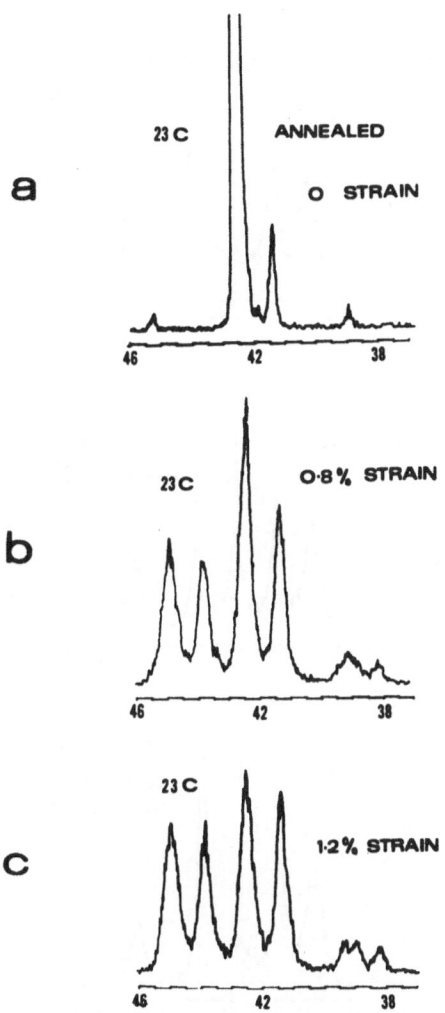

Fig. 5. Effect of an unaxial stress on the x-ray diffraction pattern of Ti-Ni alloy at 23°C. (a) zero strain, (b) 0.8% strain, (c) 1.2% strain.

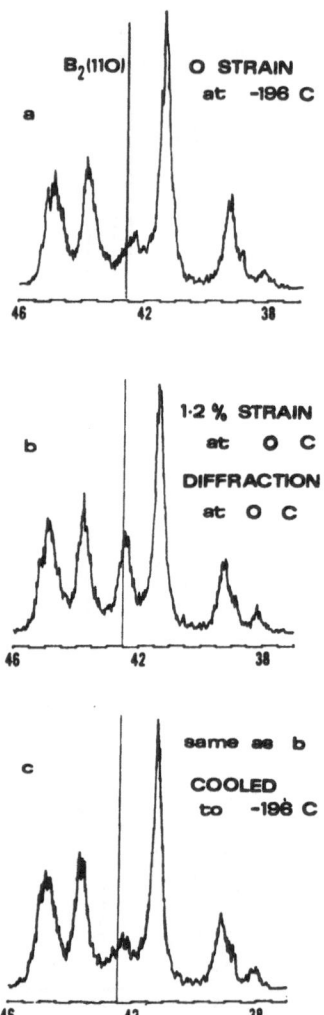

Fig. 6. Effects of an uniaxial stress on the x-ray diffraction pattern of Ti-Ni at 0°C. (a) zero strain at -196°C, (b) after 1.2% strain at 0°C, (c) cooling down to -196°C.

Every x-ray diffraction peak that results, either due to the application of an external stress as discussed above or due to cooling below the Ms temperature, can be indexed on the basis of a
monoclinic unit cell as reported by one of us in an earlier publication[28].

It has been mentioned before that stacking modulated martensites
are quite common in alloys which have a high temperature B_2
lattice. These alloys also manifest SME and the martensite is
thermoelastic in nature. We have studied a number of such alloys
as splat quenched thin foils. Splat quenching usually suppresses
the Ms temperature and the high temperature -phase is stabilized.
This metastable -phase gives rise to diffuse electron diffraction
effects such as relplanes and relrods which are similar in nature
to those associated with premartensitic instabilities in bulk alloys. In the next section we will briefly summarize some of these
recent results.

VII. LATTICE INSTABILITY AND MARTENSITIC TRANSFORMATION IN SPLAT QUENCHED SME ALLOYS

In the splat quenched Au-Cd alloys the initial manifestation of the
martensite nucleation is observed as a disturbance on the 110
type planes[29,30]. Masson[62] has suggested that in Au-Cd the
transformation can occur by alternate shear on 110 planes.
Similarly Duggin et al.[63,64] have proposed a similar mechanism
for the formation of thermoelastic martensites in the Au-Cu-Zn alloy
and Nishiyama et al.[65] have proposed a 110 shear mechanism for
the Cu-Al martensite. It has also been proposed[63,64] that certain stacking faults might act as nucleation sites, however the
formation of such faults in terms of fault energy must be carefully considered [66,67].

In the Au-Cd system the ordering forces are high and therefore
stacking fault formation by dislocation dissociation is unlikely.
The same argument also applies to other B_2 alloys. However, it
may be possible for certain types of faults to occur in this
system. The dislocation reaction proposed by Cohen et al.[68]

$$\frac{a}{2} \ 111 \qquad \frac{a}{8} \ 101 \ + \ \frac{a}{4} \ 121 \ + \ \frac{a}{8} \ 101$$

is compatible with the observed modes of martensite nucleation in
splat quenched Au-Cd alloys[30], as applied to the 101 plane.

Of particular interest is the stacking fault associated with the
$\frac{a}{8}$ 101 dislocation. In Fig. 7(a) is shown the 101 plane prior
to the dissociation and 7(b) shows the atomic displacement after
the   101 fault. It is seen in this figure that the atomic
displacements are small and unlike atom contacts are not severely
disturbed.

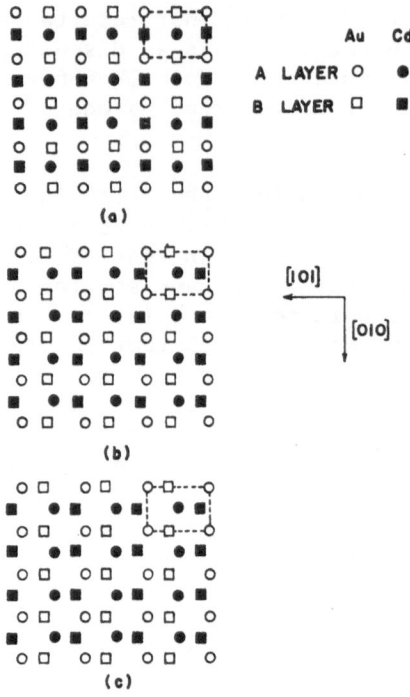

Fig. 7.   Atomic arrangement on the $(\bar{1}01)$ plane
          of the B_2 structure.
     (a) Atomic arrangement before the $\frac{a}{8}[101]$
          fault.
     (b) After the $\frac{a}{8}[101]$ fault.
     (c) $\{111\}$ type basal plane of the 2H mar-
          tensite.

If we consider the 2H martensite in the splat quenched Au-Cd, the
atomic coordinates are: (000), (0 2/3 1/2) for Cd and (1/2 1/2 0)
(1/2 1/6 1/2) for Au. The $\{111\}$ type plane of the 2H structure,
which is derived from the $\{110\}_A$ planes, has the atomic configuration as shown in Fig. 7(c). It must be noted that this configuration almost exactly coincides with the fault, as shown by the
dotted lines, in Fig. 7(b). It can be shown that the above dislocation reaction also gives rise to the orientation relationship
observed experimentally[69]. The observed[30] growth direction
for the martensite laths is [010] and since the displacement of
atoms for nucleating the structure is in the [101] direction
(Fig. 7(c)) which is normal to [010], it will appear that the transformation dislocation in the wedge shaped lath tip has predominantly screw characteristics. Warlimont[70] has discussed the possi-

bility that the screw dislocations might act as the transformation
dislocation.

Martensitic product phases in splat quenched Ag-Mg system have
also been observed by us[70]. Selected area electron diffraction
shows the presence of untransformed β and a mixture of β_2' and γ_2'
martensites. The β_2' has a 3 R structure with a stacking order
ABC BCA CAB/ABC .... and the γ_2' is 2H with a hexagonal symmetry
and an AB/AB/AB stacking. The analysis of such structures has
been carried out by Toth and Sato[71]. In the Au-Cd, Ag-Mg, and Ni-Al
alloys studied by us, the stacking modulated structures could be
derived by appropriate stacking shifts on the basal plane of the
Cu-Au type lattice. The basic Cu-Au type lattice derived from the
Cs-Cl type structure and the lattice correspondence is shown in
Fig. 8.

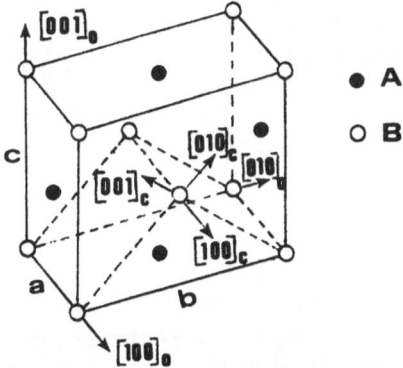

Fig. 8. Lattice correspondence between the
         Cs-Cl and Cu-Au types of unit cells:
         The Cs-Cl lattice is shown in dotted
         lines.

A typical $\langle 110 \rangle$ zone electron diffraction pattern of the 3R martensite in the splat quenched Ag-43.1 at.% Mg alloy is shown in
Fig. 9. Diffuseness of intensity and slight shift of spots from
ideal 3R position are noticeable in this figure.

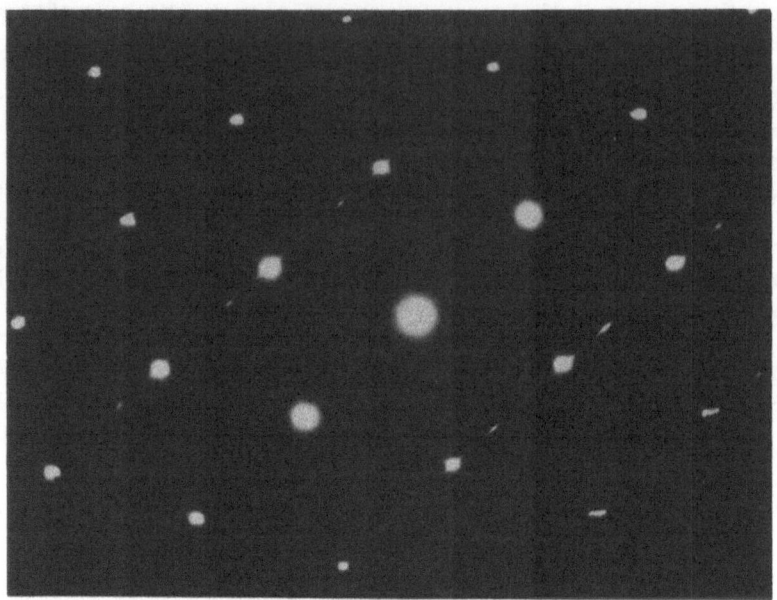

Fig. 9. Electron diffraction pattern from a
         splat quenched Ag-43.1 at.% Mg alloy.
         The structure is 3R. $\langle 110 \rangle$ zone axis.

In figure 10 is shown an electron diffraction pattern from a
selected area where retained β phase is present in the Ag-43.1 at.%
Mg alloy. Extra reflections in this diffraction pattern are due to
a ω type transition phase.

A key diagram illustrating the $\langle 110 \rangle$ zone pattern with the ω reflections is shown in Fig. 11. The orientation relationships be-
tween the β and ω are found to be $(111)_{\beta} \| (0001)_{\omega}$ and
$[110]_{\beta} \| [2\bar{1}\bar{1}0]_{\omega}$

A bright field electron micrograph of the monoclinic martensite
(distorted B19 or 2H) in the splat quenched Al-63.5 at.% Ni
alloy is shown in Figure 12. The striated substructure is identified as $\{ 111 \}$ twins in the Cu-Au type ordered structure.

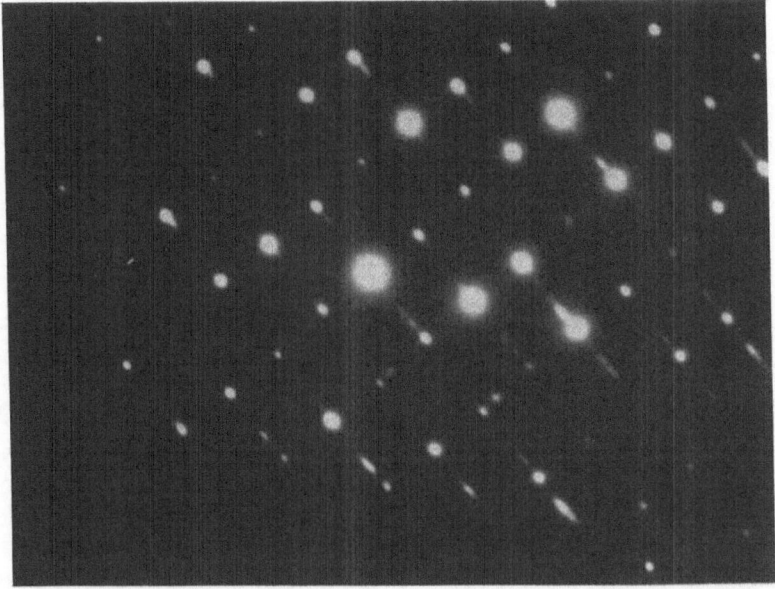

Fig. 10. ⟨110⟩$_\beta$ zone axis electron pattern of splat quenched
metastable β-phase in Ag-43.1 at.% Mg alloy. Extra
reflections are due to ω type transition phase.

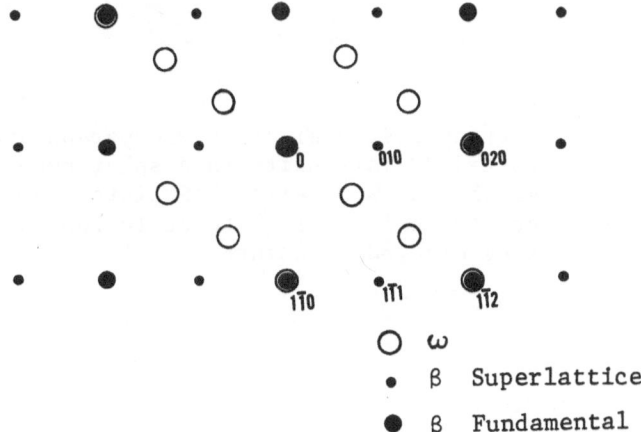

Fig. 11. Schematic pattern showing β fundamental and superlattice reflections. ⟨110⟩ zone axis, extra reflection due to ω-phase are shown.

Fig. 12. A transmission electron micrograph dis-
          torted 2H martensite in a splat quenched
          Al-63.5 at.% Ni alloy. Striated sub-
          structure is {111} twins in the Cu-Au
          type ordered structure.

Fig. 13.   <100> zone pattern of the metastable re-
           tained β-phase in the splat quenched Ni-Al
           alloy. Satellite spots around fundamental
           and super lattice reflections are observed.
           Also noticeable is diffuse <110> streaks.

In figure 13 is shown a $\langle 100 \rangle$ zone patterns of β NiAl in a splat quenched foil. Satellite spots around fundamental and superlattice reflections and diffuse $\langle 110 \rangle_\beta$ streaks are observed in this figure. It is proposed that the diffuse intensity and satellite reflections are associated with a metastable transition structure. The rel-planes correspond to a one dimensional chain of atoms with slight deviation of atomic positions from the ideal B_2 sites.

Similar microstructural and diffraction effects in splat quenched Ag-Cd and Cu-Ga have also been observed by us.[12] In all of these alloys, the presence of diffuse diffraction streaks associated with relplanes, satellite reflections and ω type diffraction effects are observed in the metastable β-phase. Since the β-phase in these alloys are also observed to transform martensitically, such diffraction effects in the β-phase are indicative of the insta-bility of the B_2 lattice and can be associated with the premarten-sitic transformation.

VIII. DISCUSSION

In the TRR range, the parent phase of the martensitic alloy becomes unstable with respect to a change of temperature (in the direction of the transition temperature) and applied external stress. Even without an external stress the nucleation and growth of martensite, and in general the martensite reversion, involve shear displacement of atoms and atomic planes. Therefore, if an external stress is applied, the resolved component of this stress on the habit plane and in the direction of the martensitic shear, will interact with the transformation process. The exact nature of this interaction will, of course, depend on various factors such as the crystal-lographic symmetry of the phases, the orientation and the nature of stress and the temperature range at which the stress is applied, etc.[72] In order for the SME to operate, the following conditions must be satisfied:

(a) Stress induced transformation of either mar-tensite or austenite (parent phase) or re-orientation of martensite.

(b) These stress induced structural changes are stable after the removal of the external force.

(c) A complete reversion to the original structure or substructure (including the spatial orienta-tion of the crystal or crystallites before the application of stress) upon heating or cooling through a transition temperature.

The shape memory effect, therefore, involves stress induced transformation and the degree of SME is related to the ease of reverse
transformation. If irreversible slip is involved either as the
lattice invariant mode of deformation or due to stress accommodation then the ease of reverse transformation is reduced and the
SME is greatly degenerated. However, this irreversible slip must
not be confused with the periodic stacking shift involved in
generating the stacking modulated martensites such as 2H, 3R, etc.
In many of the SME alloys such stacking modulated martensites are
found. The two most common martensitic structures in B_2 alloys,
with a strong SME, are β_2' and γ_2' types. Both of these structures
can be derived by stacking on the basal plane of the Cu-Au type
lattice as shown in Fig. 8.   It must be noted that the basal plane
of the Cu-Au type lattice is derived from the $\{110\}_\beta$ plane. The
β_2' martensite has the 3R structure where the stacking sequence is
ABC BCA CAB/ABC .... and thus consists of nine close packed layers.
The γ_2' martensite has the hexagonal symmetry given by the 2H
modulation and the stacking sequence is AB/AB/AB .... . Recently
Kubo and Hirano[73] and Delaey et al.[74] have applied crystallographic theories[75,76] of martensitic transformation to martensites with long period stacking order. It has been shown[73] that
the lattice deformation in these structures consists of three parts:
(i) pure distortion, (ii) homogeneous shear and (iii) shuffling.
The homogeneous shear and shuffling associated with such stacking
modulated structures could arise from certain enhanced lattice
vibrational modes[74,77,78]. From electron diffraction studies,
the following three lattice vibration modes in β -brass type
alloys have been proposed[74]:

   (1) A transverse 1/3 $\langle 110 \rangle$ wave; $\langle 1\bar{1}0 \rangle$ polarization
        vector.

   (2) A longitudinal 2/3 $\langle 111 \rangle$ wave; $\langle 11\bar{1} \rangle$ polariza-
        tion vector.

   (3) A transverse 1/2 $\langle 110 \rangle$ wave; $\langle 1\bar{1}0 \rangle$ polarization
        vector.

Of these vibrational modes, the 1/3 $\langle 110 \rangle$ wave is believed to be
associated with the softening of $1/2 (C_{11}-C_{12})$ corresponding to a
$\{110\} \langle 1\bar{1}0 \rangle$ shear of the β -phase. Therefore, both the premartensitic lattice instability and the subsequent collapse of
the B_2 lattice to stacking modulated martensites in many of the
SME alloys are intimately related to the enhanced lattice vibrational modes of the B_2 lattice.

It is important to note that many of the stacking modulated martensites are not ideal 2H, 3R, etc. and some monoclinic or orthorhombic distortions are noted from the displacements of electron

diffraction spots[27,28,31]. Such distortions could result from the strong ordering force which tries to preserve the unlike-pair contact distance. In this regard it must be noted that no attention has been given in the past to consider the effect of non-stoichiometry on the nature of stacking modulated martensites. For example, it has been shown[79] that deviation of stoichiometry in the Au-Cd β-phase alloys give rise to structural vacancies and anti-structure atoms in preferred sublattices. Therefore the distribution of atoms on the crystallographic planes vary systematically with the nonstoichiometry. Such a variation could alter the necessary distortion due to ordering forces as discussed earlier. This might account for the variation of stacking modulation with composition and even the presence of more than one modulated structure in the same alloy composition resulting from local inhomogeneities.

The presence of relrods and relplanes in the β-phase alloy near the TTR as observed by us in this work and as reported in the literature[80] indicate that certain vibrational modes occur in the TTR range indicating the incipient instability of the bcc lattice. Analogous to the premartensitic instabilities, diffuse diffraction effects are also found in the metastable β-phase as obtained by splat quenching (some examples are shown in this paper) or solid state quenching of β-phase alloys[81-85]. Transmission electron microscopy of such alloys show a mottled microstructure. At least a part of this mottling is thought to be a result of surface rippling due to the electropolishing of samples. However, such a mottled microstructure has been observed by us in most of the splat quenched β-phase foils which did not require electro-polishing[29,30,45].

In this paper we have also shown some new results to demonstrate the systematic variation of the $(110)_\beta$ peak x-ray intensity variation in the bulk Ti-Ni alloy near the transition temperature. The systematic broadening of the peak could be interpreted as due to a lattice distortion prior to the martensitic transformation followed by the peak splitting which can be interpreted as due to the loss of cubic symmetry of the lattice. We believe that the premartensitic instability in this alloy extends up to the temperature where the integrated intensity under the split peak is still constant. Beyond this temperature the instability leads to a collapse of the transition lattice in to the new martensitic structure.

We have also shown that the effect of stress on the Ti-Ni alloy in the the TTR is to produce martensite. Stress accommodation (and therefore SME) in the fully martensitic condition of this alloy occurs by the formation of β-phase rather than by martensite reorientation.

References

1. L. C. Chang and T. A. Read, Trans. AIME, 47, 191 (1951)
2. M. W. Burkart and T. A. Read, Trans. AIME, 197, 1516 (1953)
3. F. E. Wang, W. J. Buehler and S. J. Pickart, J. Appl. Phys., 36, 3232 (1965)
4. H. K. Birnbaum and T. A. Read, Trans. AIME, 218, 662 (1960)
5. A. Nagasawa, Phys. Status Solidi, (A)8, 531 (1971)
6. R. J. Wasilewski, Met. Trans., 2, 2973 (1971)
7. K. Mukherjee, F. Milillo and M. Chandrasekaran, Mat. Sc. and Eng., 14, 143 (1974)
8. E. Hornbogen and G. Wassermann, Z. Metallk., 47, 427 (1956)
9. C. M. Wayman, Scripta Met., 5, 489 (1971)
10. R. V. Krishnan and L. C. Brown, Met. Trans., 4, 432 (1973)
11. R. V. Krishnan and L. C. Brown, Met. Trans., 4, 1017 (1973)
12. A. K. Misra and K. Mukherjee, to be published.
13. A. Nagasawa and K. Kwachi, J. Phys. Soc. Japan, 30, 296 (1970)
14. H. Tas, L. Delaey and A. Deruytterre, Scripta Met., 5, 1117 (1971)
15. I. A. Arbuzova and L. G. Khandros, Fiz. Met. Metalloved, 17, 390 (1964)
16. K. Otsuka and K. Schimizu, Scripta Met., 4, 469 (1970)
17. K. Oishi and L. C. Brown, Met. Trans., 2, 1971 (1971)
18. N. Nakanishi, Y. Murakami and S. Kachi, Scripta Met., 5, 433 (1971)
19. S. Miura, S. Maeda and N. Nakanishi, Phil. Mag., 30, 565 (1974)
20. K. Enami and S. Nenno, Met. Trans., 2, 1487 (1971)
21. M. Chandrasekaran and K. Mukherjee, Mat. Sc. and Eng. 14, 97 (1974)
22. A. Nagasawa, J. Phys. Soc. of Japan, 30, 1505 (1971)
23. K. Enami, S. Nenno and Y. Minato, Scripta Met., 5, 663 (1971)
24. N. Nakanishi, T. Mori, S. Miura, Y. Murakami and S. Kachi, Phil. Mag., 28, 277 (1973)
25. L. Delaey, R. V. Krishnan, H. Tas and H. Warlimont, J. Mat. Science, 9, 1521 (1974)
26. C. M. Wayman and K. Shimizu, Met. Sc. J., 6, 175 (1972)
27. S. P. Gupta, A. A. Johnson and K. Mukherjee, Mat. Sc. and Eng., 11, 29 (1973)
28. S. P. Gupta, A. A. Johnson and K. Mukherjee, Mat. Sc. and Eng., 11, 43 (1973)
29. P. Ferraglio, K. Mukherjee and L. S. Castleman, Acta Met., 18, 1067 (1970)
30. P. L. Ferraglio and K. Mukherjee, Acta Met., 22, 835 (1974)
31. M. Chandrasekaran and K. Mukherjee, Mat. Sc. and Eng., 13, 197 (1974)
32. M. Born, Proc. Cambridge Phil. Soc., 36, 160 (1940)
33. C. Zener, Phys. Rev., 71, 846 (1947)
34. S. Zerinsky, Acta Met., 4, 164 (1956)
35. D. B. Novotny and J. F. Smith, Acta Met., 13, 881 (1965)

36. H. Pops and T. B. Massalski, Acta Met., 15, 1770 (1967)
37. K. Salama and G. A. Alers, J. Appl. Phys., 39, 4857 (1968)
38. W. D. Robertson, Iron and Steel Inst. Report, London, 93, 28 (1965)
39. M. Jovanoic, M. E. Brooks and R. W. Smith, Scripta Met., 4, 193 (1970)
40. G. D. Sandrock, A. J. Perkins and R. F. Hehemann, Met. Trans., 2, 2769 (1971)
41. L. Kaufman and M. Cohen, Trans. AIME, 206, 1393 (1956)
42. L. Kaufman and M. Cohen, Prog. Metal Phys., 7, 165 (1958)
43. A. J. Perkins, Ph.D. Thesis, Case Western Reserve Univ., 1969
44. F. E. Wang, S. J. Pickart and H. A. Alperin, J. Appl. Phys., 43, 97 (1972)
45. K. Mukherjee and P. Ferraglio, Fizika, 2, Sup. 2, 27.1 (1970)
46. H. Livingston and K. Mukherjee, J. Appl. Phys., 43, 4944 (1972)
47. P. Jongenburger, Conf. Phys. Basis Temp., Paris, 47, 41 (1955)
48. J. M. Ziman, Electrons and Phonons, (Clarendon Press, Oxford), 1963
49. D. P. Dautovich and G. R. Purdy, Can. Met. Quart., 4, 129 (1965)
50. C. M. Wayman, I. Cornelis and K. Shimizu, Scripta Met., 6, 115 (1972)
51. R. J. Wasilewski, Trans. AIME, 233, 1691 (1965)
52. R. R. Hasiguti and K. Iwasaki, J. Appl. Phys., 39, 2182 (1968)
53. S. Spinner and A. G. Rozner, J. Acoust. Soc. of Amer., 40, 1009 (1966)
54. R. F. Bunshah and R. F. Mehl, Trans. AIME, 197, 1251 (1953)
55. K. Mukherjee, Trans. AIME, 242, 1495 (1968)
56. M. J. Marcinkowski, A. S. Sastri and D. Koskinaki, Phil. Mag., 18, 945 (1968)
57. R. Scholl, D. J. Larson and E. J. Freise, J. Appl. Phys., 39, 2186 (1968)
58. F. E. Wang, W. J. Buehler and S. J. Pickart, J. Appl. Phys., 36, 3232 (1965)
59. A. Nagasawa, T. Maki and J. Kakinoki, J. Phys. Soc. Japan, 26, 1560 (1969)
60. A. Nagasawa, J. Phys. Soc. Japan, 29, 1386 (1970)
61. S. P. Gupta, A. A. Johnson and K. Mukherjee, J. Phys. Soc. Japan, 31, 605 (1971)
62. D. B. Masson, Trans. AIME, 218, 94 (1960)
63. M. J. Duggin and W. A. Rachinger, Acta Met., 12, 529 (1964)
64. M. J. Duggin and W. A. Rachinger, Acta Met., 12, 1015 (1964)
65. Z. Nishiyama and S. Kajiwara, Jap. Jour. of Appl. Phys., 2, 478 (1963)
66. P. R. Swann and H. Warlimont, Acta Met., 11, 511 (1963)
67. S. Dash and N. Brown, Acta Met., 14, 595 (1966)

68. J. B. Cohen, R. Hinton, K. Lay and S. Sass, Acta Met., $\underline{10}$, 894 (1962)

69. P. Ferraglio, Ph.D. Thesis, Polytechnic Inst. of Brooklyn, N. Y. (1972)

70. M. Chandrasekaran and K. Mukherjee, to be published.

71. R. S. Toth and H. Sato, Acta Met., $\underline{16}$, 413 (1968)

72. H. Tas, L. Delaey and A. Deruyttere, J. Less-Comm. Met., $\underline{28}$, 141 (1972)

73. H. Kubo and K. Hirano, Acta Met., $\underline{21}$, 1669 (1973)

74. L. Delaey, J. Van Paenel and T. Struyve, Scripta Met., $\underline{6}$, 507 (1972)

75. M. S. Wechsler, D. S. Lieberman and T. A. Read, Trans. AIME, $\underline{197}$, 1503 (1953)

76. J. S. Bowles and J. K. Mackenzie, Acta Met., $\underline{2}$, 199 (1954)

77. A. J. Perkins, G. Sandrock and R. F. Hehemann, Met. Trans., $\underline{2}$, 2769 (1971)

78. R. F. Hehemann and G. Sandrock, Scripta Met., $\underline{5}$, 801 (1971)

79. D. Gupta, D. S. Lieberman and D. Lazarus, Phys. Rev., $\underline{153}$, 863 (1967)

80. L. Delaey, A. J. Perkins and T. B. Mossalski, J. Mat. Sci., $\underline{7}$, 1197 (1972)

81. H. Warlimont, Proc. Sixth International Congress for Electron Microscopy, Kyoto, 437 (1966)

82. J. A. Malcolm and G. R. Purdy, Trans. AIME, $\underline{239}$, 1391 (1967)

83. E. B. Howbolt and T. B. Massalski, Met. Trans., $\underline{1}$, 2315 (1970)

84. Z. Nishiyama, H. Morikawa and K. Shimizu, Jap. J. of Appl. Phys., $\underline{6}$, 815 (1967)

85. H. Morikawa, K. Shimizu and Z. Nishiyama, Trans. Jap. Inst. of Met., $\underline{8}$, 145 (1967)

FERROELASTIC "MEMORY" AND MECHANICAL PROPERTIES IN GOLD-CADMIUM

D. S. Lieberman, M. A. Schmerling*, R. W. Karz+
Department of Metallurgy and Mining Engineering and
Materials Research Laboratory, University of Illinois
at Urbana-Champaign, Urbana, Illinois 61801

"Round about the accredited and orderly facts of every
science there ever floats a sort of dust-cloud of exceptional observations." W. James, The Will to Believe

I. INTRODUCTION

During the past dozen years, the interest, research, development, and application of alloys exhibiting mechanical "memory"
behavior have steadily increased--the particularly sharp increase
over the past few years in part stimulated this timely symposium. Most of the attention has been focused on TiNi (1) with less
on alloys such as CuAlNi (2), CuZn (3), CuZnSi (4), CuZnSn (5), NiAl
(6), InTl (7, 8), AuCd (9), AgCd (10)--although the amount of fundamental research effort devoted to these latter alloys is substantial,
as can be seen from these proceedings.   Much of this work has
been reviewed quite recently (11) and/or is treated and referenced
rather completely in this volume. This paper will consider relevant phenomena in the near-equiatomic ordered alloys of AuCd, a
system which has been so important in contributing to our understanding in such ordered alloys and other metal systems of: point
defects on quenching (12), self-diffusion (13), elastic constants
behavior and mode softening (14), martensitic phase transformations
(15, 16), twin boundary motion (17), and the unusual "memories"
described here (9).   Indeed, many important theories and definitive
experiments associated with these phenomena--including the "shape
memories"--were devised first for or on this alloy. A theory of
one of the "memories" is presented here which satisfactorily
accounts for all of the experimental observations and which provides some insight into the others.

---
*Present Address: Materials Science and Engineering, 433 Engineering
 Science Building, University of Texas at Austin, Austin, Texas 78712
+Present Address: Xerox Center for Technology, Webster, New York 14580

II. THE "SHAPE MEMORIES" IN AuCd

A. The "Piezomorphic Memory"--Ferroelastic Behavior

It was over 40 years ago that Ölander first reported the remarkable "rubber-like" behavior of an alloy of gold and cadmium at
a meeting of the Swedish Metallographic Society and in subsequent
publications (18). Although Benedicks (19) and others (20) did
work on it (and dubbed it "Ölander's Alloy"), it was not until the
late T. A. Read, Research Associate L. C. Chang (9) and graduate
students at Columbia began intensive work on this system and InTl
in the late 1940's and '50's that the behavior was characterized
and some understanding of it attempted.

At room temperature a specimen of an $Au_{1.05}Cd_{.95}$ alloy can be
severely deformed, but when the stress producing the deformation
is removed, it immediately returns to its original shape in a remarkable "rubber-like" manner. These recoverable strains can be
of the order of 8 pct. in AuCd. The original crystal structure (as
revealed by X-rays) is unchanged and the external shape is recovered, provided the crystal is not held in its deformed state too
long (see below). Hence this mechanical "memory" is characterized
by complete crystallographic and morphological recovery when the
applied stress is removed; it is observed at constant temperature.
It is only exhibited in one other alloy, InTl, but not until it is
cooled to dry ice temperature. In this section, descriptive terms
reflecting the macroscopic manifestations of the "memories" are
employed. Hence since the specimen shape is changed by the application of stress and is recovered when it is removed, the term
"piezomorphic memory" is suggested. Why the mechanism-suggesting
term "Ferroelastic" is also appropriate for the "rubber-like memory"
will be discussed below where the phenomenon is examined micro-
scopically and an atomic model developed which leads to the theory
of this "memory".

B. The "Thermoelastic Memory"

This alloy exhibits other unusual "memory" effects which are
associated with a first order martensitic phase transformation.
The room temperature crystal structure of the alloy $Au_{1.05}Cd_{.95}$
is orthorhombic. As can be seen from Fig. 1, upon heating the
specimen from room temperature, there is change in resistivity at
~75°C of ~20 pct. in going from the orthorhombic phase to a single
crystal CsCl (from X-rays). On cooling, a decrease of the same
amount occurs at about 60 degrees when the specimen again reverts
to the orthorhombic martensitic phase, (the Ölander phase) designated
as β' to distinguish it from the CsCl β (fully ordered) parent.

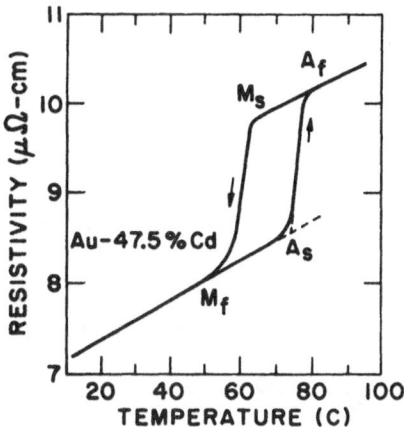

Fig. 1–Resistivity vs. temperature curve for $Au_{1.05}Cd_{.95}$. Note the narrow hysteresis width associated with the small transformation strains in this alloy (63) and the easily accessible temperature range facilitating experimental research on it.

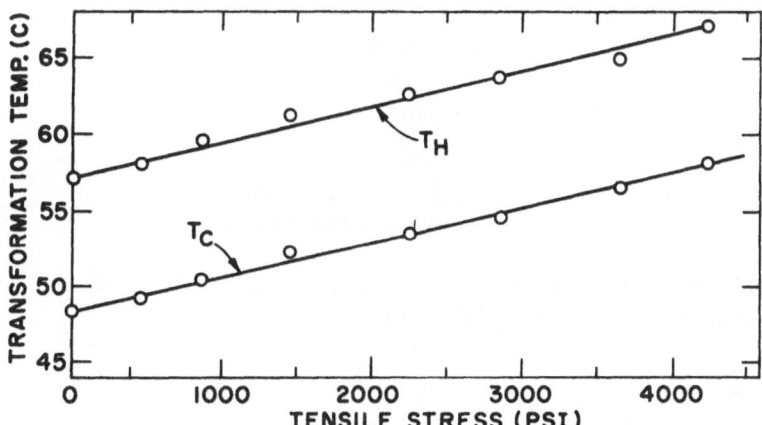

Fig. 2–The effect of stress on the transformation temperatures in a crystal reported as nominally $Au_{1.05}Cd_{.95}$. However, the lower transformation temperatures at zero stress and narrow hysteresis width (compared with Fig. 1) usually associated with compositions closer to exact stoichiometry indicate this crystal was probably $Au_{1.04}Cd_{.96}$. $T_C \approx \frac{1}{2}(M_s + M_f)$ and $T_H \approx \frac{1}{2}(A_s + A_f)$.   (21)

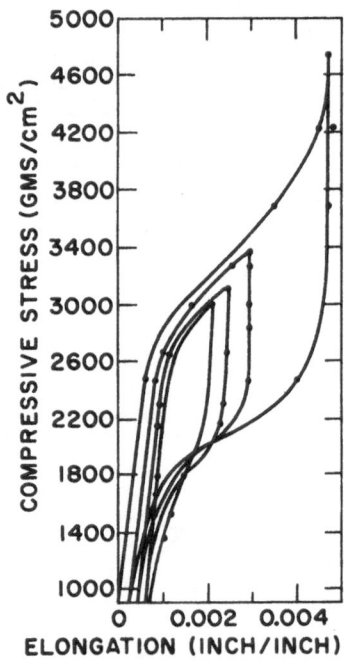

Fig. 3–Diffusionless phase transformation in InTl produced by the application of stress at 73°C. (7)

Note that the transformation is not isothermal, i.e. the specimen transforms only when the temperature is changed; the transition region on heating and cooling is about 2 or 3 degrees, while the hysteresis width is about 15 degrees. When a stress is applied, e.g., tension on a rod shaped specimen, the whole hysteresis curve is found to move toward higher temperatures on successive thermal cycles as the stress is increased (21), as shown in Fig. 2. Thus in a temperature range above M_S ($\sigma=0$), martensite is produced by the application of stress at constant temperature --"stress induced martensite," SIM--and when the stress is removed, the specimen reverts to the phase stable at that temperature. This is an ideal example of _thermoelastic_ martensite, where there is a balance or trade-off between the chemical free energy and transformation elastic strain energy, as first discussed by Kurdumov(22). In AuCd, as in InTl, the hysteresis width remains essentially constant as the stress is increased on successive thermal transformation cycles, as seen in Fig. 2, until well above the stress level where M_S under stress is above A_S under zero stress.* Thus, in AuCd, the complete transformation cycle between β and β' was accomplished _isothermally_ by the application and removal of stress--shown for the similar transformation in InTl (7) in Fig. 3--as well as by changing the temperature at constant stress (Fig. 1), as is implicit in the term _thermoelastic_. Again the original crystal structure of the parent phase and the original specimen shape are recovered when the stress is removed (or temperature is increased above A_F at constant stress) and hence this memory displays complete _crystallographic_ and _morphological_ recovery. It is tempting to invent a macroscopic term similar to "piezomorphic" above, but since the term _thermoelastic_ has already been used for more than a generation for this phenomenon, "_thermoelastic_ memory" will serve. All of the alloys mentioned in the introduction (at least) apparently transform to _thermoelastic_ martensite and thus exhibit this "memory"; they all also demonstrate the following "memory".

C. The "_Thermomorphic_ (or _Thermorphic_) Memory"

The last "memory" effect is perhaps the most dramatic. Immediately following the transformation from the parent CsCl structure to the twinned orthorhombic martensite, the specimen can be severely deformed and will not be "rubber-like" upon the removal of the stress since establishment of this state is dependent on time and temperature through a relaxation process (see below). However, upon heating the specimen up through the transformation temperature to the parent phase, the crystal structure of the parent as well as

---

*This behavior permitted the latent heat of a solid state-solid state transformation to be determined in a novel mechanical way (without calorimetry) for the first time (21, 7); it is delineated in the appendix for AuCd.

its original geometric shape are recovered.* This memory also exhibits <u>crystallographic</u> and <u>morphological</u> recovery. It has also
been observed to operate in both directions in some cases, i.e. the
same shape and crystal structure are produced on the subsequent
coolings to the martensite and hence the memory may be two-way.
This <u>shape</u> recovery is obviously observed only by varing the <u>tem-
perature</u> and (usually) at constant stress. Therefore the term
"<u>Thermomorphic</u> (or <u>Thermorphic</u>) Memory" seems most descriptive.
It is somewhat unfortunate that the term "shape memory effect" is
generally applied only to this last one when it clearly applies
equally well to all the "memories" discussed here. (There may
only be two independent ones since apparently materials which are
<u>thermomorphic</u> are also <u>thermoelastic</u>; indeed these may be different
manifestations of same material property). In any event, most of
the effort has been on this "memory" with by far the largest fraction on the TiNi (Nitinol) system and its possible device applications. Although this "<u>thermomorphic</u> shape memory" will not be
dealt with in detail here, it is interesting to note that it too
was first observed in AuCd by Chang and Read (9) in ~1950 and the
first "device" to demonstrate a "shape memory" material operating
repeatedly--and performing useful work--was constructed by Lieberman
in 1957. In the spring of that year, Read was invited to arrange
an exhibit of this "memory metal" for the 1958 Brussels Universal
and International Exhibition and asked Lieberman to prepare it. His
apparatus to cool and heat an AuCd specimen cyclically through its
transformation is shown operating in Figs. 4a, b, c, and incorporated in the exhibit in Fig. 4d. Preliminary studies of the free
energy changes in the $\beta \rightleftharpoons \beta'$ transformation were made on a similar
simple weight bending apparatus (23). Unfortunately no observa-
tions or reports were made of the behavior, life-times, etc. of the
several crystals used in the exhibit; such studies are now in
progress.

    In this paper, attention will be focused primarily on the first
mentioned "<u>piezomorphic</u> memory" occurring in AuCd and InTl, which
form a subset of those alloys demonstrating the "<u>thermomorphic</u>
memory". First, <u>ferroelasticity</u> will be explained and the theory
of the origin of the "rubber-like" restoring force in $Au_{1.05}Cd_{.95}$
will be developed and presented with experimental evidence for its
validity. Then the cyclic mechanical behavior and long fatigue life
of AuCd will be described and the model developed further to account
for these and the time-temperature dependent relaxation from
<u>ferrodisplacive</u> to <u>ferroelastic</u> behavior. Finally the "<u>thermomorphic</u>

---

*Since the "memory" alloys listed other than AuCd and InTl do not
exhibit the <u>ferroelastic</u> "<u>piezomorphic</u> memory" described above, they
can be deformed any time following the transformation to the martensitic phase on cooling and will recover their respective crystal
structures and shapes when heated back up to the parent phase(s).

(a)   (b)   (c)

Fig.4-(a) A single crystal of the high temperature β phase 10 cm
long and 3 cm in diameter is shown with the heater below. In (b),
the heater current has been turned off and the fan (lower right)on,
cooling the crystal from its reference state (a), through the transformation temperature (Fig. 1) to the product β' martensite which
deforms to relieve the stress. When the heater is turned on again
(and the fan off) in (c), the specimen (weight ~10 gms) transforms
to its initial structure and rod shape in (a), exhibiting the "Thermomorphic shape memory", and performing work on the weight (~50 gms).
Pictures (a)-(c) courtesy Eric Schaal, Time-Life (<1954). (d) shows
the apparatus automated as part of the Brussels exhibit.

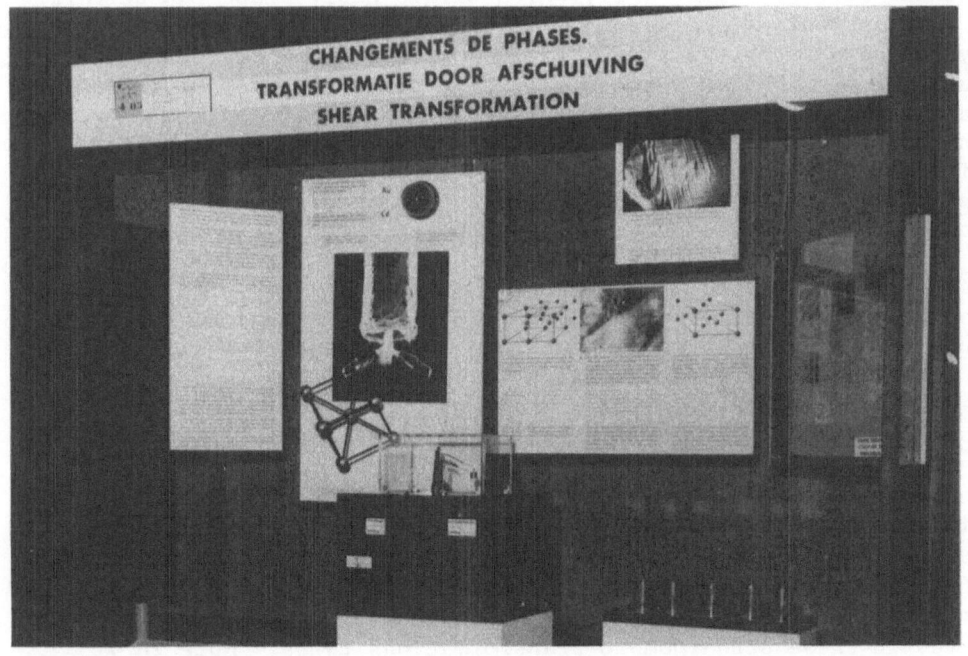

(d)

memory" will be discussed briefly, particularly with reference to
what this model can contribute to its understanding.

III. THE PHENOMENLOGY OF FERROELASTICITY

The room temperature crystal structure of this alloy is orthorhombic [(a=3.1540 Å, b=4.7645 Å, c=4.8644 Å) (24)] with four atoms per
unit cell. Photomicrographs of polished surfaces reveal striations
as in Fig. 5a, which X-rays corroborate as twins (25, 15). In
further X-ray and metallographic studies, it was shown that the
deformation is accomplished by the motion of these $(111)_0$ twin
boundaries so as to convert some twins to others more favorably
oriented to relieve the stress (17); the "rubber-like" behavior
upon the removal of the stress was observed to be associated with
the return to the original twin configuration. This is most simply
and directly studied when only twin boundaries are present in a
specimen, and preferably, only one of the four crystallographically
equivalent twin planes. Fortunately, it is possible to produce a
specimen staisfying these requirements. First, a grain boundary
free single crystal of the parent CsCl-type β phase can be grown
without too much difficulty by a modified Bridgman technique (9).
Next, although on normal cooling through the transformation temperature such a crystal transforms by several of the 24 crystallographically equivalent habit planes to many regions of differently
oriented sets of twins, (Fig. 5a) a single habit plane can be initiated at one end of the specimen (9,25,15,26) which moves along the
crystal converting the single crystal parent to an orthorhombic
product twinned on only one $(111)_0$ twin plane (15, 26) as shown in
a partially transformed specimen in Fig. 5b. That the whole specimen consists of these $(111)_0$ transformation twins--upon whose
existence the ferroelastic and other "memories" depend--is completely accounted for by the generally accepted phenomenological
theory of martensite formation (15, 16).

When such a specimen consisting of a single set of orthorhombic twins is stressed beyond a small elastic region, the twin
boundaries will start to move and one twin orientation will grow
at the expense of the other, as shown schematically in Fig. 6 and
in more detail in Fig. 7. For example, since 2 a is 7.94% smaller
than $\sqrt{b^2 + c^2}$, the twin with this shorter dimension closest to the
specimen stress axis would be expected to grow in compression at
the expense of the other to relieve the stress. In tension, the
twin with the long diagonal of the unit cell $\sqrt{b^2 + c^2}$, closest to
the stress axis would tend to increase its volume; if one twin is
not more favorably oriented, the specimen will not respond to applied
stress as described. Fig. 7 shows the geometrical relations involved in these orthorhombic transformations twins. Note in 7c
that since the angle $\angle BOB'$ is only 4.4°, OA and B'C are essentially

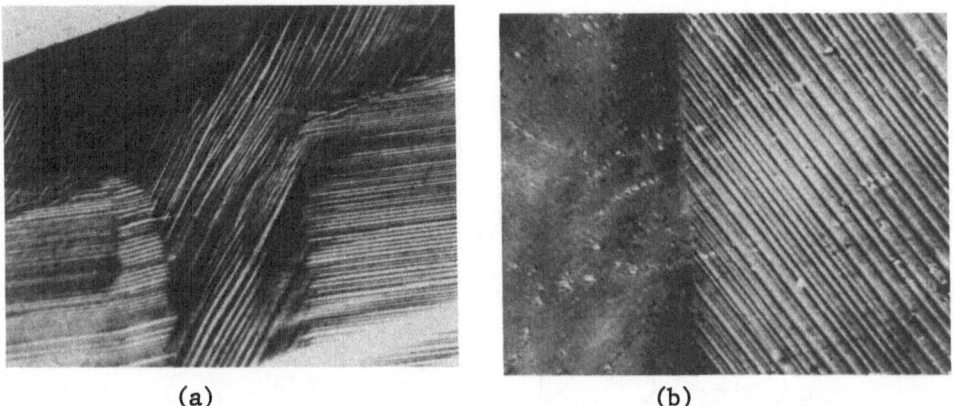

(a)   (b)

Fig. 5-Surface of AuCd specimen polished in the high temperature β
phase and transformed by multiple interfaces to (a), regions of different orthorhombic twin orientations (~150X) and (b), by a single
interface separating the high T β cubic region (left) from the low
T β' twinned orthorhombic region (right). After (26). ~150X.   The
first example of a "two-way memory" was probably an AuCd crystal
"educated" to transform by a single interface (9) and which continued
to do so by that interface to that particular set of twins.

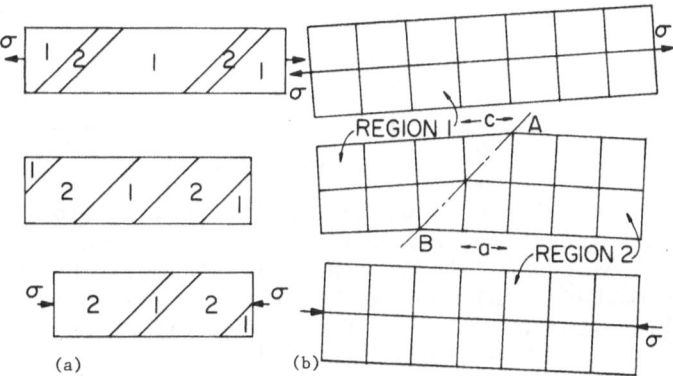

Fig. 6a-Schematic of specimen transformed by a single interface to a
single set of orthorhombic twins (Fig.5b); center, no applied stress;
above, applied tensile stress-twin 1 has grown at the expense of twin
2 to relieve the stress; and below, compressive stress-twin 2 has
grown at the expense of twin 1. (b) Center, twinned 2-D tetragonal
lattice with c/a=1.11. Tensile stress moves twin boundary, AB, to
the right producing a single crystal of region 1 above; compressive
stress moves AB to the left resulting in a single crystal of region
2 below. Maximum strain in tensile-compressive cycles is seen to be
$\frac{c-a}{a}$.

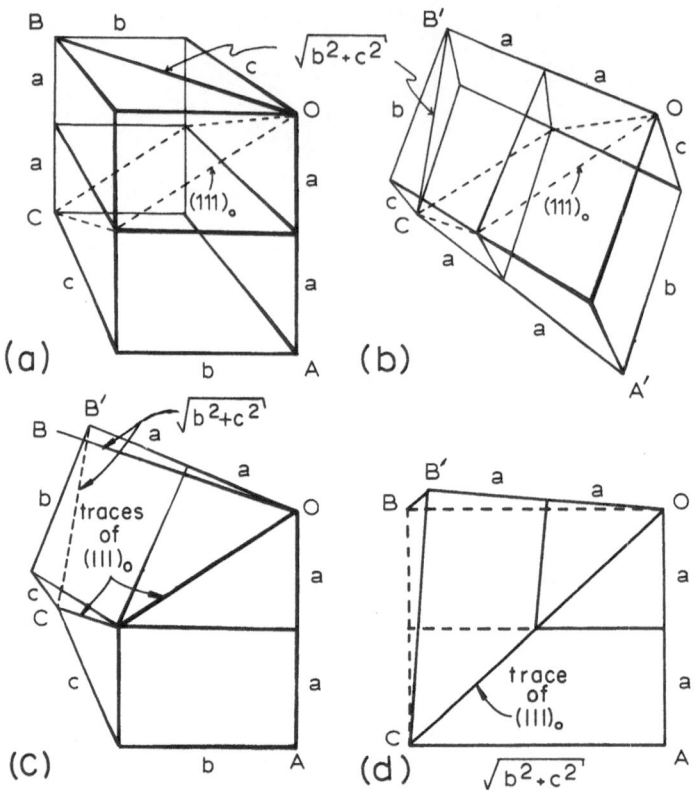

Fig. 7- Twinning of the β' orthorhombic lattice. (a)-Two unit cells placed together to form a block $2\underline{a} \times \underline{b} \times \underline{c}$ where $\underline{a}$ = 3.1540 Å, $\underline{b}$ = 4.7645 Å, $\underline{c}$ = 4.8644 Å. The twin plane is the $(\bar{1}11)_o$ plane shown dashed. (b)-Another block of two cells oriented to have a $(111)_o$ plane parallel to the one in (a) so that it is twin related to it. (c)-The part of (a) below the $(111)_o$ plane and the part of (b) above it are put together at this common $(111)_o$ twin plane; drawn to the same perspective as (a) and (b). (d)-Plane OAC in (c)-approximate since the twinning direction is irrational and OB' lies 0.55° out of it. After (26,16).

parallel (and would appear more so in an unexaggerated figure).
Thus a close approximation to the maximum strain possible if a
specimen is converted from all of one twin to all of the other
during a cycle of tensile-compressive stress would be 7.94%.*
This may be more easily seen in the two-dimensional approximation
shown in Fig. 7d which is the OAC plane ignoring the fact that
OB' actually lies 0.55° out of this plane because the structure
is slightly orthorhombic (c/b = 1.02) and the twinning direction
is actually irrational (15,26). The two dimensional analogue in
Fig. 6b illustrates this more clearly.

Use of the term <u>ferroelasticity</u> for the unusual behavior in
AuCd can be understood by examining its similarity to ferromagnetism.
Motion of twin boundaries under stress is likened to the motion of
magnetic domain boundaries when a magnetic field is applied. A particular unique vector in the twin, say the short 2<u>a</u> direction, Fig. 7,
plays a role analogous to the direction of magnetization, <u>M</u>, in
the magnetic domain. Note that each twin has a morphological or
shape vector with a "saturation" value of 2<u>a</u> in the absence of any
stress just as each domain has its value of <u>M</u> in the absence of <u>H</u>;
if such a state of strain or displacement were induced only on the
application of a stress but absent under zero stress, then it would
correspond to paramagnetism, i.e. ordinary elastic behavior is in
this sense "paraelastic". The growth of some twins (domains)
at the expense of others when the stress (magnetic field) is applied
lowers the total strain (magnetic, etc.) energy of the crystal in
accordance with a generalized LeChatelier principle; when the stress
(magnetic field) is removed, the original low energy twin (domain)
configuration is essentially recovered. The use of "ferro" here
in the absence of any iron is as good (or bad) as its use in "ferro-
electricity" in $BaTiO_3$ (no iron), for example, where an electric
field moves domain walls separating two directions of (permanent)
polarization.

The analogy can be taken further. As an applied <u>H</u> field is
increased on a ferromagnetic material with a domain structure, the
set of domains aligned with the field will grow at the expense of
the others until the material becomes saturated and then the <u>B</u>
field in the material will increase more or less linearly with <u>H</u>.
The elastic deformation after one twin orientation has been removed
from the β' phase is analogous to this increase after "saturation".
In early experiments stabilized β' crystals were strained in compression to a point just after one set of twins had apparently
been removed. On releasing the stress, it was noticed that the
strain did not completely disappear; however, on subsequent compression cycles the new strained reference state was recovered as

_____

*This compares favorably with the more exact calculation (27) and with
actual stress-strain hysteresis loops described below.

in Fig. 8. This is analogous to simply cycling an $\underline{H}$ field between
zero and some value, thus never removing the remanence. It is
therefore possible to say that in a ferroelastic material, the first
release after compression left a remanence suggesting that the twin
volume whose $\underline{a}$ axis was closer to the compression axis was left
slightly larger than its original equilibrium value after transformation from the cubic. The "saturation" is seen in Fig. 8.

Complete tension-compression stress-strain hysteresis cycles
such as shown in Fig. 9a further justify the analogy and use of
the term ferroelasticity. If the data in this figure is replotted
as strain vs. stress as in Fig. 9b (as one would plot $\underline{M}$ or $\underline{B}$ vs. $\underline{H}$),
the remanence ε_r at zero stress, the coercive force σ_c, and the
general shape of the hysteresis curve complete the analogy, which
is even more apparent in the same specimen after 100,000 cycles,
as shown in Fig. 9c. It is tempting to go further and comment
that just as the transformation in iron base alloys is from a
paramagnetic cubic phase to a ferromagnetic product of lower
symmetry, so the transformation in AuCd is from a (para)elastic
cubic parent to a ferroelastic product of lower symmetry, but this
will not be persued further here. Note "saturation" clearly in Fig. 9c.

Historically the term ferroelasticity* (28, 16, 29) was applied
only to the rubber-like "piezomorphic memory" in $Au_{1.05}Cd_{0.95}$ (and
InTl) which involved both (i), the uncommon large non-plastic deformation by the motion of twin boundaries when a stress is applied
and (ii), the even more extraordinary restoring force which brings
the twin boundaries back to their initial positions when the stress
is removed. However, the term has been used in the literature
when only condition (i) obtains, as in some of the alloys mentioned
in the introduction exhibiting only the "thermomorphic shape memory"
and in some other compounds. Ferrodisplacive may be more appropriate
for these with ferroelastic reserved when both (i) and (ii) exist;
this will be discussed further later. In any event, these terms
are much preferred to other expressions which do not convey the
magnetic analogy, give little insight into the nature of the phenomena delineated in sections II and III, and have contributed to
confusion in the literature as discussed below.

IV. NATURE OF THE RESTORING FORCE AND FERROELASTIC "MEMORY"

A. Theoretical Considerations

In III, the large dimensional shape change ($\lesssim 5\%$) in response

---

*The word ferroelasticity was first used by F. C. Frank when he
experienced the "rubber-like" behavior of AuCd in the early 1950's.

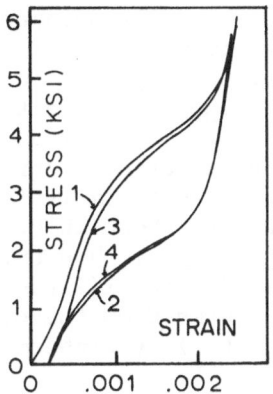

Fig. 8–Compression only σ vs ε cycles. 1 and 2 – 1st cycle of
well annealed sample. Note remanence of ε at zero σ. 3 and 4 –
2nd and subsequent cycles(s), which close completely.

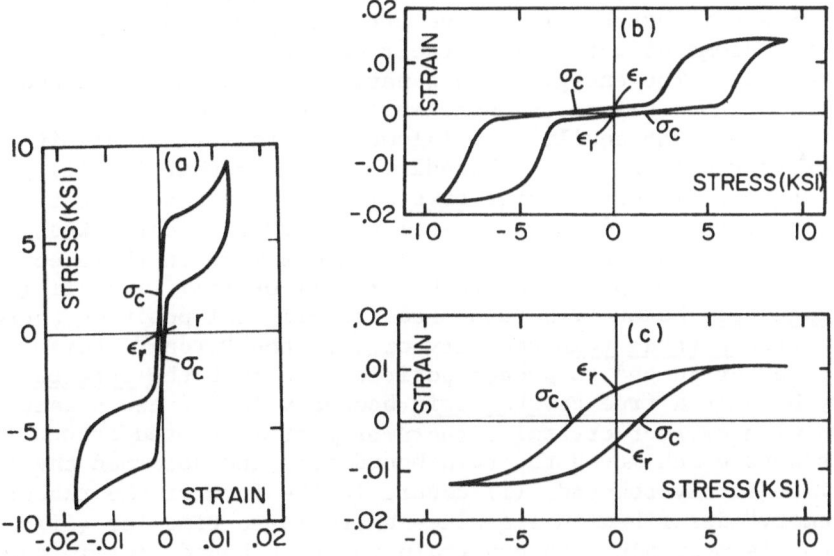

Fig. 9–Tension–compression σ–ε hysteresis loop of an hourglass
shaped crystal with one set of twins. (a)–The coercive force, σ_c,
and strain remance, ϵ_r, are shown on the hysteresis loop. (b)–Same
as (a) but with stress, the intensive parameter, now as abscissa and
strain, the extensive parameter, as ordinate. (c)–Same crystal as (b)
after 100,000 tension–compression cycles. Hysteresis loop shape,
σ_c, and ϵ_r, are now quite analogous to the corresponding ferromagnetic quantities. The transition from the behavior in (b) to that
in (c) is explained in the text, section V–A.

to relatively small stress was easily accounted for from a simple
consideration of the twin geometry using block models of the orthorhombic unit cell. However, the explanation of the rubber-like
behavior associated with the return of the twin boundaries to their
original positions when the stress is removed requires a more detailed examination of the actual crystal structure. The orthorhombic phase of $Au_{1.05}Cd_{.95}$ is not simply face-centered* but the
more complex B-19 (30) structure shown in Fig. 10a which will be
referred to as Ölander orthorhombic. How this "shuffled" structure
with the Cd atoms at 0,0,0 and 0,5/8,1/2 and the Au atoms at 1/2,
1/8,1/2 and 1/2,1/2,0 (20) forms from the CsCl parent in Fig. 10b
has been discussed in some detail by Lieberman (16).

The salient feature for the model proposed here is that any
movement of a twin boundary, viewed as a homogeneous shear in any
small volume, moves some atoms to positions not those of the perfect
Ölander unit cell since this is a structure with more than one atom
per lattice point, i.e. a lattice with a basis. Atoms are, in
effect, moved into the "wrong" places. This is understood more
easily from the examination of the two dimensional analogue in
Fig. 11. It can be seen that an homogeneous twinning shear results
in a lattice twin but not in a true crystal structure twin since
half of the atoms do not go into their correct positions to be
mirror related. The energy of this part of the specimen is now
higher than it was before the homogeneous twinning shear since it
does not possess the equilibrium (=lowest energy) structure it (i)
had prior to the lattice twin boundary motion or that it (ii)
would possess if it were a true crystal twin in its new position
with a mirror related correct crystal structure. Clearly the
specimen energy would be reduced if the atoms were in their correct (lowest energy) positions either in (i) or (ii), i.e. if (i)
the lattice twin boundary returned to its original position where
it was a true crystal twin boundary or (ii) the "wrong" atoms
"reshuffled" or jumped to proper positions so that the lattice twin
boundary becomes a true crystal twin boundary in its new strained
position with correct crystal structures mirror related across it.
If the stress which moved the twin boundaries and deformed the crystal is immediately removed, (i) occurs in the form of the rubber-
like "memory" described in IIA since there is no time for (ii); if
the stress is maintained to constrain the twin boundaries in their
new positions, (ii) occurs with temperature and time dependent relaxation kinetics. Upon complete stabilization, the specimen will
remain in the deformed configuration when the stress is removed and
will be "rubber-like" from there. Note that the driving force to
restore the planar lattice twin boundaries to the initial crystal
twin positions in a "rubber-like" manner in (i), σ_r, and to force the

---

*That is, face-centered if the distinction between gold and cadmium
atoms is ignored. See Figs. 10a and 10b.

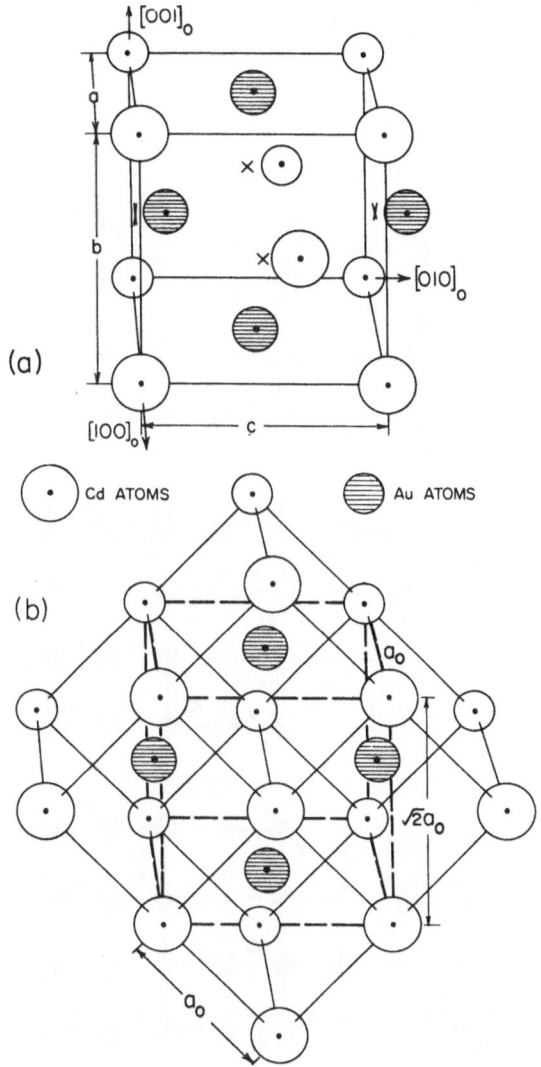

Fig. 10-(a)-c/8 shuffles on alternate (001)$_o$ planes produce the β'
Ölander cell from the f.c.o. cell (atoms at X) transformed by Bain
distortion from the 'tetragonal' cell delineated in 4 β CsCl cells,
(b)-β' cell constants: a=0.949 a$_o$, b=1.0138($\sqrt{2}$a$_o$), c=1.0550($\sqrt{2}$a$_o$)
(16).

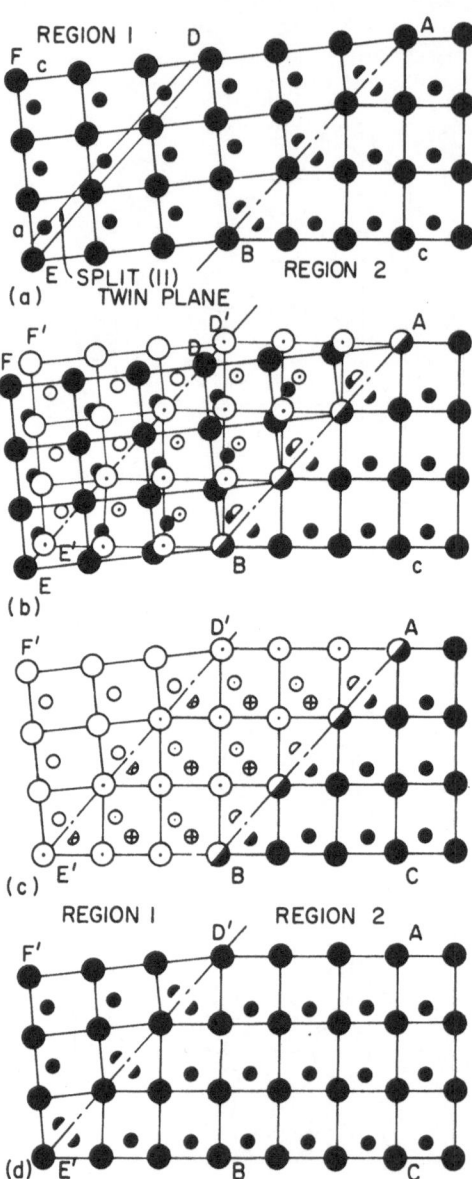

Fig. 11-Lattice of Fig. 6b with 2 atoms/unit cell. (a)-(11)$_0$ twin
plane AB mirrors ABC & ABD. (b)-Compression moves AB to DE, twinning ABED to ABE'D' and translating DEF to D'E'F'. (c)-<u>Lattice</u> of
⊙ ABE'D' is lattice 2 and a twin to lattice 1 D'E'F', but <u>crystal</u>
ABE'D is not crystal 2 since O are at ⊙ from twin shear (b), not
at ⊕ (from d). Resultant higher energy is reduced if ⊙ go to correct positions in either (i) region 1 by restoring the twin boundary to AB in (a) or (ii) region 2 by 'reshuffling' to ⊕ in (c) to
produce (d).

atoms to reshuffle to correct positions in (ii) is the result of
a <u>volume</u> free energy minimization process to produce the correct
structure and hence is a <u>volume</u> force. Other suggestions about
the mechanism will be critically discussed elsewhere (27).

The question arises as to why there should be a time dependence to the reshuffling process in AuCd as evidence by the observed
stabilization effects. $Au_{1.05}Cd_{.95}$ is close to being a hexagonal
close packed structure in the β' phase (30, 16), as shown in Fig. 12
If the atoms were the same size and kind and alternate $(001)_o$
planes were displaced $\underline{c}/6$ instead of $\underline{c}/8$ determined by Olander (18),
it would be hcp; Fig. 13 shows the relation between orthorhombic
and hexagonal coordinates used here. Since hcp also has more than
one atom per lattice point, not all atoms are moved to their final
positions by a simple homogeneous twinning shear (31, 32). The
twinning process is usually described as a homogeneous shear and
a shuffling of atoms to regain the hcp structure which must occur
simultaneously since no observations are available that indicate
any time dependence in hcp twinning. However, $Au_{1.05}Cd_{.95}$ does
appear to have a time dependent reshuffling process associated with
twinning and, therefore, reshuffling is more than just a way to
describe the process.

The significant differences between the AuCd Ölander structure and hcp materials, which must be intimately associated with
this unusual time-temperature dependent "reshuffling" and its extraordinary consequences, are:

(i) $Au_{1.05}Cd_{.95}$ is an <u>ordered alloy and the right type of</u>
<u>atom must be twinned (and shuffled) to the right place</u>; that is
it must remain ordered. This alloy is highly ordered even up to
the melting point (12); the ordering energy is so high that the
material shows an ionic character in the mode of slip in both the
cubic (33) and orthorhombic (34) phases. This high degree of order
seems to dictate a twinning mode probably not found in single component materials.

(ii) The transformation <u>twinning system in AuCd is not the</u>
<u>twinning system observed in normal hcp metals</u>. While the hexa-
gonal representation of the $(111)_o$ twin plane in AuCd is seen
from Fig. 13 to be $(10\bar{1}1)$, twinning of normal hcp metals is on the
$(10\bar{1}2)$ plane. $(10\bar{1}1)$ is sometimes a twinning plane in such materials, but with a $[10\bar{1}2]$ twinning direction, not the $[\bar{9}\ 13\ \bar{4}\ 5]$
hexagonal representation of the accepted twinning direction in
this alloy. As has already been mentioned in III, the $(111)_o$
twins in AuCd are <u>transformation</u> twins formed during the martensitic transformation from the CsCl parent to produce the habit
plane of zero average distortion in Fig. 5b (15, 16). Hence there is
really no reason to expect them to be the same, in the hexagonal

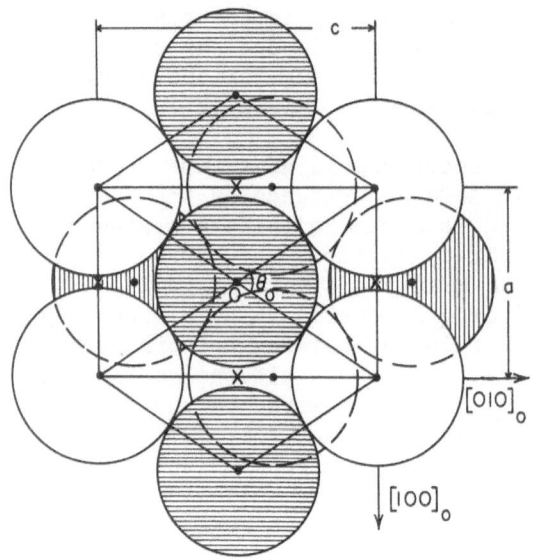

Fig. 12—The (001)$_o$ plane of the β' structure of Fig. 10a. The "shuffles" in the planes below and above produce a structure very close to hcp. (16).

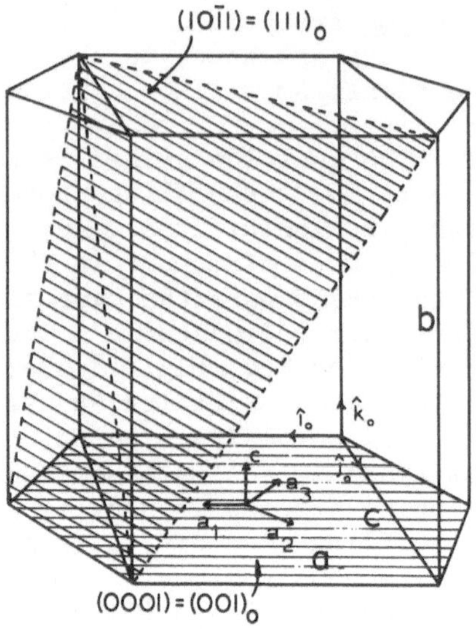

Fig. 13—The relationship between hexagonal and orthorhombic coordinates for the Ölander structure.

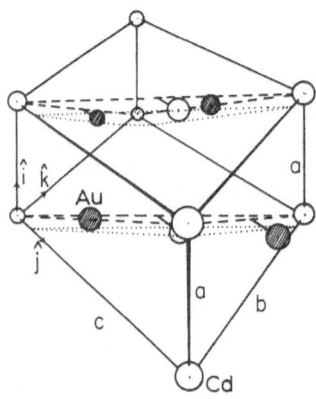

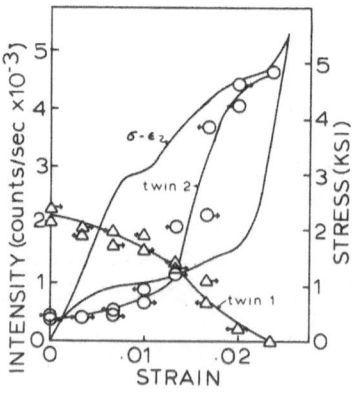

Fig. 14—The β' cell tilted to show the splitting of the (111)$_o$ planes. The "shuffle" system planes (001)$_o$ and (100)$_o$ in Fig. 10a are not split. See text.

Fig. 15—X-ray intensity vs ε. Arrows indicate the direction ε was being changed when intensity was recorded. The zero on the ε axis is the offset remanence of Fig. 8 since the σ vs ε cycle shown was not the first.

representation, as the twins normally observed in hcp metals during plastic deformation, etc.; the salient point is that they are indeed different Calculation of the small shear in the Ölander structure compared with what it would be if hcp, and further analysis, will not be presented here (27).

(iii) Except for the shuffle (shear) plane and the plane of shuffle (shear) in Fig. 10a, all others are split into two planes. In particular the $(111)_o$ atomic planes are split into two planes about .3A apart, as shown in Fig. 14, and in this alloy, the atoms in each plane of a pair can move through the spaces in the other plane. This reshuffle in the real Ölander orthorhombic structure is analyzed in detail elsewhere (27); here it will suffice to note from the two dimensional analogue of Fig. 11 that the main component of the reshuffling to form a true crystal twin is essentially perpendicular to the twin plane just as in the Ölander alloy where the paired $(111)_o$ planes can penetrate each other. This is not true of the corresponding planes in an hcp structure-- the holes are not large enough to let this interpenetration happen, as can be easily calculated. Assuming that the atoms act like hard spheres and that the size of the gold atom is 49% of the close packed interatomic distance in Fig. 12 and the cadmium 51% (16), the ratio of the hole radius to the radius of the atom which tries to pass through it can be calculated for AuCd as

	Au	Cd	hcp
r_{hole}/r_{atom}	1.028	1.005	.984

Thus the type of restoring force, σ_r, and time dependent relaxation process observed in AuCd would not be expected in pure hcp metals.

B. Some Experimental Results and Discussion

The experimental procedure essentially involved stabilizing a specimen (transformed by a single interface to a single set of twins) so that it was rubber-like, straining the specimen to move the twin boundaries, holding the specimen at constant strain, and measuring some property which was expected to change as the atoms reshuffled in the lattice twin as it became a true crystal twin according to the model developed in IVA above. Since the reshuffling of atoms changes the crystal structure in the direction of the Ölander B-19 structure, it should be observable as changes in X-ray peak intensity with time. Preliminary measurements made using resistivity as the observed parameter as the sample was deformed and allowed to restabilize were inconclusive. Furthermore, since this type of experiment cannot lead to a detailed understanding of atomic processes, the primary experimental techniques employed were X-ray diffraction and stress relaxation.

X-Ray Diffraction. Only crystals with twin planes oriented approximately midway between the long axis of the sample and a diameter were used since only these could deform easily by twin boundary displacement when compressed along the long axis (27). Although this is a necessary condition for displacement, it is not sufficient--the short a (unique) axis of one twin must also be considerably closer to the compression axis than the short axis of the other twin(27). In order to avoid plastically deforming the specimen, it was important to know the maximum amount of displacement expected from twin boundary motion for each crystal, i.e. the strain at the end of the low effective modulus part of the stress-strain curve in Fig. 9a just before the sharp upturn indicating that one lattice twin had been completely converted to the other. This was calculated from the orientation of the twin plane relative to the stress axis system and the orientation of one of the ortho-rhombic unit cells (27).

A special compression jig was constructed for use on an horizontal X-ray diffractometer. One sample was compressed by small strain increments while positioned for reflections from the twin related (113)$_0$ planes. From the intensities plotted in Fig. 15 and the accompanying stress-strain curve, it is evident that the change in twin ratio is continuous from that of the unstressed set of twins to the compressed single lattice orientation specimen and in the reverse direction upon removing the compressive stress. The disappearance and reappearance of twins was also observed when a specimen was stressed in an SEM. Several sections from another long single crystal prepared as delineated above were held on the diffractometer at various temperatures after applying sufficient stress to strain them to the predetermined strain at which a single lattice (twin) thus formed in each section could become a single crystal (twin) by reshuffling, according to the model developed here. See Figs. 6 & 11.

The intensity of a Bragg peak was monitored as a function of time at the several temperatures; the total change in peak intensity was < 6%. A least squares fit to the equation $f = \exp(-t/\tau)$ was made where f is the fractional intensity change from the extrapolated final slope at time t and τ is the time constant for the process. From the values of the time constants at different temperatures the constants in the equation $\tau = \tau_0 \exp(Q/RT)$ could be obtained from a semi-log plot of $\tau \ln \tau$ versus 1/T. The results for the (022)$_0$ and (0$\bar{1}$2)$_0$ reflections are shown in Figs. 16-18. τ_0 is calculated to be 8.3 X 10^{-13} seconds and Q=24,000 cal/mole. A similar activation energy is obtained using the rate equations $\delta f/\delta t = F(f)\exp(-Q(f)/RT)$ or $1/t = G(f)\exp(-Q(f)/RT$ as suggested by Hillert (35) for the first stages of an activated process. Chang (25) reported

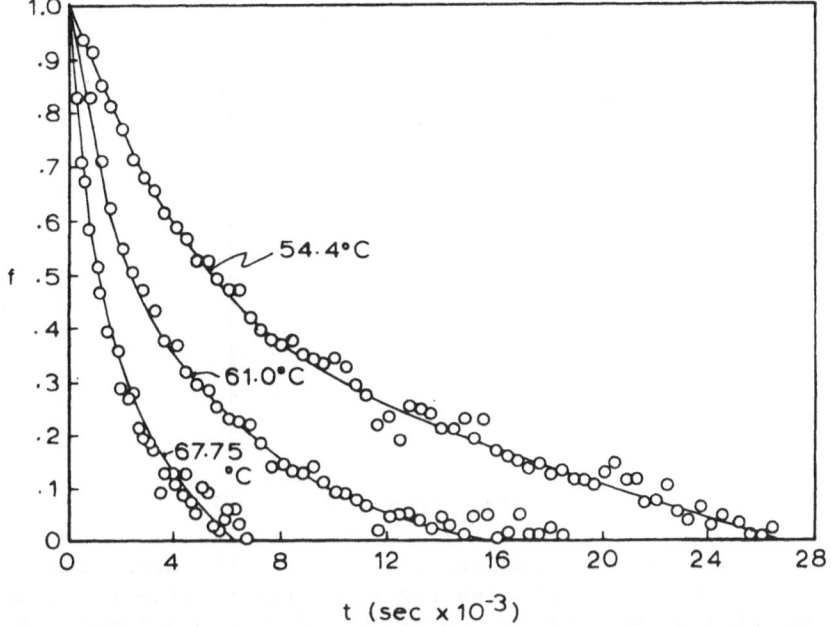

Fig. 16—Fraction of total intensity change remaining at time t, f, vs t for 3 segments of one long crystal at 3 different temperatures. The upper test temperature is limited by A_s in Fig. 1.

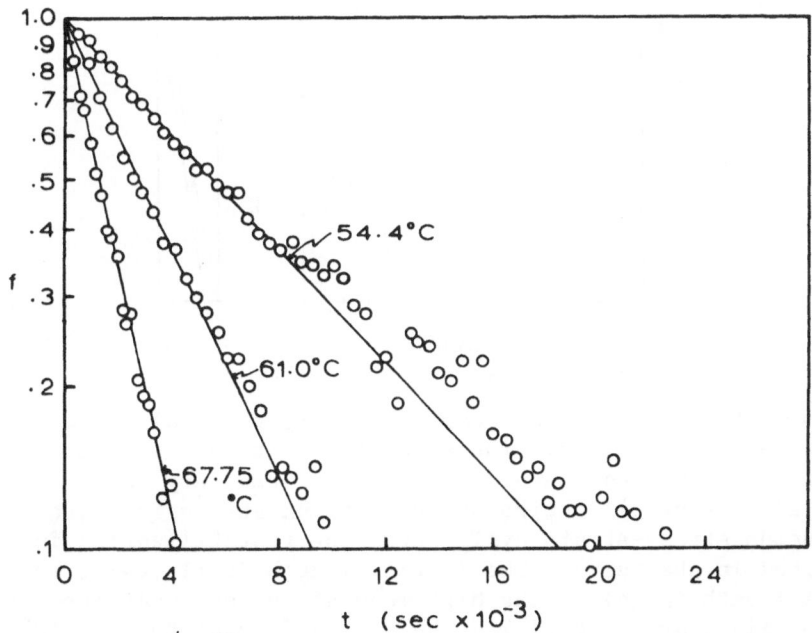

Fig. 17—Ln f (defined in Fig. 16) vs t; same data as Fig. 16.

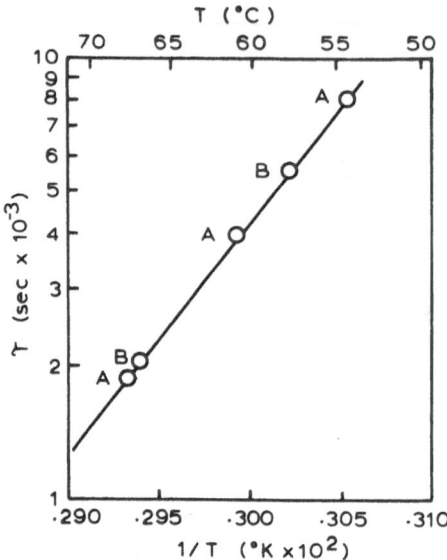

Figure 18-Ln of the time constant τ vs 1/T for the three sample
runs of Figs. 16 and 17 (A) plus two runs from a different specimen
(B). From this an activation energy, Q, of 24,000 cal/mol and a
time constant, τ_o, of 8.3×10^{-13} seconds were determined for the
atom "reshuffling" process proposed.

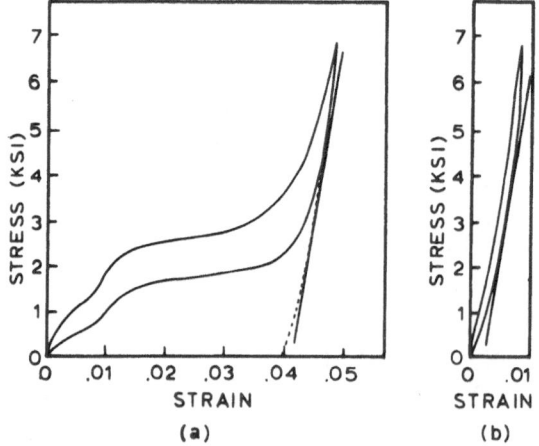

Fig. 19-Restabilization of a compressed crystal. (a)-A compression
only stress-strain cycle. The zero of ε is the remanence offset in
Fig. 8. Superimposed (dashed line) is the recovery curve for the
sample after restabilization at the maximum strain indicated. (b)-
Compression stress-strain cycle after the restabilization in (a)
(described in the text). The initial length is (1-.04) times the
initial length in (a). The high modulus regions indicated by
straight lines are essentially the same and equal to $\sim 8 \times 10^5$ psi.
Testing temperature was $\sim 60^o$C.

an experimentally quite different behavior in single crystal AuCd
with an activation energy of 24,000 cal/mole; this relaxation process and the one reported here may be related.

τ_o is approximately the reciprocal of the Debye frequency,
and therefore τ can be interpreted as the time per jump. Since
the intensity change is completely over in a time of 2 or 3 time
constants, it can be assumed that atoms make very few jumps before
they are permanently trapped. This is consistent with the model
of atoms in the "wrong" places jumping to the nearby "correct"
positions and the activation energy is also reasonable for an
energy of motion; no energy of vacancy formation is necessary in
this model since the correct spaces are already there in the structure. The experimental data, however, are not consistent with
the collapsing of dislocation loops proposed for the restoring
force-relaxation process by Birnbaum and Read (17) unless it is
assumed that after a few jumps, all the loops snap to zero size.
Nor do they support Wasilewski's suggestion (36) that the displacement of the domain boundaries is due to a double stress-
assisted transformation of one twin related region from its martensite orientation to (possibly) the same martensite orientation
as the undisturbed region by way of stress-induced transitional
parent cubic β structure in a temperature region $A_d < T < M_f$, where A_d
is the minimum temperature at which the stress-induced phase can be
formed. Such parameters as the extension due to twin boundary
motion (from load-extension curves), slope during boundary motion,
and stress necessary to move the boundaries, were observed to vary
smoothly and continuously with no indication of any discontinuous
or anomalous behavior from room temperature down to liquid nitrogen
temperatures. A_d would have to be more than 200°C below M_f in this
alloy for this proposal to apply; more   complete arguments against
these and other proposals will be found elsewhere (27).

Stress Relaxation. Since the X-ray diffraction data were from
specimens held at constant strain in a "hard" machine, it was important to determine if there were any stress relief. A crystal
treated and strained to a single lattice twin exactly as the X-ray
specimens, was maintained at constant temperature in an MTS electro-
hydraulic closed-loop test system set to control the strain and
monitor any change in stress required to keep the strain constant
and thus keep the second twin from reappearing. No stress relaxation was observed as a specimen was held in this strained state
at 60°C. for over 14,000 seconds. When the compressive load was
removed, the specimen recovered only slightly. Fig. 19a shows a
compression only hysteresis stress-strain curve, compression to the
maximum strained state, and--as a dashed line--the recovery curve
following complete restabilization in that strained state. Further
evidence that the crystal has completely restabilized in this time
at this temperature is contained in Figs. 16 & 17. The applied

stress necessary to keep the lattice twin boundaries in their
strained positions, $=\sigma_r$, can only be measured by unloading the specimen
after various holding times ($<$ the time for complete stabilization at that temperature--Figs. 16-18) and noting where the stress-
strain curve on such unloading departs from the dashed curve in
Fig. 19a from a completely restabilized specimen. This stress
should monotonically decrease from its maximum value to zero in
the same way with time and temperature as the excess X-ray intensities; both indicate the atomic reshuffling in response to the
volume force and decreasing crystal energy as, in this case, the
single lattice become a single crystal with all atoms correctly
in the Ölander structure. See Figs. 6b & 11.

The completely restabilized single crystal, Fig. 19a, can no
longer exhibit ferroelastic twin boundary motion in a compression-
release cycle but only the elastic response, shown in Fig. 19b.
Hence the strain recovered in Fig. 19a when the load is removed is
elastic strain with no twin boundary motion component. It can be
seen that the slope of the recovery curve (modulus) is essentially
the same as that which occurs when the stress is first released
in the earlier cycle and as that in the subsequent restabilized
cycles. Thus not only did the experiment demonstrate that (i), the
volume restoring force (and therefore the applied force to prevent
twin boundary return) "relaxed" to zero, but equally important,
that (ii), the stress necessary to keep the Ölander structure at
constant elastic strain did not relax irrespective of twin boundary
motion, atom reshuffling, etc., and hence the X-ray diffractometer
results above require no reconsideration on that score. Further
critical information for developing the model of ferroelasticity
will now be sought in an examination of dynamic mechanical experiments rather than in additional static observations.

V. FERROELASTIC ⇄ FERRODISPLACIVE* BEHAVIOR

A. The Effects of Cyclic Stress

β' AuCd specimens can deform by five mechanisms (37): (i) deformation by transformation twin boundary motion discussed in detail in this paper; (ii) nucleation of mechanical twins (38) in
crystals so oriented that neither of the transformation twins can
relieve the applied stress by growth; (iii) nucleation and growth
of twins in single crystals (e.g., the single crystal produced by
restabilization in compression discussed in IVB can deform by this
mechanism under tensile stress); (iv) movement of the regional
boundaries in Fig. 5a separating sets of crystallographically
equivalent twins; and (v) conventional plastic deformation. These
have all been studied and analysed in considerable detail (37); only
(i) is of immediate interest here.

*See note added in proof, p. 242.

Cyclic tension-compression (fatigue) testing was performed on the $\pm$20 Kip MTS closed loop machine mentioned earlier. Most specimens were tested in load control primarily because of the abrupt transition to the high modulus region terminating deformation by twin boundary motion(Figs. 9a, 20a) which could result in highly variable peak loads and a certain amount of non-twin related (plastic) deformation if stroke (ram displacement) control were used. Since the primary purpose of this research was to study deformation by twin boundary motion and not by classical plastic mechanisms, the load limits were generally set far below that for measurable plastic deformation but sufficiently high to observe the end of twin boundary motion. Testing frequencies were generally 6 to 7 Hz but were occasionally as high as 20 Hz and as low as 2×10^{-4} Hz. A simple temperature control system was devised to allow testing between -126°C and +65°C on the MTS machine. A crystal containing a single set of transformation twins was predicted to deform by twin boundary motion in both tension and compression by an algorithm developed for this purpose (37). The tension-to-compression stress-strain hysteresis loop (henceforth called the hysteresis loop or "loop") is shown in Fig. 20a. As expected from the discussion above, when the tensile stress was increased from zero, a narrow, high modulus region was first observed followed by a rather abrupt transition or threshold to a low modulus region as twin boundaries moved and one twin grew at the expense of the other. Because of its shape, this portion of the stress-strain loop will be referred to as a "stress step." When twin boundary motion ceased (i.e., when the specimen was completely converted to one <u>lattice twin</u> orientation), another high modulus region was observed. There was a third high modulus region on unloading followed by another low modulus regions as the twin orientation that previously grew under stress contracted and the twin boundaries returned to their original positions in response to the restoring force discussed above. The compressive portion of the loop showed similar behavior to that in tension.

The loop showed less than 0.001 strain hysteresis at zero stress and remained essentially unchanged for the first 10,000 cycles. Over the next 20,000 cycles, the loop gradually changed to that shown in Fig. 20b and then remained essentially unchanged as long as cycling was continued. The stress step at zero strain (fig. 20a) disappeared but the two high modulus segments at the strain limits after the specimen was converted to a single orientation (at the start of unloading) were still observed and were essentially unchanged; there was also a substantial strain remnant. Loops similar in shape to that in Fig. 20b showing no stress steps would be called "stable" loops in conventional fatigue terminology but to prevent confusion with the volume "stabilization process" and "well-stabilized specimens", they will be called "terminal loops" here since there was no further change in their

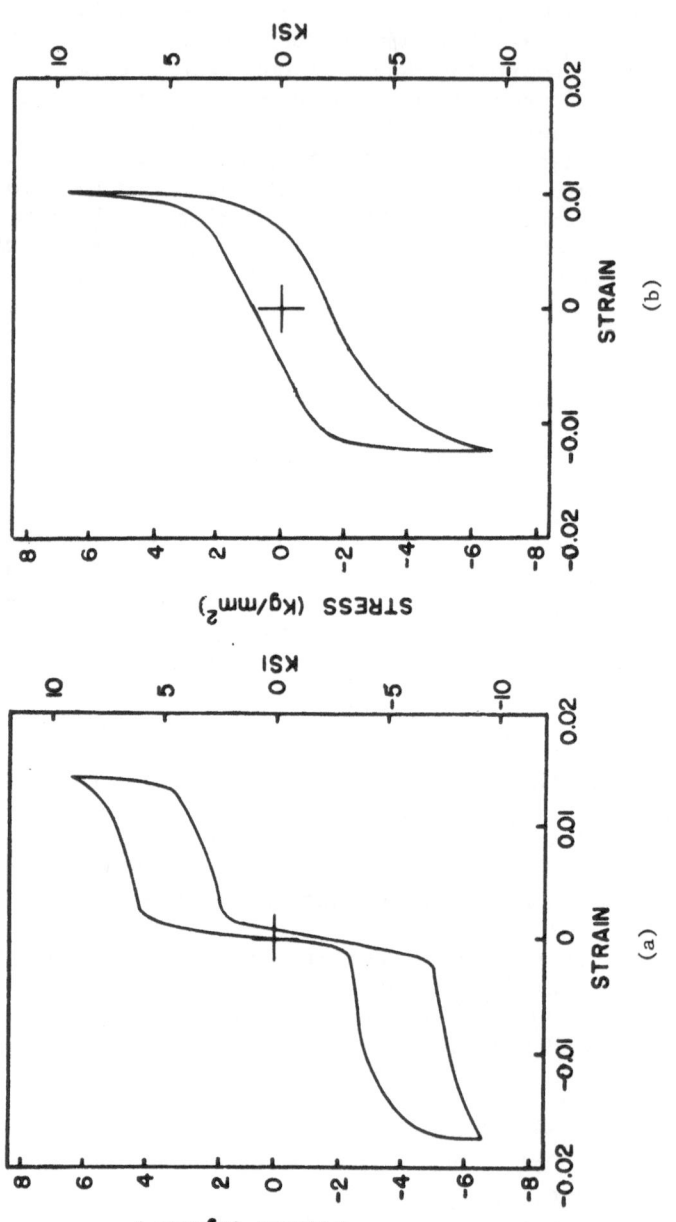

Fig. 20–Hysteresis loop for a specimen transformed by a single interface, as in Fig. 5, to a single set of twins and which deforms by the motion of the twin boundaries (37). (a)–After 25 cycles, the specimen still exhibits abrupt "stress steps" near zero strain in both tension and compression, as in the first cycle. See text. (b)–After 100,000 cycles, the "stress steps" have disappeared, but the stress hysteresis width is essentially unchanged; there is no further change in the loop shape with cycling–even if the test is continued until failure.

shape with continued stress cycling. The testing temperature and
cycling frequency were found to have significant effects on the
shape of the "terminal" hysteresis loops, especially at frequencies
below 10^{-1} Hz. Loop shapes for three frequencies traced at four
temperatures are shown in Fig. 21; the loop hysteresis generally
decreased with increasing frequency and decreasing temperature.

The formation of a stress step whenever cycling was stopped
for more than a few seconds was a common feature of all "terminal"
loops of specimens deforming by twin boundary motion. Fig. 22a
shows the effect on the shape of the "terminal" loop of holding
at zero load at room temperature. One side of the loop becomes
displaced vertically relative to the other with increased holding
time, as can be seen. The "step height" is defined to be the
vertical displacement of the center C (in the "terminal" loop
at left) in the two displaced portions of the loop, i.e. between
the two stress midpoints C' and C" of the two half loops in
Fig. 22a. As will be developed below, this is twice the volume
restoring force, σ_r, and is plotted versus holding time at
several temperatures in Fig. 22b. Steps may be created anywhere
in the low modulus regions of the loop (i.e., not only at zero
stress), and several steps may be introduced into one loop. When
cycling is resumed, the steps shrink and disappear, as shown in
Fig. 22c for the same specimen as in Fig. 22a; the rates are
also dependent on frequency and temperature.

Cyclic stresses clearly have a profound effect on the stress-
strain relationships in AuCd crystals as can be seen from
Figs. 20-22. The volume restoring force $\sigma_r = \frac{1}{2}$ "stress step"
disappears and the strain recovered at zero stress essen-
tially vanishes also. However, deformation is still primarily
by the reversible motion of twin boundaries (low slope region)
with a small amount of elastic strain at the extremities (high
slope region). It has already been suggested in III that ferrodisplacive would be appropriate when (i), deformation is by the
reversible motion of twin boundaries and ferroelastic when (i)
obtains and (ii), there is a restoring force which returns the twins
to their initial configuration in a rubber-like manner when the
stress is removed. Evidence has been presented here that on stopping the cyclic stress, the "stress step" grows and approaches its
original magnitude at a temperature and time dependent rate and
the ferroelastic restoring force returns. Thus in AuCd, the stable
equilibrium state would seem to be ferroelastic and therefore that
term would seem to be proper for the complete phenomenon and behavior in AuCd, and perhaps InTl. However, those "shape-memory"
alloys which deform in the low temperature martensite phase by the
reversible motion of twin boundaries but do not exhibit the rubber-
like "piezomorphic memory" could well be described as ferrodisplacive.
In any event, the model of ferroelasticity partially developed in

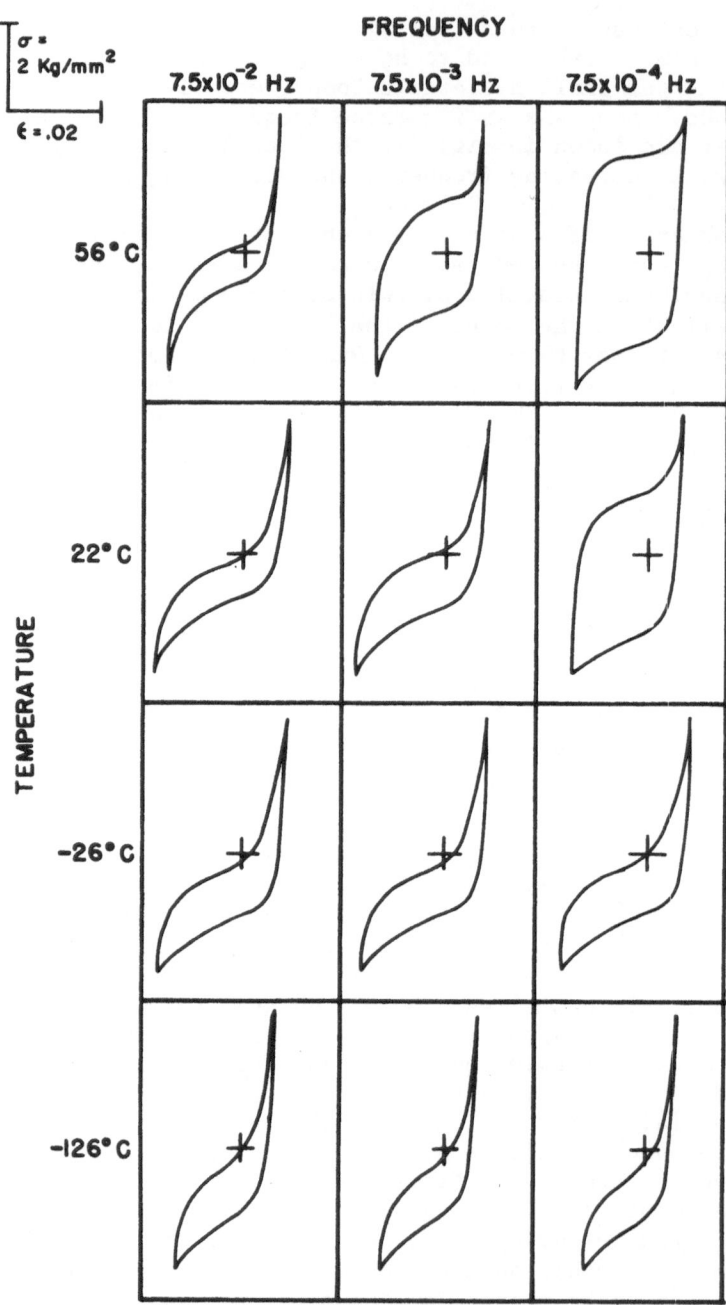

Fig. 21–The effect of cycling frequency and temperature on the "terminal" loop shape; frequency decreases to the right and temperature decreases toward the bottom of the page.

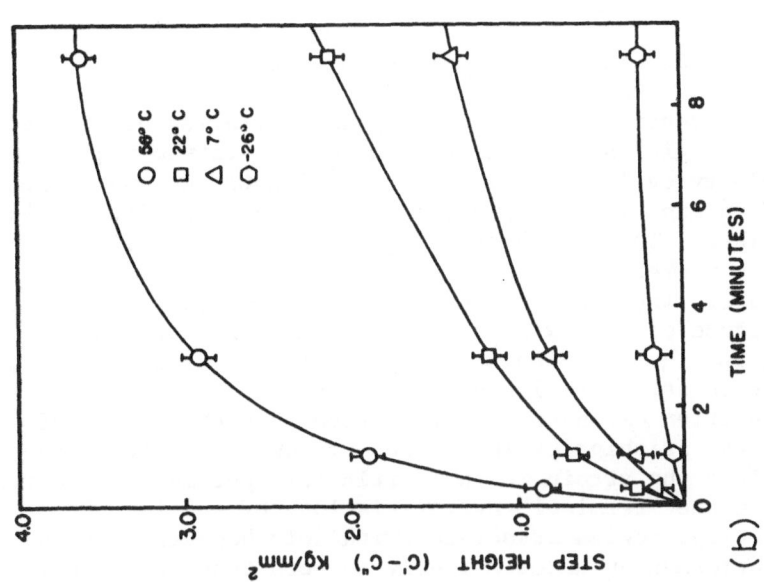

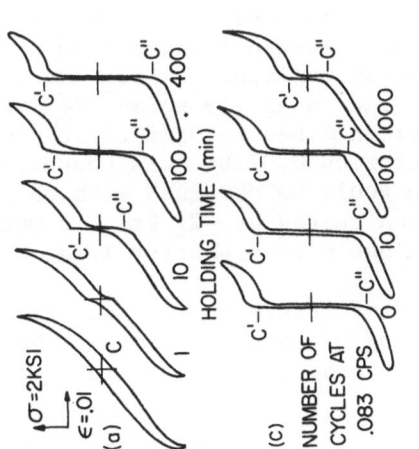

Fig. 22 (a)–The growth of a "stress step", C–C', with time in a terminal loop at no load and at room temperature. (b)–The effect of temperature on the growth rate of the "stress step". (c)–The shrinkage of the "stress step" with resumed cycling and disappearance as the "terminal" loop is again exhibited.

IV, may   have to be modified further to account for the effects
of cyclic stress (including variations in frequency, temperature,
interruptions, etc.) on the stress-strain behavior; this will be
examined in VI.

B.   Fatigue Properties of AuCd

Tension-compression fatigue testing was performed on several
specimens with single sets of twins, some with multiple regions
Fig.5a and polycrystalline specimens. The β' AuCd fatigue behavior
is summarized in Fig. 23 as log total strain amplitude ($\Delta \epsilon$) vs log
cycles to failure. The very broad range of fatigue behavior illustrates how extremely orientation dependent such phenomena are
in AuCd crystals. Thus specimen A was tested 1,005,000 cycles at
a strain amplitude of approximately 0.07 without failure while
specimen B failed in 32 cycles at a strain of approximately 0.05.
The difference in strain amplitude for these two specimens was
primarily a function of specimen orientation, but the spread of
fatigue lives was principally influenced by the stress level required to achieve the strain in the particular specimen as observed,
for example, in the "terminal" hysteresis loop of specimen C
where the low slope region associated with twin boundary motion
plus elastic deformation ended abruptly in both tension and compression. Hence raising the load limits (in load control) would
greatly raise the stresses without significantly raising the strain.
Likewise raising the strain limits (in strain control) would also
cause greatly increased stresses. Since no more deformation by
the motion of twin boundaries is possible in either mode of control, these increases can only raise the plastic component of the
deformation and thus shorten fatigue life without significantly
increasing the strain amplitude. Specimen B failed in 32 cycles
because it was tested under $\pm$19 kg/mm^2 ($\pm$27 ksi) of longitudinal
stress while specimen A had not failed at over 1,005,000 cycles,
not because it deformed 0.07 by twin boundary motion, but because
this large strain could be obtained with only $\pm$2.8 kg/mm^2 ($\pm$4 ksi).
Had specimen A been tested at $\pm$27 ksi, it would have shown negligible
additional strain but a much shorter life.

The influence of stress on fatigue life is shown in Fig. 24 ,
as log stress vs log cycles to failure. The data appears to fit
the linear relation proposed by Basquin (39) which corresponds to
the equation

$$\sigma_a = \sigma_f^i (2N_f)^b$$

where σ_a is the half life stress limit, σ_f^i is the fatigue strength
coefficient, $2N_f$ is the number of reversals to failure and b is the
fatigue strength exponent (40). The best fit line through the

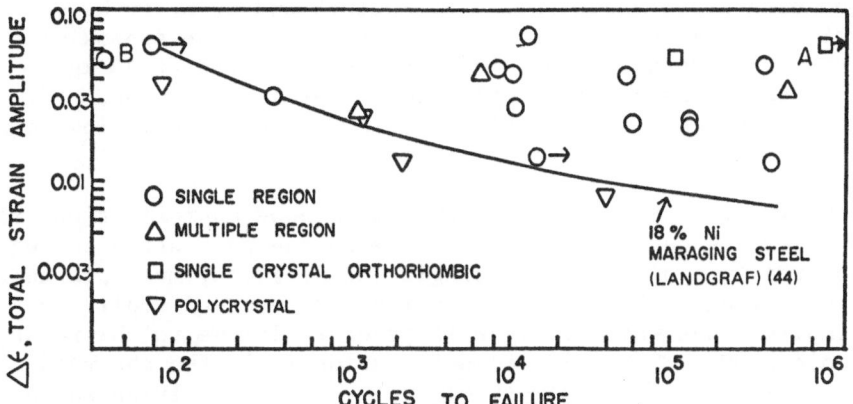

Fig. 23–Tension-compression fatigue data plotted as log total
strain amplitude (Δε) vs log cycles to failure. Because of the
extreme orientation dependence of the fatigue life (discussed
in the text), no line through the data would have any meaning
even if it could be drawn. Data from an 18% Ni maraging steel
facilitate comparison with a strain fatigue resistant steel.
The values plotted are half-life strains; an arrow indicates
no failure. Lettered crystals, etc., are discussed in the text.

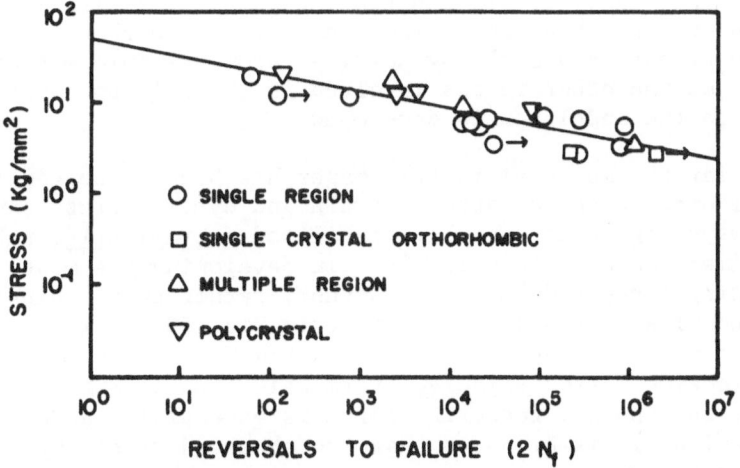

Fig. 24–Tension-compression fatigue data of stress vs number
of reversals to failure on a log log scale. AuCd follows
Basquin's stress relation (39) regardless of orientation. The
stresses given are half life values; an arrow = no failure.

fatigue data gives values for b of -0.1875 and σ_f^i of 49 kg/mm^2 (69 ksi). It is interesting to note that σ_f^i may be approximated by the true fracture strength which has been measured at 44 kg/mm^2 (63 ksi) and 38 kg/mm^2 (54 ksi) on two tensile specimens. Fig. 24 indicates that the stress fatigue resistance of AuCd is low compared with most metals.

The fatigue data indicates that to a good approximation, fatigue lives are determined principally by the peak stresses regardless of the type of specimen (single set of twins, multiple twin sets, or polycrystalline), orientation, or observed amplitude. This implies that while fatigue is essentially due to the relatively isotropic {100} <100> plastic deformation (34), the observed strain amplitude is due almost entirely to the highly anisotropic deformation by twin boundary motion which apparently effects the actual fatigue life very little. Thus the long fatigue lives at high strain amplitudes are possible because the large cyclic strains can be obtained at stress levels far below that which cause measurable fatigue damage by conventional means (even) in this material

VI. MODEL OF THE "PIEZOMORPHIC MEMORY" AND FERROELASTIC ⇄ FERRODISPLACIVE* BEHAVIOR

The results of the dynamic experiments in V-B reinforce the conclusion of the "static" experiments in IV-B concerning the time-temperature restabilization process and disappearnce and reappearance of the volume restoring force and "ferroelastic memory" or rubber-like behavior. In addition, careful consideration of the transition from the hysteresis loop in Fig. 20a, to that in Fig. 20b (by cyclic stress) and back (by holding) provides information about the nature of the other forces involved in the twin boundary motion and leads to the model herein developed.

Thus far the material in this paper has been presented in historical order with the stress-strain and cyclic stress experiments always beginning on fully stabilized ferroelastic, rubber-like specimens, such as in Fig. 20a. In developing the model of the restoring force and "memory" further, rather than beginning with the complex ferroelastic behavior in Fig. 20a it is easier-- and perhaps more instructive--to begin with the simpler and more familiar looking ferrodisplacive "terminal" loop of Figs. 20b & 22a since here the unusual restoring force is essentially absent. For this situation, it is proposed that the observed stress hysteresis is due to a Peierls-Nabarro stress σ_{pn} for twin boundary motion. This is based on Sumino's recent theory (41) of twin boundary motion in which the twin plane itself is considered to be a two dimensional dislocation that moves by the expansion of pill box

*See note added in proof, page 242.

shaped kinks. The Burgers vector of the twinning dislocation is the twinning shear vector, and Sumino pointed out that since no bonds are broken in the propagation of twinning dislocations, their energy of motion would be lower than that for slip dislocations. He applied Peierls-Nabarro calculations to his model and reported the calculated stress for twin boundary motion to be in reasonable agreement with that measured in several materials including β' AuCd (42, 43). A hypothetical hysteresis loop with the Peierls-Nabarro plus elastic stresses acting is shown schematically in Fig. 25a; its slope is discussed and explained below.

The near absence of the volume restoring force in specimens after many cycles can be explained by the continuous movement of a twin boundary back and forth past a given volume converting it from one twin orientation to the other. While the jumping process of shuffled "wrong" atoms to their proper positions is undoubtedly occurring, it does not contribute to the restoring force because the volume cannot be associated with a particular twin. This is especially true in the center of the twin boundary travel, as shown in Fig. 26 since a volume there spends half the time in each twin. It is less true near the limits of twin boundary motion where a volume is in one twin over most of the cycle thus allowing more stabilization (or properly positioned atoms) with respect to that twin than the other, and hence a restoring force. Therefore the "terminal" loop shape is seen to be determined by the Pierls-Nabarro + elastic stresses, with increasing restoring force components as the strain limits (and thus limits of twin boundary motion) are approached. Fig. 25b combines these with Fig. 25a; compare with experimental "terminal" loop of Fig. 22a repeated in 25c.

When cycling is interrupted and the twin boundaries are held stationary, every volume can henceforth be associated with a particular twin orientation, and the "wrong" atoms can reshuffle to their correct positions unambiguously. Thus there is a growing discontinuity in stabilization at the twin boundary positions with volumes on each side of the boundaries being stabilized as differentiable twin-related Ölander crystals. Hence displacement of the boundaries into either twin causes an appropriately directed restoring force, and thus a stress step--but no yield point-- because the restoring force is a volume force essentially independent of displacement. A stress step can be put anywhere in a stable loop simply by stopping twin boundary motion (by stopping the machine) and thereby removing the ambiguity in stabilization caused by twin boundary motion. Ultimately the volumes in each twin will be completely stabilized, and, except for the large discontinuity at the holding point, the restoring force will remain constant as idealized in Fig. 25d. From Fig. 25h for example, it can be seen that in a well stabilized rubber-like specimen, the Ölander structure is elastically strained to ϵ_c by the applied stress

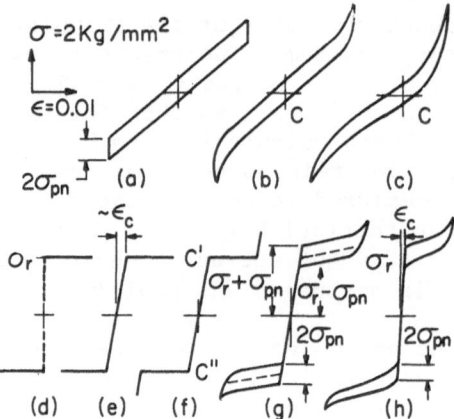

Figure 25–Synthesis of (above) a <u>ferrodisplacive</u> "terminal" loop
in a specimen following many cycles of stress, and (below), a <u>ferro-
elastic</u> loop typical of a stabilized rubber-like specimen. Note
comparison is with the same crystal and its behavior as in Fig. 22
a&c, and to the same scale. Terms defined and steps described in
text.

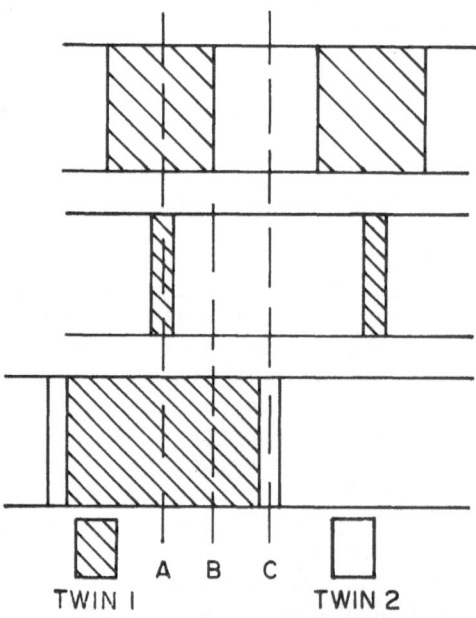

Fig.26-Schematic of twin boundary motion showing twin boundary configurations at <u>center</u>-midpoint of travel, <u>above</u>-one extreme, and
<u>below</u>-other extreme. Over most of the cycle position A is in twin
1 and C is in twin 2; B is equally shared by both twins. See Text.

as it is increased to the critical restoring force, σ_r, before the stabilized <u>crystal</u> <u>twin</u> boundaries can be moved into the other, twin related, Olander crystal structure and produce a <u>lattice</u> <u>twin</u> with the "wrong" structure--and must be kept at this elastic strain to prevent the twin boundaries from returning, as suggested by the stress relaxation experiments discussed in IV-B. This initial high modulus elastic response is shown schematically in Fig. 25e and combined with the restoring force (of Fig. 25d) in Fig. 25f. To it have also been added the elastic responses, with substantially the same modulus, at the extremeties of the twin boundary motion when only one twin variant (single crystal) remains. The Peierls-Nabarro stress is (still) the primary source of stress hysteresis associated with the twin boundary motion according to Sumino's theory (42) invoked above. When it is added (from Fig. 25b) to Fig. 25f, Fig. 25g results--the ideal stress-strain curve for a fully stabilized, "rubber-like", crystal--to be compared with that of the real experimental specimen in Fig. 25h. Note that in order to deform such a crystal by twin boundary motion, the applied stress must be increased to $\sigma_r + \sigma_{pn}$ to overcome the frictional force, while on releasing the load, the applied stress has to be reduced to $\sigma_r - \sigma_{pn}$ for the same reason before the twin boundaries can be restored. The "terminal" loop of Fig. 25b & c has thus been converted to that of a well stabilized rubber-like crystal, such as at the beginning of the cyclic testing, Fig. 25g & h, primarily by the time-temperature addition of the restoring force, σ_r, to the Peierls-Nabarro force, $\pm \sigma_{pn}$, as in Fig. 22a.

When cycling is resumed, the ambiguity of twin orientation is reintroduced and the "terminal" loop is recreated as the restoring force stress step is reduced (Fig.22c). Some discrepancies between the $\sigma - \epsilon$ behavior of the model thus far developed and that observed as stress cycling is continued can be understood, at least qualitatively, in terms of Sumino's model (41). Since the pill box shaped kinks by which the twin boundaries move always propagate outwards towards the specimen's surface, any fatigue generated lattice defects (indicated by the marked degeneration of Laue spots in X-ray photographs of tested specimens) would be swept towards the surface. Twin lamella are thousands of times longer ($\sim$mm.) than they are wide ($\sim \mu$) and thus it is likely that most defects would diffuse out of the paths of the kinks and into areas not swept by twin boundary motion long before they reach the surface. As these defects accumulate with cyclic stress, they would reduce the amplitude of the twin boundary motion and thus account for the reduction in strain amplitude observed, for example, in going from Fig. 20a to Fig. 20b. They could also be contributing to the increased slope--hardening--in the low effective modulus region associated with the twin boundary motion as in Figs. 20, 22 & 25.

VII. DISCUSSION AND CONCLUSIONS

The major features of the "memory" and mechanical behavior of β' AuCd described here can now be accounted for by the effects of (i), the proposed time-temperature dependent volume restoring force (responsible for the rubber-like behavior) and (ii), the Peierls-Nabarro stress (producing the stress hysteresis) on the motion of twin boundaries: the shape of the stress-strain loops for well stabilized rubber-like specimens, the absence of stress relaxation in strained specimens, the transition to the "terminal" loop as stress cycling proceeds, and the development and disappearance of a "stress step" when cycling is suspended and then resumed. The temperature and frequency dependence of the loop shape (Fig. 21) can also be understood in terms of continuing stabilization during cycling: if the frequency is sufficiently low and the temperature sufficiently high, a given volume will have enough time to begin to stablize in one twin before the passage of the twin boundary. Thus this passage will be resisted by the volume restoring force in addition to the Peierls-Nabarro + elastic forces always present, and the stress hysteresis will alter in accordance with the model and as observed. Some effects and details require for their understanding an additional stress such as that generated by the accumulation of defects produced as delineated above and which would be expected to increase with increasing strain and increasing cycling: the reduction in strain amplitude with cycling (Figs. 20 a & b), the increase in slope--hardening--in the region where deformation is by twin boundary motion (Figs. 20 a & b, 22 c, 25) and its decrease with holding time (Figs. 22 a, 25) as the debris anneals out, and the details of the effects of frequency and temperature (Fig. 21).

The sensitivity of the mechanical behavior to crystal orientation relative to the stress axis, specifically mentioned in discussing the fatigue data in Fig. 23, at least partially account for the broad range (evident in Figs. 8, 9, 15, 19, 20, 21, 22, and 25) in the initial elastic moduls; σ_r, the stress to initiate twin boundary motion in a well stabilized specimen; the effective modulus for twin boundary motion both in the ferroelastic and ferrodisplacive states; the rate of change with cycling and/or holding time in the later; the maximum strain amplitude before the rapid increase in stress, usually signifying single crystal elastic response, etc. Note that the area within the hysteresis loop representing the energy lost per cycle remains essentially constant for a given crystal whether operating in the ferroelastic or ferrodisplacive mode- witness Fig. 9 b & c and Fig. 20 a & b. Fatigue lives were shown to follow Basquin's stress relation (39) regardless of orientation, type of specimen (single region, multiple region or polycrystal) or principle deformation mode (transformation twin boundary motion, mechanical twinning, or plastic deformation). This strongly suggests that relatively isotropic plastic deformation (as compared with twin boundary motion) was mainly responsible for fatigue. Failure showed brittle characteristics typical of ordered alloys (45).

The premonitory pretransition phenomena (terms first introduced by Ubbelohde two decades ago (46)) are discussed by Varanayan and Hehemann in their article on premartensitic behavior in β phase alloys. (47) To this rather comprehensive treatment of AuCd will only be added a comment especially relevant to the "shuffles" and volume restoring force model developed here. Zener (48, 49) first proposed (three decades ago) that the bcc and CsCl phases in metal systems decrease in relative stability as the $(110)_c [1\bar{1}0]_c$ shear modulus decreases with decreasing temperature and that this could result in a martensitic shear transformation to a more closely packed structure, usually of lower symmetry. $Au_{1.05}Cd_{.95}$ does indeed exhibit a softening of this $1/2 (C_{11}-C_{12})$ elastic constant as the temperature decreases toward M_s, as is seen in Figure 27. However, what is more interesting is the observation that the plane and direction in both the twin system and the "shuffle" system in the product Ölander structure came from this $(110)_c [1\bar{1}0]_c$

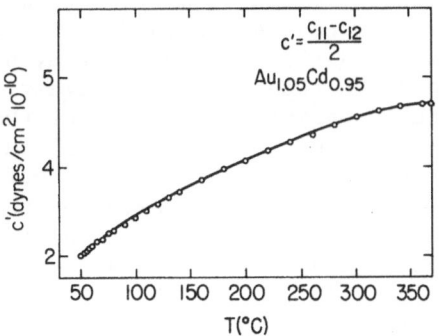

Fig. 27-"Softening" of the $(110)_c [1\bar{1}0]_c$ shear constant as the transformation temperature is approached indicating an instability. Both the twin and "shuffle" systems in the Ölander produce phase come from this system in the parent. (After ref. 14).

shear system in the parent cubic phase, as can be seen from (16) for the former and Fig. 10 for the latter. Both of these systems have previously been shown to be salient components of the total transformation distortion in AuCd, the twin system as the lattice invariant shear (15) and the "shuffle" producing the hcp-like B19 Ölander structure (16). And here it has been concluded that these particcular two systems are necessary conditions for the existence of the extraordinary ferrodisplacive $\rightleftarrows$ ferroelastic behavior in this alloy; perhaps it is not surprising that they are both so intimately related to that system in the parent which is already so 'active' well above M_s, forewarning of the impending transformation and preparing for the key roles it will play when that temperature is reached.

There are many other important points which should be made about the "memory phenomena, phase changes, and related items discussed in this volume. Unfortunately, space limitations permit only a few comments and questions here; more complete discussions will be developed elsewhere (27, 37, 50). It is urged that more care and responsibility be taken-by authors and referees- to prevent statements typified by the following from appearing in the literature: "The terms superelasticity (51), stress-induced pseudoelasticity (5), ferroelasticity (16), and rubber-like behavior (17) have been used to describe the unusually high restorable strains, which are obtained in certain alloys undergoing a stress induced transformation" (52); the reference numbers have been changed to conform to this paper and the underlining has been added. While (17) does refer to the rubber-like behavior in AuCd as "pseudo-elastic"- a decade before the thermoelastic application (5)-both (16) and (17) make it very clear that ferroelastic behavior is (and must always be) observed below M_f; at no time is the specimen heated into the parent CsCl phase*. As for the first two terms, since all they do is label an aspect of thermoelastic martensite as part of a thermoelastic phase change generally accepted as included in its operational definition as long ago as 1953 (53, 54), could they both not be replaced by the unequivocal "martensitic thermoelasticity"+? Finally, although Aizu was probably unaware of the use of the word ferroelasticity for the rubber-like behavior when he (re) introduced it in 1969 (55), it is clear from his definition, his defining idealized Fig. 1, etc. that he is describing what here has been called ferrodisplacive deformation by the reversible motion of twin boundaries, as in Fig. 20b, and not the rubber-like behavior due to the volume restoring force, Fig. 20a. It can also be seen that Fig. 2 of (56), for example, is very similar to Fig. 20b and not at all like Fig. 20a- but it is labelled "ferroelastic" (56). To avoid further confusion, it is strongly urged that the response to stress by the reversible displacement of coherent twin boundaries be termed ferrodisplacive to emphasize the analogy so strongly suggested by all

---

*Indeed it would be possible (although not very fruitful) to study the phenomenon without knowing that a phase transformation existed.
+A term independently and persuasively argued for by Perkins et al (62)

the quantities in Fig. 20b, while _ferroelastic_ be reserved for those
few (so far) cases in which in addition involve a volume restoring
force. That AuCd can be "educated" to behave in the _ferrodisplacive_
mode as described herein is interesting since it may furnish some
insight into the "thermomorphic shape memory" (also exhibited by
AuCd) because those systems discussed in this volume which display
this unusual behavior also deform in the low temperature phase by
the _ferrodisplacive_ growth of coherent twins - or, if not, by an
equivalently conservative process involving stacking faults - but
never by the disapative motion of slip dislocations producing plastic deformation.

VIII.   ACKNOWLEDGEMENTS

The authors acknowledge partial support of this work by the
US AEC through Contract AT (11-1)-1198 and by NASA through Grant
No. 14-005-162. The generous aid of the staff of the Lawrence
Berkeley Laboratory, University of California, during the final
preparation of the manuscript is appreciated. It is a pleasure
to recall many helpful and stimulating discussions with colleagues;
in particular, Dr. J. Christian should be mentioned for first seeing
that in _ferroelastic_ rubber-like behavior, the restoring force on
the planar twin boundaries must be a volume force. This paper is
primarily based on theses submitted by M. A. Schmerling and R. W.
Karz to the University of Illinois at Champaign-Urbana in partial
fulfillment of the requirements of the degree of Doctor of Philosophy;
sections III and IV are essentially the work of M.A.S. and V and VI
of R.W.K.

APPENDIX

On the Latent Heat of Thermoelastic Martensites

The AuCd (21) and InTl (7) systems were the first solid state-
solid state transitions for which the latent heat of transformation
was determined by application of the Clausius Clapeyron equation to
a mechanical measurement, as also discussed a few years later (57).
Following (7), it can be seen that these transformations are first
order since with L the specimen length and F the force on it, the
derivative of the free energy G can be written as $L = (G/F)_T$ as the
uniaxial force analogue of $V = (G/P)_T$. From Fig. 2 it can be seen
that transforming the specimen under uniaxial load increases the
transformation temperature both on heating and cooling; these
temperatures, T_H and T_C, are defined in the caption. In the case
of AuCd (and InTl (7)), these transformation temperatures vary
linearly with stress and further, in the same (parallel) way, as
can be seen. Hence the equilibrium temperature, T_o, which cannot
be observed because of the superheating and undercooling in such a
hysteretic transformation, can safely be assumed to lie somewhere
between T_H and T_C and to vary with stress as they do; Rodriguez

and Brown (58) have questioned the approximate nature of this
general treatment, but at least for these alloys, it seems to be
reasonable. Thus the uniaxial analogue of the Clausius Clapeyron
equation dT/dP=TΔV/ΔH becomes, with a little manipulation,

$$\lambda = \frac{\Delta H}{V\rho} = \frac{T_o}{V\rho} \frac{d\sigma}{dT} \varepsilon$$

where λ is the latent heat in cal/gm, σ is the uniaxial stress,
ε is the strain on transformation, and ρ is the density of the
alloy.

For the experiment of Fig. 2, the following were obtained:
ρ, the density of $Au_{1.05} Cd_{.95}$ =14.11 gms/cm^3; T_o=3.4x10^2 ^{o}K;
ΔT=10° C, $\Delta\sigma$=3x10^3 gms/mm^2, and ε =0.025. The specific heat was
then calculated to be λ= 0.485 cal/gm. The measured values using
calorimetry were in surprising agreement with each other
(59, 60)$\overset{\sim}{=}$ .5 cal/gm; the close agreement with that calculated may
be fortuitous. However for InTl, the agreement is also well within
experimental error: the calculated value from plots such as Fig. 2-
but carried out in tension and compression- was 0.36 cal/gm. at. (7)
and the calorimetric value reported 0.49±0.03 cal/gm.at. for the
fcc $\overset{\rightarrow}{\leftarrow}$ fct transition in an 80In-20Tl alloy (61); again the agreement
is favorable. Thus for these two alloys, both of which exhibit
ferroelasticity and martensitic thermoelasticity, the extension of
the Clausius-Clapeyron equation to uniaxial stresses seems to work,
in spite of some theoretical reservations (50).

Note added in proof: To differentiate between deformation by the reversible motion of twin boundaries (as generally exhibited by "memory"
metals in the martensite) and 'ordinary' plastic deformation by the
irreversible motion of slip dislocations, the term ferrodisplacive was
introduced for the former. However ferroplastic deformation or ferro-
plasticity describes the phenomenon and parallels ferroelasticity
better. Ferro- (still) conveys the close analogy to ferromagnetism
of Fig. 9c and -plastic that the deformation remains after the stress
producing it is removed. An operational definition is offered here:

Ferroplasticity- Deformation by the reversible motion of boundaries
between transformation twins, and/or mechanical twins induced by
stress, (and possibly stacking faults) in thermoelastic martensite
with ε-σ behavior completely analogous to B-H in ferromagnetism. The
deformation remains after the stress is removed, is reversible to relatively high strains, and occurs at quite low stress levels with a
very low effective modulus.

In this paper, please read ferroplastic for ferrodisplacive; this
will be discussed further elsewhere (50).

REFERENCES

1. W. J. Buehler, J.V. Gilfrich, and R.C. Wiley, J.A.P. $\underline{34}$, 1475 (1963).

2. K. Otsuka and K Shimizu, Scripta Met. $\underline{4}$, 469 (1970).

3. C.M. Wayman, Scripta Met. $\underline{5}$, 489 (1971).

4. D.V. Wield and E Gillam, Scripta Met. $\underline{6}$,1157 (1972).

5. H. Pops, Met. Trans. $\underline{1}$, 251 (1970).

6   K. Enami and S. Nenno, Met. Trans. $\underline{2}$,1487 (1971).

7. M. W. Burkart and T.A. Read, Trans. AIME $\underline{197}$,1516 (1953).

8. Z.S. Basinski and J.W. Christian, Acta Met. $\underline{2}$,101 (1954).

9. L.C. Chang and T.A. Read, Trans. AIME $\underline{191}$, 47 (1951).

10. R.V. Krishnan and L.C. Brown, Met. Trans. $\underline{4}$, 423 (1973).

11. L.Delaey, R.V. Krishnan, H. Tas, and H. Warlimont, J. Mat. Sci., $\underline{9}$, 1521,1536, 1545 (1974).

12. M.S. Wechsler, Acta Met. $\underline{5}$, 150 (1957).

13. D. Gupta, D. Lazarus, and D.S. Lieberman, Phys. Rev. $\underline{153}$, 863 (1967).

14. S. Zirinsky, Acta Met. $\underline{4}$, 164 (1956).

15. D.S. Lieberman, M.S. Wechsler, and T.A. Read, J. Appl.Phys. $\underline{26}$, 473 (1955).

16. D.S. Lieberman, Phase Transformations, ASM. p. 1, Metals Park, Ohio, (1970)-ASM-AIME Meeting, Detroit, 1968.

17. H.K. Birnbaum and T.A. Read, Trans. AIME $\underline{218}$, 662 (1960).

18. A. Olander, Ztsch. Krist.$\underline{83A}$, 145 (1932).

19. C. Benedicks, Arkiv for Matematik, Astronomi och Fysik $\underline{27A}$,No. 18, (1940-1941).

20. A. Bystrom and K. E. Almin, Acta Chemica Scandinavica $\underline{1}$, 76 (1947).

21. J. Intrater, L.C. Chang, and T.A. Read, Phys. Rev. $\underline{86}$, 598 (1952).

22. G. V. Kurdjumov and L.G. Khandros, Doklady Akab. Kaak, SSSR, $\underline{66}$, 211 (1949).

23. D.S. Lieberman, T.A. Read, and L.C. Chang, Phys. Rev. $\underline{82}$, 340 (1951).

24. L.C. Chang, Acta Cryst. $\underline{4}$, 320 (1951).

25. L.C. Chang, J. Appl. Phys. $\underline{23}$, 725 (1952).

26. D.S. Lieberman, T.A. Read, and M.S. Wechsler, J. Appl. Phys. $\underline{28}$, 532 (1957).

27. M.A. Schmerling and D.S. Lieberman, to be published.

28. D.S. Lieberman, USAEC Report NYO-3968, Oak Ridge Tennessee, August 31, 1954.

29. D.S. Lieberman, Friday Evening Discourse, The Royal Institution, London, May 21, 1971- to be published.

30. R.S. Toth and H. Sato, Acta Met. $\underline{16}$, 413 (1968).

31. J. W. Christian, The Theory of Transformations in Metals and Alloys, Oxford, Pergamon Press (1965).

32. B.A. Bilby and A.G. Crocker, Proc. Royal Soc.$\underline{288}$,240 (1965).

33. W.A. Rachinger and A. H. Cottrell, Acta Met. $\underline{4}$, 109 (1956).

34. H.K. Birnbaum and W. Class, Acta Met. 6, 609 (1958).
35. M. Hillert, Acta Met. 7, 653 (1959).
36. R. J. Wasilewski, Met. Trans. 2, 2973 (1971).
37. R.S. Karz and D.S. Lieberman, to be published.
38. H. K. Birnbaum and T.A. Read, Trans. AIME 218, 381 (1960).
39. O.H. Basquin, Proc. ASTM 10, 625 (1910).
40. C.E. Feltner and M.R. Mitchell, Manual on Low Cycle Fatigue Testing, ASTM Special Technical Publication No. 465,pp. 27-66 (1969).
41. K. Sumino, Acta Met., 14, 1607 (1966).
42. K. Sumino, Phys. Stat. Sol. 33, 327 (1969).
43. T. Aoyagi and K Sumino, Phys. Stat. Sol. 33, 317 (1969).
44. R.W. Landgraf, Cyclic Deformation and Fatigue Behavior of Hardened Steels, T.& A.M. Report No. 320, University of Illinois, Urbana, Ill. p. 73 (1968).
45. Ordered Alloys, B.H. Kear, C.T. Sims, N.S. Stoloff and J.H. Westbrook,eds.,Baton Rouge,Claitor's Pub. Div.,pp 259-331(1970).
46. A.R. Ubbelohde, Br. J. Appl. Phys. 7, 313 (1956).
47. S.Vatanayon and R.F. Hehemann, this volume, page
48. C. Zener, Phys. Rev. 71, 846 (1947)
49. C. Zener, Elasticity and Anelasticity of Metals, University of Chicago Press, Chicago 1948
50. D.S. Lieberman, to be published.
51. W.A. Rachinger, Br. J. Appl. Phys. 9. 250 (1958).
52. H. Tas, L. Delaey, and A. Deruyttere, J. Less Common Metals 28, 141 (1972).
53. R.W. Cahn, Nuovo Cimento 10, 350 (1953).
54. L. Kaufman and M. Cohen, Progr. Metal. Phys.7,165 (1958).
55. K. Aizu, J. Phys. Soc. Japan 27, 387 (1969).
56. N.Nakanishi, Y Murakami, S. Kachi, T. Mori, and S.Miura, Phys. Let. 37A, 61 (1971).
57. D.S. Lieberman,The Mechanism of Phase Transformations in Metals Br. Inst. of Metals Monograph No. 18, 337 (1956).
58. C. Rodriguez and L.C. Brown, this volume, page
59. S. Fishman, R. Karz, and D.S. Lieberman, to be published.
60. N. Nakanishi,T.Mori, S. Miura, Y. Murakami,and S. Kachi, Phil. Mag. 18, 77 (1973).
61. B. Predl, Z. Metallkunde 55, 117 (1964).
62. J. Perkins, G.R. Edwards, C.R. Such, J.M. Johnson, and R.R. Allen, this volume, page 273.
63. D.S. Lieberman, Trans. Japanese Iron and Steel Inst. 11, 1149 (1970).

THE SHAPE MEMORY EFFECT IN TiNi: ONE ASPECT OF STRESS-ASSISTED

MARTENSITIC TRANSFORMATION

R.J. Wasilewski

National Science Foundation

Washington, D.C.   20550

INTRODUCTION

The objective of this paper is to discuss the (irreversible) shape memory effect in TiNi which was first reported by Buehler et al.,some 12 years ago (1). Considerable efforts at developing some applications of this remarkable phenomenon have been made since, and a number of these will be discussed in the subsequent papers at this session. Most of this developmental research had been carried out on an empirical basis, in absence of a fundamental understanding of the phenomenon involved. With hindsight it is clear that this posed a severe handicap both to the engineers involved with this persnickety material, and to the metallurgists or materials science addicts baffled by the apparently irreproducible behavior reported by the engineers. Today it may appear surprising that this effect had been originally ascribed to a stress-assisted compositional change (1,2), and that the first attempt to rationalise it in terms of the martensitic transformation was reported by Zijderveld et al., in 1966 (3). Considering the scanty experimental data then available that hypothesis showed considerable insight.

Today we know that there exists a direct relationship between the shape memory effect and the martensitic transformation, and that this behavior is far more common than once believed. The precise nature of the relationship is yet to be firmly established, but considerable progress has been, and is being, made. One of the factors initially contributing to the slow rate of progress was the tendency to view the shape memory behavior as unique, or as limited to materials undergoing some special kind of transformation. Nagasawa and

Kawachi seem first to have suggested (4) that such behavior should
be expected of all materials undergoing a martensitic transformation.
If this suggestion is correct then the shape memory effect must be
merely one of the aspects of the effects of stress on the course of
a martensitic transformation. The same conclusion was reached by Wasilewski (5) on the basis of extensive observations on TiNi. Most of
the experimental work on TiNi through mid-1970 has been summarized
in an early review (6). The more recent research on the effects of
stress on the martensitic transformation has been directed mainly
at observations on single crystals, particularly of Cu-Al-Ni alloys.
An up to date summary has been reported in a three-part review(7-9).

     In the present paper these effects of stress on the transfor-
mation, both forward and reverse, will be more fully developed,with
particular reference to those conditions under which shape memory
behavior can be induced. Experimental evidence for some of the proposed mechanisms will be presented. I shall concentrate on the basic
mechanisms involved. The discussion of phenomena other than the shape
memory will be limited to the extent necessary to understand why apparently trivial changes in the experimental conditions can cause
major differences in the observed shape memory behavior. The mechanical property anomalies which represent the effects of stress on the
transformation will then be discussed in some detail, and experimental results on both TiNi and on some other materials exhibiting shape
memory effect will be selectively considered. Finally the conditions
under which optimum shape memory behavior may be expected will be
suggested.

                              DEFINITIONS

     A great many different terms have been used in conjunction with
the mechanical property anomalies referred to in the following. To
avoid confusion the following definitions will be used:

Shape memory effect: A material apparently plastically deformed at
a suitable temperature recovers, fully or partially, its original
shape on subsequent heating to a moderately higher temperature.
Strains of the order of up to 8 - 10% can be thus fully recovered,
the magnitude of the recoverable strain varying with the material,
and for a given material varying with grain size, texture, and the
conditions under which the strain is effected. Subsequent cooling
to the temperature at which deformation took place does not restore
the specimen to its deformed configuration.(1-17).

     It should be noted that a complete shape recovery has been usu-
ally implied in most of the prior work. In fact such full recovery
is seldom experimentally attained. In the discussion to follow we
shall concentrate on the ideal, full recovery conditions, but we

shall also consider the reasons why these are seldom realised, and
why a partial shape recovery is more generally observed.

Two-way shape memory effect: Under suitable prior deformation conditions in either the martensitic or the parent structures a "reversible" expansion/contraction may be developed to accompany the forward transformation; and a change of equal magnitude but opposite
in sense will then be present in the reverse transformation. Typically the magnitude of this effect is a factor of 5-10 lower than
that attainable in the "irreversible" shape memory effect, and
amounts to a dimensional change of the order of 1% (5,18-22).

Superelasticity: A material deformed well past its apparent yield
point fully recovers its initial shape on removing (rather than reversing) the load. Significant stress-strain hysteresis, indicative
of energy absorption in the lattice, is usually observed. This behavior is due to stress-assisted martensite formation during loading,
this martensite being thermally unstable except under stress. Thus
unloading is accompanied by the reversion to the initial structure
(and orientation). The temperature range within which this can be
observed is $A_f < T < M_d$ (5,8,23-26).

Ferroelasticity: A stress-strain behavior indistinguishable from
superelasticity but occurring in material stressed while in fully
martensitic condition. It has been frequently observed to involve
a stress-assisted growth of martensite lamellae of one orientation
at the expense of similar lamellae of other orientation(s),widely
interpreted as a reversible twin-boundary motion. An alternative
mechanism proposed is that of stress-assisted double transformation
(5), within the temperature range $A_d < T < M_f$ (8,27,28).

PRIOR WORK

Stress-assisted Martensite Formation

   The first explicit suggestion that the shape memory effect in
TiNi is associated with the martensitic transformation seems due to
Zijderveld et al.,(3). Subsequently de Lange and Zijderveld proposed
a mechanism for this effect (29). According to their hypothesis the
TiNi investigated contained a significant amount of untransformed
parent structure, i.e. the initial shape deformation was effected
at a temperature $M_f < T < M_s$. The martensite already present was assumed to be a random distribution of all the equivalent variant orientations. On application of stress, however, the residual parent structure would further transform to a martensite variant of some preferred
orientation, the particular orientation(s) formed being determined by
the nature of the applied stress. On the subsequent heating, and thus

thermal reversion to the parent structure, "the latter transformation
causes a deformation which annihilates the former mechanical deformation."(29). In conjunction with this shape recovery resulting from
the reverse transformation the authors also noted the need for the
reversion "to the original cubic crystal", and qualitatively confirmed this for TiNi by examination of the diffraction patterns of
repeatedly transformed specimens.

Twin Boundary Motion in Thermoelastic Martensite

The mechanism proposed above can only account for the shape memory effect being induced in a partially transformed material. It can
not, however, apply to a fully martensitic TiNi structure. The early
observations were reviewed by Wayman and Shimizu in an attempt to determine the characteristics required of the material, and/or of the
transformation it undergoes, which are necessary to develop shape memory behavior (30). In these considerations they assumed that any
partial shape memory behavior is likely to be caused by factors other
than those leading to a complete shape recovery, and concerned themselves only with the latter. It must be stressed that much of the information now available on the effects of stress on the transformation had not been reported at that time.

Wayman and Shimizu concluded that the "...shape memory effect
can be universally correlated with a martensitic transformation that
is thermoelastic in nature, the thermoelasticity being attributed to
ordering in the parent and martensitic phases." They further stated,
following Otsuka and Shimizu (12) and Otsuka (13), that the shape memory effect is only possible when the martensite is internally
twinned; and that, therefore, twinning rather than dislocations must
characterise the lattice invariant component of the transformation
deformation required by the phenomenological approach.

These conclusions are logically derived from the point of view
of the formal crystallography of the transformation, as was the stated objective of the authors. However, they are of necessity limited
to considerations of the geometry involved, and give little insight
into the physical mechanism -- therefore into the significant factors -- involved. Specifically, as discussed elsewhere (31),the following explicit or implied conclusions may be questioned:

(a) The thermoelastic transformation is defined (30) as one in which
"..a given plate or domain of martensite grows or shrinks as the temperature is lowered or raised, and the growth rate appears to be governed only by the rate of change in temperature.." This definition
implies a fundamental distinction between the thermoelastic and other
(burst? isothermal?) types of transformation; therefore also the
existence of at least two, and possibly more, different types of martensitic transformation.

The first reservation to this argument is experimental: It is quite straightforward to induce a "burst" transformation in TiNi, in which between 50 and 80% of the specimen transforms apparently instantaneously with a loud click reminiscent of twin formation. Therefore thermoelastic behavior is not characteristic of TiNi, a material exhibiting very striking memory behavior.

The second reservation is more fundamental. It has become common over the years to describe differing experimental observations as due to Umklapp, Schiebung, burst, isothermal, thermoelastic,etc., transformation. This usage may imply - or may lead to explicit assumption of - the existence of different kinds of martensitic transformations. Further, it may lead to attempts to account for each set of different observations on the basis of basic differences in the transformation mechanisms involved. Since I believe in the inherent simplicity of nature I do not share this view. On the contrary, I suggest that the fundamental nature of the martensitic transformation remains the same <u>regardless</u> of the material, or of the conditions under which the transformation takes place. Consequently the experimentally observed differences in behavior must be due to causes other than those directly and necessarily involved in the transformation itself (31). According to this hypothesis <u>all</u> the transformations are basically thermoelastic; therefore the shape memory effect should be inherently attainable in any material undergoing a martensitic transformation. If so, then it is only necessary to determine the conditions under which a complete shape recovery occurs.

(b) The requirement that the martensite be internally twinned is also subject to reservations on both experimental and logical basis. As pointed out by the authors, Cu-Zn alloys are not internally twinned, yet some of them exhibit shape memory behavior (16). In the substructures of Cu-Al (14) and Cu-Al-Ni (12) alloys some of the lamellar structures do, but others do not, exhibit twin/matrix relationship. Furthermore, the reversible twin boundary motion is certainly not characteristic of twins in most metallic materials: on the contrary, twinning is viewed as a mode of plastic (i.e. irreversible) deformation. For these reason the present author has argued that the lamellar substructures consist of different but equivalent variants of martensite; that the twin/matrix relationship between neighboring lamellae may be a desirable, but is not a necessary, feature of the transformation; and that no plastic deformation need be involved in the transformation (5,31). A similar argument for the structure of Cu-Al martensites was presented in some detail by Tas et al.(32).

Inasmuch as the distinction between a "twinned" and an "equivalent martensite variant" substructure need not involve any geometrical changes this distinction may be considered trivial. It is not. Twinning is widely considered as involving a dislocation mechanism. Therefore, if its existence is invoked to satisfy the geometrical requirements of the invariant lattice strain then some dislocations will have to be introduced into the resulting "twinned" martensite. This is difficult to reconcile with the authors' requirement that the lattice invariant deformation be "twinning rather than disloca- tions", as dislocations clearly must be involved.

Stress-Assisted Double Transformation

The alternative approach proposed by Wasilewski (5) was deve- loped from considerations of the effects of stress applied to TiNi in the vicinity of its (thermal) martensitic transformation range. The following assumptions have been made:

(a) Plastic deformation, defined as the formation and flow of dislo- cations, is not necessarily involved in either the forward or the reverse transformation. This is assumed regardless of whether the transformation takes place thermally (on cooling or heating), or under the action of an applied stress.

(b) At any suitable temperature $T < A_s$ the effect of an applied stress is to cause one or more of the following stress-assisted transformations to take place:

(i) $M_s < T < A_s$:
$$\beta \overset{\sigma}{\rightleftharpoons} M'$$

(ii) $A_d < T < M_f$:
$$M \overset{\sigma}{\rightleftharpoons} (\beta') \overset{\sigma}{\rightleftharpoons} M''$$

(iii) $M_f < T < M_s$:
$$\beta + M \overset{\sigma}{\rightleftharpoons} M' + M \overset{\sigma}{\rightleftharpoons} (M' + \beta') \overset{\sigma}{\rightleftharpoons} M'$$

Here the characteristic transformation temperatures are denoted conventionally. The primes and the double primes indicate the pro- ducts, respectively, of stress-assisted transformation or of double transformation (5).

Case (i) is straightforward, and the stress-assisted martensite formed remains stable on removal of stress. Under suitable condi- tions it should be possible to obtain a martensite single crystal, as had been reported for AuCd by Birnbaum and Read (27). Case (ii) represents the onset of stress-assisted double transformation mecha- nism. Here the β' formed by the stress-assisted reverse transforma-

tion is <u>both</u> thermally unstable <u>and</u> subjected to an applied stress;
Therefore it instantaneously transforms to a martensite variant
which is equivalent to, but of an orientation different from, that
of the original martensite. Case (iii) is a combination of the
above. Under the simplest conditions possible this will lead also
to a final single orientation of the martensite, and this orientation will in general be different from that (those) originally
present.

The shape changes involved in these stress-assisted transformations are schematically shown in Fig.1. The two equivalent variants
of martensite which can form from the original parent structure (β)
orientation are denoted by M' and M. (The orientation of M" is identical with that of M'). In (i) an applied compressive stress will facilitate the formation of M' and oppose that of M. In (ii) for simplicity's sake we show a single crystal of M as the initial condition,
and a complete reversion to β' as the intermediate step in the double
transformation to M". In fact the initial thermal martensite will
consist of both M and M', but only the former will be affected by
the stress; And the transition to β' is probably limited to a layer
between the transforming M and the product of the double transformation M". Such layers have been experimentally observed (33), although
their structures have not been determined.

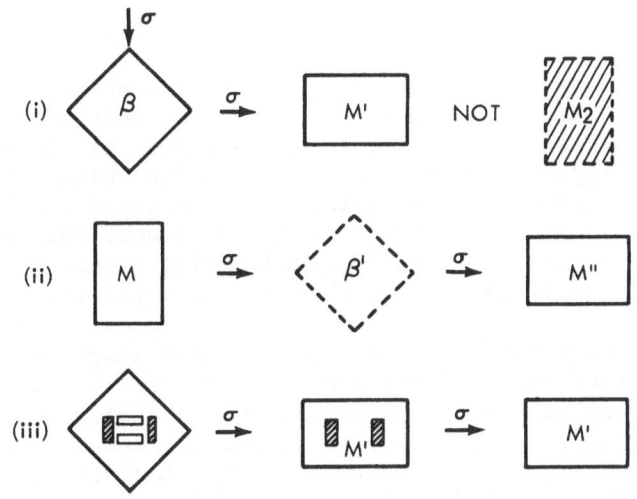

Fig. 1. Schematic representation of the shape changes involved in
stress-assisted transformations (i) - (iii) leading to optimum
shape memory effect.

I also doubt whether the simple conditions expressed by (iii)
are likely to be attained experimentally. A more probable final condition will consist of both the variants of the model of Fig.1, the
retention of the variant underlined unfavorably oriented with respect to the
stress being compensated by elastic stresses resulting from the
transformation. In real transformations, where the number of equivalent variants may be as high as 24, the final condition will be
that of several martensite variants, rather than the single orientation most favorably oriented with respect to the applied stress.
This factor introduces some complications, but does not affect the
validity of the argument.

The above are the conditions most favorable for inducing the
shape memory effect inasmuch that -- given the postulated absence
of slip deformation -- all of the residual strain remaining after
the removal of the applied stress is directly derived from the
stress-assisted transformation of the original structure orientation
to the martensite orientation(s) most favorably oriented with respect
to the applied stress. Therefore all of this strain can be fully recovered on the reverse transformation to the parent structure,
provided that

(c) There is a unique geometrical relationship between any martensite variant and the orientation of the parent structure to which
it transforms on heating.

Complete vs Partial Shape Memory Effect

A more detailed discussion of the assumptions (a) - (c) will
be presented in the following sections. Now we only stress that the
above conditions are optimal for the development of the shape memory effect, and appear to be the only ones under which a full and
complete shape recovery can be attained. Nevertheless a partial shape
memory effect can be much more generally obtained, and is probably
characteristic of all the materials undergoing a martensitic transformation. This conclusion is reached by the following argument:

If the assumption (a) is relaxed and some slip deformation does
take place, either in the initial stress-assisted transformation or
in the subsequent reverse transformation on heating, some residual
strains will remain in the reverted parent structure. Unless some
significant recovery does concurrently take place these will result
in residual stresses, elastic in nature, which then act on any subsequent thermal transformation on cycling across the transformation
temperature range. Such residual stresses, due to the stress fields
of the dislocations introduced during the plastic deformation, lead
to the "two-way" memory effect on subsequent transformations, whether
forward or reverse, in absence of any external stress (5,18-22).

Regarding now assumption (b) we first note that of the three alternative initial structures only (ii) consists entirely of martensite prior to the transformation. If the initial martensite, M, is the product of a purely thermal transformation on cooling, it consists of all the possible variant orientations which can be obtained from the original structure orientation, whether single crystal or polycrystalline. In absence of any directional constraints, such as applied stress, thermal gradients, or texture/shape effects (31) the resulting martensite will contain all the possible variants in self-accomodating groups of single variant domains (14,31,32) and therefore will have a grain size effectively smaller than that of the parent structure. It seems reasonable to suggest that the yield stress of this structure should be higher than that of its parent; Therefore the material may be expected to be the least susceptible to slip under any given value of an applied stress up to the plastic yield stress of the martensite. Conversely, one may expect the fully or partially β structures of (i) and (iii) to undergo some plastic deformation at stresses exceeding the β phase yield strength but below the yield of martensite.

Consequently all the above may be expected to exhibit perfect shape memory up to some limiting stress value $\sigma_y(\beta)$, and the corresponding maximum transformation strains reached. For applied stress $\sigma_y(\beta) < \sigma < \sigma_y(M)$, and the correspondingly higher strains attained, only (ii) will continue to retain perfect shape memory. In the remaining two cases some plastic strain will be involved at these stress levels in addition to the transformation strain. Only the latter is recoverable on subsequent reverse transformation. Therefore only partial shape memory effect can be expected.

It is important to note that the temperature at which the deformation is effected need not be uniquely significant. The prior thermal history of the material, even when involving such seemingly trivial factor as whether the specimen is cooled or heated to this temperature, can be of importance (5).

EXPERIMENTAL

Extent of Plastic Flow

We shall now examine the experimental observations pertinent to the assumptions made. Regarding first the incidence of slip deformation concurrent with transformation we note the data of Rozner and Wasilewski (34) indicate an anomalously low tensile yield stress of TiNi over the temperature range spanning the transformation. If we accept the yield strength at the higher temperatures, well away from the transformation, to be representative of the true plastic yield, then the usual extrapolation to the lower temperatures would follow approximately the dashed line in Fig.2. It is reasonable to conclude, therefore, that at very much lower apparent yield stresses observed the strain effected is largely due to stress-assisted transformation, and that true plastic strain can form at most a minor part of the total strain. This conclusion is further confirmed by a more detailed analysis of the stress-strain curve in Fig.3. Details of the technique are given in the original paper (5). We note that the total strain attained under maximum load consists of the following individual components:

(i) An <u>elastic</u> strain, ε_e, with a modulus apparently different from that observed on initial loading;

(ii) An <u>anelastic</u> strain, ε_{an}, which results in some recovery on unloading, most markedly at low stresses. This represents ferroelastic behavior;

(iii) A <u>transformation</u> strain, ε_{tr}, which is recovered on subsequent reverse transformation and represents the shape memory effect. And

(iv) A residual <u>plastic</u> strain, ε_p, the extent of which depends markedly on the type (tensile,compressive) of applied stress.

Similar observations were reported also for compressive loading by Sastri and Marcinkowski (35). They assumed, however, elastic recovery on unloading.

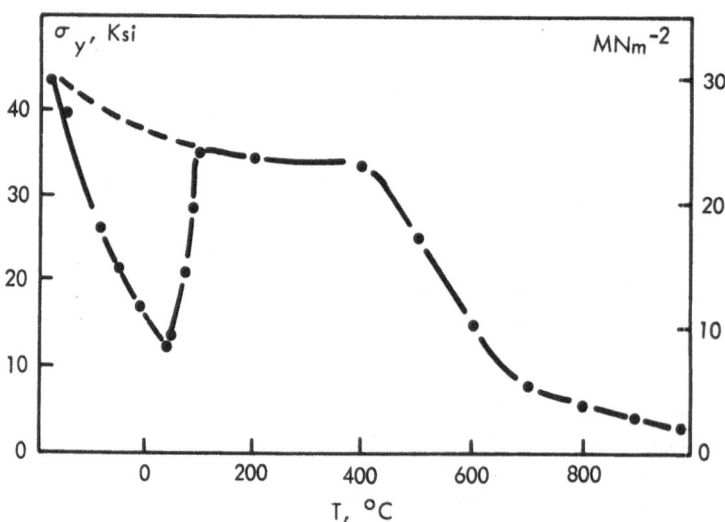

Fig. 2. The variation of the tensile yield stress of polycrystalline TiNi with temperature (34). Dashed line: extrapolated "true plastic" yield stress.

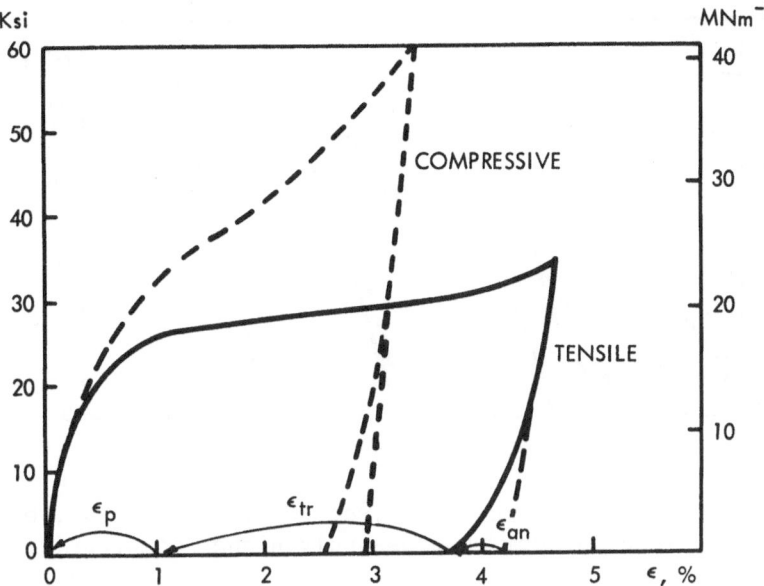

Fig.3. Stress-strain plots for tensile and compression loading of polycrystalline TiNi. Strain rate $\dot{\epsilon}$ = 10^{-4} per sec.

ϵ_{an} : anelastic recovery on unloading
ϵ_{tr} : transformation strain
ϵ_p : residual plastic strain

        The relatively minor amount of the residual plastic strain for the given value of applied stress in compressive - as opposed to tensile - loading is presumed due to some texture being present in the specimens tested. This texture has not been determined, and in the present work we shall discuss primarily compressive deformation in which the plastic strain is conveniently minimised. Because of the relatively low applied stress level - the apparent yield being well below the extrapolated yield strength of Fig.2 - it is believed that this plastic deformation must be here directly caused by the high local stresses arising from the transformation. The effect of loading to progressively increasing stress levels in absence of intermediate reversion anneals is shown in Fig.4. To what extent the difference between the stress-strain behavior in this, and in the preceding, figures is due to the interrupted loading, or to some slight compositional/texture differences, is not known. Nevertheless it is clear that the anelastic recovery on unloading decreases

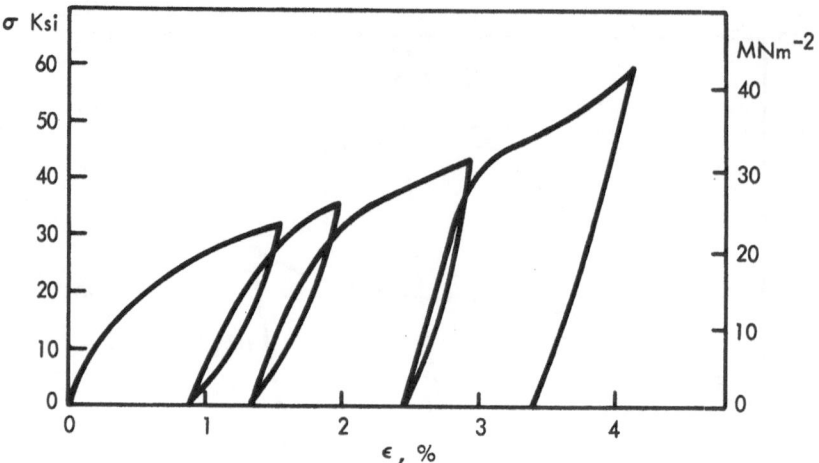

Fig.4. Total strain and anelastic recovery in Ti-50.2%Ni, on loading to and unloading from progressively higher compressive stress levels.

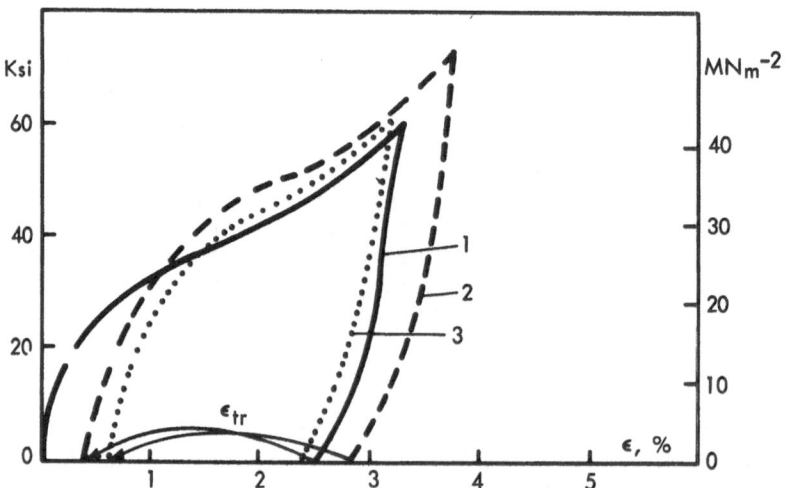

Fig.5. The effect of intermediate reversion anneals at 163°C on stress-strain behavior of TiNi on subsequent loading.
    1 -- "as received", max. stress 60 ksi; ε_{tr}=2.5%; ε_p=0.4%
    2 -- as reverted,   max. stress 73 ksi; ε_{tr}=1.85%; ε_p=0.2%
    3 -- as reverted,   max. stress 60 ksi; ε_{tr}=1.75%; ε_p<0.1%

markedly relative to the total strain attained. In the following Fig.5 we show the effect of intermediate reversion anneals between successive loadings of the TiNi rod of Fig.3. The reversion was effected by a two-minute anneal in boiling glycerol (163°C),i.e. at a temperature too low to permit significant relaxation of any work hardening caused by the plastic deformation. The only effect of the reversion anneal, therefore, was the annihilation of the pure transformation strain.

The apparent yield strength is little affected, and the stress strain plot at low strains remains smoothly curved on successive loading. The overall shape of the plot also remains similar. In each successive step, however, the total strain for any given stress level decreases; So does the anelastic recovery, the transformation strain, and in particular the plastic strain.

One concludes that the plastic deformation of TiNi results in work hardening, which then raises the effective plastic yield stress of this material, just as it does in other materials; Therefore prestraining TiNi to a high stress level should inhibit plastic flow on subsequent loading to a lower stress. Therefore the extent of plastic deformation on such subsequent loading should be minimized; The fraction of the overall deformation due to the transformation strain alone should be maximized; And thus the optimum shape recovery should be obtained. However, such prestraining will also affect the extent of the anelastic strain recovery (springback?), the two-way shape memory (22), and the magnitude of the irreversible shape memory effect.

To check this conclusion the effects of prior prestrain followed by a reversion anneal on the subsequent stress-strain behavior and on the memory behavior were investigated. The type of stress strain behavior generally observed is schematically shown in Fig.6a. The approximately elastic slopes below Y_1, and particularly above Y_2, as well as the slope and the extent of strain between Y_1 and Y_2, varied considerably depending on whether tensile or compressive stress was applied, as already shown in Fig.3. Nevertheless the overall character of the curves remained similar. On successive loading to well above Y_2 with intermediate reversion anneals between the tests, the values of Y_1 showed a progressive decrease for both tensile and compressive loading (Fig.6b). The same trend was observed for Y_2 in tensile loading, but in compression a small increase in Y_2 occurred initially. In all these tests, however, the extent of strain observed between Y_1 and Y_2 decreased very markedly between tests #1 and #2, and then at a much lower rate in the following tests.

We interpret these observations as follows. Y_1 indicates the onset of stress-assisted transformation. In the relatively soft "as received" TiNi this is accompanied by a significant amount of

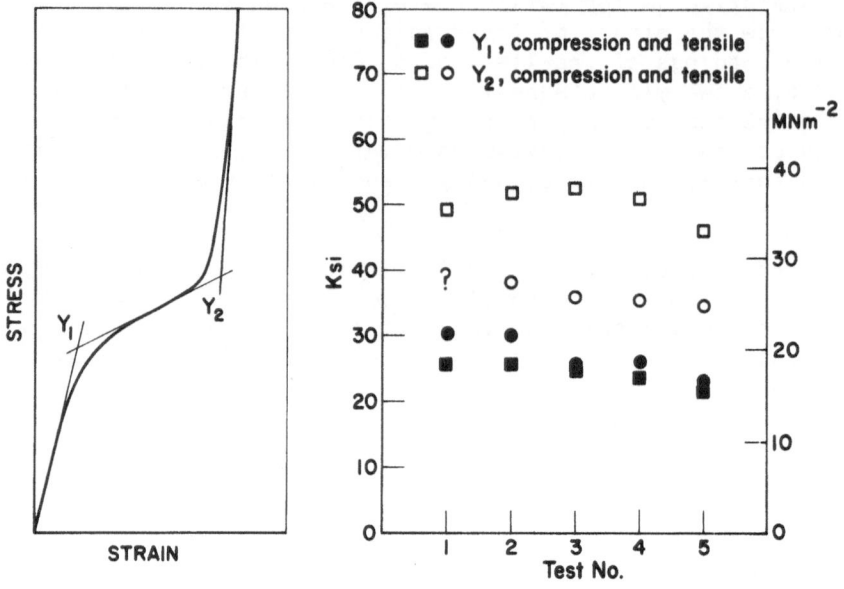

Fig. 6. Stress-strain behavior on successive loading of TiNi with intermediate reversion anneals.
   a. Schematic, shape of stress-strain plots.
   b. Variation of Y_1 and Y_2 with successive loading in tension and in compression.

concurrent plastic deformation. Initial stress-assisted transformation occurs in those martensite domains which are most favorably oriented with respect to the applied stress. As transformation proceeds, however, the continuing stress-assisted transformation must set in in other, progressively less favorably oriented, domains; Therefore its initiation requires progressively higher stresses. It follows that the relative proportion of the plastic deformation to the transformation deformation progressively increases. Finally, at Y_2, essentially all the favorably oriented martensite domains have undergone transformation. In a random polycrystalline martensite specimen this will occur at approximately half of the domains present having transformed, as the others will be stabilized by the applied stress (5,31). Therefore any further deformation must occur by plastic flow. It appears that the work-hardening manifests itself more in the extent of rounding of the curve at Y_2, rather than in increasing the actual yield value of Y_2.

The corresponding strain and shape memory effect were determined as follows. Several compression specimens 25.4 mm long by 6.4 mm diameter were prestrained to different stress levels (between 0 and 180 ksi -- (0 to 126 MNm^{-2}), reverted at 163°C, and cooled back to martensite. They were then deformed in compression at 60 or 90 ksi (42 or 63 MNm^{-2}). The strain components determined at each of these stages are summarized in Table I.

Table I. Strain components observed in polycrystalline TiNi on prestraining to different stress levels, and in subsequent memory-inducing deformation. All in compressive loading. (10 ksi = 7 MNm^{-2}. All strain values in percent)

Prestrain, ksi	0	60	120	180
total strain	0.0	2.84	3.66	4.66
anelastic recovery	0.0	0.24	0.31	0.25
transformation strain	0.0	2.15	2.63	3.60
plastic strain	0.0	0.45	0.72	0.81
two-way memory	< 0.02	0.13	0.81	1.30
After reversion and loading to 60 ksi				
total strain	2.89	2.20	1.92	1.85
anelastic recovery	0.26	0.17	0.10	0.11
transformation strain	2.26	1.90	1.75	1.74
plastic strain	0.37	0.13	0.07	0.0
two-way memory	0.16	0.20	0.92	1.22
After reversion and loading to 90 ksi				
total strain	3.18	3.08	2.81	2.68
anelastic recovery	0.27	0.22	0.14	0.12
transformation strain	2.51	2.68	2.58	2.54
plastic strain	0.40	0.18	0.09	(0.02)
two-way memory	0.20	0.37	1.10	1.36

It is concluded that the prestrain indeed can suppress the onset of the plastic flow, and thus cause the strain in the memory-inducing deformation step to be largely limited to the transformation strain. Thus, although the absolute value of the recoverable strain (i.e. the extent of the recoverable shape deformation) decreases, complete recovery is facilitated. Concurrently, with an increasing prestrain the two-way memory effect progressively increases.

In order to affect the true yield stress the initial prestraining must clearly be carried out in the cold-working temperature range. It does not have to be necessarily restricted to temperatures below M_f. Inferring from the observations on the two-way memory recently reported (22) the prestraining of TiNi can be effective even when done well above the transformation temperature range. If effected below the transformation range, however, it must be followed by a reversion anneal and thermal transformation back to martensite. A subsequent deformation of the martensite will then induce the shape memory effect under optimum conditions.

It should be noted that prestraining to a high stress level need not be the only way of work-hardening TiNi (or any other similar material), and thus of reducing the extent of true plastic flow. I believe that repeated thermal cycling over the transformation range may result in similar strengthening due to the progressively larger densities of accomodation dislocations (31) introduced in the successive transformations. There may be an additional effect on the yield strength due to the decreasing crystallite (domain? sub-grain?) size, because of the fragmentation of the original grains during the repeated forward and reverse transformations. As far as I know, however, this aspect has not been investigated, except perhaps for the observed modest increase in hardness on thermal cycling of TiNi (36). The other means of increasing the yield strength -- by alloying -- will be considered later.

Stress-assisted Transformations

The formation of stress-assisted martensite and the associated shape memory effect corresponding to case (i) proposed here was reported by Oishi and Brown for Cu-Al-Ni crystals (37). These authors investigated the extent of shape recovery on heating crystals deformed to a comparable residual strain (of 3 - 4%) at a temperature 10° above M_s, as well as that of comparable crystals deformed in martensitic condition at 15° below M_f. The latter corresponds to our case (ii).

Their observations indicate that for deformation temperatures $M_s < T < A_f$ the recovery on heating begins close to the original temperature A_s, and is essentially completed at A_f except for a small amount of residual strain (Fig.7). For deformation temperatures below M_f the recovery does not begin until a temperature well above the original A_s is reached. However, once initiated the recovery is rapid, and completed over a narrow temperature interval with very little residual strain. The observed reversion temperatures were stated to depend on the amount of the deformation effected. Further observations were reported by Eisenwasser and Brown on Cu-Zn-Sn alloys (38), in which a systematic variation of the shape memory effect (called by the authors "strain-memory") and the anelastic recovery (called by them "elastic recovery") with temperature was observed

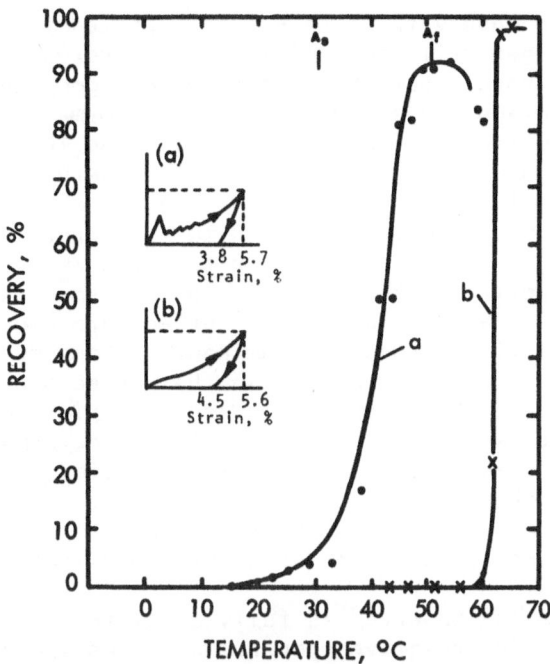

Fig.7. Shape memory effect in Cu-14Al-3Ni single crystals deformed at: (a) 10°C above M_s, and (b) 15°C below M_f. Inset: the respective stress-strain plots. (Oishi and Brown, ref.37).

and discussed. Similar observations on Ag-45%Cd alloys were reported by Krishnan and Brown (25).

In these investigations (25,37,38) the authors were primarily concerned with the formation of stress-induced martensite and the associated effects. They noted that the anelastic recovery observed in polycrystalline materials is less than that attainable in single crystals (37,38); That shape memory effect is observed in both poly-crystals and in single crystal material (37); And that there exists anisotropy of anelastic recovery with the crystallographic direction of the applied stress (25). Krishnan and Brown further observed that the "memory-inducing" deformation results in the formation of a new variant of the martensite. This new variant is equivalent to the other, already present, variants, but forms preferentially under the action of the applied stress (25).

No comparable observations have been reported for TiNi, the work on which has generally been limited to polycrystalline specimens. Nevertheless the above observations are in very good agreement with the effects of the applied stress on the transformation in TiNi as discussed by the writer (5) and further developed in the preceding section. In particular the "pseudoelastic" recovery of Brown et al., is clearly identical with the anelastic recovery observed here. This is the effect often previously referred to as "superelasticity",or "ferroelasticity", depending on whether the temperature was above A_f or below M_f, respectively. Brown et al., have also shown that the anelastic recovery and the shape memory effect are closely connected, and that the transition between predominantly shape memory and predominantly anelastic recovery occurs in their specimens in the vicinity of A_f. They also showed that some residual plastic deformation is usually present, and that the extent of this deformation is less for specimens deformed in the martensitic condition. This lower residual strain was obtained in spite of the stress required to attain the same total strain level was some 50% higher for the martensite than for the untransformed parent structure specimens.

Thus the behavior observed is fully consistent with that shown for TiNi in Fig.3. The relative proportion of the individual components of the total strain effected by the applied stress, as well as the specific values of the stress and strain observed, will vary from specimen to specimen, depending on the particular experimental conditions. Nevertheless whenever a specimen of TiNi is subjected to an applied stress in the vicinity of the transformation, and particularly at $A_d < T < M_d$, the total strain resulting will involve: The transformation strain (inducing the shape memory effect on subsequent thermal transformation); The anelastic strain (leading to super-, or ferroelastic effect); And the plastic strain (which induces the two-way memory effect on subsequent thermal transformation). We neglect here the truly elastic component of the total strain. This is, of course, always present under load, but its existence does not affect the phenomena discussed.

We conclude that the stress-assisted transformations corresponding to cases (i) and (ii) can be fully accounted for, and at least the trends expected of shape memory behavior, anelastic recovery, and the transition from one to the other can be predicted. The only additional complication in case (iii), i.e. for transformation temperature $M_f < T < M_s$, is that the transformation strain no longer corresponds to the simple transformation either from β to M', or from M to M''. Rather it is now composed, even in the simplest and most idealized case, of a more complex combination of the strains involved in the transformation of a part of the specimen β →M', and for the rest of the specimen M → M'' (=M'). Apart from this additional degree of complexity, however, there is no fundamental difference in the overall behavior.

The additional factor to note is that the use of any fixed values for the characteristic transformation temperatures is misleading. The values determined for thermal transformation, whether forward or reverse, under "no-stress" conditions (31) no longer apply when the transformation takes place either under an applied stress, or following a prior deformation. The observations of Oishi and Brown (37) clearly indicate that a relatively small shift in A_s and A_f results from deformation effected above M_s; This suggests that the lattice damage involved in the formation of stress-induced martensite is relatively slight. When the deformation is effected below M_f, however, the reverse transformation sets in some $30^{\circ}C$ above the original "no-stress" A_s temperature, which is already some $10^{\circ}C$ higher than the original A_f. Therefore the lattice damage due to the stress-induced double transformation (5,31) must be quite extensive to cause such a shift. No observations concerning a probable concurrent shift in M_s and M_f have been reported. One may expect that these will be shifted down with increased lattice damage.

Effects of Alloying

Alloying commonly increases the yield strength. It can also affect the transformation behavior. To check possible effects on the transformation of TiNi a number of specimens were prepared of composition $TiNi_{0.95}X_{0.05}$, where X was an element of the group Ti,V, Cr, Mn, Fe, Co, Ni, and Cu. These were prepared by non-consumable arc melting of 200 g buttons, homogenizing for 24 hours at $1000^{\circ}C$, then forging and hot-rolling at $850^{\circ}C$ to a strip 3.2 mm thick. The as-rolled surfaces were then machined (0.2 mm deep on each surface), and polished to a metallographic finish. Finally the alloys were heat-treated for 1 hour at $650^{\circ}C$ and furnace cooled to room temperature.

With the exception of the two non-stoichiometric binary compositions all the specimens appeared single phase, no discernible second phases having been observed on microscopic examination. All the ternaries showed an equiaxed grain structure, except for those containing V and Cu, the structure of which indicated at least partial transformation having taken place.

The transformation behavior of the ternary compositions was investigated briefly by determining the thermoelectric power (TEP) at room temperature (39,40), and by x-ray diffraction scans during cooling to $-180^{\circ}C$ and subsequent heating back to RT. The TEP data obtained are shown in Fig.8. These indicate that the (p-type) TEP is lowered by substitution of Ni with each of the elements from V through Co, maximum effect having been observed for Cr. Conversely, the addition of either Cu, or of excess of Ti in the binary, increases the TEP at room temperature. This suggests that the sub-

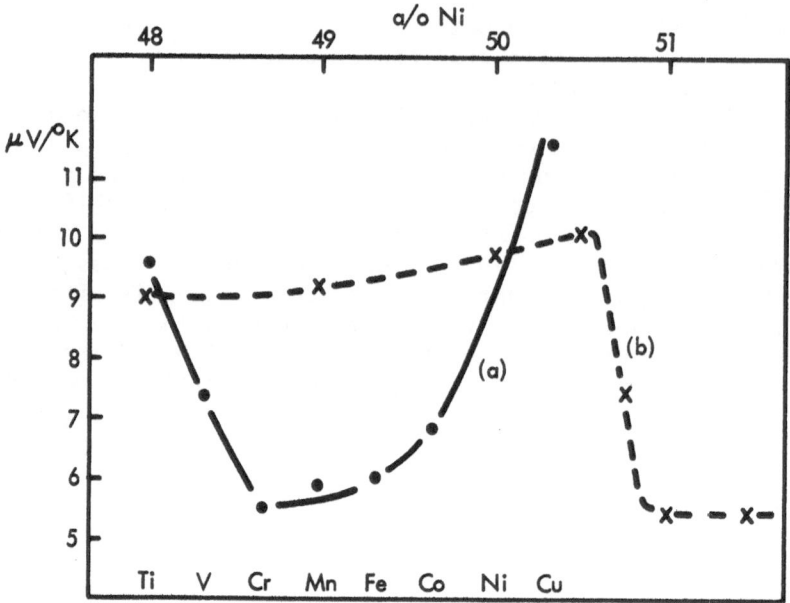

Fig. 8. Thermoelectric power (p-type) at 25°C, in (a) ternary alloys of composition $TiNi_{0.95}X_{0.05}$, and (b) non-stoichiometric TiNi.

stitution of Ni by V through Co stabilizes the β-structure, and lowers the M_s.

X-ray diffraction scans on continuous cooling and heating permitted further to establish the approximate values of the characteristic transformation temperatures. These are summarised in Table II. It is noted that the effect of alloying is not only in depressing the forward transformation temperatures, but also in increasing the transformation intervals $(M_s - M_f)$ and $(A_f - A_s)$ as compared with the behavior of stoichiometric TiNi. To what extent this may have been due to incomplete homogenization of the alloys has not been established. However, the spreading-out of the transformation range on alloying is an expected result of introducing solute atoms into the lattice (31).

One can expect alloying to increase the resistance to plastic deformation. Though no mechanical property data were obtained on this series of specimens we did make hardness measurements on all the compositions at room temperature. These were of the same order as the hardness of stoichiometric TiNi (180-190 VHN,1kg load) for the ternaries with V and Cr; But,surprisingly, varied between 140

and 150 VHN for those containing Mn,Fe,Co and Cu. This suggests that
the hardness impressions were anomalously large in these materials
due to the indenter load having been applied in the proximity of
the transformation temperature range, and specifically at some temperature $M_s < T < A_f$, M_d. Thus the size of the hardness impression
is representative of the transformation strain effected by the indenter, rather than of the extent of plastic deformation under it.
This conclusion is further supported by earlier observations of
"giant" hardness impressions at room temperature in Ti-51%Ni heat
treated to an M_s of $0°C$: in some of the β-grains indentations of a
size corresponding to hardness of the order of 20-30 VHN were observed (41).

Table II. Characteristic transformation temperatures, in degrees C,
on continuous cooling and heating of $TiNi_{0.95}X_{0.05}$ alloys. (ND --
not determined).

X	M_s	M_f	A_s	A_f
V	> 25	<-140	<-64	> 25
Cr	-100	<-180	<-58	> 25
Mn	-116	<-180	<-63	> 10
Fe	ND	<-180	-30	> 25
Co	ND	ND	0	> 25
Cu	> 25	<-100	?	> 25
$TiNi_{0.95}$	70	60	108	113
TiNi	60	52	71	77
$Ti_{0.95}Ni*$	-50	<-180	?	20(?)

* In supersaturated condition: as furnace cooled from $650°C$ (42).

Geometry of Reverse Transformation

In the preceding sections we discussed the reasons why the
strain attributable directly to stress-induced transformation should
be maximized for optimum shape memory behavior. Some conditions best
suited for maximizing this strain, and also the means for minimizing
the concurrent plastic deformation, have also been outlined. The final necessary condition concerns the geometry of the reverse transformation.

It can be stated quite generally that the symmetry of the parent structure is higher than that of the product martensite, even
though the structure of the latter may well be at times uncertain
or ill-defined (31). Consequently there are a number -- in some cases

as many as 24 -- of equivalent martensite orientations, or variants, which can result from the transformation of a given orientation of the parent structure. The transformation strain involved in the formation of each individual variant is the same when determined relative to the corresponding crystal directions in both structures; But the strain components of these individual variants as determined with reference to a fixed set of coordinates will differ. Thus the overall deformation in bulk transformation, in which all the possible variants form, is the sum total of the individual variant deformations, and does not exhibit any preferred direction. Rather, the overall deformation amounts to a macroscopically isotropic volume change, if we neglect the effects of the local strains and the elastic stresses resulting from the transformation. A reverse volume change accompanies the reverse transformation, again neglecting the residual lattice damage resulting from the formation of accomodation dislocations (31). Thus the material undergoing such idealized transformation may be considered to possess shape (volume) memory, regardless of the geometry of the reverse transformation, provided no external stress is applied during either the forward or the reverse transformation, and further provided that all the possible variants form to a comparable extent throughout the specimen.

When the forward transformation takes place under the action of stress, however, only some variants of martensite are likely to form, while the formation of others will be suppressed. Therefore the overall deformation of the specimen will no longer be isotropic, but will be directly related to the direction and sense of the applied stress. This applies both to the formation of stress-assisted martensite from the parent structure, and to the formation of the more favorably oriented variants of martensite by stress-assisted double transformation mechanism. However, this formation of the preferred variants of martensite in parts of the specimen leads directly to elastic stresses in the surrounding, untransformed, matrix. Therefore, unless these stresses are relieved by reverse transformation upon removal of the stress any subsequent reverse transformation on heating will be subject to these internal stresses. It is immaterial whether the untransformed matrix consists of the parent structure, or of martensite: In each case some internal stresses will have been introduced, and only the extent to which they may be lowered by reverse transformation on removal of stress (i.e. by superelastic or ferroelastic behavior) will vary. Therefore the difference between the behavior of the matrix of different structure is one of the order, rather than of the kind.

The assumption of residual elastic stresses in the untransformed matrix is, I believe, correct only to a first approximation. Even if we postulate -- as is done here -- that no plastic flow need accompany the transformation, considerable lattice damage is effected by the formation of accomodation dislocations (31). These serve to accomodate, in part, the local strains arising from the transformation, therefore their formation will affect the residual stresses in the surrounding matrix. In consequence these residual stresses are not necessarily those directly opposing the local transformation strains: Their magnitude must be smaller, and the sense of the local stress may also be affected. Therefore we can not assume that these residual stresses will simply assist the reverse transformation by reversing the strain involved in the forward transformation. Furthermore, there are some observations indicating that the product of the double transformation is relatively stable. This is suggested, for example, by the shape recovery usually occurring towards the end of the reverse transformation, rather than in its initial stages. It follows that the residual stresses acting on the reverse transformation of the "deformation" martensite are even more complex, and if the transformation geometry permits variants of the parent structure other than the original orientation may be expected to form.

The effects of transformation cycling, both with and without an applied stress, in materials which can form several equivalent parent structure orientations have been reported for Co (43-45), Co-Ni alloys (46), and for some ferrous martensites (47,48). In all of these the formation of parent structure orientations other than the original one has been observed. Under "no-stress" conditions this can simply lead to the formation of a polycrystalline aggregate from the initial single crystal specimen. With an applied stress acting on both the forward and the reverse transformation, however, a shape change resulted from a full transformation cycle. At sufficiently high tensile stress levels both the forward and the reverse transformation in Co were accompanied by elongation. This suggests that, just as one or more preferred martensite variants form in the forward transformation, so a preferred variant of the parent structure, which was different from the original orientation, formed on reverse transformation.

We conclude that, given a variety of alternative variants of the parent structure orientation resulting from the transformation of martensite, the reverse transformation occurring under the action of stress is unlikely to result in complete recovery of the initial shape. Conversely, a complete shape recovery is possible if a unique geometrical relationship exists between any of the equivalent martensite orientations and the parent structure orientation from which it was derived, and to which it can revert.

It is evident that in the above argument we neglect the possible effects of accomodation dislocations. The formation of these is certainly irreversible, and probably cumulative (49). The extent to which these are likely to affect the shape memory is probably minor. They do, however, indicate the amount of energy stored in the lattice, and this no doubt affects the level of the residual stresses present.

DISCUSSION

Although we are yet far removed from satisfactory quantitative understanding and predictive capability of the complex effects of stress on the transformation, we are now in a position to determine both the fundamental aspects to be investigated and the most promising application potential.

Concerning, first, the most pertinent research directions, the basic problem posed is in determining -- and ultimately being able to predict -- the relative proportions of the transformation, anelastic, and plastic strain components for any given material subjected to a given stress and at a given temperature. These relative proportions are likely to vary, for the same material, with:

Structural perfection: single crystal vs polycrystal, and with grain size in polycrystalline materials; Deviation from the stoichiometric composition in compounds, or the solute concentration in alloys; The concentration of thermal defects (vacancies, interstitials) or of dislocations; Inclusions.

Prior thermal history: In particular within the range $M_f < T < A_f$ the initial structure of the specimen will depend on whether it is cooled or heated to the test temperature. If the transformation temperature range lies close to room temperature, therefore, considerable differences in behavior may be caused by apparently trivial pre-test temperature excursions, as the result of which the test specimen may be either in parent structure or in martensitic condition during the test.

Orientation: In single crystals, or

Texture, (and probably also specimen shape) in polycrystals.

Stress level: in particular the effects of the resulting lattice damage in shifting the effective transformation characteristic temperatures from those of "no-stress" transformation levels. It is clear that, in order to investigate the effects of stress on transformation behavior the concentration on relatively low stress levels is indicated. At higher stresses the incidence of plastic flow becomes increasingly significant, and the effects on the transformation itself thus more difficult to determine.

Stabilization: The mechanism responsible for suppression of the anelastic recovery (whether "superelastic" or "ferroelastic"), which thus leads to an increase of the deformation recoverable by the shape memory effect.

The above seem to be the most promising directions for fundamental work. As regards the application potential I believe that the shape memory effect, though unquestionably most spectacular when almost full recovery is attained, may not be of primary importance. This belief is based on the fact that a complete shape recovery is very difficult to attain in practice, and that the conditions which favor a complete recovery simultaneously limit the extent of maximum recoverable distortion.

It seems more probable that the high stress levels accompanying the "memory" reverse transformation, even when the shape recovery is incomplete, offer a much better application potential. These have been already briefly investigated in TiNi, and stresses of the order of 100 ksi (70 MNm^{-2}) are readily attainable (50,51). The other, and perhaps even more attractive, aspect of stress-assisted transformation is that of two-way memory effect. This reversible change is, presumably, also accompanied by potentially useful stress levels. In TiNi a dimensional change of ± 1% can be readily induced, and thus in a TiNi couple a relatively large differential length change of 2% can be obtained over a temperature range of a few tens of degrees. It would be of interest to determine the useful stress levels accompanying this behavior, the effects of the necessary prior plastic deformation on the temperature range of the transformation, and finally the persistence of the reversible change on continuing cycling through the transformation.

CONCLUSIONS

1. It is concluded that the shape memory effect in TiNi, and in other materials undergoing a martensitic transformation, is merely one of the possible effects of stress on the course of the transformation. In the specific case of the shape memory effect the stress acting on the (reverse) transformation is internal, and arises from the elastic stresses introduced by prior deformation. The cause of this effect is the same regardless of whether a complete or a partial shape recovery results.

2. Anelastic strain, recoverable on removal of stress, is another common effect of applied stress on the transformation.

3. Plastic strain, though not a necessary element of stress-assisted transformations, is usually observed at higher applied stress levels. Plastic deformation results in residual elastic stresses, probably largely localized at the interfaces between the transformed and untransformed regions. These lead to the "two-way" memory effect.

4. Shape memory effect can be induced by a suitable deformation of either the parent structure or the martensite. Depending on the conditions of the transformation, however, the total strain may con-

sist of varying proportions of the above three components, even in a given material and at a given temperature.

   5. Optimum conditions facilitating a complete shape recovery are:
      (a) A low homologous temperature M_s, preferably well below the recovery temperature.
      (b) Absence of plastic flow during the memory-inducing deformation. This requires
         (i) high yield strength of the initial matrix, whether martensite or parent structure, and
         (ii) Low stress levels required to effect the stress-induced transformation, or double transformation. This in turn suggests that
      (c) The optimum temperature for memory-inducing deformation is close to M_f, where both the requirements of (b) should be optimized.

   6. Increasing the plastic yield strength of TiNi by prestraining leads to more complete shape recovery, but at the same time lowers the maximum recoverable strain levels.

   7. The two-way memory effect, though perhaps less spectacular, offers a more reproducible and more reliable application potential than the one-way shape recovery. The two-way memory is readily induced by suitable cold working in either structure.

REFERENCES

1. W. J. Buehler, J. V. Gilfrich, and R. C. Wiley, J.appl.Phys.,1963, v.34, 1475
2. J. V. Gilfrich, Proc.XI Ann.Conf. on X-Ray Analysis, Plenum Press. New York. 1963. p.74
3. J. A. Zijderveld, R. G. de Lange, and C. A. Verbraak, Mem.sci.Rev.Met.,1966, v.63, 885
4. A. Nagasawa and K. Kawachi, J.Phys.Soc.Japan,1971,v.30,296
5. R. J. Wasilewski, Met. Trans.,1971, v.2, 2973
6. C. M. Jackson, H. J. Wagner, and R. J. Wasilewski, NASA Special Report NASA-SP 5110, 1972
7. L. Delaey, R. V. Krishnan, H. Tas, and H. Warlimont J. Mat.Sci.,1974, v.9, 1521
8. R. V. Krishnan, L. Delaey, H. Tas, and H. Warlimont,Ibid,1536
9. H. Warlimont, L. Delaey, R. V. Krishnan, and H. Tas,Ibid,1545
10. L. C. Chang and T. A. Read, Trans. AIME, 1951, v.191, 47
11. Z. S. Basinski and J. W. Christian, Acta Met.,1954, v.2,101
12. K. Otsuka and K. Shimizu, Scripta Met.,1970, v.4, 469
13. K. Otsuka, Japan J.appl.Phys.,1971, v.10, 571
14. H.Tas, L. Delaey, and A. Deruyttere,Scripta Met.1971,v.5,1117
15. I. A. Arbuzova and L. G. Khandros,Fiz.Met.Metalloved.1964,v.17,390

16. C. M. Wayman, Scripta Met.,1971, v.5, 489
17. K. Enami and S. Nenno, Met.Trans.,1971, v.2, 1487
18. E. Hornbogen and G. Wassermann,Z.Metallk.,1956, v.47, 427
19. H. Tas, L. Delaey, and A. Deruyttere,J.less common Metals,
                        1972, v.28, 141
20. T. Saburi and S. Nenno, Scripta Met.,1974, v.8, 1363
21. J. Perkins, Scripta Met.,1974, v.8, 1469
22. R. J. Wasilewski, Scripta Met.,1975,v.9, 417
23. W. A. Rachinger, Brit.J.appl.Phys.,1958,v.9, 250
24. H. Pops, Met.Trans.,1970,v.1, 251
25. R.V.Krishnan and L. C. Brown, Met.Trans.,1973,v.4, 423
26. R. J. Wasilewski, Scripta Met.,1971,v.5, 127
27. H. K. Birnbaum and T. A. Read,Trans.AIME,1960,v.218, 662
28. R. E. Busch, R. T. Luedeman, and P. M. Cross, U.S.Army Mater-
                        ials Res.Rept.AD 629726, 1966.
29. R. G. de Lange and J. A. Zijderveld,J.appl.Phys.1968,v.39,2195
30. C. M. Wayman and K. Shimizu, Met.Sci.J.,1972,v.6, 175
31. R. J. Wasilewski, Met. Trans.,1975,v.6A,in press
32. H. Tas, R. V. Krishnan,and L. Delaey,Scripta Met.,1973,v.7, 183
33. E. Hornbogen, A. Segmüller,and G. Wassermann,
                        Z. Metallk.,1957,v.48, 379
34. A. G. Rozner and R. J. Wasilewski,J.Inst.Met.,1966,v.94, 169
35. A. S. Sastri and J. M. Marcinkowski,Trans.AIME,1968,v.242,2393
36. K. Mukherjee, F. Milillo,and M. Chandrasekaran,
                        Mat.Sci.Eng.,1974,v.14, 143
37. K. Oishi and L. C. Brown, Met.Trans.,1971,v.2, 1971
38. J. D. Eisenwasser and L. C. Brown, Met.Trans.,1972,v.3,1359
39. R. J. Wasilewski, S. R. Butler,and J. E. Hanlon,
                        Met.Sci.J.,1967,v.1,104
40. J. E. Hanlon, S. R. Butler,and R. J. Wasilewski
                        Trans.TMS-AIME,1967,v.239, 1323
41. R. J. Wasilewski, unpublished work
42. J. E. Hanlon, S. R. Butler,and R. J. Wasilewski,unpublished work
43. R. T. Johnson,Jr.,and R. D. Dragsdorf, J.appl.Phys.,1967,v.38,618
44. J. O. Nelson and C. Altstetter,Trans.TMS-AIME,1964,v.230, 1577
45. R. Adams and C. Altstetter, Trans.TMS-AIME,1968,v.242, 139
46. E. de Lamotte and C. Altstetter,"The Mechanism of Phase Transfor-
                        mations in Crystalline Solids",Monograph #33,
                        The Institute of Metals, London,1969,p.189
47. G. Krauss,Jr., Acta Met.,1963, v.11, 499
48. H. Kessler and W. Pitsch, Acta Met.,1967,v.15, 401
49. R. J. Wasilewski, Scripta Met.,1971, v.5, 207
50. V. F. Beuhring, C. M. Jackson, and H. M. Wagner, Final Report
                        BMI-X-579,Battelle Memorial Institute,1969
51. W. J. Buehler and W. B. Cross, Wire J.,1969,v.2, 41

THERMOMECHANICAL CHARACTERISTICS OF ALLOYS EXHIBITING MARTENSITIC THERMOELASTICITY

Jeff Perkins, G. R. Edwards, C. R. Such, J. M. Johnson, and R. R. Allen
Materials Group, Department of Mechanical Engineering, Naval Postgraduate School
Monterey, California, USA, 93940

ABSTRACT

Successful application of unique mechanical effects related to thermoelastic martensite transformation, deformation, and reversion is discussed in terms of the requirements to characterize various unique thermomechanical parameters, such as reversion stress, strain limits, stress and strain stability, and cycling effects. The dependence of reversion stress on such factors as prestrain, partial relaxation during reversion, temperature, and time is evaluated for Ni-Ti-base alloys and recent data are correlated with existing data for NiTi and other "shape-memory effect" alloys. The kinetics of martensitic deformation and reversion is discussed and related to proposed microstructural models for martensitic thermoelastic effects. Practical considerations such as stress relaxation, consequences of exceeding strain limits, and conditions for development of reversible effects are discussed. The general prerequisites for thermoelastic martensite effects are summarized, and implications regarding applications are discussed.

INTRODUCTION

Shape memory effect (SME) behavior is defined when a material is deformed beyond an apparent "yield point", yet recovers all or nearly all strain upon stress release and/or on heating. A number of distinct effects fitting this general definition have been observed, and these unique phenomena have acquired various descriptive titles, as described in several recent reviews (1-5). For example, immediate strain recovery on unloading a deformed alloy

is now widely known as the "pseudoelastic" (6) or "superelastic"
(7) or "rubberlike" (8) effect, while a strain reversion requiring
heating to some temperature above the deformation temperature is
generally termed the "shape memory effect" (SME), per se. In many
instances, a combination of immediate and heating-induced strain
reversion is observed; in fact, this may be considered the general
case.

The first report of the SME (per se) was by Chang and Read in
1951 in the alloy AuCd (9). Since that time, SME has been discovered in a number of alloy systems. The most widely known and
developed alloys are the NiTi-based "Nitinols", in which the SME
was publicized by development work at the Naval Ordnance Laboratory (NOL, therefore, NiTiNOL) in the early 1960's (10). Other systems in which the SME is now known to occur include InTl (11),
CuZn (12), CuAl (13), NiAl (14,15), Fe_3Pt (12), AgZn (16), AgCd
(17,18), and other alloys based on some of these systems, such as
CuAlNi (19), CuZnSi (20), CuZnSn (21), and CuZnAl (22). The immediate strain recovery effect is also now known to occur in a
number of systems, including TiNi (23,24), AuCd (25), In-Tl (8),
CuZn (26), Cu-Al-Ni (7), CuZnSn (6,21), AgCd (17,27), and AuCuZn
(28). Such effects have now been established to be associated in
all cases with the reversion of stress-induced martensite and/or
deformed martensite. A number of common structural features have
been determined for SME occurrence. A primary requirement apparently is that a thermoelastic martensitic transformation occurs,
as established by the extensive work of Wayman and co-workers
(1,15,16,29,32). Indeed, a more sound scientific terminology for
SME would be "thermoelastic martensitic reversion" or "martensitic
thermoelasticity". This general terminology avoids the use of lay
terms such as "shape" and "memory" (all definitions of which imply
mental capacity), yet is descriptive, and will be adopted in this
paper, along with the abbreviated general terminology SME, or
simply "these effects".

Other contributors to this International Symposium on Shape
Memory Effects and Applications will deal extensively with mechanisms and microstructural features of SME. The present paper
will consider the general thermal-mechanical characteristics of
alloys which exhibit strain reversion associated with thermoelastic
martensites, based on recently obtained data and a review of earlier
work. Delineation of a number of unique mechanical properties
(thermoelastic martensite parameters) will be made, including martensitic prestrain, strain recovery, strain limits, reversion
stress, and so on. A review will be made of the general dependence of these parameters on controllable variables. To this end,
reference will be made to earlier work, especially the extensive
data deriving from NASA-sponsored projects in the 1960's (33-35),
the work of Pops (6), Brown and co-workers (17,21), and others, as
well as recent work in the authors' laboratories (36-38). Also,

some effort will be made to correlate phenomenological observations with current microstructural models for SME, in order to develop some scientific methodology for use in application of these effects.

EXPERIMENTAL

New data included in the present paper is based on recent work performed in the authors' laboratories; experimental details, reported elsewhere (36-38), will be briefly summarized here. Several TiNi-base alloys were obtained from Raychem Corporation, Menlo Park, California. Alloy compositions utilized were 49.2 Ti-49.8 Ni-1.0 Al and 49.5 Ti-49.5 Ni-1.0 Fe (all at. %), with Ms of about 0°C and -70°C, respectively (39). Alloys were prepared by electron beam melting of a 1.25 in. dia. ingot, hot swaging to 0.5 in., cold drawing, in 15% passes with intermediate anneals, to 0.0968 inches ±0.0002 in. and vacuum annealing at 900°C for thirty minutes (39). Samples were tested directly in rod form in an Instron testing machine. Specimens 2" long were held in collet-type grips with a 1" gage length between grips. Thorough checks were made to assure uniform deformation along the gage length and absence of slippage or creep in the grips. Further details concerning strain measurement, thermal environment control, data recording, and calibration and characterization of the apparatus are discussed elsewhere (36-38).

THERMOELASTIC BEHAVIOR: PHENOMENOLOGY

Several unique parameters may be usefully defined when dealing with SME. Discussion of these will be assisted by reference to Figures 1 and 2. The parameters of interest include: ε_i, the "initial martensitic strain" (produced by deformation $A \xrightarrow{\sigma} M(d)'$ and/or $M(+) \xrightarrow{\sigma} M(d)''$, where M(d) is stress-induced martensite and M(+) is athermal martensite). (Note: $\varepsilon_i = \varepsilon - \varepsilon_e$, where ε_e is the elastic strain component); ε_r, "reversion strain"; ε_r(SE), "immediate (superelastic effect) reversion strain"; ε_r(SME), "heating-activated (shape memory effect) reversion strain"; ε_f, "residual strain" after heating above A ($\varepsilon_f = \varepsilon_i - \varepsilon_r$); ε_L, the "strain limit" for $\varepsilon_f \approx 0$ (Note: $\varepsilon_f \approx \varepsilon_i - \varepsilon_L$ when $\varepsilon_i > \varepsilon_L$); "% strain recovery" (= $\varepsilon_r/\varepsilon_i$ x 100); σ_r , "reversion stress" (developed on heating above Af when all or part of ε_r is constrained against recovery); ε_r(FREE), "unconstrained or free portion of reversion strain" (that portion of ε_i or ε_r which reverts prior to constraint); ε_r(Con), "constrained reversion strain" (ε_r(Con) = $\varepsilon_i - \varepsilon_r$(FREE) if $\varepsilon_i < \varepsilon_L$; ε_r (Con) = $\varepsilon_i - \varepsilon_r$(FREE) - ε_f, if $\varepsilon_i > \varepsilon_L$)

More common mechanical properties such as yield stresses and moduli are also useful to describe the properties of the various

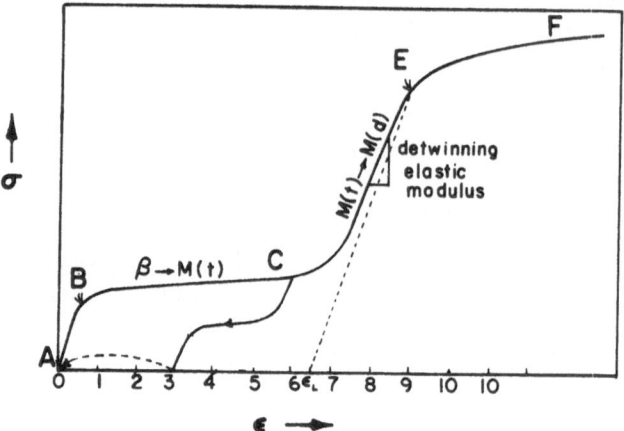

Figure 1:   Typical stress-strain curve for SME alloy
            at As<Td<Af<Md.   Point B: $\sigma_{A \to M}$.   Point E:
            true plastic deformation begins.

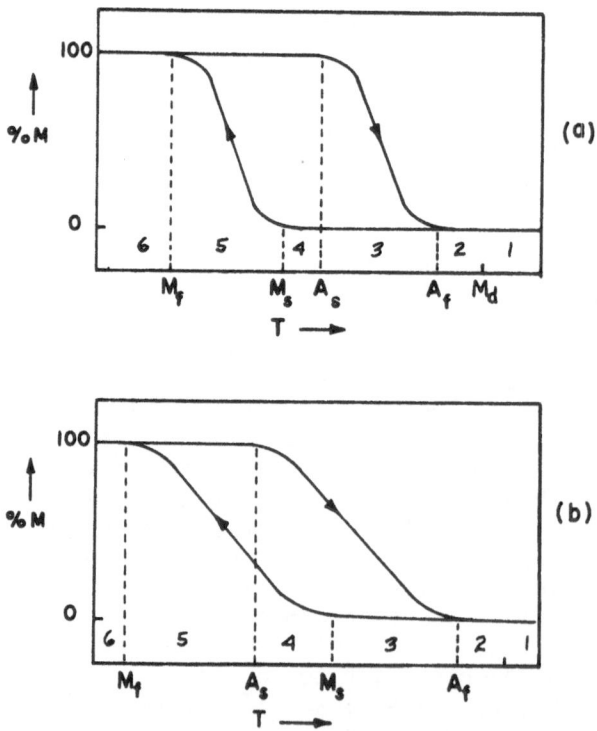

Figure 2:   Schematic thermal transformation hysteresis
            loops for (a) TYPE I and (b) TYPE II thermo-
            elastic martensites, defining unique deforma-
            tion temperature ranges.

phases and deformation reactions. However, even some of these properties exhibit unusual behavior; for example, certain apparent yield
points correspond to the onset of reversible transformation plasticity rather than to classical deformation mechanisms and therefore exhibit unexpected temperature dependencies. Referring to Figure 1, a
typical $\sigma-\varepsilon$ curve for an SME alloy with Type 1 thermoelastic behavior
(Ms<As) at Td (deformation temperature) where for this example
As<Td<Af<Md: $\sigma_{A \to M}$ is the flow stress for the stress-induced reaction
$A \overset{\sigma}{\to} M$; $\sigma_{m \to A}$ is the flow stress for the reversion reaction M→A on reversal of stress sense starting in the stress-induced martensite
state; $\sigma_{m \to m(P)}$ is the flow stress for true plastic deformation of
stress-induced martensite; ε_e is the elastic strain component, which
may generally be attributed to one or to both of the two phases A and
M, as $\varepsilon_{e(A)}$ and $\varepsilon_{e(M)}$.

Referring to Figure 1, $\sigma_{A \to M}$ decreases with decreasing temperature above Ms due to increasing instability of the parent A phase,
becoming approximately zero at Ms (6,40). Similarly, $\sigma_{m \to A}$ decreases
with decreasing temperature above As, becoming approximately zero at
As (6,17) with the explanation that when Td is closer to As, stress-
induced thermoelastic martensite is less unstable (more stable), so
that stress must be reduced to a lower level before reversion to
parent A phase is possible; in the limit, at Td < As, there will be
no immediate (unloading) reversion at all, and thermal activation of
the M→A reaction by heating will be required, i.e., only SME behavior
will occur. Temperature dependencies of these stresses are represented in Figure 3. Of course, temperature dependencies of the
yield points for true plastic deformation reaction (e.g., $M \overset{\sigma}{\to} M(p)$ and
$A \overset{\sigma}{\to} A(p)$) show a classical increase with decreasing temperature. Only
if $\sigma_{A \to A(P)}$ is sufficiently high will it be possible to stress-induce
martensitic transformation prior to plastic deformation of parent
phase.

In practical terms, two of the most interesting parameters relative to application of thermoelastic martensite effects are reversion
stress and strain, σ_r and ε_r, respectively. These will be discussed
in more detail in following sections.

Shape Recovery (Reversion Strain), ε_r

Temperature-stress effects on attainment of ε_r. One of the primary parameters useful for application of thermoelastic martensite
effects is ε_r , the ability to return toward $\varepsilon \approx 0$. Assuming that
$\varepsilon_i < \varepsilon_L$ and no external constraint, then $\varepsilon_i \approx \varepsilon_r$. Referring to
the schematic hysteresis loops of Figure 2, a number of distinct de-
formation ranges can be defined. The deformation and reversion mechanisms are uncertain for several of these ranges, although the phenomenological behavior is known. It is clear that if Af < Td < Md,

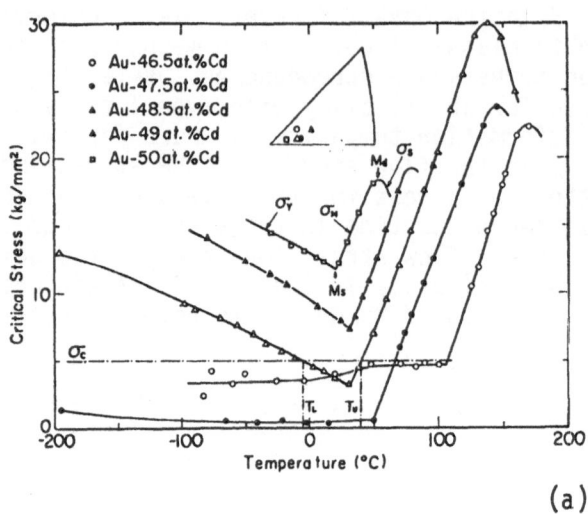

(a)

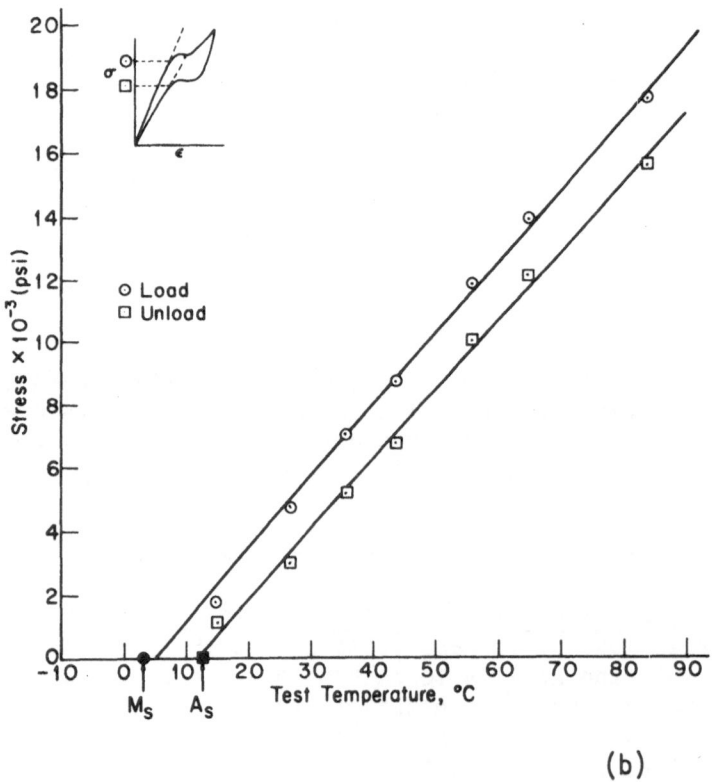

(b)

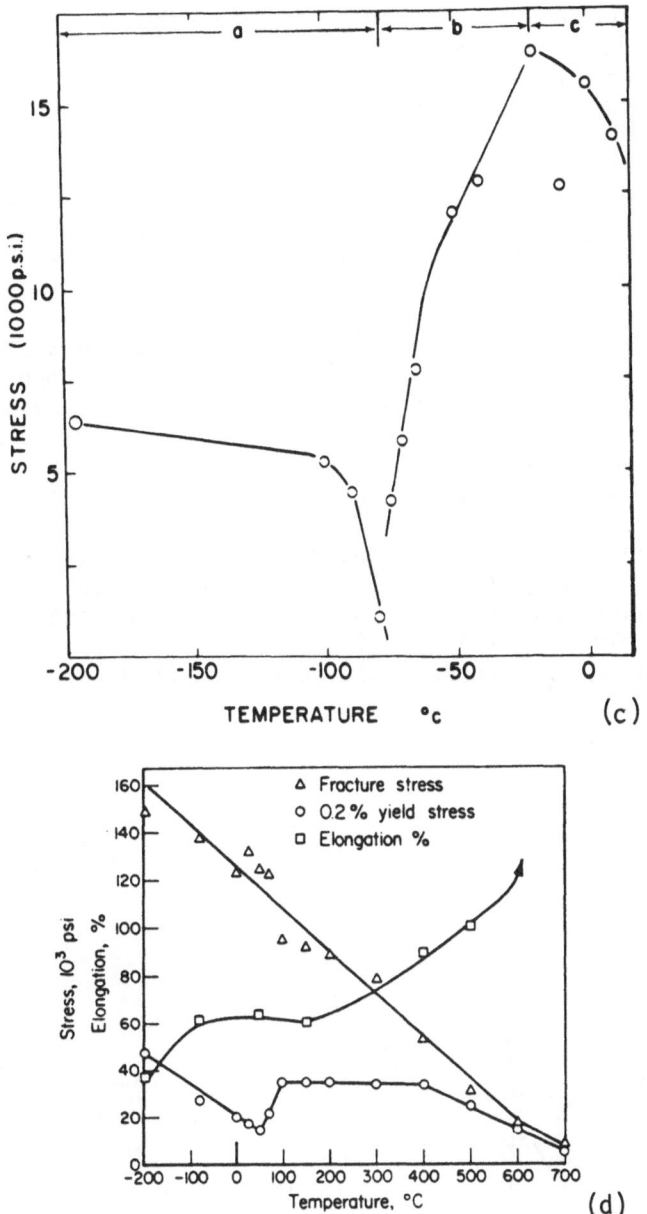

Figure 3: Critical yield stresses for transformation and true plastic deformation (first deviation from linearity of stress-strain curve) versus test temperature, for (a) Au-Cd single crystals (from Ref. 50), (b) Cu-Zn-Si coarse-grained polycrystals (from Ref. 6), (c) Ag-45a/o Cd polycrystals (from Ref. 17), (d) NiTi poly-crystals (from Ref. 35).

$\varepsilon_r = \varepsilon_r$ (SE); on the other extreme, if Td<As<Md, $\varepsilon_r$will be totally realized on heating above Af and, $\varepsilon_r = \varepsilon_r$(SME). In a more general case, if As<Td<Af<Md, $\varepsilon_r$will be a mixture of ε_r(SE) and ε_r(SME). If the deformation reaction in this range (Range 3 in Figure 2a) is A→M(t), then, for $\varepsilon < \varepsilon$ on release of the stress to zero, to a reasonable approximation (1-f)(M(t)→A (superelastic effect), and on heating above Af, (f)M(t)→A (shape memory effect, per se), where f=(Af-Td)/(Af-As), i.e., when Td=Af, f=0 (complete SE) and at Td=As, f=1 (complete SME). This interrelation of SE and SME is shown in Figure 4.

In practical terms, it is important to realize that martensitic transformation takes place over a range of temperature or stress, for athermal and stress-assisted reactions respectively, and that similarly, reversion takes place over a range of temperature or stress. In the case of stress-induced reaction, there is an essential equivalence of stress and temperature for thermoelastic martensites, so that if the $\sigma - \varepsilon$ curve of Figure 1 is adjusted such that strain is the dependent variable and elastic strains are removed, the resulting curve describes the kinetics of stress-induced martensite formation and reversion, analogous to Figure 2 for athermal transformations. This is shown in Figure 5 for several representative deformation temperatures.

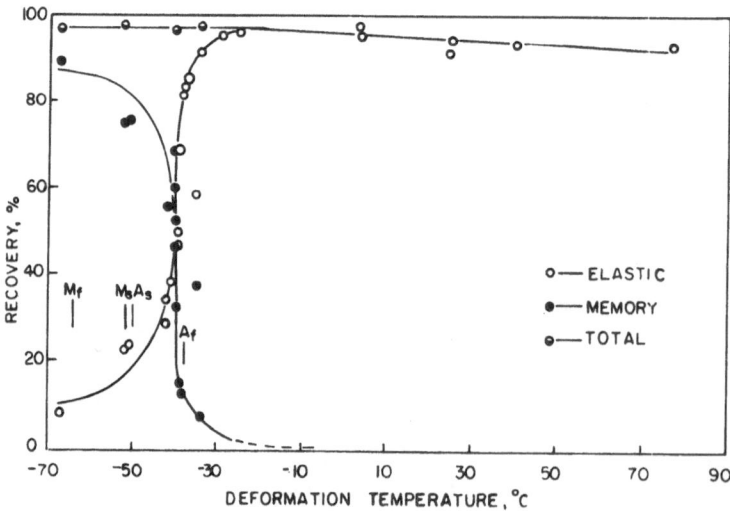

Figure 4. Percentage strain recoveries for the superelastic effect and shape memory effect versus deformation temperature, for single crystal Cu-Zn-Sn, showing that $\varepsilon r_{SME} + \varepsilon r_{SE} \approx 1$ in the vicinity of the thermoelastic range (from Ref. 21)

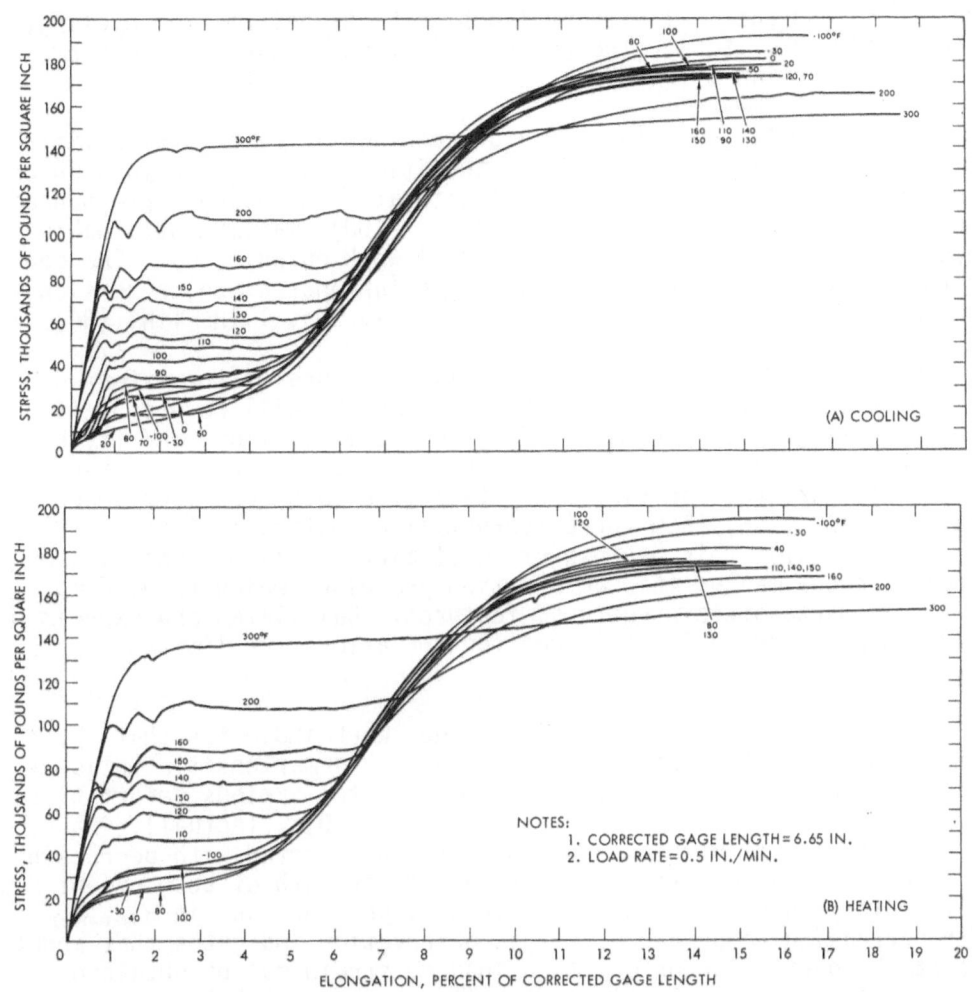

Figure 5: Stress-elongation curves as a function of
temperature for NiTi, during cooling and
heating cycles (from Ref. 34).

It is important to realize also that an (A+M) microstructure is
an irreversible path function. Also, the characteristic temperatures (Ms, Mf, As, Af) are a function of the stress state, and conversely, the characteristic streses for stress-induced reaction
are a function of temperature. The absolute values of the characteristic temperatures and stresses can also be modified by alloying

and by thermomechanical history (such as by transformation cycling, prior cold work, etc.). The general kinetics of martensitic formation and reversion will be discussed in a later section.

Strain limit, ε_L. There is for each SME alloy a "strain limit, ε_L, a critical value of initial martensitic strain (ε_i) beyond which many of the desirable parameters deteriorate, including a pronounced decrease in % strain recovery($= \varepsilon_r / \varepsilon_i \times 100$), maximum obtainable reversion stress, σ_r (max), and potential work output (34). The completeness of strain recovery during free (unconstrained) reversion decreases beyond ε_L, as in Figure 6 for polycrystalline TiNi (34).

Of course, ε_L is microstructurally a measure of the extent of martensitic transformation strain, and as such, will be dependent on the crystallographic parameters for the specific alloy, and may be calculated from such data (5, 21). Experimentally, ε_L is always potentially greater for single crystals (the value realized depending on orientation) than for polycrystalline material of the same alloy (5, 6, 21), this because of constraint of grain boundaries on stress-induced martensitic deformation processes which require mobility of martensitic structural features. Calculated and experimentally observed strain limit values for SME alloys are limited at present.

It is interesting to note that other work indicates that ε_L, which represents the potential for recoverable martensitic "plasticity," seems to be approximately constant for various deformation temperatures below Md for a given alloy (34, 40). (Actually, there is in general a slight increase in ε_L to be expected with decreasing Td below Ms (34), due to increasing yield strength of both parent and martensite phases, this reflected in a slight increase in $\sigma_{m \rightarrow m(P)}$ with decreasing temperature.) This observation indicates that about the same amount of transformation-induced strain can be obtained throughout the thermoelastic martensite range of stability, and suggests, for example, that purely stress-induced martensite formation (Ms<Td<Md) realizes about the same strain as deformation ("reorientation") of a fully martensitic microstructure (Td<Mf).

Cycling effects on ε_r and repeatability. Surprisingly little information has been obtained from most SME alloys regarding thermal and/or SME (strain-heat-cool) cycling on SME parameters. NiTi has been most extensively studied in this regard, largely in connection with the NASA investigations (34, 35). Figure 7 indicates the typical behavior on SME cycling (deformation to a given strain ε_i each cycle, heating to just above Af each cycle), which shows a gradual decrease in % recovery($= \varepsilon_r / \varepsilon_i \times 100$) with increasing cycles until an apparently "stabilized" value is attained, this value decreasing with increasing ε_i (and requiring increased cycles to stabilization). (34).

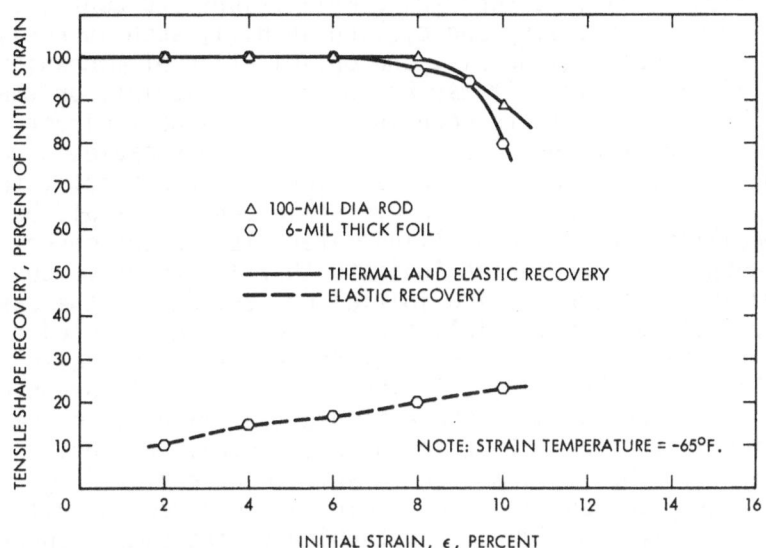

Figure 6: Percentage strain recovery [%R = $(\varepsilon_r/\varepsilon_i)$ x 100] versus ε_i for NiTi, showing a decrease in %R beyond ε_L (from Ref. 34).

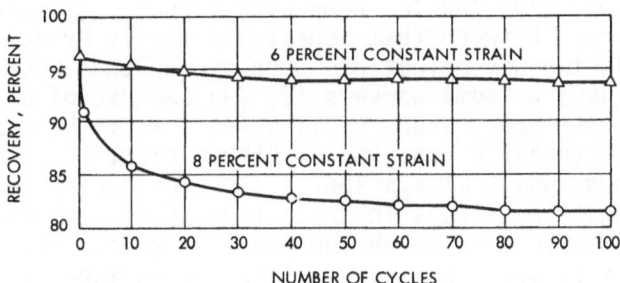

Figure 7: Percentage strain recovery versus number of SME cycles (deform below Mf, heat above Af) for NiTi, showing the attainment of perfect repeatability with cycling (from Ref. 34).

In view of various reported complications and controversies associated with annealing and cycling in NiTi, such as resistivity peaks, microstructural debris, and optimization of mechanical behavior (34, 35, 41-44), it may be unwise to speculate on the origin of the SME cycling effects seen in Figure 7 without direct microstructural correlations. However, it seems reasonable, since in most SME alloys the martensitic behavior is not perfectly thermoelastic, that some degree of irreversible behavior would be expected, especially for polycrystalline materials; recent observations in the authors' laboratories indicate that irreversible behavior is realized and exhausted during initial cycles, and in the case of SME-type cycling, irreversible behavior would be expected to be greater for higher ε_i (even when ε_i is less than crystallographically calculated values of ε_L.) The saturation of irreversible behavior can be envisoned as the attainment of microstructural stabilization in the form of "transformation strain hardening" to a point where further deformation is able to be resisted; when this condition is obtained, imposed ε_i can be recovered in full on each subsequent cycle (in Figure 7, returning to the same proportion of the initial length). Microstructural evidence of the buildup of dislocation substructure via thermal cycling has been recently reconfirmed (38), as reported earlier by Sandrock and coworkers (3, 42, 43). A correlary observation is that thermal cycling tends to raise the transition temperature range, i.e., tends to stabilize the low temperature structure (38). Note that the mechanism of substructure generation by transformation cycling implies that rather good high-strain fatigue behavior would be realizable for many SME alloys; such favorable fatigue properties have to date been characterized for NiTi (45) but for few other SME alloys.

Reversible (two-way) effects. Reversible shape memory effects have been reported (2, 5, 10, 23, 34, 46) in which an alloy "remembers" both high and low temperature shapes (strain positions), and may be cycled between them repeatedly merely by heating and cooling. This two-way effect has been termed the "reversible linear change" (RLC) by some workers (2, 23) because of a lack of hysteresis during temperature cycling; this is to be contrasted with the characteristic behavior of "thermoelastic" martensitic transformations, which always show hystereses during temperature or stress cycling. The alloys in which two-way effects have been reported are all alloys in which the usual (one-way) SME occurs, and it is likely that most known SME alloys can be expected to exhibit such behavior. The most pronounced two-way effects are apparently observed in cases where an alloy is deformed severely and nonuniformly, such that some fibres exceed the strain limit for complete recovery, ε_L, and others do not, and has been explained in this case in terms of residual stress effects (47). Such nonuniform deformation may occur commonly in fabrication operations such as bending, rolling, swaging, tube-drawing, and so on. If one

considers that deformation may develop nonuniformly on a microscopic
scale, such as from one martensite plate to another, the two-way
effect may also occur during operation with macroscopically uniform
stress application. Delaey and coworkers have advanced a similar
microscopic argument (5) rationalizing two-way memory behavior on
the basis of retention of residual defects after oriented nucleation
of martensite plates during the initial stress-induced transformation, these defects acting as preferred nuclei for subsequent thermal martensite arrays. This explanation does not account for reported two-way effects where the initial deformation is of fully marten-
istic alloys (34), although it is clear that the microscopic structure associated with the residual stress arguments (47) is consistent
with the induced defects proposed by Delaey and coworkers.

    In any case, where certain regions in a piece have exceeded
ϵ_L, while others have not, residual stresses will develop on heating above Af, since some fibres will be constrained from full reversion. These fibres will develop a residual reversion stress
(σ_r) proportional to and in the sense corresponding to the retained
unreverted strain ϵ_r (con). If the alloy is subsequently cooled below Md, this residual stress pattern will stress-induce martensite
in a pattern of appropriate orientation to accommodate the stress
(47). Fibres in regions of residual compression will shorten, while
fibres in regions of residual tension will elongate, giving rise to
a net shape change that will tend toward the initial deformed low
temperature shape (47). This sequence is represented schematically
in Figure 8. In general there will not be complete recovery of

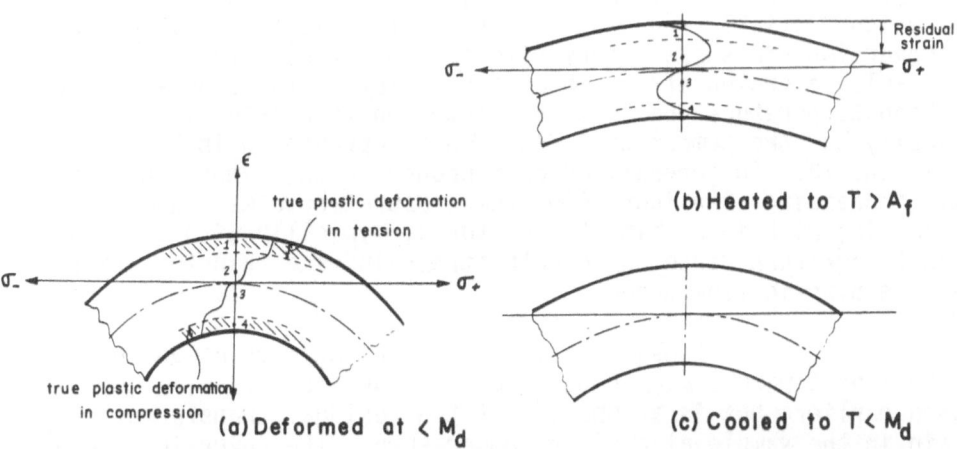

Figure 8.   Residual stress as an origin of two-way shape memory
             effects (a) Stress-strain distribution in sample bent
             below Md, with ϵ_L exceeded at outer and inner fibres;
             (b) residual stress pattern in sample heated to above
             Af; (c) shape change on cooling below Md (from Ref. 47).

either the initial low temperature deformed shape or the original
annealed high temperature shape (2, 46), but the piece will exhibit
some ability to cycle repeatedly between two strain positions between these limits (2, 46), unless the temperatures reached during
the heating cycles are sufficient to relieve the residual stresses
(anneal the residual defects), in which case the two-way effect will
deteriorate (5). Thermomechanical parameters associated with the
two-way effect are not yet well characterized for various SME alloys.

Reversion Stress, σ_r

Kinetics and limits of σ_r. If the initial strain ε_i is constrained to prevent reversion, significant internal stress will be
developed on heating through the As-Af range. σ_r increases in sig-
moidal fashion over this range (see Figure 9). Note that the As-Af
range is generally broadened and displaced to higher temperatures
by the stress (34). The maximum reversion stress attained, $\sigma_{r(max)}$
is a function of ε_i, increasing with ε_i to a peak (at $\varepsilon_i \approx \varepsilon_L$), then
decreasing (Figure 10) for the fully constrained case. It has been
observed that so long as ε_L is not exceeded, at a given deformation
temperature, $\sigma_{r(max)}$ is a linear function of the final constrained
strain value (34, 36). This partial free reversion prior to constraint does not alter the capacity of the material to develop a
reversion stress value characteristic only of the final degree of
constrained transformation strain. Further details on the development of σ_r will be discussed in a following section on the stability
of thermal and stress-induced transformation and reversion.

Note that $\sigma_{r(max)}$ may be substantially higher than the value of
$\sigma_{A \to m}$ which initially induced ε_i; indeed this is the basis of energy conversion via shape memory effect alloys (48). The limit on
attainable reversion stress in a given alloy is the flow stress of
the high temperature phase at the reversion temperature; $\sigma_{A \to A(P)}$
typically has the temperature dependence represented in Figures
3, 11, and 13. In general, Md corresponds to the temperature where
$\sigma_{A \to m}$ (increasing with increasing temperature above Ms) equals $\sigma_{A \to A(P)}$
(decreasing with increasing temperature), typically with a trans-
ition temperature range where both stress-induced transformation
and true plastic flow occur.

The reversion stress at any temperature between As and Af dur-
ing buildup toward $\sigma_{r(max)}$ is not a direct function of ε_i, as mentioned earlier, but is a function of the residual transformation
strain in the sample at a given temperature. The reversion stress
at any temperature can be approximated by the flow stress of the
alloy at that temperature as an upper bound (see Figure 11). The
stress differential between the flow stress vs. temperature plot
and the reversion stress kinetics (for the same strain value) is

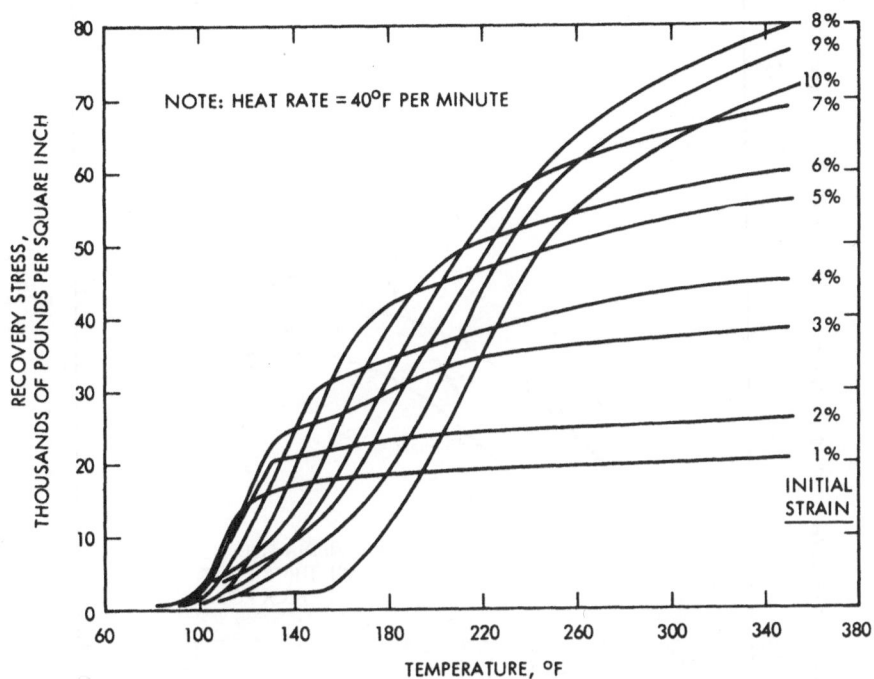

Figure 9: Reversion stress, σ_r, versus temperature during heating for various ε_i for NiTi, showing increasing $\sigma_{r(max)}$ (below ε_L) and upward-displaced reversion temperature range with increasing ε_i (from Ref. 34.)

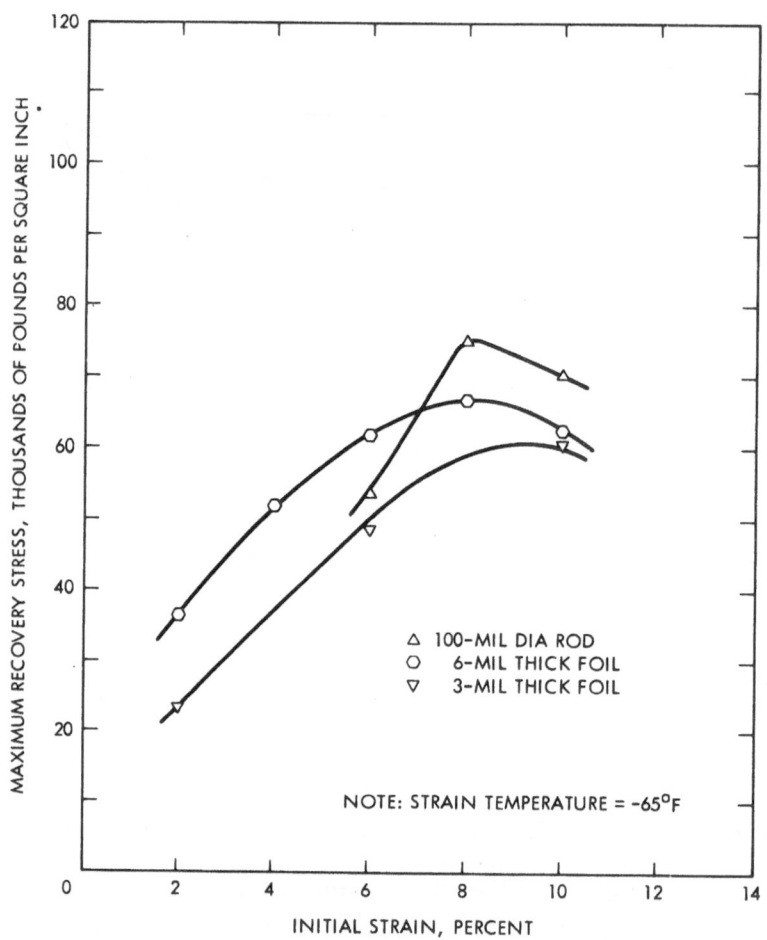

Figure 10: Maximum reversion stress, $\sigma_{r(max)}$ versus ε_i , for NiTi, showing a peak in attainable $\sigma_{r(max)}$ at $\varepsilon_i \approx \varepsilon_L$ (from Ref. 34).

real and is believed to be due to variations in the dislocation
substructure between the two types of deformation, and to differences in the dynamics of the two test methods (37). Flow stress
vs. temperature data are plotted as a function of strain in Figure 13, forming a surface which represents the upper bound on
reversion stress and can be utilized to consider strain-tempera-
ture parameters to optimize thermomechanical performance in a
given application. For a given ε_i, σ_r (ε_i, T) during cooling is
most closely associated with $\sigma_{A \to m}(\varepsilon)$; i.e., in general, the reversion stress exhibited for a given prestrain at a given tempera-
ture, $\sigma_r(\varepsilon_i$, T), is a measure of the stability of the martensite
phase at that temperature, so is directly related to the stress
required to induce the same amount of transformation strain at
that temperature $\sigma_{A \to m}(\varepsilon)$, where $\varepsilon_i = \varepsilon$, as seen in Figure 11.

Stability of σ_r. The ability to maintain a significant and
constant level of reversion stress may be of practical interest
in many applications. Two primary situations can be envisioned
in which, given a constant value of ε_r(con), σ_r might deteriorate:
(1) during elevation or prolonged maintenance of temperature above
Af; (2) during subsequent reduction of temperature to below Md or
Ms. In the first case, yielding, due to a decrease in the high
temperature phase flow stress with increasing temperature (see
Figure 12), or classical stress relaxation (creep) processes, may
occur, even at moderate operating temperatures, or relaxation
processes unique to the martensitic alloys might operate. In any
case, dramatic decreases in reversion stress may be observed (34,
37).

As mentioned earlier, Md corresponds to that temperature
where $\sigma_{A \to m}$ equals $\sigma_{A \to A(p)}$; the former is the "transformation yield
point", the latter the "true plastic yield point." Work in the
author's laboratories has shown that if the stress generated by
the constrained reversion reaction M→A(σ_r) exceeds $\sigma_{A \to A(p)}$ during
reversion heating, the high temperature phase will yield, limiting
further stress elevation to movement along the A phase flow curve
(37). If the temperature subsequently is further elevated, additional stress relaxation will result as the flow stress of the
high temperature phase decreases. By this process, the initially
induced martensitic "plasticity" will be partially converted into
true plastic strain, manifested as a permanent set when constraint
is removed. Limited comment on this phenomena in the early NASA
work on NiTi indicated that moderate heating above Af while in a
constrained condition (ε_r(con), σ_r) could seriously degrade subsequent shape recovery capability (34). (It was mentioned that in
alloys with Ms = 35°F and 45°F and Af = 160°F and 170°F, respectively, elevation of temperature to 300°F and 212°F - while constrained with on the order of 6% strain - gave rise to severe loss
of shape recovery when subsequently released and heated above Af,

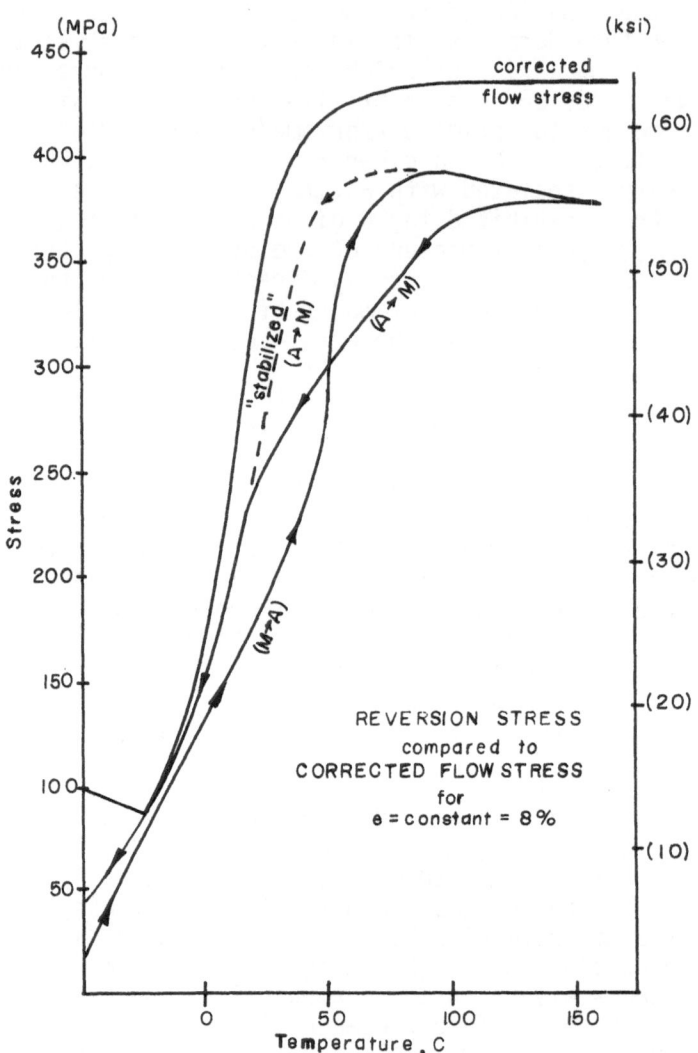

Figure 11: Reversion stress development in NiTi on
            heating and cooling (ϵ_i = 0.08)
            compared with temperature dependence of
            $\sigma_{A \to m}$ flow stress at ϵ = 0.08 (from Ref.
            37).

indicating a conversion of thermoelastic transformation strain to true plastic strain).

In the case of decreasing temperature into the martensitic range, reversion will exhibit a reversible behavior with hysteresis, as typified in Figures 11 and 12 (Figure 11 corresponds to an initial cycle; Figure 12 corresponds to a condition of stabilization produced by multiple cycling). In cases where an external dead load is applied in opposition to the reversion stress, and this load is maintained as the internal σ_r decreases on cooling, true transformation plasticity may occur (34,5). This results from the drastic reduction in yield stress (internal stress capability) as the alloy approaches Ms. True transformation plasticity should not result in cases such as Figures 11 and 12, where, as temperature decreases, the applied stress decreases as well.

These discussions lead to the realization that a limited temperature range of σ_r stability exists for any alloy. This range has an upper limit at the temperature where $\sigma_r = \sigma_{A \to A(P)}$, $T_L(\sigma_r)$, and a lower limit at Ms ($\sigma_{r(max)}$) σ_r will drop off beyond both of these limits. Below Ms ($\sigma_{r(max)}$)σ_r will drop reversibly (can be recovered by reheating), while above $T_L(\sigma_r)$, σ_r will diminish irreversibly with time. In cases where the external load does not decrease as the internally-generated σ_r decreases (such as when a dead load is attached), irreversible behavior will be accentuated, whether above $T_L(\sigma_r)$ or below $Ms(\sigma_{r(max)})$.The range of σ_r stability can be extended by several means, involving both metallurgical and design considerations. These include increasing $\sigma_{A \to A(P)}$ (via thermomechanical treatment or cycling), decreasing $\sigma_{r(max)}$(decrease ε_i or increase the section size to exert the same force), etc.

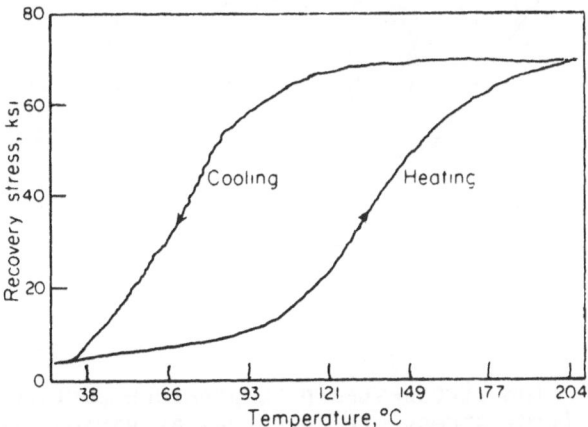

Figure 12.   Reversion stress, σ_r, versus temperature during heating and cooling of NiTi (pre-stabilized by transformation cycling) (From Ref. 35).

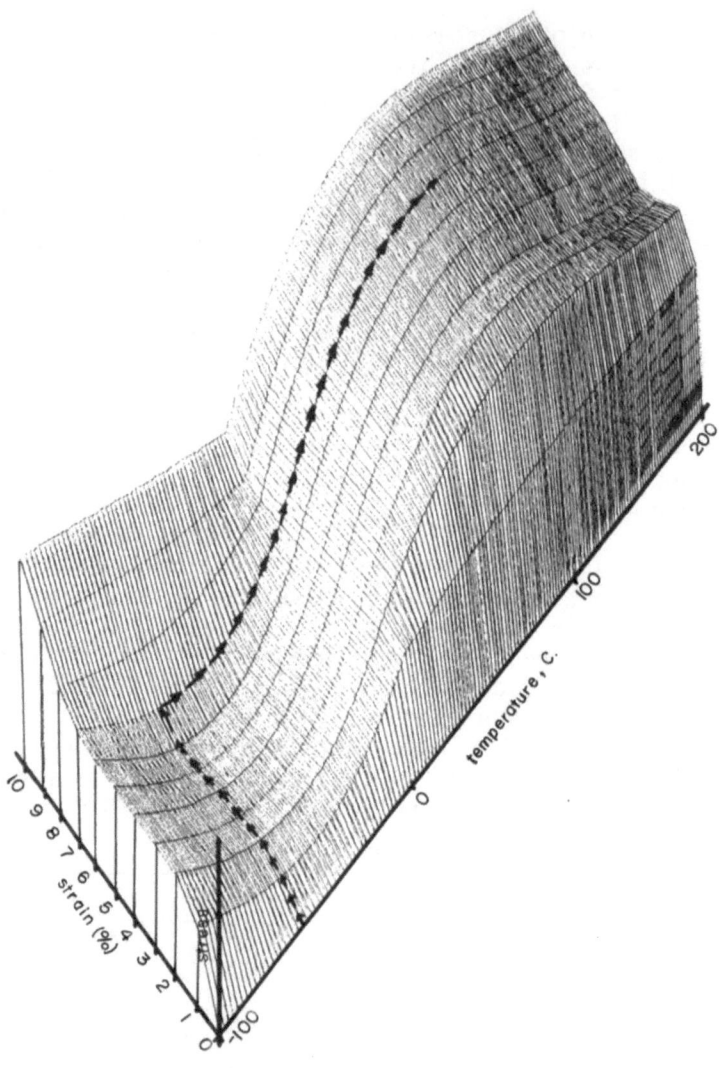

Figure 13: Flow stress-strain-temperature surface for NiTi
(from stress-strain curves at various temperatures)
(from Ref. 37).

MARTENSITIC THERMOELASTICITY: PRINCIPLES

Thermodynamic Conditions

A primary requirement for SME behavior is that a thermoelastic (t-e) martensitic transformation (MT) occur, as established by
Wayman and coworkers (1, 15, 16, 29-32). A characteristic of t-e
MT is a balance between the chemical driving energy for transformation, ΔG^c, and the nonchemical opposing energy, ΔG^{nc}, (essentially
elastic strain energy) built up during the formation of martensitic
structure A→M. There is for t-e MT an essential equivalence of temperature and stress, such that a change in either will tend to ad-
just the microstructural state. One can refer to a generalized
temperature or stress, such as $Y = (T-C\sigma)$, to be used to determine
the net transformation tendency (when Y = Ms, A→M; when Y = As,
M→A and similarly, with respect to transformation completion temperatures. (49)

Two "classes" of t-e MT are commonly defined: Class I, with
Ms<As and a relatively narrow transformation temperature range,
and Class II, with As<Ms and a relatively broad range. The two
classes for various alloys reflect differences in the magnitude of
ΔG^{nc}, with larger values for Class II alloys; in a sense these are
less perfectly thermoelastic alloys. The effect of ΔG^{nc} on transition temperatures is represented by the relations developed by
Tong and Wayman (29):

$$\lim_{\Delta G^{nc} \to 0} Mf = Ms \text{ and } \lim_{\Delta G^{nc} \to 0} As = Af$$

where ΔG^{nc} refers to the total accumulated nonchemical opposing
energy in the system at a given point; therefore, clearly the effective forward and reverse transformation temperatures depend on
the percentage martensite already present (2). For low specific
values of ΔG^{nc} (relative to ΔG^c) the forward and reverse reactions
will be completed over narrow temperature ranges and As will tend
to be higher than Ms. Note also that the percentage martensite
formed in a given temperature interval generally decreases in
cooling from Ms to Mf (31). Note that the specific value of ΔG^c
increases with decreasing temperature, while the specific value of
ΔG^{nc} is relatively unaffected (although the accumulated opposition
energy will change significantly, of course.)

Two characteristic temperatures are defined for the forward and
reverse reactions: To, where $\Delta G^c = 0$, and To', where $\Delta G^c + \Delta G^{nc} = 0$.
Tong and Wayman (29) have pointed out that To and To' are properly
bracketed as follows: To = 1/2 (Ms + Af), To' = 1/2 (Mf + As).
True t-e equilibrium exists only between To and To'. In considering hysteresis loops as in Figure 2, it is found that, for a thermal

transformation at $\sigma = 0$, the path is reversible above Ms and below As when describing the complete (full transformation) loop, whereas, in general (recalling the continuous nature of the transition temperatures described above on the basis of accumulated nonchemical energy ΔG^{nc}) the path will not be retraced if there is reversal in "Y" during ongoing microstructural changes. This hysteresis reflects the typical imperfect thermoelastic behavior and heterogeneous defect-related dependence of these transformations; a related effect is the common "stabilization" if Y is temporarily arrested. The first t-e plate to form (at Ms) will be the last to revert (at Af); the last to form (at Mf) will be the first to revert (at As); this is why Ms and Af become the more fundamental reference points on the temperature scale (29).

Because of the essential equivalence of temperature and stress, $\sigma_{A \to m}$ is observed to decrease as temperature is decreased below Md, to a minimum at Ms (34, 6, 50, 21, 37), as seen in Figures 3, 11, and 13. The temperature coefficient of the critical stress for martensite formation above Ms can be calculated from the Clausius-Clapeyron equation:

$$\frac{d\sigma_{A \to m}}{dT} = \frac{\rho \Delta G^c}{\varepsilon_L T_0}$$

where ε_L is the maximum macroscopic strain crystallographically realizable in the transformation, ΔG^c is the heat of transformation, and ρ is the density of the material.

Microstructural-Crystallographic Conditions

Factors governing strain limits. On a microscopic scale, a degree of thermoelastic martensitic behavior is an evident requirement for SME. Examining this requirement in greater detail, it is clear that in many cases, mutual accommodation of transformation strains between neighboring microstructural or submicrostructural units can be an effective means of obtaining a low net shape change (51). When these self-accommodating structures are subjected to stress, or are formed under conditions of stress-induction, the overall arrangement of units will be altered to produce the appropriate sense of strain. This suggests an available mechanism of mobile boundary movement between units, or alternately, that there is a reversion-retransformation mechanism; the former idea seems more plausible and has been supported by many workers. The motion of twin boundaries within martensite plates as a means of stress-accommodation is well established in twinned martensites (11, 52, 53). Such motion can give rise to substantial strains, but the introduction of irreversible defects, such as through "tapering" effects (11, 5) is likely at high strains. Similarly, the motion of inter-plate boundaries under stress in self-accommodating plate groups has been advanced (5, 51); this mechanism has the advantage of requiring no change in the plate substructure, so that invariant plane strain (IPS) requirements are not taxed, and irreversible deformation is more likely to be avoided.

The strain limit, ε_L, for reversible martensitic deformation is governed by the crystallographic features of the martensitic transformation, principally the magnitude of the macroscopic shape change (5, 21, 17). Many of the martensitic transformations in SME alloys (including TiNi, AuCd, CuZn, AgCd, CuAlNi, and CuZnX) (5) can be regarded as essentially body-centered cubic (BCC) to face-centered orthorhombic (FCO) transformation if slight distortions and atomic species are ignored, with a lattice correspondence $[100]_{BCC} \,||\, [0\bar{1}0]_{FCO}$ and $[010]_{BCC} \,||\, [110]_{FCO}$ (unit cells as given in Ref. 43.) The lattice transformation can be regarded simply as accomplished by a homogeneous lattice transformation (Bain strain), B, given by:

$$
B = \begin{vmatrix} n_1 & 0 & 0 \\ 0 & n_2 & 0 \\ 0 & 0 & n_3 \end{vmatrix}
$$

where n_1, n_2, n_3 are the principal distortions referred to the FCO axes, as

$$
n_1 = \frac{a_{FCO}}{a_{BCC}}, \quad n_2 = \frac{b_{FCO}}{2a_{BCC}}, \quad n_3 = \frac{c_{FCO}}{2a_{BCC}}.
$$

The maximum strain theoretically realizable from a given lattice transformation can be obtained by comparison of a reference sphere representing the cubic phase with an ellipsoid (with principle axes in the ratio n_1: n_2: n_3) representing the martensite structure. Of course, the crystallographic theories of martensitic transformations lead to superposition of other (lattice invariant) strains on B, leading to an invariant plane (the habit plane) strain for the total shape deformation. This is accomplished by the additional action of an inhomogeneous strain, S, and a rigid body rotation, R. For the cubic to orthorhombic case, the lattice invariant deformation, S, can be associated with a $\{110\}_{BCC} <110>_{BCC}$ type shear of magnitude $\bar{g}$. The magnitude of g can be calculated by the expression given in the WLR theory as:

$$
\bar{g} = \frac{n_1^2 - n_3^2}{n_1 n_3} \left\{ \frac{1}{2} + \frac{1}{2} \frac{(1-n_1^2 n_2^2)}{n_1^2 - n_2^2} \left(1 - \left[\frac{(n_1^2 - 1)(1 - n_2^2)}{1 - n_1^2 n_2^2} \right]^2 \right)^{1/2} \right\}
$$

The macroscopic shape change (surface tilt), F, is given by
the expression F = RBS. Generally, the observed macroscopic shape
change of a piece is minimized by self-accommodation within and
between individual microstructural units. However, it can be seen
that the realizable transformation strain for a given martensitic
alloy will be limited by the crystallographic parameters. For
stress-induced transformation or deformation of thermal martensite,
total strain-minimizing accommodations in the microstructure will
be removed as much as possible by what may generally be termed
"reorientation" or "stress-accommodation" processes. In some
cases, stress-induced martensite may even be of a different crystal
structure than thermal martensite in the same alloy (54), or different lattice invariant shear systems may operate, for deformation of thermal martensite, variations in the relative amounts of
internal deformation or of particular plate orientations may occur;
stress-induced transformation from one martensite structure to
another may even occur. (See other papers in this conference).
In general, the upper limit on realizable strain for a given crystallographic change is represented in the Bain strain, B. Good
correlation has been shown between theoretical crystallographic
calculations and experimental strain limits in cases where data is
available.

   Submicrostructural features.   It has been shown in certain
SME alloys that the lattice invariant deformation is described
by internal faulting rather than twinning (55-60). This "faulting"
is not to be regarded as slip in the classical sense of displacement, but a fault in the sequence of placement (of close packed
planes with respect to one another) in martensite formation. The
fact that "faulted" martensites have been regarded as essentially
"FCC structures with slip as inhomogeneous deformation" is a result of a purely phenomenological assumption and not reflective of
the actual mechanism by which these structures are formed. Therefore, these structures cannot be regarded as deformed in the sense
that glissile dislocations have traversed the microstructural units.
The mechanism of formation of the microdomains within faulted martensite plates is not clear at this time (59). It has been suggested that the atom motions of a premartensitic lattice wave stage
(43, 44, 61-70) could exhibit local variations in orientation from
place to place, giving rise to the observed irregular fine structure
within plates (57, 59); this scheme is fully consistent with the
concept that martensitic thermoelasticity is only realized if there
is a reversible path for A↔M (59).

   Recent observations in the authors' laboratories have suggested that in certain SME alloys, such as NiTi, several martensitic
structures may form simultaneously in the same sample (38). It
also appears that certain martensites may be defined as "imperfectly

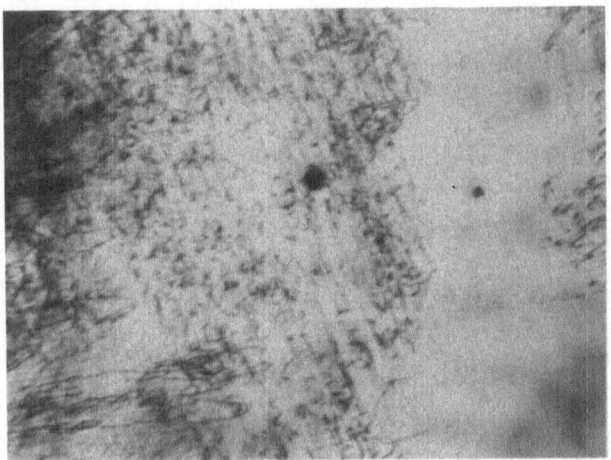

Figure 14:   Dislocation debris remaining above Af after three
             full thermal cycles (below Mf, above Af) in NiTi,
             showing non-uniform distribution of thermal trans-
             formation-generated substructre (from Ref. 38).

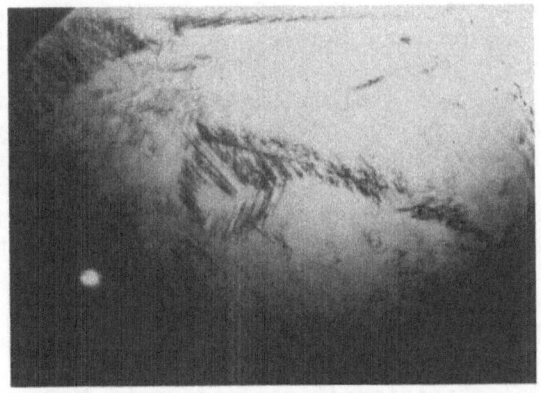

Figure 15:   Dislocation substructure (existing at about Af
             on heating) associated with strained (≈ 4%)
             thermal martensite in NiTi (from Ref. 38).

thermoelastic", at least in terms of formation from a fully annealed matrix, and therefore, these martensites tend to generate dislocations simply as a result of thermal formation (see Figure 14). Further thermal cycling increases the dislocation substructure as previously reported by Sandrock and coworkers (3, 61). Eventually, the alloy attains transformation behavior that is "truly thermoelastic", i.e., there is no further formation of dislocations on cycling, the elastic limit of the alloy having been effectively increased. In terms of thermo-mechanical performance, this effect of transformation cycling serves to explain the gradual attainment of complete shape recovery on (thermal or SME) cycling as seen in Figure 7. Note that detailed observation of dislocation formation by cycling processes in NiTi indicates that dislocations are primarily (A:M) interfacially generated in thermally cycled material, but that strain (below ε_L) imposed on martensitic alloys gives rise to extensive dislocations internal to M plates, including plate-centered dislocation tangle mid-ribs, as seen in Figure 15.

    <u>Lattice stability considerations</u>. Lattice softening and lattice oscillations prior to thermoelastic martensitic transformation appears to be a common characteristic (43, 50, 61, 62, 64, 70, 71). Due to a drastic reduction in the A phase shear eleastic constant, $C' = \frac{1}{2} (C_{11} - C_{12})$ as T_d is decreased toward Ms, martensitic transformation can be induced by very low values of resolved shear stress due to externally applied force (6, 17, 21, 50). The positive temperature coefficient of $\sigma_{A \to M}$ (Figure 3) is directly related to the variation of C' above Ms. On a submicroscopic lattic dynamics scale, low values of this critical stress may be related to a lowering of nucleation energy barriers by the coupling of specific lattice oscillation modes (62, 65, 67, 68). Experimental data also reflects the existence of premonitory lattic oscillations for both forward and reverse thermoelastic shear transitions (34, 75, 76); for example, resistivity curves exhibit discontinuities just above Ms and just below As on cooling and heating, respectively (34). This correlates with the existence of pseudoelastic behavior (spontaneous reversal on stress release) only above some critical temperature (approximately at As) where nucleation barriers for M→A are lowered by the action of lattice oscillations in the M structure.

SUMMARY: CONTROL AND UTILIZATION OF MARTENSITIC THERMOELASTICITY

    1. The general characteristics leading to thermoelastic martensitic transformation are small values of ΔG^c, the transformation volume change, $\Delta V_{A \to M}$, and the shear component of the macroscopic shape strain, and a high matrix phase elastic limit.

    2. From standpoint of utility of shape memory effects, large transformation strains and heats of transformation are favored;

since these features are not compatible with thermoelasticity requirements, compromises must be made in practice.

3. For each shape memory effect alloy, there is a strain limit, ε_L, beyond which the desirable SME parameters, such as % strain recovery, reversion stress, and work output, deteriorate; this limit is determined by the crystallographic parameters of the martensitic transformation.

4. Crystallographic and thermodynamic parameters are fixed for an alloy in a given alloy system, but can typically be manipulated via composition control.

5. The upper bound on attainable reversion stress (internal stress, σ_r), at any temperature and for a given martensitic prestrain, is the flow stress of the alloy at that temperature and strain value.

6. Matrix strength can be changed in a given alloy by thermomechanical treatment, such as by prestraining the high temperature phase above Md, or by transformation cycling.

7. There is a distinct temperature range over which reversion stress, σ_r, is relatively stable, with Ms(σ_r) as a lower bound and an upper bound at a critical temperature above which creep activation or substantial true plastic flow of the matrix phase takes place.

ACKNOWLEDGEMENTS

This work was sponsored in part by Office of Naval Research (ONR), Naval Postgraduate School ONR Research Foundation, NSWL, Dahlgren, VA, and NAVSEA Ship Silencing Program Office.

REFERENCES

1. C. M. Wayman and K. Shimizu, Met. Sci. J. 6 (1972) 175.

2. L. Delaey, J. Van Paemel, and T. Struyve, Scripta Met. 6 (1972) 507.

3. J. Perkins, Met. Trans. 4 (1973) 2709.

4. D. Dunne, J. Austral. Inst. Met. 19 (1974) 28-34.

5. L. Delaey, R. V. Krishnan, H. Tas, and H. Warlimont, J. Matl. Sci 9 (1974) 1521.

6. H. Pops, Met. Trans. 1 (1970) 251.

7. W. A. Rachinger, Brit. J.A.P. 9 (1958) 250.

8. M. W. Burkart and T. A. Read, Trans. AIME 197 (1953) 1516.

9. L. C. Chang and T. A. Read, Trans. AIME 191 (1951) 47.

10. W. J. Buehler, J. V. Gilfrich, and R. C. Wiley, J.A.P. 34 (1963) 1475.

11. Z. S. Basinski and J. W. Christian, Acta Met. 2 (1954) 101.

12. C. M. Wayman, Scripta Met. 5 (1971) 489.

13. A. Nagasawa and K. Kawachi, J. Phys. Soc. Japan 30 (1971) 296.

14. K. Enami and S. Nenno, Met. Trans. 2 (1971) 1487.

15. Y. K. Au and C. M. Wayman, Scripta Met. 6 (1972) 1209.

16. I. Cornelis and C. M. Wayman, Scripta Met. 8 (1974) 1321.

17. R. V. Krishnan and L. C. Brown, Met. Trans. 4 (1973) 423.

18. H. C. Tong and C. M. Wayman, Scripta Met. 7 (1973) 215.

19 K. Otsuka and K. Shimizu, Scripta Met. 4 (1970) 469.

20. D. V. Wield and E. Gillam, Scripta Met. 6 (1972) 1157.

21. J. D. Eisenwasser and L. C. Brown, Met. Trans. 3 (1972) 1359.

22. H. Pops and N. Ridley, Met. Trans. 1 (1970) 2653.

23. R. J. Wasilewski, Scripta Met. 5 (1971) 127.

24. S. Spinner and A. G. Rozner, J. Acoust. Soc. Am. 40 (1966) 1009.

25. A. Olander, J. Amer. Chem. Soc. 54 (1932) 3819.

26. E. J. Suoninen, R. M. Genevray, and M. B. Bever, Trans. AIME 206 (1956) 283.

27. S. Miura, T. Mori, and N. Nakanishi, Scripta Met. 7 (1973) 697.

28. N. Nakanishi, Y. Murakami, and S. Kachi, Scripta Met. 5 (1971) 433.

29. H. C. Tong and C. M. Wayman, Acta Met. 22 (1974) 887.

30. K. Otsuka, K. Shimizu, I. Cornelis, and C. M. Wayman, Scripta Met. 6 (1972) 377.

31. D. P. Dunne and C. M. Wayman, Met. Trans. 4 (1973) 137, 147.

32. K. Otsuka, T. Sawamura, K. Shimizu, and C. M. Wayman, Met. Trans. 2 (1971) 2583.

33. H. V. Schuerch, NASA CR-1232, November 1968.

34. W. B. Cross, A. H. Kariotis, and F. J. Stimler, NASA CR-1433, September 1969.

35. C. M. Jackson, H. J. Wagner, and R. J. Wasilewski, NASA-SP 5110, (1972).

36. C. R. Such, M.S.M.E. Thesis, Naval Postgraduate School, June 1974.

37. J. M. Johnson, M.S.M.E. Thesis, Naval Postgraduate School, March 1975.

38. R. R. Allen, M.S.M.E. Thesis, Naval Postgraduate School, March 1975.

39. J. D. Harrison, Raychem Corporation, Menlo Park, California (private communication)

40. K. Oishi and L. C. Brown, Met. Trans. 2 (1971).

41. A. Ball, S. G. Bergerson, and M. M. Hutchison, Proc. Int. Conf. Str. Met. and Alloys, Trans. Jap. Inst. Met. 9 (1968) (Supp.) 291.

42. G. D. Sandrock, Ph.D. Thesis, Case Western Reserve Univ. 1971.

43. R. F. Hehemann and G. D. Sandrock, Scripta Met. 5 (1971) 801.

44. C. M. Wayman, I. Cornelis, and K. Shimizu, Scripta Met. 6 (1972) 115.

45. R. J. Wasilewski, Scripta Met. 5 (1971) 207.

46. A. Nagasawa, K. Enami, Y. Ishino, and S. Nenno, Scripta Met. 8 (1974) 1055.

47. J. Perkins, Scripta Met. 8 (1974) 1469.

48. H. C. Tong and C. M. Wayman, Met. Trans. 6 (1975)

49. H. C. Tong and C. M. Wayman, Scripta Met. 8 (1974) 93.

50. N. Nakanishi, T. Mori, S. Miura, Y. Murakami, and S. Kachi, Phil. Mag. 28 (1973) 277.

51. H. Tas, L. Delaey, and A. Deruyttere, Met. Trans. 4 (1973) 2833.

52. H. K. Birnbaum and T. A. Read, Trans. AIME 218 (1960) 662.

53. T. Aoyagi and K. Sumino, Phys. Stat. Sol. 33 (1969) 317.

54. K. Otsuka and K. Shimizu, Phil. Mag. 24 (1971) 481.

55. H. Pops and L. Delaey, Trans. AIME 242 (1968) 1849.

56. W. Arneodo and M. Ahlers, Scripta Met. 7 (1973) 1287.

57. J. Perkins, Metallography 7 (1974) 345.

58. J. Perkins, Phil. Mag. 30 (1974) 379.

59. J. Perkins, Scripta Met. 9 (1975) 121.

60. T. Tadaki, M. Tokoro and K. Shimizu, Scripta Met. 8 (1974) 1077.

61. G. D. Sandrock, A. J. Perkins, and R. F. Hehemann, Met. Trans. 2 (1971) 2769.

62. L. Delaey, J. Van Paemel, and T. Struyve, Scripta Met. 6 (1972).

63. L. Delaey, A. J. Perkins, and T. B. Massalski, J. Matl. Sci. 7 (1972) 1197.

64. I. Cornelis, R. Oshima, H. C. Tong, and C. M. Wayman, Scripta Met. 8 (1974) 133.

65. P. C. Clapp, Phys. Stat. Sol. 57 (1973) 561.

66. J. Perkins, Scripta Met. 8 (1974) 31.

67. J. Perkins, Scripta Met. 8 (1974) 439.

68. J. Perkins, Scripta Met. 8 (1974) 975.

69. F. E. Wang, S. J. Pickart, and H. A. Alperin, J.A.P. 43 (1972) 97.

70. K. Chandra and G. R. Purdy, J.A.P. 39 (1968) 2176.

71. N. Nakanishi, Memoirs Konan University, Science Series, No. 15, Art. 77 (1972).

72. S. Zirinsky, Acta Met. 4 (1956) 164.

73. D. B. Novotny and J. F. Smith, Acta Met. 13 (1965) 881.

74. P. L. Young and A. Bienenstock, J.A.P. 42 (1971) 3008.

75. E. S. Fisher and C. J. Renken, Phys. Rev. 135 (1964) A482.

76. S. C. Moss, D. T. Keating, and J. D. Axe, BNL Report 18098 (1973).

SHAPE MEMORY EFFECTS AND APPLICATIONS: AN OVERVIEW

Walter S. Owen

Department of Materials Science and Engineering

M.I.T., Cambridge, Massachusetts

ABSTRACT

An attempt is made to identify those deformations, microscopic mechanisms and thermal stabilizing and recovery processes which appear to be operative in potentially useful shape memory alloys. The basic physical phenomena controlling these mechanisms are discussed briefly.

Some applications of shape-memory alloys have been developed to the commercial stage. Many more are being actively explored. Some of the problems relating to alloy properties likely to be encountered are reviewed. It is clear that the engineering of these alloys is at a very early stage. The potential for development appears to be very great.

INTRODUCTION

It is impossible to summarize in the space available the vast amount of new information about shape-memory alloys which has been presented at this Symposium. Instead, in this paper an attempt is made to evaluate our present understanding of the basic mechanisms, to point to some problems for which no convincing answers exist at present and to assess the prospects for exploitation of some of the effects described in other papers in this volume. The term 'shape memory' has been used to cover a wide variety of effects, but in all of them the shape at a high temperature is recovered after deforming the specimen at some lower temperature to a different shape and reheating to the original temperature (1). The two

305

major groups of alloys which exhibit this effect are binary alloys
of transition metals, one component to the left and the other to
the right of chromium (1) and β-phase alloys of the precious metals
(2). Most of them have a CsCl structure at the high temperature
and all transform martensitically to a low-temperature phase with
a lower symmetry. Most, but not all, of the higher-temperature
phases are ordered. The most extensively studied alloy of the
first group is NiTi. AuCd is typical of the second. Uranium and
some of its alloys also have a shape memory under suitable conditions. In these alloys the high-temperature phase is disordered
b.c.c.

DEFORMATION MODES

Martensitic transformation is a deformation process. The
total change in shape of the crystal lattice is usually assumed to
be an invariant-plane strain combining a pure shear with a change
in volume. The latter is probably unimportant in most shape-
memory effects. If, as is usual, a plate of martensite grows within the parent crystal, this change in shape must be accommodated
by strains in the untransformed crystal; the only exception being
the transformation of a single crystal by the movement of a single
interface. In addition the martensite must develop a lattice-
invariant strain as it grows. These strains are achieved by the
operation of one or more conventional deformation modes within the
martensite crystal. If now the partially or completely transformed
alloy is deformed by the application of external stress the
martensite, and the parent phase if any remains, will be strained
further. If the resulting total macroscopic strain is to be recovered on heating all the strains must be produced by deformation
modes which are mechanically reversible. If some mechanically-
irreversible deformation occurs the shape can be recovered only
partially.

Some mechanically reversible deformation processes which may
be operative in some shape-memory alloys are: elastic and anelastic
deformation, reversible-growth of martensite, transformation or
deformation twinning, movement of stacking-fault partials, slip by
superlattice dislocation in crystals with long-range order. It is
generally agreed that the interface between a parent and a martensite phase is coherent or semicoherent (Frank interface) and thus
there is no reason why the movement of such an interface should not
be reversible. It is the other deformations associated with the
transformations which determine that, for example, the interfacial
movement in Fe-30Ni is irreversible but in ordered Fe_3Pt is reversible. Some deformations which are irreversible mechanically
are: non-planar slip, irreversible growth of martensite, high-
temperature creep, various processes which result in relaxation

of dislocation configurations. It has been stated that the only deformation mode introducing the lattice invariant strain into the martensitic phase of shape memory alloys is twinning (1), but recent work has shown that in some Cu-Zn, Cu-Zn-Si and Cu-Zn-Al alloys which have a substantial, if not always perfect, shape memory the lattice invariant strain is produced by stacking faults (1,2,3). They have the untwinned "modified" 9R structure consisting of a long-period stacking structure modified by shuffles.

The only known mode of reversible deformation accommodating the lattice shape-change on forming martensite is elastic strain. Electron microscopic evidence for accommodation by the movement of superlattice dislocations has been found in ordered Fe_3Pt alloys (4), but it is not expected that these dislocations reverse direction when the alloy is cooled through the M_s temperature. Because the reversible accommodation strains are elastic all mechanically reversible growth of martensite is thermoelastic. The elastic energy stored in the specimen during the forward transformation contributes to the driving force for the reverse transformation leading to interesting effects which will be mentioned later.

There is a second important property which deformations involved in shape memory mechanisms must posses; the product of the deformation must be capable of being stabilized at the temperature of deformation and of being released at the higher temperature at which the original shape is recovered. If no such shape stabilizing mechanism operates at the temperature of deformation the deformed specimen simply springs back to its undeformed state when the applied stress is removed.

To produce an unrestricted change of shape of a single-crystal or polycrystalline specimen requires that at least five non-parallel independent shears operate (5). Several participants in the Symposium reported the "growth" of a "single crystal of martensite" by deformations which activated only one variant of the martensite. The macroscopic deformation which produced this situation must have been specific and restricted in such a way that in the final stages only one shear operated. This is not the general case and such specific deformations can never occur in randomly oriented polycrystalline specimens.

MECHANISMS INVOLVED IN SHAPE MEMORY EFFECTS

The following is a list of structural mechanisms reported at this Symposium or previously thought to operate separately or in combinations in many shape-memory alloys.

Inter-twin Growth

Macroscopic deformation by change in the thickness of two twin variants under an applied stress was first reported by Ölander (6) and studied in detail in AuCd and InTl alloys by Chang and Read (7). When discussing the mechanism Lieberman (8) emphasized that no movement of martensite-martensite interfaces is involved as can be shown by experiments with a "single crystal" of martensite formed by a single-interface transformation of a single crystal of the parent phase. The room temperature structure is orthorhombic. Twin striations in parallel array are traces of $\{111\}_O$ twin boundaries. On applying a stress these boundaries move so as to thicken those twins favorably oriented to the stress. On removing the stress the twin boundaries return, relatively slowly, to their original position and the specimen recovers its original shape. Recoverable strains may be as large as about 0.08 in AuCd. This "rubber-like" behavior has been called "ferroelastic deformation" and the memory has been described as a "piezomorphic memory" (8). If the deformed shape is held for some time, the specimen does not recover its original shape when the strain is released but it may do so on being heated to some higher temperature. Clearly, a thermally activated stabilization process is operative. Lieberman has proposed one such process which might operate in the AuCd alloys (8,9). It is based on the observation that although within the volume of crystal affected by a moving twin interface the lattices of the two variants are twin-related the crystal structures are not. When a $\{111\}$ interface moves under stress converting one variant to the other, local atomic rearrangements akin to shuffles are necessary to put the atoms in the proper twin positions. Until these rearrangements occur that volume of the crystal which has been twinned by the applied stress is in a higher energy state than the unaffected parts of the crystal and, consequently, when the stress is removed the direction of the movement of the interfaces is reversed. The atomic rearrangements, described in detail by Lieberman, require thermal assistance; the activation energy is estimated to be 26 K cal/mole. After the rearrangments have taken place the whole crystal is in its lowest energy state and on removing the stress no restoring force exists.

InTl has not been studied in the same detail as AuCd, but because InTl alloys are not ordered and there appears to be no distinction between the twinned lattice and the twinned atomic structure it is improbable that Lieberman's stabilization process operates in this alloy. Ferroelasticity has been observed, usually in conjunction with at least one other mechanism, in many shape-memory alloys. Large strains, up to 0.10, which can be fully recovered anelastically are usually possible by this mechanism. However AuCd seems to be the only alloy for which thermal stabilization has been clearly demonstrated. It is likely that the ferroelasticity behavior is easily reversible in most cases, stabilization, when it occurs, being provided by the other mechanisms which take place along with the ferroelasticity.

Stress or Strain-induced Reversible-growth Martensite.

The essential feature of this mechanism is that plates of martensite with orientations and shear directions favored by the dominant shear required by the imposed macroscopic change in shape grow, while other variants, less favorably oriented, shrink. The growth must be mechanically reversible and therefore, as indicated earlier, the growth process must be thermoelastic if the shape is to be completely recoverable. The martensite plates which participate in this mechanism may be nucleated under the action of the applied stress, in which case only the favorably-oriented variants will be produced, or they may be plates which have been formed athermally before the external stress is applied. In most alloy systems stress and thermally induced martensites are identical phases which exhibit the same thermoelastic behavior. Martensites formed in Cu-Al-Ni alloys are exceptional. Shimizu (10,11) reported that in a Cu-14.2 Al-4.3 Ni the parent β_1 phase transforms athermally to γ_1' but when the martensite is stress-induced at a temperature above A_f a different structure, β_1', is formed. Both γ_1' and β_1' grow reversibly thermoelastically.

The shape is stabilized when the martensite is formed at, or cooled under stress to, a temperature at which it is thermodynamically stable in an unstressed specimen. Thus, if the martensite is nucleated and grows under the action of a stress applied at a temperature above A_s, the growth will be reversed on removing the stress. If, however, the stress-induced martensite is formed from parent phase retained below A_s it will be stable and after the stress is removed the plates will remain unchanged. Reverse growth will start on reheating above A_s. The condition for thermal stability is complicated, in this case, by the necessity to include the stored elastic-strain energy in the energy balance which determines the A_s temperature.

The only condition which must be satisfied if growth is to be thermoelastic is that it should not be accompanied by irreversible deformations. Generally, this means that the extent of accommodation slip should be negligible.

When NiTi is thermally cycled between temperatures above A_f and below M_f, with no external stress applied, the area of the hysteresis loop decreases continuously over the first ten or twenty cycles and therafter is stable. The initial decrease is accompanied by a decrease in M_s on each successive cycle. For example,

Hehemann (30) showed some data for Ti-50.5 Ni demonstrating that M_s is reduced from 22°C to 0°C after 100 cycles. In Fe_3Pt the hysteresis loop decreases on cycling as does the loop for NiTi, but the M_s temperature increases with each successive cycle up to about five cycles (4,17). In specimens subjected to such a cycling treatment it is found, by electron microscopy, that a substantial density of dislocations has developed in the parent phase. This accumulation of dislocation debris has been described by Perkins (13). The shape recovery of a specimen subjected to only one or two cycles is imperfect. It improves as the number of cycles is increased and becomes nearly perfect in specimens which have been cycled a sufficient number of times to establish a stable hysteresis loop. It is suspected that the dislocation debris is the accumulation of sessile dislocations formed either by the dissociation of interfacial dislocation or the interaction of these dislocations with annealed-in dislocations during the forward or backward motion of the interface. The debris contains the dislocation arrangements which are the sites for the heterogeneous nucleation of athermal martensite on cooling (4). Thus, M_s increases as the accumulation increases the density of available sites up to some saturation limit beyond which the high density of sessile dislocations inhibits growth and thereby lowers the M_s temperature observed experimentally.

Reorientation of Martensite Plates Under an Applied Stress

Many examples were presented at the Symposium of macroscopic deformation by an externally applied stress through the growth of martensite plates favorably oriented in the stress field at the expense of less suitably oriented plates (see, for example, 1,10, 11,13,14). There appears to be common agreement that when the interface moves the habit plane is unchanged, suggesting that the ratio of the thickness of alternate transformation twins within the plates remains unchanged. The twins may thicken, however, by inter-twin growth (mechanism 1) at the same time as the reorientation of martensite plates is occuring. Thus the "rubber-like" behavior described earlier may be an important component of the total strain produced by the reorientation mechanism.

This mechanism has been studied extensively in fully-transformed alloys deformed below A_s. It is possible, in principle, that in partially transformed specimens it could operate in competition with the formation of stress-induced martensite, although no examples seem to have been reported. Delaey (15) described the behavior of Cu-Zn-Al in which the interfaces between stress-induced martensite plates are glissile under the action of an applied stress whereas interfaces with plates of martensite formed athermally are not.

The process responsible for stabilizing the shape of specimens deformed below M_f is unclear. Warlimont (14) suggested that "interfacial frictional forces" may be sufficient to prevent the interface returning to its original position after the stress is removed. The origin of this "interfacial friction" is not known. Presumably, it is the negative of the stress required to move the interfacial transformation-dislocations through a martensite lattice. Since the interfaces appear to be glissile, moving easily when only a very small stress is applied, and since the magnitude of the friction stress cannot depend upon the direction of motion, it would seem that frictional stresses of this kind are very small. However, the restoring force may be very small also. Possibly, there may be none. Since below M_s all variants of the martensite are stable thermodynamically, the only ways in which the total energy of the system can increase is by the interfacial area per unit volume being increased by the interfacial boundary migration or by the storage of additional elastic strain as a result of the deformation. It is unlikely that either of these terms will be of significant magnitude. The former term may be negative since the grain boundary area per unit volume may be decreased by the preferential growth of less than five variants. Thus, the configuration of martensite plates produced by the imposed macroscopic shape change may, in fact, be mechanically stable.

A special and specific mechanism whereby martensite plates can be reoriented under stress has been proposed by Wasilewski (16). He describes it as a "stress-assisted double transformation" and proposes that, at temperatures between A_d and M_f, when a stress is applied martensite transforms to β' by the "stress-assisted reverse transformation" and then, because it "is both thermally unstable and subject to an applied stress" it transforms instantaneously to a martensite variant more favorable to the applied stress field than was the original plate. It must be admitted that, at present, the experimental evidence for the transitory existence of an intermediate β' phase is far from conclusive.

Martensite-to-Martensite Transformations under Stress

Shimizu (10) showed that in Cu-Al-Ni alloys not only could two different martensites, β_1' and γ_1', form from the parent β_1 depending upon the transformation temperature but specimens completely transformed to β_1' can be partially transformed to γ_1' by applying a stress at a temperature below M_s. This $\beta_1' \rightarrow \gamma_1'$ transformation does not reverse on unloading. The shape stabilization mechanism is unknown. Considerations similar to those relating to the stabilization of microstructures developed by reorientation of only one form of martensite probably apply. Delaey (15) demonstrated a similar stress-induced martensite-to-martensite transformation in a Cu-Zn-Al alloy.

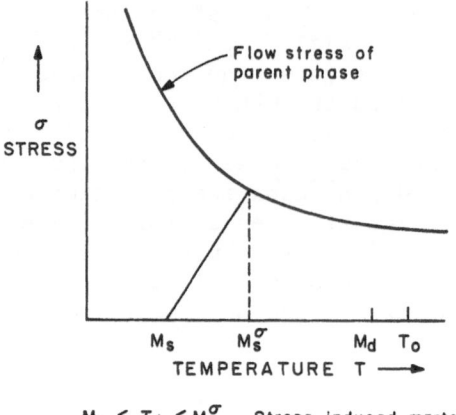

$M_s < T_d < M_s^\sigma$ Stress induced martensite

$M_s^\sigma < T_d < M_d$ Strain induced martensite

Figure 1.   Schematic representation of the stress to cause the first deformation as a function of temperature.

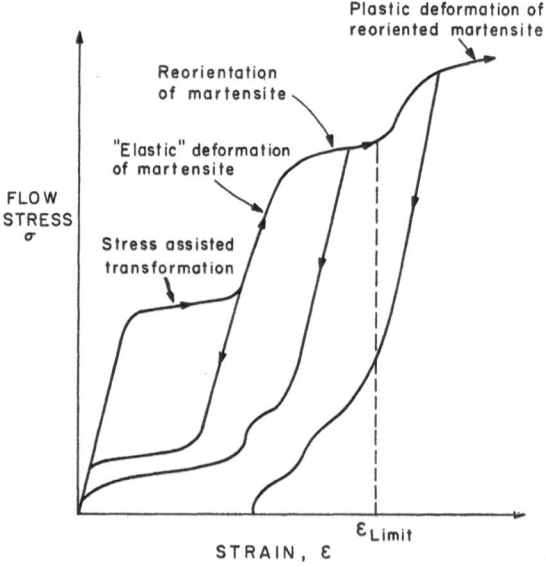

Figure 2.   Schematic representation of isothermal deformation at $A_s < T_d < M_s^\sigma$.

THE MAGNITUDE AND TEMPERATURE OF THE DEFORMATION

The relationships between the different transformation temperatures on cooling and the applied stress, σ, are represented schematically in Figure 1. At T_0 the Gibbs free-energies of the parent and martensitic phases are equal. At M_s, martensite starts to form spontaneously in a stress-free specimen. On cooling to a temperature between M_s^σ and M_s the transformation can be induced by the application of a stress smaller than the flow stress of the parent phase and the deformation is by martensite formation. However, at high temperatures, between M_d and M_s^σ plastic flow by slip in the parent phase precedes the formation of martensite. Between M_s^σ and M_s the martensite is stress-induced. Between M_s^σ and M_d it is strain induced. The martensite formed in the latter temperature range may still be reversible but only if no significant additional slip occurs during the transformation.

Stress-strain relationships on loading a typical shape-memory alloy at a temperature above A_s but below M_s^σ are shown schematically in Figure 2. At small strains the deformation is by stress-induced martensite. On removing the stress the growth direction is reversed and the initial shape of the specimen is recovered. At medium strains, the alloy is completely martensitic and further strain is produced by the reorientation of the plates of martensite. The former is sometimes described as "superelastic" deformation and the latter as "ferroelasticity" but, unfortunately there is no consensus at present about the application of these and similar terms used to describe the deformation behavior of shape-memory alloys. The deformation resulting from the reorientation of martensite under an applied stress is also recoverable completely on unloading up to some limiting strain, ε_{limit}, beyond which slip occurs thereby introducing some irreversible deformation into the specimen. The limiting strain is usually near 0.10. It varies with grain-size, texture, alloy compositions and "the conditions under which the strain is effected" (16).

If, however, the temperature at which the stress is applied is below A_s but above M_f, although the successive deformation modes introduced with increasing magnitude of the deformation are unchanged, growth is not reversed on removing the stress and the initial shape is not recovered until the specimen is reheated to A_f or a higher temperature. Of course, if it has been strained too far, the irreversible deformation introduced will not be recovered at any temperature.

The sequence of events of stressing, removing the stress, and reheating are similar when the temperature of deformation is below M_f, except that the first step, the formation of stress-induced martensite, is absent because the specimen in its initial condition consists of martensite formed athermally above M_f.

REVERSIBLE SHAPE MEMORY

The most commonly studied form of the shape-memory effect is the one by which the shape of a specimen deformed at a temperature below M_f is recovered on reheating to above A_f. The magnitude of the shape change that can be recovered completely is limited by the requirement that the maximum strain at any location must be less than ε_{limit}. This sequence can usually be repeated a large number of times provided the stress is reapplied at the appropriate temperature in each cycle. There has been reported in the literature and at this Symposium examples of a much more surprising shape memory effect whereby the movement from the initial to the deformed shape occurs simply by thermally cycling between A_f and M_f without any application of external stress after the first cycle. This has been described as the reversible, repeatable or multiple shape memory effect. For it to occur an additional basic mechanism, not discussed in detail in the Symposium, must operate; the alloy must possess a microstructural memory. That is, each time the martensite forms on cooling the parent phase below M_s plates must nucleate at the same locations and grow in the same orientations as the martensite plates present in the specimen after straining below M_f on the first cycle. Further, the growth development of each plate must be the same each time the specimen is cooled. This probably means that the sequence in which the plates disappear on heating is exactly reversed on cooling and both sequences are repeatable. These conditions are necessary because the shape change is arbitrary and consequently five or more shear modes must operate during the deformation on the first cycle; each shear being determined by the growth development of martensitic plates of a single orientation. There is a negligible probability of obtaining the same shape change in subsequent thermal cycles by a different combination of more than five shears.

BASIC PHENOMENA

There are a small number of basic phenomena involved directly in one or more of the mechanisms responsible for the various shape memory effects.

Stress and Strain-induced Nucleation and Stress-Directed Growth

Interactions of the applied stress field with the nucleation and growth processes are, presumably, through interactions of elastic strain fields. The nucleation sites are probably the same,

or similar to, those activated during athermal nucleation. During
growth, the variant favored is that on which the shear stress,
determined by the Schmidt product, is a maximum. If the applied
stress produces slip before transformation, then the nucleation
sites might well be produced by this slip and the sites will be un-
related to the athermal sites. Microstructurally, it is usually
easy to distinguish between stress- or strain-induced martensite
and that formed athermally.

Thermoelastic Reversible Growth

To reverse the direction of growth of a plate of martensite
embedded in a matrix of parent phase by reversing the direction
of the applied stress, or by changing from cooling to heating, requires that the interface between the two phases be glissile and
stable. It is also necessary for all the strains involved in the
transformation to be reversible, which usually means that the
shape change has to be accommodated elastically. Consider the
growth of a lenticular plate, axial ratio c/r, growing in a matrix
of parent crystal. The c/r ratio during free growth will be determined by kinetic restrictions and according to Raghavan and Cohen
(20) will be constant. When radial growth is blocked by other
grains or plates or some other obstacle, the plate will continue to
thicken as the temperature is decreased. If all the accommodation
strain is elastic, the ratio c/r will reach a maximum value at each
temperature and all growth will stop when the decrease in chemical
free energy, $\Delta G^{P \to M}$, is balanced by the increase in stored elastic
strain energy and interfacial energy. Except when the plate is
very small, the latter term is probably negligible. An analysis
similar to the continuum elastic calculations of Eshelby (21) shows
that, after radial growth of the plate is blocked, the size of the
zone within which slip is expected to occur (that is, within which
the flow stress σ_f is exceeded) increases with decreasing value of
σ_f^P/μ_p^P (22) assuming that the shear modulus of the parent crystal,
μ, and of the martensite, μ^M, are nearly equal. This indicates,
not surprisingly, that elastic accommodation is greater the higher
the flow stress and the lower the modulus.

The magnitude of the chemical free-energy difference, $\Delta G^{P \to M}$,
which is available for conversion into mechanical work at the M_s
temperature is a nearly linear function of the temperature differ-
ence $\Delta T = M_s - T_o$. Thus, it has usually been assumed that $\Delta G^{P \to M}$
at M_s must be small because it has also been usually assumed that
the energy stored is relatively small (3,4,17). The problem is to
decide how small $\Delta G^{P \to M}$ really is. Experimentally, it is necessary
to distinguish between stored elastic energy and a change in chemical free energy; both forms of energy being reduced on heating.
Measurements in a differential calorimeter of the "latent heats" of

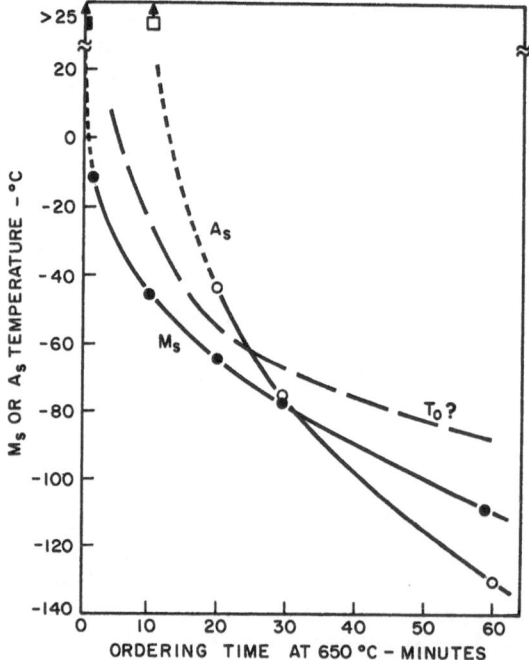

Figure 3.   The M_s and A_s temperatures of Fe-25 at pct Pt alloy as
a function of ordering time at 650°C.

transformation indicate that the total enthalpy change is small in thermoelastic martensites. In Fe_3Pt, Tong and Wayman (24) found values of about 80 cal/mole for the total energy change and similar values (50-80 cal/mole) were found by Djuric (25) in Fe-27Pt alloys. These data should be compared with a value of about 600 cal/mole for irreversible martensites, such as Fe-30Ni, in which there is no significant contribution from the release of stored elastic energy. Wayman (3) assumes that in the thermoelastic alloys the measured energy is nearly all elastic energy, but the basis of this assumption is questionable. There is clearly a need for independent measurements of the stored elastic energy.

From a theoretical point of view there is a major problem which must be solved before an acceptable model of the thermodynamics of a thermoelastic martensitic transformation can be developed; the position of T_0 in relation to M_s and the other transformation temperatures. When an ordered parent phase transforms martensitically, the Bain strain translates the atoms into their new positions in the martensite but atomic order is preserved. However, the imposed order in the martensite is not necessarily the same as the order which may exist in the low-temperature phase in thermal equilibrium. Since there is little or no change in the relative positions of the atoms, the additional energy of the martensitic non-equilibrium phase cannot be due to a change in the configurational entropy and must, therefore, be a result of changes in the vibrational entropy No estimates of the magnitude of these entropy changes have been made and, consequently, no estimate of the value or the temperature dependence of the Gibbs free energy of the ordered martensitic phase can be made at the present time. Thus, reliable T_0 values are not available as yet for the thermoelastic case.

It is usual, for irreversible martensites, to assume that $T_0 = (M_s + A_s)/2$, but, as pointed out by Tong and Wayman (3,24), this experimental approach to the determination of T_0 cannot be applied to thermoelastic martensites. Some experimental measurements of M_s and A_s for Fe_3Pt as a function of a parameter (annealing time at 650°C) which increases with the degree of perfection of the order are shown in Figure 3 (4). When long-range order is established in the parent phase, the transformation becomes thermoelastic and M_s is higher than A_s. The change is due, of course, to the release of stored elastic-strain energy contributing to the driving force for the reverse transformation. Tong and Waymen have assumed that stored elastic energy is of no significance in the energy balance determining the nucleation of martensite at M_s on cooling. Thus, T_0 must be greater than M_s. On heating, the last martensite plate to transform to the parent phase would then be influenced by an elastic strain field of negligible magnitude compared with that experienced by plates transforming earlier and, consequently, it can be assumed that A_f is affected very little by the release of stored elastic energy. This led Tong and Wayman (3,24) to suggest that $T_o = \frac{1}{2} (M_s + A_f)$. This cannot be true, however, because, as consideration of the reverse kinetic path shows, A_f must be lower than T_o. The difference may be small because it represents the contribution of the elastic strain energy and the interfacial energy of the last plate to the reverse driving-force.

There exists now a number of theoretical models which account satisfactorily for the main feature of the athermal heterogeneous nucleation of irreversible martensites with a driving force of the order of 300-350 cal/mole (14,26). In each of these models the driving force is approximately linearly related to the shear modulus of the parent phase. Thus, in the relatively elastically-soft phases in which thermoelastic martensites form it might be expected that the driving force at M_s might be reduced by a half or a little more. Until either credible thermodynamic models, or experimental methods of measuring separately the contributions of the chemical energy and the stored elastic energy to the total transformation energy are developed, the validity of these nucleation models when applied to thermoelastic martensites must remain in doubt.

In most, but not all thermoelastic martensites, the plates of martensite grow in stacks of nearly parallel plates alternating in shear direction so that the total assembly of plates is self-accommodating (23). Consequently, after the first few plates of the stack have formed the strains to be accommodated during the growth of the martensite are those surrounding and within the self-accommodating stack. There is evidence that many plates of one variant grow first and that the second variant then grows to fill in the interstices. The complex strain fields which develop during the growth of self-accommodating plates have not been analyzed. Consequently, estimates of the magnitude of the elastic strain-energy stored during growth cannot be made at the present time. However, it is clear that the strength and shear modulus of the growing plate become increasingly important, relative to the corresponding para-meters for parent phase, as the transformation proceeds.

Rubber Elasticity

There seems little doubt that the phenomenon commonly called rubber elasticity involves the motion of coherent twin interfaces, as was clearly demonstrated by Chang and Read (7) more than twenty years ago. There is much greater uncertainty about the nature of the driving force which reverses the direction of the strain when the applied stress is removed and the relaxation processes which occur in those cases in which a permanent set can be given to a deformed specimen by annealing for some time before the stress is removed. The Lieberman model (8,9) appears to require the twinned phase to have a complex, but ordered, crystal structure. It provides a plausible explanation of the reverse driving force and the relaxation process in AuCd. However, InTl exhibits exactly parallel rubber-like behavior, but it is not ordered and when twinned the atoms and the lattice points are both in twinned positions. Thus, the Lieberman model does not appear to be capable of extensive generalization.

Microstructural Memory

A microstructural memory has been observed directly on thermally cycling unstressed specimens of partially-ordered, and of ordered, Fe_3Pt (4,17). The plates do, in fact, start to shrink on heating in the reverse order to that in which they nucleate and grow on cooling (17). Crystals of the parent phase formed by the reverse growth of martensite plates contain aggregates of dislocation loops at locations in the lattice at which the last vestiges of martensite plates disappeared (18). The dislocations within these aggregates are sessile and have a Burgers vector which is the same as that of the usual glide dislocation in the parent lattice. It is very probable that the aggregates identify the sites at which martensite plates are heterogeneously nucleated on cooling (4). This specific distribution of the dislocation debris formed by the reverse transformation provides the basis of an explanation of the constancy of nucleation site locations on repeated thermal cycling, but the reason why on each cycle the newly formed plate at a specified location is the same variant as that formed in all the preceding cycles is still obscure.

In some alloys, debris from the reverse-martensite transformation can nucleate plates of bainite on subsequent cooling to a suitable temperature. This is perhaps not surprising since the bainitic and martensitic nucleation processes must be similar to each other. What is more remarkable is that the bainite formed by the decomposition of the β-phase Cu-Zn-Si alloy described by Pops (19) must have developed the same shears as those produced by the martensite formed by the cooling step in the preceding thermal cycle since a "reverse shape-memory" developed. The study of shape-memory effects depending on bainitic transformations clearly needs attention.

SOME GENERAL EFFECTS

Since Christian (27), in 1965, first tried to identify some common features of thermoelastic martensitic transformations there have been several attempts to specify conditions necessary for the transformation to proceed thermoelastically. Three of the most popular conditions proposed are that the parent phase must have a complex crystal structure, be ordered and be subject to lattice softening as the temperature is reduced. All three conditions can, in appropriate circumstances, be important factors but none, alone or combined with others, are both necessary and sufficient conditions. The only condition that is appropriately so described is that no irreversible strains shall be produced by the transformation.

In this Symposium, Foos, Frantz and Gautois (28) challenged the assumption by Wayman (1) that for the transformation to occur thermoelastically it is necessary for the parent phase to be ordered. They described the results of some experiments with quenched specimens of Fe_3Pt in support of their contention, but on close examination it appears that these specimens contained substantial short-range order based on the Fe_3Pt superlattice and so the result was not significantly different from those reported earlier by Dunne and Wayman (17) and by Kajiwara and Owen (4). However, it did draw attention to the fact that some degree of ordering, usually long-range ordering, is present in all parent phases which are known to transform thermoelastically. Slip in ordered structures occurs by the movement of superlattice dislocation pairs and, consequently, it is confined to a single glide plane. This might be a factor ensuring that the accommodation deformation is reversible or that the transformation interface is stable and reversible, but no direct evidence of either of these effects is known. Probably a more important role of ordering is to strengthen the parent and the martensite phases without affecting significantly the value of the shear modulus, thereby increasing the ratio σ/μ and reducing the probability that irreversible slip will occur. Unlike other forms of strenghtening (by precipitates, interstitials, vacancies, dislocations, etc.) order strengthening does not introduce disturbances in the lattice which are likely to disorganize the transformation interface when they interact with it. It has been demonstrated experimentally (29) that even the short range order introduced into Fe_3Pt on quenching in a manner similar to that adopted by Foos, Frantz and Gautois is sufficient to increase the flow stress to three times its value in nearly completely disordered specimens.

It has been suspected for some time that the parent phases of all thermoelastic alloys soften elastically on cooling. In fact, in most cases, the temperature derivative of the elastic shear constant $(C_{11} - C_{12})/2$, $\partial C'/\partial T$, is strongly positive. This suspicion has been reinforced by much data presented at this Symposium (30-33). No exceptions have been reported, although in $CuSn_{15}$ the slope of $\partial C'/\partial T$ is nearly zero (31), and this generalization can now be considered to be well established. In some alloys of the transition metals the lattice softening is accompanied by a magnetic transformation, the Invar effect, but during thermoelastic martensite transformation these alloys do not appear to behave differently from the β-phase alloys which do not change magnetically. In all thermoelastic alloys significant lattice softening of the parent phase has been measured at temperatures above M_s, in many cases over a wide temperature range. Further, recent measurements by Nakanishi (31) indicate clearly that in those alloys examined the lattice softening extended into the martensitic phase, a result suggested also by some of Hehemann's results (30).

The exact nature of the lattice softening phenomena is not well understood. Some of the problems were discussed by Rusović and Warlimont in terms of a simple lattice potential model. Many mysteries are likely to persist until much more experimental work using sophisticated techniques, such as electron scattering, has been completed. There is, however, strong evidence that the softening revealed by a change in sign of $\partial C'/\partial T$ is accompanied by soft phonon instabilities similar to those involved in ω-transitions (30,32,33). This has led several contributors to the Symposium to suggest that the assumed anomalously small thermodynamic driving force at M_s is due to the nucleation of martensite in a soft lattice being assisted by long-wavelength phonon instabilities. Clapp (34) has suggested the possibility of the existence of a strain spinodal at a free surface and Delaey and Warlimont (14) have attempted to combine this concept with a detailed crystallographically mechanistic mode of the β to 3R- or 3H- martensite transition to develop a model of the nucleation of these martensites. The model uses the point-group matching concept developed by Mendelson (35). Energetically the model is qualitative and so it is difficult to decide whether or not this concept of nucleation through "premartensitic shear instabilities" does, in fact, predice an unusually small value of $\Delta G^{P \to M}$ at M_s. The nucleation of a second phase through soft-mode instabilities is the subject of much interest at the present time (36) and quantitative developments of these and related ideas can be expected in the next few years.

As with ordering, it is probable that the major, and certainly most direct, effect of lattice softening on the nature of martensitic growth is on the ratio σ_f/μ. A decrease in the shear modulus increases the magnitude of the shear deformation which can be accommodated elastically. The softening of the shear modulus is a strongly anisotropic effect. Using the inosotropic elastic constants, estimates of the size of the zone within which slip occurs during the growth of a plate of martensite are smaller than calculated following Eshelby's analysis which uses isotropic elasticity (21, 22). It should be noted also that no attempts to incorporate the anisotropic lattice-softening effects with recent defect models of the heterogeneous nucleation of martensite have yet been reported. When this is done, the discrepancy between the driving force at M_s predicted by the nucleation models and the assumed anomalously low energy may be reduced.

PRACTICAL APPLICATIONS

Impressive developments were reported at the Symposium in
the practical application of shape memory alloys in three areas;
devices depending on shape recovery, heat engines and acoustical
damping materials. Shape-recovery devices were probably the earli-
est application of NiTi. Harris and Hodgson's (37) description
of commercial electrical and mechanical connectors operating on
a temperature difference of more than $100^{\circ}C$ demonstrated that
satisfactory engineering solutions have been found to the anticipated problems of strength, fatigue life and loss of memory. Nor
are applications confined to NiTi as was shown by the use of a
copper-base alloy in the application of the shape memory effect
to the manufacture of an integrated circuit package (19). As a
wider range of practical and economical alloys are developed it
will become possible to tailor the alloy to the optimum temperature
cycle for the design and to select alloys and thermo-mechanical
treatments to give, within limits, the optimum combination of
strength, ductility, fatigue resistance, corrosion resistance,
creep resistance, etc.

The feasibility of constructing a heat engine using the shape
change on thermal cycling as the motive force was shown clearly
by the prototype engines demonstrated by Banks and Weres (38)
which use NiTi. Delaey (15) showed moving pictures of a Banks
engine which uses Cn-Zn-Al as the shape memory alloy. NiTi has
severe limitiations in heat-engine applications; the most important
being cost, difficulty of fabrication and relatively low mechanical
strength. The possibility of developing engines using low-cost
copper alloys is attractive. Wayman (3,39) and Banks (38) independently estimate that efficiencies greater than 20 per cent are
theoretically possible. Wayman points out that a high efficiency
is favored by a large latent heat, a high recoverable strain, a
low operating temperature and a large temperature difference in
the themal cycle. With the rapid development in understanding shape-
memory alloys which is evident in this Symposium in the near future
it should be possible to design an alloy to optimize the thermal
efficiency of a shape-memory driven engine. Such engines might
prove to be of considerable importance in connection with the
utilization of large, low-temperature heat sources such as solar
energy.

Finally, Kaufman, Kulin, Neshe and Salzbrenner (40) described
efforts currently underway to understand the unusually high
acoustical damping exhibited by some thermoelastic martensitic
alloys. NiTi increases markedly in damping capacity on cooling into
the temperature range in which premartensitic lattice-softening
effects are detected. The high damping presists after the alloy
has become martensitic, but some fall-off is observed below M_s.

The magnitude of the corresponding decrease in damping in other thermoelastic alloys after they have transformed to martensite is not known. A practically useful damping alloy must have acceptable strength, ductility and stiffness. Thermoelastic martensites formed athermally from annealed parent-phase are usually weak and the elastic modulus decreases markedly due to lattice-softening at temperatures above M_s. If solutions to these two problems can be found thermoelastic martensites may well prove to be important damping materials.

ACKNOWLEDGEMENTS

Many of the ideas and opinions evident in this short review developed in numerous discussions with my colleagues at M.I.T.: particularly, J. W. Cahn, M. Cohen, L. Kaufman, G. Olson and R. Salzbrenner. It is hard to identify the origin of each thought and so I apologize if I have omitted to give proper credit. On the other hand, obviously my colleagues should not be held responsible for any of the opinions I have expressed.

The work on Fe-Pt alloys referred to in the text was sponsored by the National Science Foundation under contracts numbered H-33635 and DMR-74-22719.

REFERENCES

1. C. M. Wayman and K. Shimizu, Metal Sci. J., 6, 175 (1972).

2. H. Warlimont and L. Delaey, Progress in Materials Science, Pergamon Press, 18 (1974).

3. C. M. Wayman, this Symposium.

4. S. Kajiwara and W. S. Owen, Met. Trans., 5, 2047 (1974).

5. R. von Mises, Z. angew. Math. u. Mech., 8, 61 (1928).

6. A. Ölander, Z. Krist., 83A, 145, (1932).

7. L. C. Chang and T. A. Read, Trans. A.I.M.E., 191, 47 (1951).

8. D. S. Lieberman, M. A. Schmerling and R. W. Karz, this Symposium.

9. D. S. Lieberman, Phase Transformations, A.S.M., 1 (1970).

10. K. Shimizu, this Symposium.

11. K. Otsuka, H. Sakamoto and K. Shimizu, this Symposium.

12. G. Olson, unpublished calculations.

13. A. J. Perkins, Met. Trans., 4, 2709 (1973) and this Symposium.

14. L. Delaey and H. Warlimont, this Symposium.

15. L. Delaey, this Symposium.

16. R. J. Wasilewski, Met. Trans., 2, 2973 (1971) and this Symposium.

17. D. P. Dunne and C. M. Wayman, Met. Trans., 4, 137 (1973).

18. S. Kajiwara and W. S. Owen, Met. Trans., 4, 1988 (1973).

19. H. Pops, this Symposium.

20. V. Ragavan and M. Cohen, Acta Met., 20, 779 (1972).

21. J. D. Eshelby, Proc. Roy. Soc., A241, 376 (1957).

22. G. Olson, M.I.T., unpublished calculations.

23. J. W. Christian, <u>Physical Properties of Martensite and Bainite</u>, Iron and Steel Inst. Spec. Report 93, 1 (1965).

24. H. C. Tong and C. M. Wayman, Acta Met., <u>22</u>, 887 (1974).

25. B. Djuric, M.I.T., unpublished data.

26. G. Olson, Sc.D. Thesis, M.I.T., 1974.

27. J. W. Christian, <u>The Theory of Phase Transformations in Metals and Alloys</u>, Pergamon Press, Oxford (1965).

28. M. Foos, C. Frantz and M. Gautois, this Symposium.

29. W. S. Owen and J. L. Nilles, Symposium on Mechanical Behavior of Materials, Kyoto (1974).

30. R. F. Hehemann and S. Vatanayon, this Symposium.

31. N. Nakanishi, this Symposium.

32. K. Mukherjee, this Symposium.

33. N. Rusović and H. Warlimont, this Symposium.

34. P. C. Clapp, Phys. Stat. Sol. (b), <u>57</u>, 561 (1975).

35. S. Mendelson, this Symposium.

36. For example, P. C. Clapp, J. W. Cahn, M. Green, S. Moss, M.I.T. Conference on Martensite, unpublished (1975).

37. J. D. Harrison and D. E. Hodgson, this Symposium.

38. R. Banks and O. Weres, this Symposium.

39. H. C. Tong and C. M. Wayman, Met. Trans. <u>6A</u>, 29 (1975).

40. L. Kaufman, S. A. Kulin, P. Neshe, R. Salzbrenner, this Symposium.

DIRECT OBSERVATION OF MARTENSITIC TRANSFORMATION

BETWEEN MARTENSITES IN A Cu-Al-Ni ALLOY

K. Otsuka*, H. Sakamoto and K. Shimizu

Institute of Scientific and Industrial Research

Osaka University

Yamadakami, Suita, Osaka 565, Japan

(*Presently with the Department of Metallurgy and

Mining Engineering, University of Illinois

Urbana, Illinois 61801, U.S.A.)

I.   Introduction.

The martensitic transformations in metals and alloys usually occur from matrix to martensite by cooling below some critical temperature, that is, from a loosely packed structure to more close packed structure in order to lower the internal energy, except for ferrous alloys where the magnetic energy is more significant.   These transformations are usually characterized by the presence of the habit plane, the lattice invariant shear and the lattice rotation, and their crystallographies are well accounted for by the phenomenological theory under the invariant plane strain assumption (1, 2).   Now, if stress is couppled as the second variable, even the martensitic transformation between martensites may become possible, in cases where several martensitic structures with similar free energies are available (3).   In fact, there have been a few reports on the crystal structure change from one martensite to another by deformation in such alloy systems, Cu-Al ($4 \sim 7$) and Au-49(at%)Cd (8), although the transformation process itself is not fully investigated. Further, the present authors recently found that the martensitic transformation occurs from one martensite to another under certain conditions in a Cu-Al-Ni alloy (9).   The purpose of the present paper is to describe in detail on how the transformation between martensites proceed morphologically and crystallographically in such an idealized situation that the starting specimen is a martensite single crystal.

It is well established that Cu-Al-Ni alloys with composition near Cu-14.2Al-4.3Ni (wt%) transform martensitically from the β_1 matrix phase (DO_3 type ordered structure) to the γ_1' martensite phase (2H type stacking order structure) when cooled below the Ms temperature (10, 11). However, the structure of the martensites stress-induced above Ms is temperature dependent (12~14). That is, if the temperature is higher than Af, the β_1' martensite (18R type long period stacking order structure) is stress-induced, while if the temperature is lower than Af, the γ_1' martensite is stress-induced. The crystal structures of the two martensites are shown in Fig. 1. As is seen from the figure, both structures are quite similar in a sense that both are kinds of long period

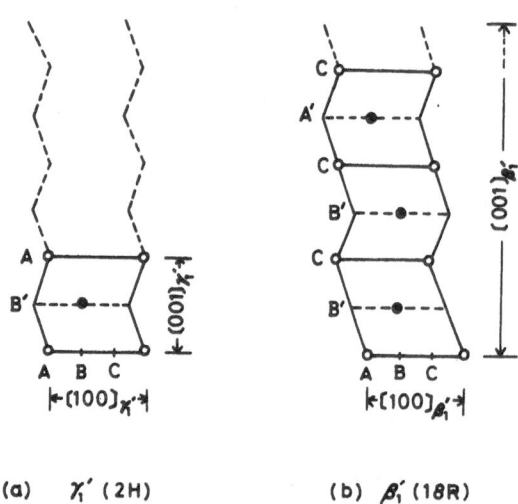

(a) γ_1' (2H)   (b) β_1' (18R)

Fig. 1. Crystal structures of γ_1' (a) and β_1' martensites (b) in Cu-Al-Ni alloy as viewed from $[010]_{\gamma_1}$' or $[010]_{\beta_1}$' direction, respectively. Open circles represent Al atoms, and closed circles Cu or Ni atoms. (After Nishiyama and Kajiwara (20))

stacking order structures with the common basal plane. Thus, the structural change from the β_1' martensite to the γ_1' martensite, which is described in this paper, may be accomplished only by shuffling in the $[100]_{\beta_1}$' direction on the $(001)_{\beta_1}$' basal plane.

II   <u>Experimental Procedure</u>

The methods of specimen preparation and single crystal fabri-

cation have been described elsewhere (11, 15). Two single crystals with the composition Cu-14.1Al-4.2Ni(wt%) have been used and
their orientations are shown in
Fig. 2. The final heat treatment of the specimens was such
that

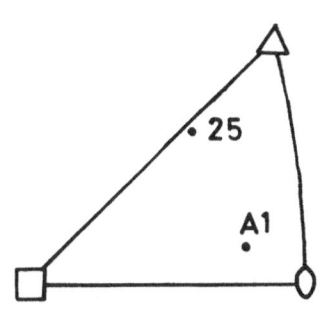

$$1000°C \times 2hr \longrightarrow W.Q.$$

The characteristic temperatures
of Specimen Al were Ms = -18°C,
Mf = -30°C, As = 2°C and Af =
21°C. Those of Specimen 25 were
not measured, but they are to be
quite close to those of Specimen
Al, since the single crystal was
made from the same lot. After
shaping by a Servomet Machine
into tensile test specimens with
the gauge length of 20 mm, they
were electropolished for suffi-

Fig. 2. Orientations of
specimens used in the present investigation.

cient time to remove strains introduced by the spark cutting.
  Then the specimens were tensile tested with an Instron Machine
of TT-CM-L type, which is equiped with a temperature control chamber. The macroscopic morphological change during the tests were
recorded by a camera with a telescope lense. The special attach-
ment has been made to take the back Laue reflection pattern from
specimens under tension, and the same apparatus has been used for
the microscopic observation of the transformation with an optical
microscope by immersing the whole system into an cooled ethanol
liquid.

III   Results and Discussion

  The process of the martensitic transformation from the β_1'
martensite to the γ_1' martensite is realized by the following
three steps. First we stress-induced the β_1' martensite from the
β_1 matrix single crystal at a temperature above Af. This step is
represented by the stress-strain curve of Region I in Fig. 3. As
reported previously (3, 14), the curve between $\underline{B}$ and $\underline{C}$ is characteristic of the β_1 to β_1' transformation, and at point $\underline{C}$ the
specimen became a single crystal of the β_1' martensite. Then,
as the second step, the specimen was cooled keeping the stress
constant, as shown in Region II in Fig. 3. However, no structural or morphological change was observed during the cooling down
to the liquid nitrogen temperature. Thus, the specimen was unloaded as the third step, and the stress-strain curve of Region
III was obtained (Note that the increase of strains in Region III
simply means the contraction of specimens). Now, it is apparent
that a stage is present between $\underline{s}$ and $\underline{f}$ which is characterized by
serrations. In fact this stage corresponds to the martensitic

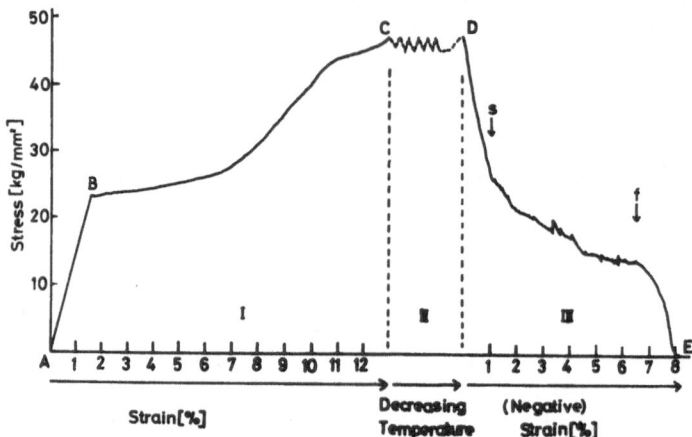

Fig. 3. Stress-strain or stress-temperature relation for the
process described in the text (Specimen 25). Region I :
Loading with constant strain rate 2.5 x 10^{-2}/min. at 49.2°C.
Region II : Cooling from 49.2°C (C̲) to -194.5°C (D̲), keeping the stress constant. Region III : Unloading with constant strain rate 2.5 x 10^{-2}/min. at -194.5°C.

transformation from the β_1' martensite to the γ_1' martensite as
will be proved in the following.

The macroscopic morphological change associated with the
stage in Fig. 3 is shown in Fig. 4.   (a) corresponds to the point
D̲ in Fig. 3, which represents the β_1' single crystal martensite
at the liquid nitrogen temperature.   (b) corresponds to the point

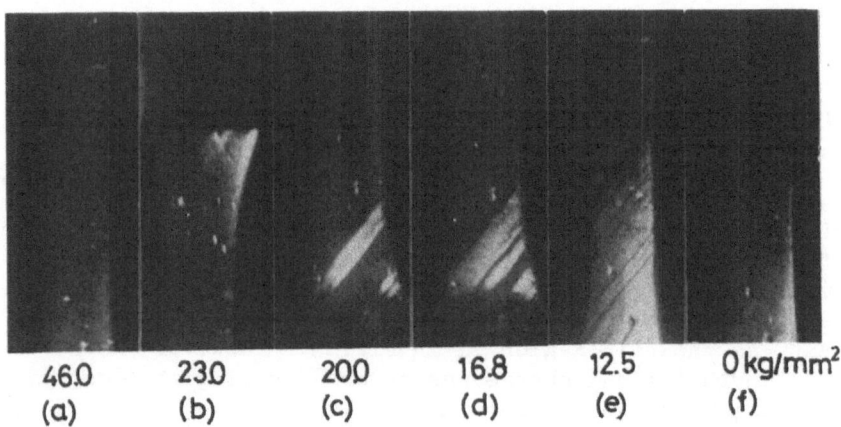

Fig. 4. Macroscopic morphological change associated with the
stress-strain curve of Region III in Fig. 3 (Specimen 25).
See text for details.

s̲ in Fig. 3, showing that thick bands appear in the region near the grips where the crosssectional area is large. Thus, with decreasing strains, the transformation starts in the lower stress region near the grips. With further decrease of strains, the transformation proceeds by the formation and growth of these thick bands as shown in (c) and (d). It is seen from these figures that the product martensite has a habit plane like those in usual martensitic transformations from matrix to martensite. (e), which corresponds to point f̲ in Fig. 3, shows that the transformation is nearly completed at the end of the stage. Further unloading to zero stress (point E̲) results in the nearly single crystal of the product martensite phase, as shown in (f).

The above process was observed more in detail by using an optical microscope, and the result is shown in a series of micrographs of Fig. 5. (a) represents the original β_1 matrix single crystal, and the arrow (↟) indicates an etch pit as a fiducial marker. By applying stress in the temperature range above Af, acicular β_1' martensite plates are stress-induced (b). With increasing strains more and more β_1' martensite plates are stress-induced (c), and eventually the specimen becomes the β_1' single crystal martensite (d). The back reflection Laue pattern was taken in this condition, and it was confirmed that the state corresponding to (d) is really a β_1' single crystal martensite, as will be shown later. Next, the specimen is cooled to some temperature below Af, keeping the stress constant, and then unloaded at the temperature*. Now, the transformation starts below some critical stress, and banded martensites appear as shown in (e) and (f). The structure of the banded martensite has been confirmed to be γ_1' phase, as will be shown later. The habit plane determined by the two surface analysis was close to $\{\bar{1}\bar{1}3\}_{\gamma_1'}$ or $\{\bar{1}\bar{1},25\}_{\beta_1'}$, although there were some scatters from place to place. Obviously, the habit plane is different from the basal plane, which is a common plane to both martensites. Fine striations inside the banded martensites have been determined by the two surface analysis to be $\{\bar{1}01\}_{\gamma_1'}$ twins ($\{\bar{1}102\}$ twins in the hexagonal indices). This twin system as a lattice invariant shear is different from that in the $\beta_1 \rightarrow \gamma_1'$ direct transformation (i.e. $\{121\}_{\gamma_1'}$ twins [$\{10\bar{1}1\}$ twins in the hexagonal indices]) (11, 16), although the product is the same in the two transformations. With further decrease of strains, the transformation proceeds by the growth and formation of these banded martensites, and the density of twins inside the martensites decreases markedly (g∼j)**. Eventually, the specimen becomes a

---

* The $\beta_1' \rightarrow \gamma_1'$ transformation occurs upon unloading not only at the liquid nitrogen temperature but also at all temperatures below some critical one near Af, as will be shown later in Fig. 9.
** Some striations different from the $\{\bar{1}01\}_{\gamma_1'}$ twins were sometimes observed in some area of the specimens as shown in Fig. 5(p). The two face analysis showed that the normal to the plane is close to (010) of β_1' or γ_1'. But the nature of the striations is not clear.

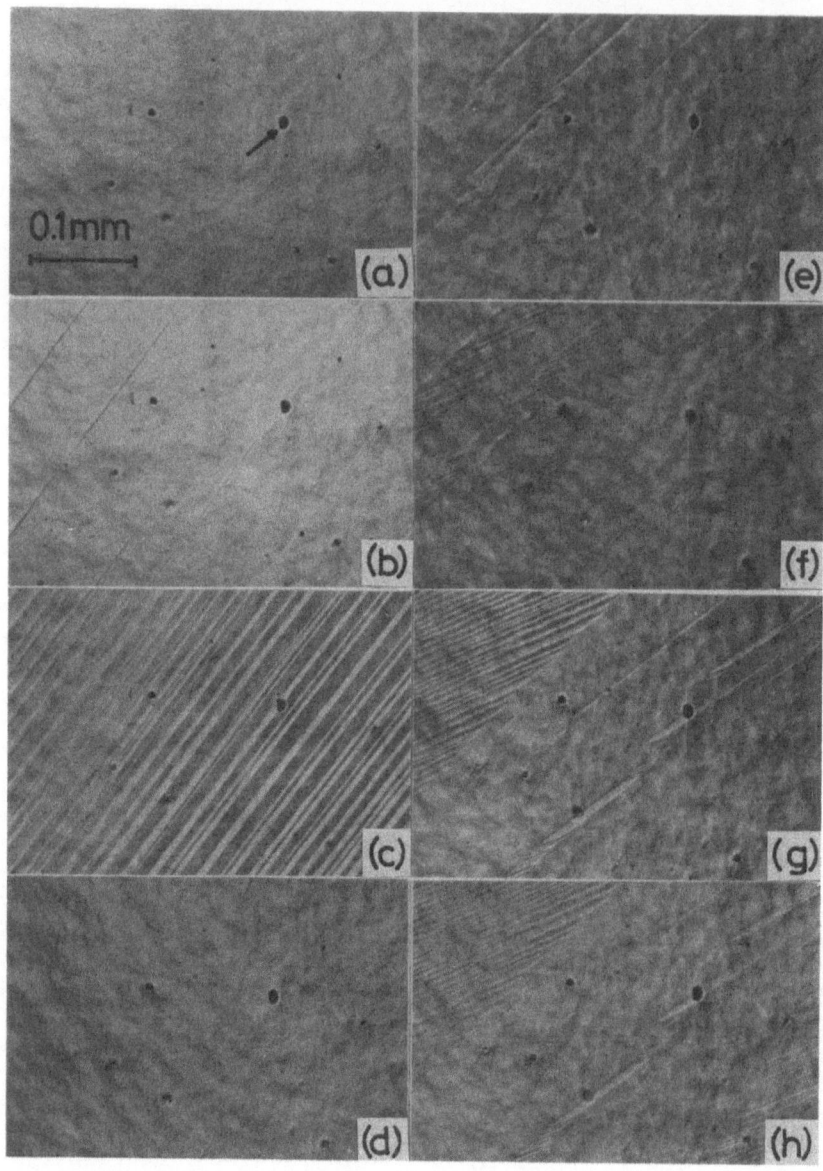

Fig. 5. Optical micrographs showing the series of the transformations from β_1 to β_1' and then to γ_1' under the similar process shown in Fig. 3 (Specimen Al). (a) β_1 matrix.
(b) ∽ (d) under increasing stress at 35°C. (d) β_1' single
crystal. (e) ∽ (j) under decreasing stress at -20°C.

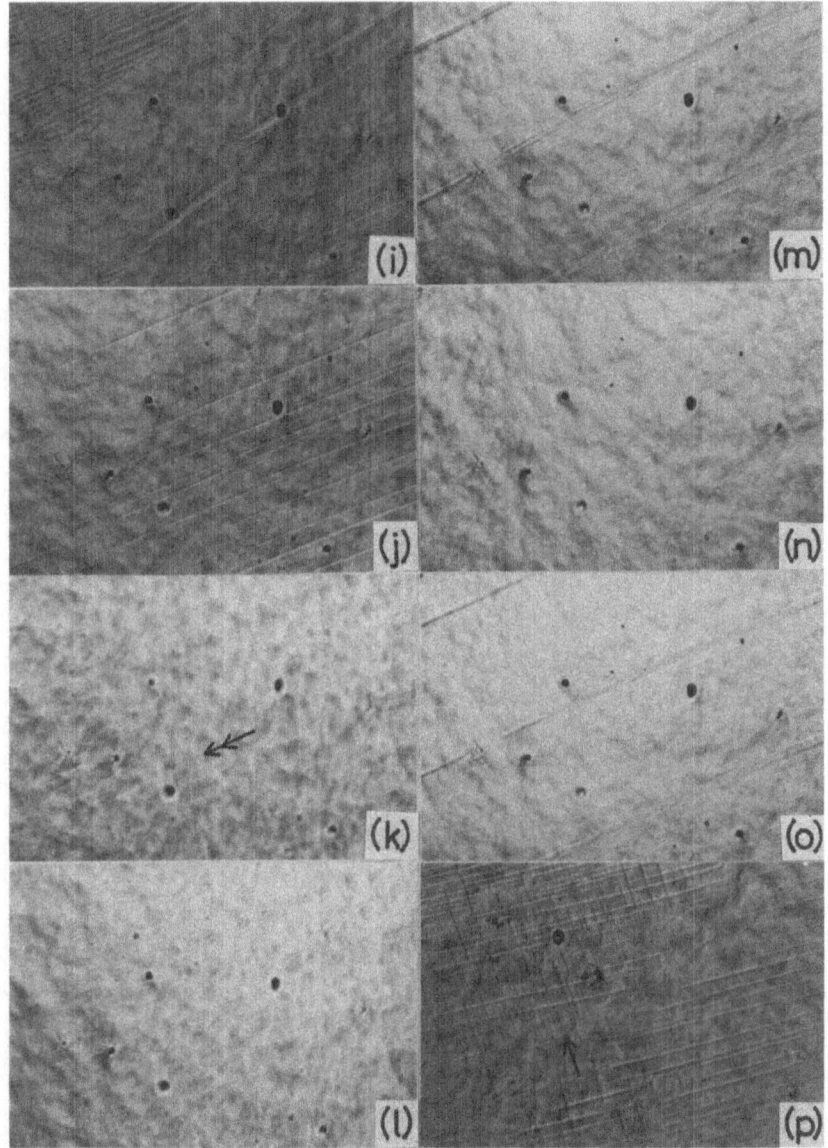

Fig. 5 (continued) (k) γ_1' martensite at zero stress at -20°
C. (ℓ) under stress again at -20°C. (m) at zero stress a-
gain at -20°C. (n) under re-loading at -20°C. (o) again
unloaded at -20°C. (p) different area corresponding to the
stage (i ∿ j) in the above figure. See text for detail.

single variant of the product martensite phase with a small number of twins inside (j). Further unloading to zero stress results in a nearly single crystal of the product phase, although a few traces (↑) of the twins are still observable (k).

The behaviour of the above mentioned twins is rather peculiar. If the specimen is pulled again, the twins disappear completely as shown in (ℓ), since the tensile stress causes detwinning. However, the behaviour observed upon re-unloading is different depending upon the stress level of the previous tensile stress. If the stress is very low, no twin appears upon re-unloading, while if the stress is high the twins appear upon re-unloading as shown in (m). If the same process is repeated, the same behaviour is observed as shown in (n) and (o). In fact, even the locations of the appearing twins are identical in (m) and (o). This behaviour is analogous to that known as "rubber-like" behaviour in Au-47.5 (at%)Cd alloy (17), although the density of twins are much lower in the present case. The reason why twins appear upon unloading, and why they appear in the same locations are not known at present.

Now, let us return to the stress-strain curve of Fig. 3. If the specimen is pulled again from the state designated as $\underline{E}$, the stress-strain curve as shown in Fig. 6(a) is obtained, which is characterized by the low flow stress region $\underline{EF}$. However, if the experiment is repeated once again, the region disappears as shown in Fig. 6(b). We believe that the low flow stress region $\underline{E}$ $\underline{F}$ in the first run is caused by the detwinning process of many mobile micro twins, which is presumably present at point $\underline{E}$, although they are hardly visible under an optical microscope. The region between $\underline{G}$ and $\underline{H}$ in Fig. 6(a) and that between $\underline{G}'$ and $\underline{H}'$ in Fig. 6(b) are almost linear. Thus, they seem to represent the elastic deformation of the γ_1' single crystal martensite. However, it is not true, because the unloading process does not trace back the loading process. The deformation between $\underline{G}$ and $\underline{H}$ or $\underline{G}'$ and $\underline{H}'$ is probably carried out by the slip of partial dislocations which are very mo-

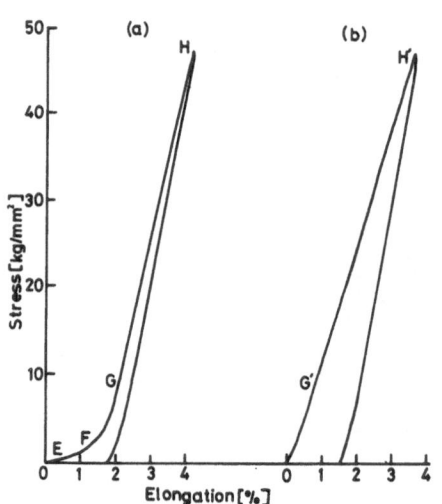

Fig. 6. Stress-strain curve of the γ_1' nearly single crystal obtained by the process of Fig. 3 (Specimen 25). (a) First cycle, (b) Second cycle.

bile and abundantly present in the γ_1' martensite. Since the slip is not recoverable by unloading in general, the hysteresis is left behind after the unloading process as shown in Fig. 6(a) and (b).

The crystal structure change associated with the above mentioned transformation was confirmed by the back reflection Laue method, and the result is shown in Fig. 7.   (a), (b) and (c) correspond to Fig. 5(a), (d) and (ℓ), respectively, and each pattern

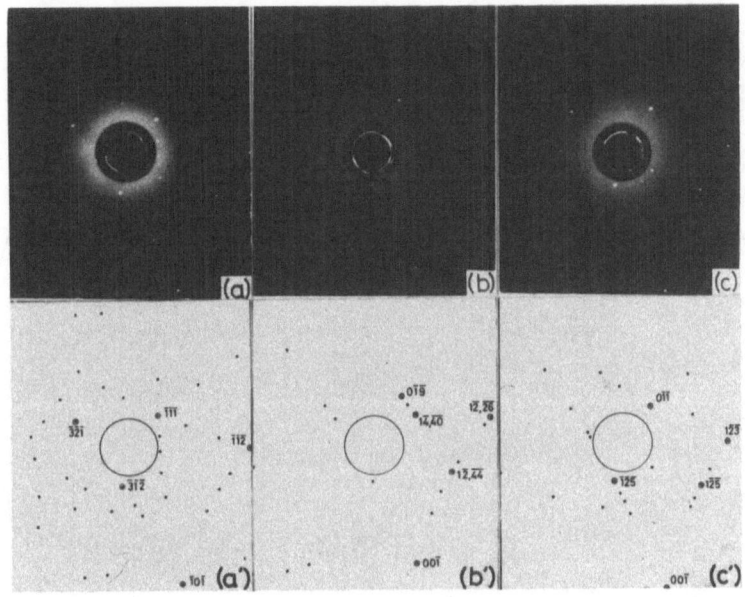

Fig. 7. Back reflection Laue patterns of the three phases
(Specimen Al).   (a) β_1, (b) β_1', (c) γ_1'.

was consistently indexed as β_1, β_1' and γ_1' phases as shown in each key diagram. Therefore, it is quite clear that the above mentioned process occurring upon the unloading process is the martensitic transformation from the β_1' martensite to the γ_1' martensite.

It is seen by the comparison of Fig. 7 (b) and (c) that the basal planes of the two martensites are pretty close to each other, i.e. about 4°. The deviation is partly caused by the lattice rotation associated with the transformation, and partly caused by the experimental error, since the two diffraction patterns were taken separately. If they were taken simultaneously from the region containing both phases, the angle between the two basal planes would be smaller*. At any rate, the above fact indicates

---

* In fact, the angle between $(001)_{\beta_1'}$ and $(001)_{\gamma_1'}$ was less than 2° in one experiment, and the result of Fig. 8 is based on the data.

that the crystal structure change from the β_1' martensite to the γ_1' martensite is carried out by the shuffling $1/3 \cdot [100]_{\beta_1'}$ on $(001)_{\beta_1'}$ plane, as described in the first section.

The crystallography of the transformation from β_1' to γ_1' is summarized in the stereographic projection of Fig. 8, which is based on the X-ray diffraction analysis and the two surface analysis.

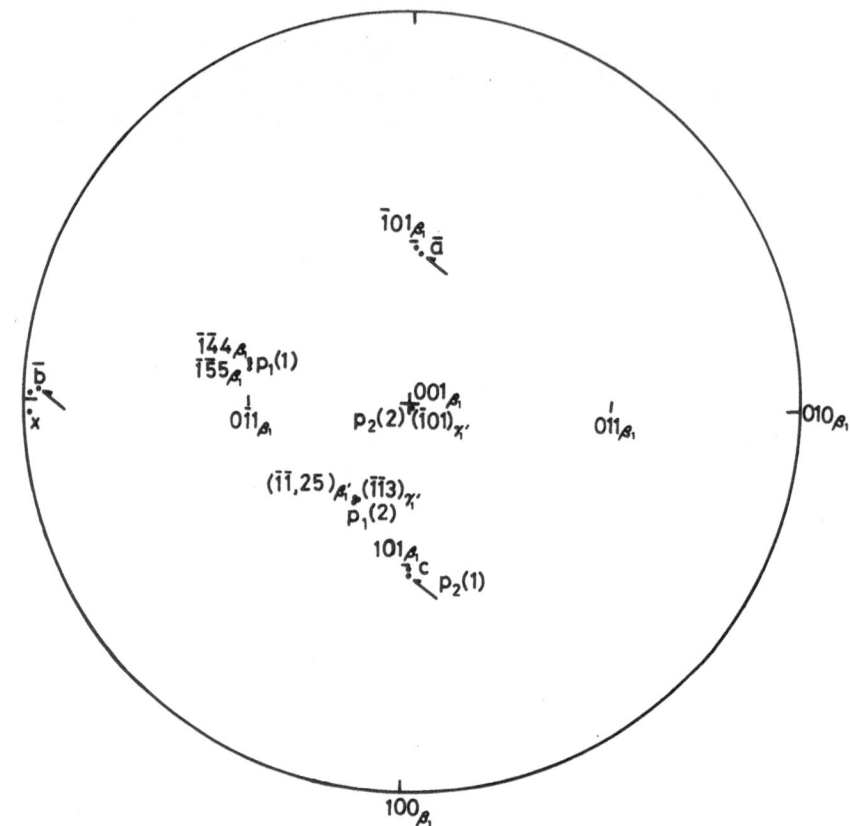

Fig. 8. Crystallography of the transformation $\beta_1 \xrightarrow{1} \beta_1' \xrightarrow{2} \gamma_1'$. $\mathbf{p}_1$ and $\mathbf{p}_2$ represent the habit plane and the plane of lattice invariant shear, respectively. The numbers in the parentheses following $\mathbf{p}_1$ and $\mathbf{p}_2$ correspond to the order of the transformations, that is, 1 for the $\beta_1 \rightarrow \beta_1'$ transformation and 2 for the $\beta_1' \rightarrow \gamma_1'$ one. $\mathbf{a}$, $\mathbf{b}$ and $\mathbf{c}$ represent the crystal axes of the β_1' and γ_1' martensites relative to the β_1 matrix axes The poles indicated by arrows correspond to β_1' poles, and those without arrows to γ_1' poles. x represents a plane corresponding to the trace shown in Fig. 5(p).

If the γ_1' single crystal martensite obtained by the above process is heated, the reverse transformation occurs directly from the γ_1' martensite phase to the β_1 matrix phase. However, the reverse transformation did not start even at the Af temperature, which was associated with the usual athermal $\beta_1 \rightleftharpoons \gamma_1'$ transformation upon cooling and heating in the same specimen. The cause of the shift of the As temperature toward higher temperature side is probably due to the difficulty in the nucleation of the β_1 matrix phase in the γ_1' single crystal martensite, since the preferred nucleation sites such as boundaries between martensites are not available in this situation. It is also to be noted that the specimen reverts to the original shape by the reverse transformation to the matrix phase upon heating, even though it was subjected to the two successive transformations.

As previously reported (3, 14), the γ_1' single crystal martensite may be stress-induced directly from β_1 in the temperature range Ms < T < Af. Thus, we have two ways to make a γ_1' single crystal martensite, either directly or indirectly via β_1' martensite. **The orientations** of the γ_1' martensites produced by the two methods were found to be identical in Specimen Al. Since the elongation obtained by the $\beta_1 \rightarrow \beta_1'$ transformation under stress is always larger than that obtained by the $\beta_1 \rightarrow \gamma_1'$ transformation (3, 14)*, specimens must shrink upon the $\beta_1' \rightarrow \gamma_1'$ transformation. The fact that the $\beta_1' \rightarrow \gamma_1'$ transformation occurs in the unloading process is in good accordance with this requirement.

In order to examine the phase relationship between the three phases, β_1, β_1' and γ_1', the experiment similar to that shown in Fig. 3 has been carried out, and the critical stress for the $\beta_1' \rightarrow \gamma_1'$ transformation has been measured for different unloading temperatures. Since the critical stress is not constant between s and f in Fig. 3, a value was chosen from the plateau region near point f. For the value represents the critical stress in the parallel portion of specimens. The result is shown by the solid line A C in Fig. 9. The critical stress is almost constant below the temperature Af, but the line has a slight negative slope. If the β_1' phase, which forms under stress, is a stable phase, the phase relationship between the three phases will be represented as shown in Fig. 9. Then the transformation from the γ_1' phase to the β_1' phase may also be possible by increasing stress. In order to check this possibility, the γ_1' single crystal (Specimen 25) obtained by the procedure of Fig. 3 has been tested at the liquid nitrogen temperature. However, the evidence of the transformation from γ_1' to β_1' was not detected up to the stress of 45 kg/mm^2 which is much higher than the line A C in Fig. 9. This is probably because a large hysteresis is associated with this transformation.

There is another interpretation for the appearance of the

---

* In case of Specimen Al, the elongation obtained by the transformation from β_1 to β_1' is 7.0 %, while that by the transformation from β_1 to γ_1' is 4.2 % (3).

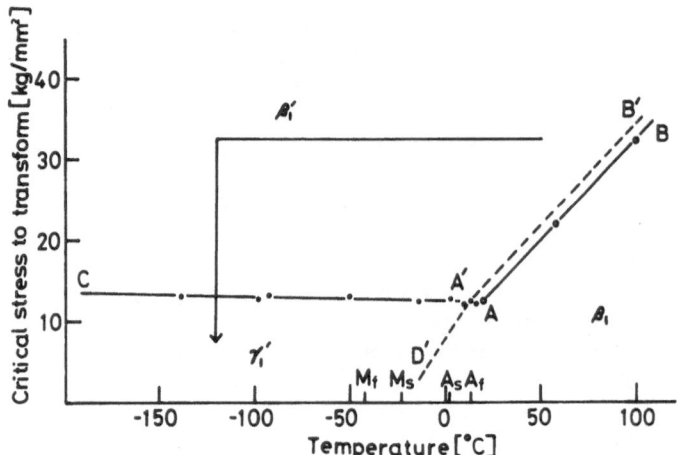

Fig. 9. Phase relationship between the three phases, β_1, β_1' and γ_1'. Solid lines $\underline{A}\,\underline{B}$ and $\underline{A}\,\underline{C}$ represent the measured crit- ical stresses for the $\beta_1' \to \beta_1$ and $\beta_1' \to \gamma_1'$ transformations, respectively (Specimen Al). Dotted lines $\underline{A}'\underline{B}'$ and $\underline{A}'\underline{D}'$ repre- sent the hypothetical critical stresses for the $\beta_1 \to \beta_1'$ and $\beta_1 \to \gamma_1'$ transformations, respectively. The path indicated by the arrow corresponds to the process described in Fig. 3.

transformation from β_1' to γ_1'. This interpretation is based on the hypothesis that the β_1' phase is a transient phase which is stress-induced for some reason. Since it is a transient phase, it tends to transform to the more stable γ_1' phase. However, the transformation can not be accomplished without the unloading process, because it is associated with the shrinkage of specimens which causes the rapid increase of stress without unloading, as described in the above. According to this view, the critical stress represented by the line $\underline{A}\,\underline{C}$ in Fig. 9 does not represent the phase boundary, and the unloading process is required simply for avoiding the rapid stress increase upon the transformation. Thus, no such a transformation from γ_1' to β_1' is expected from this view even under high stress level, which is consistent with the present experimental status. Since stress is an independent variable as well as temperature, the authors support the former interpretation. However, the latter interpretation can not be ruled out completely at the present status, since the transfor- mation from γ_1' to β_1' has not been found yet.

So far the transformation from β_1' to γ_1' upon unloading was described in the above. However, the similar transformation was also observed in the loading process in a narrow temperature- stress region, which is close to point $\underline{A}$ in Fig. 9, that is, in the boundary region between the three phases, β_1, β_1' and γ_1'. It is important to note that this transformation can occur only when the β_1 matrix phase still remains in specimens. If the β_1 matrix phase is still present in specimens, the stress increase due to the transformation from β_1' to γ_1' is compensated by the transformation from β_1 to β_1' or β_1 to γ_1' which occurs simultaneously in the remaining β_1 matrix phase region. The authors think that the observation made by Oishi and Brown in Fig. 3 (D) (but not Fig. 3 (E)) of Ref. (19) corresponds to this case.

IV   Conclusions

The martensitic transformation from the stress-induced β_1' single crystal martensite to the γ_1' single crystal martensite was realized in a Cu-Al-Ni alloy only when cooled and then unloaded, and the process of the transformation was described in detail. To summarize the result, the transformation process was found to be quite similar to that of the usual transformation from matrix to martensite in morphology and crystallography. Thus, it is expected that the transformation is also treated by the phenomenological theory. Although the crystal structures of the two martensites are simply related with the common basal plane, the habit plane was not the basal plane, but was close to $\{\bar{1}\bar{1}3\}_{\gamma_1}'$ or $\{\bar{1}\bar{1},25\}_{\beta_1}'$. The lattice invariant shear was $\{\bar{1}01\}_{\gamma_1}'$ twinning in the product martensite phase, which is in contrast to the $\{121\}_{\gamma_1}'$ twinning in the $\beta_1 \rightarrow \gamma_1'$ transformation, although the product phase is the same. Some rather peculiar behaviour of this $\{\bar{1}01\}_{\gamma_1}'$ twins was described, and the phase relationship between the three phases β_1, β_1' and γ_1' was also discussed in the temperature-stress coordinates.

Acknowledgements :   The authors would like to express their sincere appreciation to Professor C. M. Wayman, University of Illinois, for usefull discussion. This work was partially supported by the Grant-in-Aid for Fundamental Scientific Research (Ippan B, 1972-3) from the Ministry of Education of Japan.

References

(1) C. M. Wayman, Introduction to the Crystallography of Martensitic Transformation. Macmillan (1964).
(2) J. W. Christian, The Theory of Transformations in Metals and Alloys. Pergamon Press (1965).
(3) K. Otsuka, K. Nakai, H. Sakamoto, K. Shimizu and C. M. Wayman To be published (1975).
(4) A. B. Greniger, Trans. AIME, 133, 204 (1939).
(5) I. Isaitschew, E. Kaminsky and G. V. Kurdjumov, Trans. AIME, 128, 361 (1938).
(6) S. Kajiwara, Trans. Jap. Inst. Metals, 9, suppl. 543 (1968).

(7) H. Tas, L. Delaey and A. Deruyttere, Scripta Met., 5, 1117
     (1971).
(8) N. Nakanishi, Preprint for the Symposium on Thermoelastic
     Martensite and Shape Memory Effect (1974), Tokyo, Japan.
(9) K. Otsuka, H. Sakamoto and K. Shimizu, Scripta Met., 9, No. 5
     (1975), in press.
(10) M. J. Duggin and W. A. Rachinger, Acta Met., 12, 529 (1964).
(11) K. Otsuka and K. Shimizu, Jap. J. appl. Phys., 8, 1196 (1969).
(12) K. Otsuka and K. Shimizu, Phil. Mag., 24, 481 (1971).
(13) K. Otsuka, T. Nakamura and K. Shimizu, Trans. Jap. Inst. Met.
     15, 200 (1974).
(14) K. Otsuka, K. Nakai and K. Shimizu, Scripta Met., 8, 913
     (1974).
(15) K. Otsuka, M. Takahashi and K. Shimizu, Met. Trans., 4, 2003
     (1973).
(16) K. Otsuka and K. Shimizu, Trans. Jap. Inst. Metals, 15, 109
     (1974).
(17) H. K. Birnbaum and T. A. Read, Trans. AIME, 218, 662 (1960).
(18) K. Otsuka and K. Shimizu, Trans. Jap. Inst. Metals, 15, 103
     (1974).
(19) K. Oishi and L. C. Brown, Met. Trans., 2, 1971 (1971).
(20) Z. Nishiyama and S. Kajiwara, Jap. J. appl. Phys., 2, 478
     (1963).

MICROSTRUCTURAL CHANGES DURING SME BEHAVIOR

L. Delaey and J. Thienel [°]

Dept. Metaalk.Kath. Univ. Leuven, Belgium and

Inst. für den Wiss. Film [°], Göttingen, W.-Germany

Many works published on thermoelasticity, pseudoelasticity, shape memory effect and the two-way shape memory effect associated with martensitic transformations deal in part with the microstructural changes occuring during the manipulations of the individual effects. Because the four effects are very closely interrelated, a study of the microstructural changes occuring in the same area of the sample is needed and should yield important information for the understanding of the interrelationships. Because the microstructural changes, which occur during the shape recovery associated with the memory effects, are dynamic, the analysis of a cine-film, even frame by frame, will also bring some new aspects. The attention should mainly be drawn on two phenomena which need a better understanding : the microstructural changes taking place during the shape recovery stage of the shape memory effect and the comparaison of the microstructure obtained by thermoelastic martensitic transformation with that obtained during the two-way memory effect.

In order to clarify these questions the microstructural changes occurring during the four effects have been filmed. The film presented during the present symposium deals with the microstructural changes during SME behavior and more specifically during the two-way memory effect. The conclusions which can be drawn from this film will be compared with those reported in the recent publications (1-8) on two-way memory effects.

The "two-way memory effect"(1,7), also termed "reversible shape memory effect" (2) or "reversible linear change on transformation" (4), refers to a reversible shape change accompanying thermally induced martensite formation and reverse transformation, after that

the sample has been apparently plastically deformed above A_f or below M_s. The sample thus remembers both the high and low temperature shapes, and may be cycled repeatedly between these two shapes.
The phenomenon can fade out during repeated thermal cycles (1) or
shows no hysteresis even after more than twenty thermal cycles (2).

Experimental Procedure

A Cu-Zn-Al alloy, with the nominal composition 72.8 wt % Cu,
21.2 wt % Zn and 6.0 wt % Al, was induction melted, casted, hot-
rolled and heattreated. Tensile samples were heated in the β-phase
field for 10 minutes and subsequently water quenched. The tensile
samples, which are in the martensite condition at room temperature
($M_s \sim 50$ °C), were electropolished at room temperature. Consequently, upon heating the samples to a temperature above A_f, an inverse
surface relief will be visible (see for example figure 3). The microstructure was analysed by means of polarized light for the two
following reasons. The surface relief was not always sufficiently
pronounced and interfered with the inverse surface relief obtained
by the electropolishing. Each martensite plate orientation is now
characterized by a specific color, allowing us to differentiate between the various martensite plate variants. The film is in color,
but the figures of this paper are black-and-white prints of the ne-
gatives. The use of polarized light reduced seriously the intensity of the reflected light, therefore, a 2,5 KW Xenon lamp was used
as illumination source.

The tensile sample was stressed in the microscope and heating
was performed by induction.

Experimental Results

The present film is the fourth out of a serie of four films
on the martensite transformation in copper base alloys. The first
film deals with the thermoelastic martensite formation and reversion. The second and the third film deal with the pseudoelastic
martensite formation and the fourth, the present film, deals with
the shape memory effects. The four films will be published in Encyclopaedia Cinematographica (9).

The figures 1 to 16 are the prints showing the microstructural
changes occuring during the shape memory manipulations. The figures 1, 2 and 3 correspond with the first heating step from room
temperature to $T \simeq A_f$. The apparent microstructure visible on figure 3 is due to the above mentioned reverse surface relief. The
sample was subsequently stressed at $T > A_f$ (figure 4) and cooled
to room temperature while keeping the elongation constant. The
sample was then completely unloaded at room temperature (figure 5).

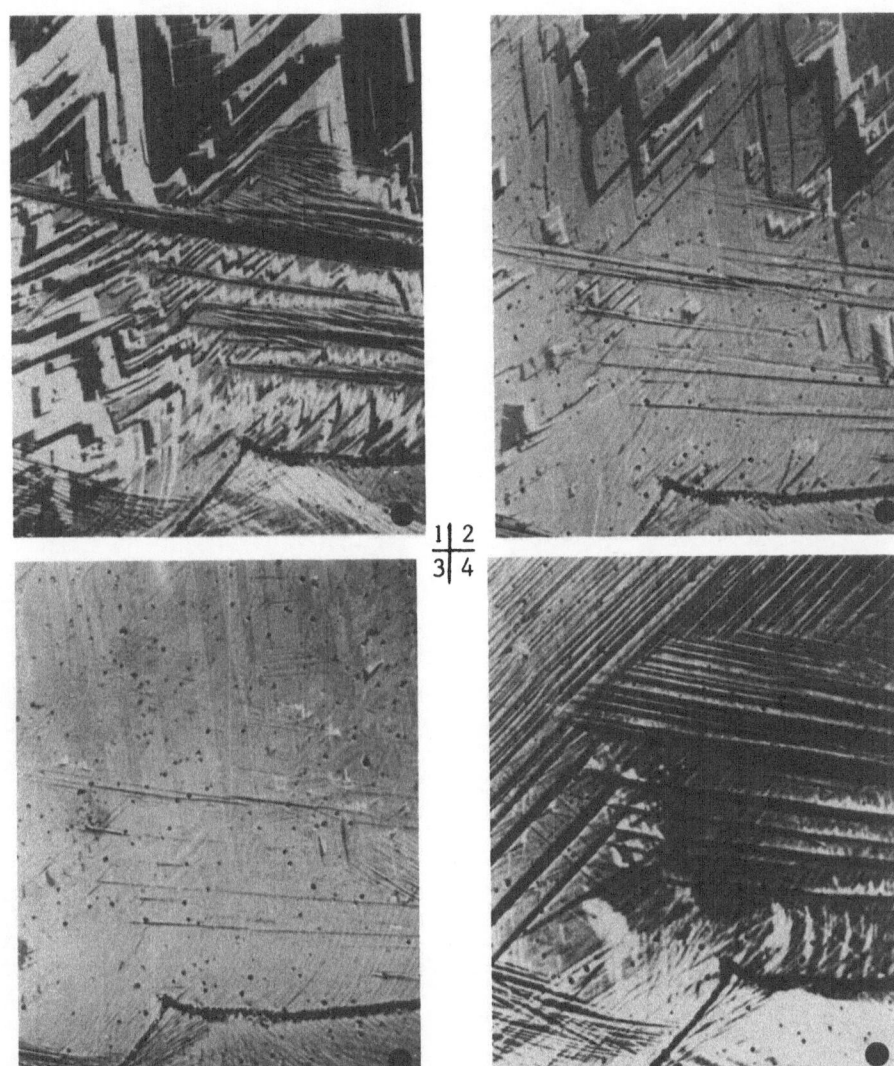

Figure 1 to 16

A series of micrographs taken from the cine-film and corresponding to various stages during the shape memory manipulations (Width of image 360 μm).

Figure 1 : Microstructure obtained after quenching, heating above A_f and subsequent cooling to room temperature but before any stress has been applied.

Figure 2 : Shrinkage of the martensite plates during heating from room temperature to $A_s < T < A_f$.

Figure 3 : Microstructure at a temperature $T \simeq A_f$ ($\pm$ 60 °C).

Figure 4 : The martensite plates induced by applying a tensile stress at a temperature $T > A_f$.

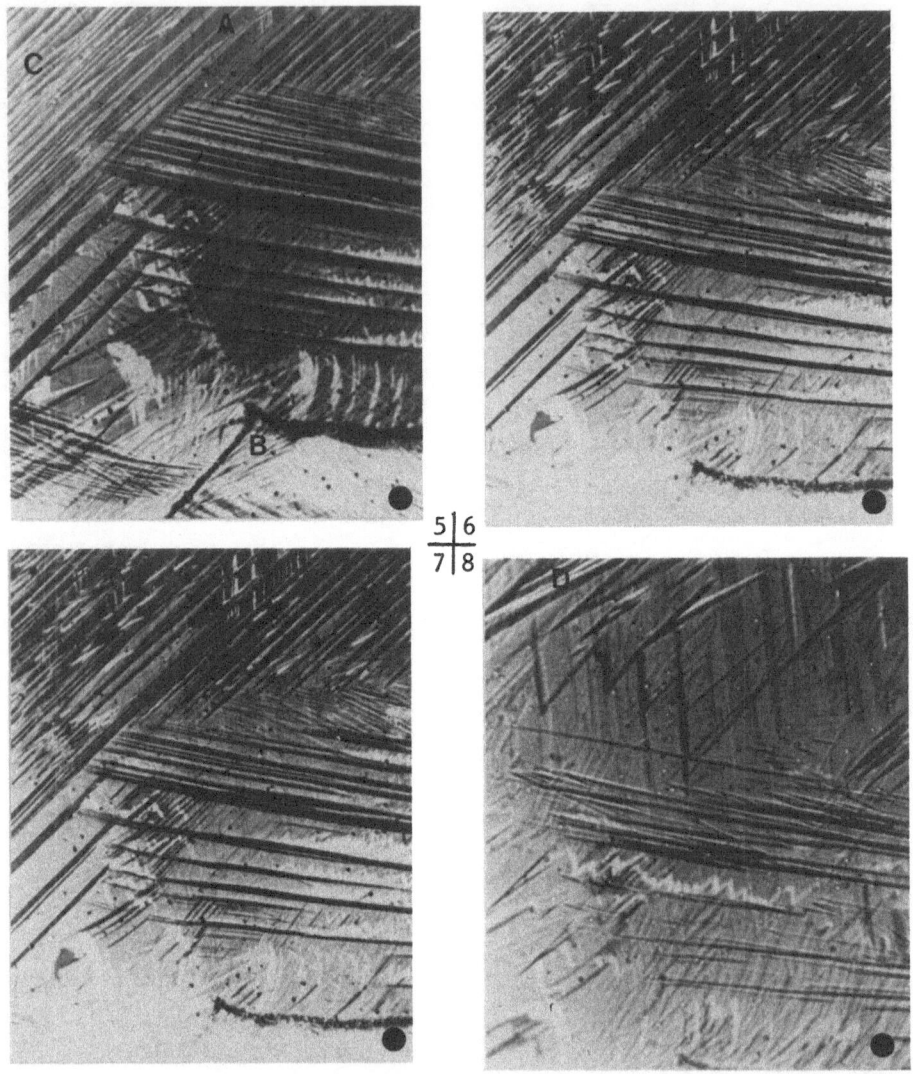

Figure 5 : Microstructure after cooling under stress to room tem-
           perature and retained after releasing the stress.
Figure 6 : Microstructure during the first stages of heating from
           room temperature to T > A_f and after the shape recovery
           has started.
Figure 7 : Microstructure appearing as an arrangement of martensite
           plates intermediate between the microstructure of figure
           6 and just before completion of the shape recovery.
Figure 8 : Microstructure obtained during the first cooling cycle
           from T > A_f to room temperature.

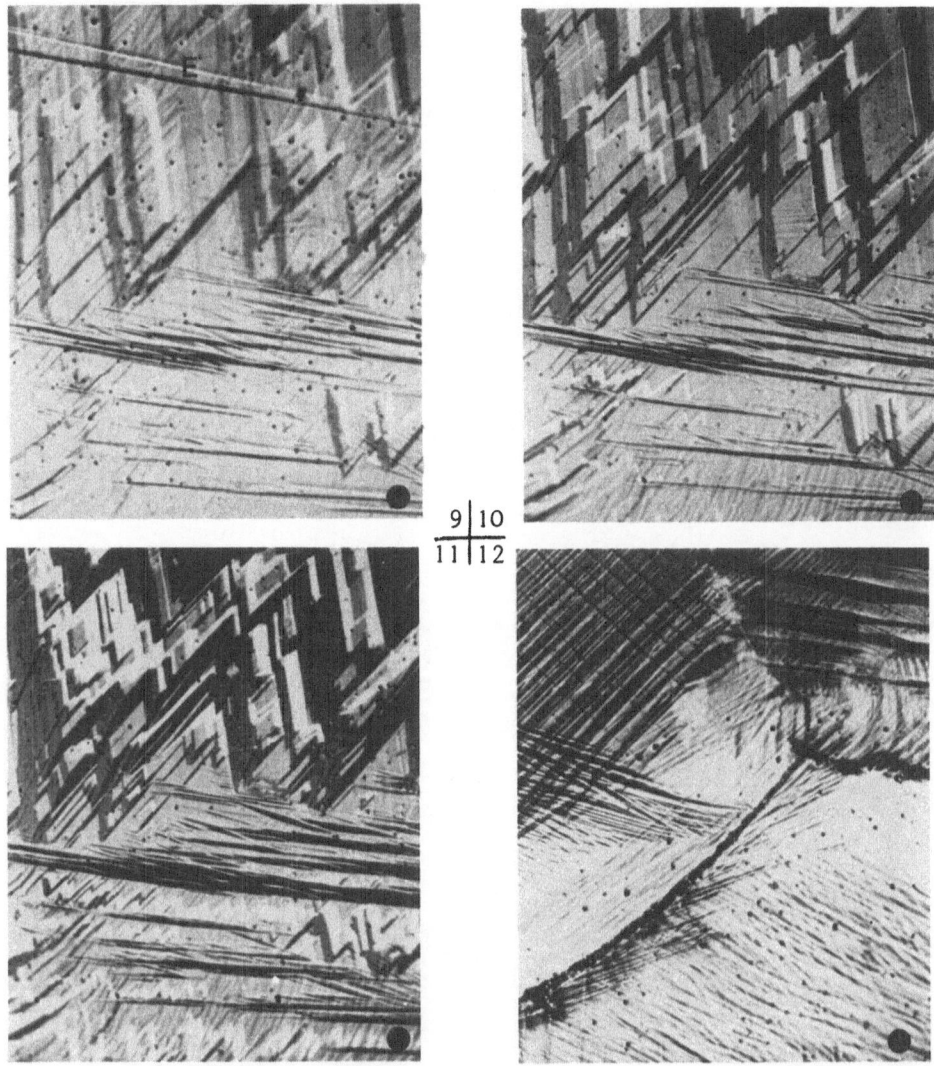

Figure 9 : Microstructure obtained during the second cooling cycle
            after the sample was heated to $T > A_f$ but without any
            stress was applied at $T > A_f$.

Figure 10 : Microstructure obtained during the third cooling cycle
            and under the same conditions as figure 9.

Figure 11 : Microstructure at room temperature after completing the
            third cooling cycle. Microstructure to be compared
            with that of figure 1.

Figure 12 : Microstructure after cooling under stress to room tem-
            perature and induced at $T > A_f$ but after the foregoing
            manipulations (figure 1 - 11) have been applied ten
            (10) times.

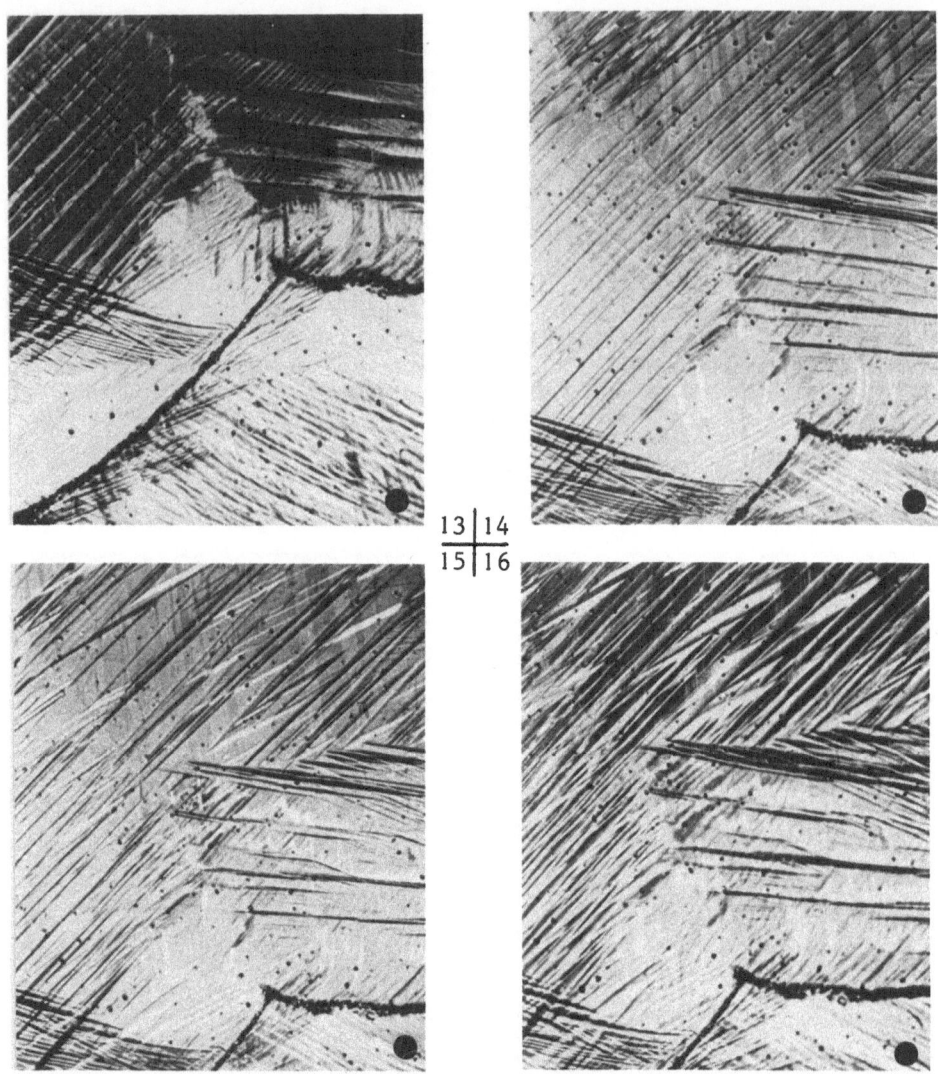

13 | 14
15 | 16

Figure 13 : Microstructure at room temperature of the stress indu-
           ced martensite, after releasing the stress and sample
           fixed at one end.
Figure 14 : Microstructure during the heating cycle of the two-way
           memory effect.
Figure 15 : Microstructure during the cooling cycle of the two-way
           memory effect.
Figure 16 : Microstructure at room temperature and after completing
           the cooling cycle of the two-way memory effect. Micro-
           structure to be compared with that of figure 8, 7 and
           1.

During this unloading and cooling step some small microstructural changes occurred in the area labelled A on figure 5. Small internal markings appeared inside the stress-induced martensite plates. Most of those stress-induced martensite plates crossed the β-grain boundary labelled BC, which is a common feature in polycrystalline material (10). The internal markings became more pronounced on figure 6, a micrograph taken after the sample was fixed at one end and heated in order to recover the shape. During this shape recovery stage an intermediate martensite plate configuration appeared, but only for a fraction of a second. As can be deduced from the color, two different martensite plate variants appeared in the V-type arrangement, which is characteristic for the strain accommodating martensite. It is just this particular martensite plate arrangement that will be responsible for the two-way memory effect.

After the martensite plates have been completely reversed to β by heating above A_f, the sample was cooled to room temperature and the microstructural changes were filmed. During the first cooling cycle part of the original martensite plates appeared (compare fig. 8 with fig. 1) but, in the area labelled D on the fig. 8, the V-shaped martensite plates also formed. Analysing the color of the martensite plates, it can be concluded that the same martensite plate variants as in figure 7 were present.

The sample was then heated above A_f and cooled to room temperature a second and a third time, but no stress was applied at T > A_f (the sample was fixed only at one end). During the second cooling cycle (fig. 9), the V-shaped martensite plates visible on fig. 7 and 8 did not appear, but nearly all the original plates (visible on fig. 1) reappeared, with the exception of the stress induced martensite plate labelled E. During the third cooling cycle, no stress-induced martensite plates formed spontaneous (fig. 10), the original microstructure was restored (compare fig. 11 with fig. 1).

After going through this cycle (heating to T > A_f, stressing at T > A_f, cooling to room temperature and releasing the stress, unfixing at one end, heating to T > A_f and cooling to room temperature without applying stress), the sample was again fixed at the two ends and the same cycle was applied 9 times. The sample now exhibited the two-way memory effect. The microstructures which appeared after cooling under stress and after unloading but just before heating are shown resp. in the figures 12 and 13. During heating, the V-shaped arrangement appeared again as an intermediate microstructure (figure 14). This V-shaped arrangement was much more pronounced during the cooling cycle (figure 15) and after completing the cooling cycle (figure 16). During the heating and cooling cycle, and while the sample was fixed at one end, the sample movement could be observed by a movement of the image. The micro-

structure shown in figure 16 is completely different from that
shown in figure 1 and 11, but shows the same V-shape arrangement
as that shown in figure 7 and 8. The latter microstructural changes
appearing during the two-way memory effect manipulations could be
repeated several times.

Conclusion and Discussion

Upon heating a sample wherein the martensite plates have been
formed by a tensile stress, the stress-induced martensite plates
disappear progressively until the balance "three dimensional stress
minimization" versus "two dimensional stress minimization" becomes
unfavorable for the single stress-induced martensite plates. As
soon as a single martensite plate ends in the middle of a crystal
in a tapered form (see f.e. fig. 8 and fig. 13 of ref. 7), either
one plate boundary or both plate boundaries deviate from the habit
plane orientation (figure 17 a).

Consequently an energy increase will be built up around the
martensite plate. Such a microstructure can only be retained

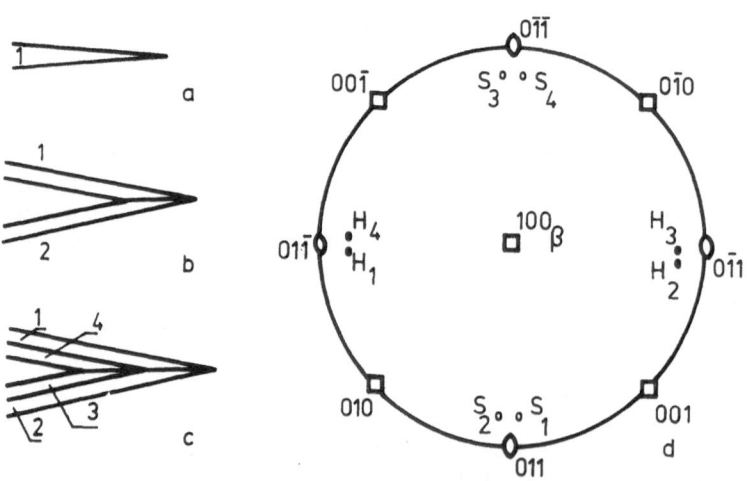

Figure 17 : Martensite plate arrangements and their associated macroscopic shape deformation. H : habit plane, S : direction of total shear.

if this increase is balanced by a decrease in free energy (thermo-elasticity) or by the application of an external stress (pseudo-elasticity). The macroscopic deformation associated with the martensite plate variant 1 is given in the stereographic projection of figure 17 d (12). From the same figure, it can be seen that the macroscopic deformation associated with variant 2 lies very close to that of variant 1. The martensite plate arrangement shown schematically in figure 17 b is characterized by plate boundaries in the exact habit plane orientation but with a total macroscopic deformation slightly deviating from that of variant 1. Depending upon the actual internal stress tensor, the energy balance may now be in favour of the arrangement of figure 17 a or 17 b. In case of a fully self accommodating martensite plate group (figure 17c), the total macroscopic deformation will be zero (13).

Once the V-shaped martensite plate arrangements have been formed, further decrease in their size can progressively occur while further increasing the temperature.

By the series of micrographs given above and by the film, it has now been proved that a preferred orientation distribution and the growth of particular martensite plate variants are responsible for the spontaneous shape change associated with the two-way memory effect while cooling the sample from $T > A_f$ to $T < M_f$. This conclusion is in accordance with ref. 1, 2, 5 and 6. It has also been shown that the two-way memory effect or the reversible shape memory effect can be obtained by several deformation cycles above A_f but below M_d but after cooling under constant load or elongation to a temperature below M_f, which is in accordance with ref. 1 and 8. The two-way shape memory effect may fade out as shown by the figures 8, 9, 10 and 11, or can be repeated even more than hundred times (5).

As a concluding remark, the authors want to draw the attention on the fact that the two-way memory effect was already reported to occur in Cu-Zn alloys in 1956 by Hornbogen and Wassermann (11), although no name was given at that time to the observed phenomena. The same report contains also the first proofs for the occurrence of SME in Cu-Zn alloys.

Acknowledgements.
The authors want to thank Ing. G. Hummel, IWF, Göttingen for his guidance and supervising while preparing the films. One of the authors, L.D., wants also to thank Prof. G. Wolf for providing the facilities in his laboratory. Part of this work was supported by the Belgian Nationaal Fonds voor Wetenschappelijk Onderzoek.

References

1. H. Tas, L. Delaey and A. Deruyttere, J. Less Common Metals, 1972, vol. 28, p. 141.
2. A. Nagasawa, K. Enami, Y. Ishino, Y. Abe and S. Nenno, Scripta Met., 1974, vol. 8, p. 1055.
3. W.J. Buehler, J.V. Gilfrich and R.C. Wiley, Journal of Appl. Phys., 1963, vol. 34, p. 1475.
4. R.J. Wasilewski, Met. Trans., 1971, vol. 2, p. 2973.
5. T. Saburi and S. Nenno, Scripta Met., 1974, vol. 8, p. 1363.
6. J. Perkins, Scripta Met., 1974, vol. 8, p. 1469.
7. L. Delaey, R.V. Krishnan, H. Tas and H. Warlimont, J. of Mat. Science, 1974, vol. 9, p. 1521, 1536 and 1545.
8. F.E. Wang and W.J. Buehler, Appl. Phys. Lett., 1972, vol. 21, p. 105.
9. L. Delaey, G. Hummel and J. Thienel, Encyclopedia Cinematographica, Ed. G. Wolf, Göttingen, to be published Winter 1975.
10. L. Delaey, F. Van de Voorde and R.V. Krishnan, Proceedings of the Int. Symp. on SME, 1975, this volume.
11. E. Hornbogen and G. Wassermann, Z. Metallk., 1956, vol. 47, p. 427.
12. H. Tas, Ph. D. thesis, Leuven 1971.
13. H. Tas, L. Delaey and A. Deruyttere, Met. Trans., 1973, vol. 4, p. 2833.

Martensite Formation as a Deformation Process in Polycrystalline

Copper-Zinc Based Alloys

L. Delaey, F. Van de Voorde and R.V. Krishnan [*]

Dept. Metaalk., Kath. Univ. Leuven, Belgium

[*] now at : Nat. Aeronautical Lab. Bangalore, India

The crystallography of the stress-induced martensite formation in
single crystals is easy accessible, both from the theory and the
experiments. Some interesting results as the critical resolved
stress necessary for the onset of the martensite formation (1) and
the orientation of the martensite obtained by unaxial stress (1,2)
have already been reported. The martensite formation in polycrystalline material is easily observed while stressing the β-phase,
but a complete explanation of the microstructure thus obtained is
rather difficult. The aim of the present work is to collect and
classify the various modes of stress-induced martensite formation
and to study the influence of the presence of the β-grain boundaries on the martensite plate arrangements.

Starting with the criteria for stress-induced martensite formation in single crystals and from the texture of the polycrystalline β-phase, one should be able to calculate the mechanical behaviour of a stressed β-phase while it transforms. However, because the deformation is achieved by a phase transformation, thermodynamic criteria should also be included in the calculations.
Factors, such as the fraction of transformed volume should be
known or measured as a function of the achieved macroscopic deformation. The whole β-phase volume does not get transformed and part
of the work done in the system is stored as elastic stored energy.
As will be shown in another (3) contribution, the elastic stored
energy enters as an important factor into the energy balance for
the growth of a martensite plate. It will be shown here, how the
transformed volume is distributed over the sample and where the
elastic energy is stored. Three different mechanisms will be discussed in detail : 1. the mechanical coupling due to the self-
accommodating character of the martensite (4); 2. the taper forma-

tion of martensite plates (5,6); 3. the irreversible plastic deformation and cracking (7).

The Self-Accommodating Character of the Martensite

The crystallography of the self-accommodating martensite plate
group formation in copper-aluminium alloys has been explained in an
earlier work (4). Optical microscopy has shown that the self-accommodating character is not only valid for copper-aluminium, but also
for the β-type martensite formation in the other copper-base alloys
and for some silver- and gold-based alloys (8). The β-type martensite is characterized by the formation of groups of martensite plates of four different variants; since there are twenty-four variants for one β-orientation, six differently oriented groups can
be observed in one β-grain. A schematic diagram showing the spatial arrangement of the four martensite plate variants is given in
figure 1 a. The figure shows the perspective, top and front views
of such a group. The macroscopic shape deformations of plates 1 and
4 and also of plates 2 and 3 are equal and roughly opposite; this
is clearly visible in the top and front views. The macroscopic deformations of plates 1 and 2 and also of plates 3 and 4 are exactly equal and opposite. The formation of a group of nearly parallel
martensite plates of variants 1 and 4 will result in a small shape
deformation which is then compensated by changing over to the other
pair of martensite plates of variant 2 and 3. We are thus dealing
not only with the plane of minimum distortion (invariant plane strain)
but also with the minimization of the three-dimensional volume distortion.

The habit planes and the directions of macroscopic shear associated with the formation of each martensite plate within one
group is given in fig. 17 of the previous paper (9). An optical
micrograph exhibiting clearly the V-shaped plate arrangement is shown
in figure 2. The thermoelastic growth and shrinkage of such a martensite plate group has been recently analysed through motion pictures (7). The polished surface was orientated such that the pictures correspond either to a front view, a top view or a view parallel to one of the habit planes.

If the sample is now slightly strained, the volume fractions
of the four variants are not equal any more, meaning that the macroscopic deformation is not completely compensated (figure 1b).
For this it is necessary that the martensite plate boundaries between the different variants are movable under the influence of
stress. Such movement of the martensite plate boundaries has indeed
been observed in earlier work (10,11).

A larger amount of macroscopic deformation can be obtained if
only two martensite plate variants form instead of four (figure 1c).

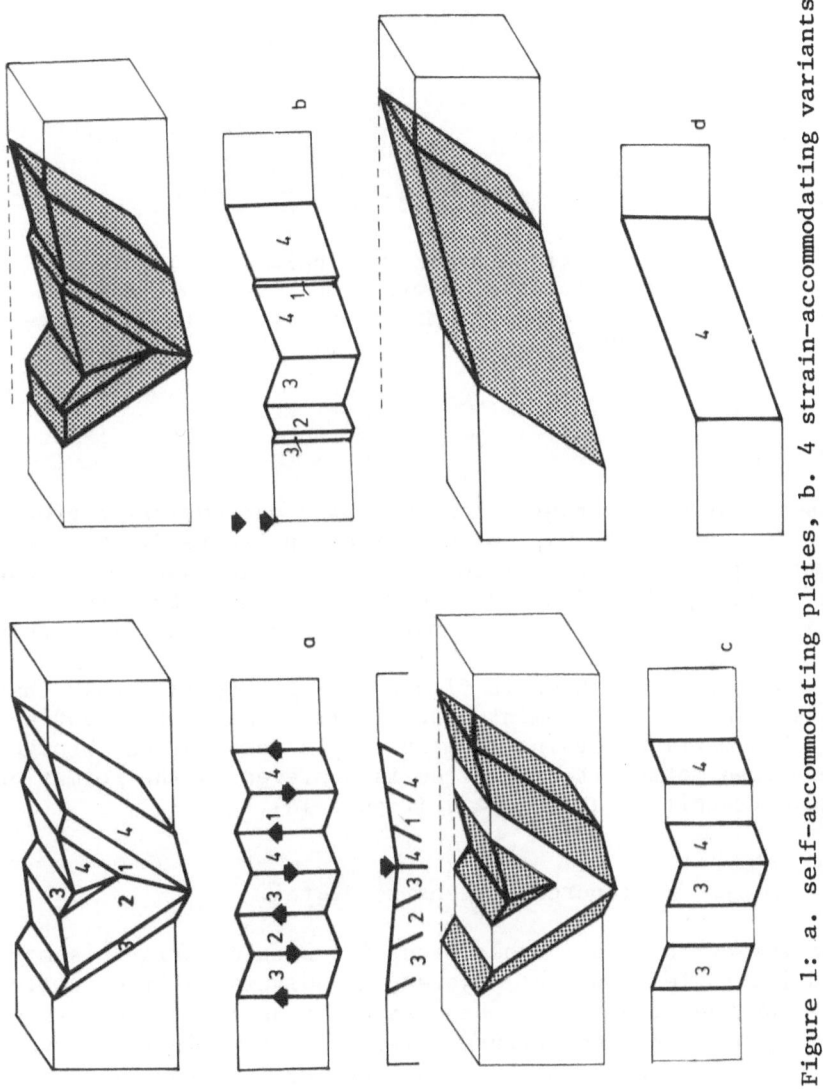

Figure 1: a. self-accommodating plates, b. 4 strain-accommodating variants, c. two plate variants and d. a single plate variant.

Figure 2 : Optical micrograph taken with polarized light of a Cu-
          Zn-Al β_1-martensite showing the V-shaped arrangement
          of the four self-accommodating martensite plate va-
          riants (400 x).

A series of optical micrographs showing the pseudoelastic growth
of two martensite plate variants is given in figure 3. The arrow
on the micrographs shows a fixed position of the sample. It can be
seen that the tip formed by the two variants moves forwards and
backwards. This movement is far more important than the thickening
of the martensite plates itself. The macroscopic deformation need
not be symmetrical as shown in figure 1c. Figure 3.1 shows that besides the V-shaped arrangement other isolated martensite variants
appear. The maximum obtainable macroscopic deformation through
stress-induced martensite formation is achieved by the formation of
a single martensite plate variant (figure 1d).

                    Tapered Martensite Plates

    The formation of one broad band of a single variant is mostly
observed in single crystals; however, in polycrystalline samples it
often is observed that such a band is built up of numerous thin parallel martensite plates (figure 4.1.). As shown in this figure,
these bands often cross the β-grain boundary. In the present case
the band extended across four β-grain boundaries. Such a group of
martensite plates has a higher energy compared with a single thick
martensite plate, the surface energy. However if the bands were
to terminate somewhere in the middle of the sample, high local elastic stresses will be built up due either to increased misfit between two β-grains or strain incompatibilities. If the elastic
stresses could be equally spread over a larger area of the sample,

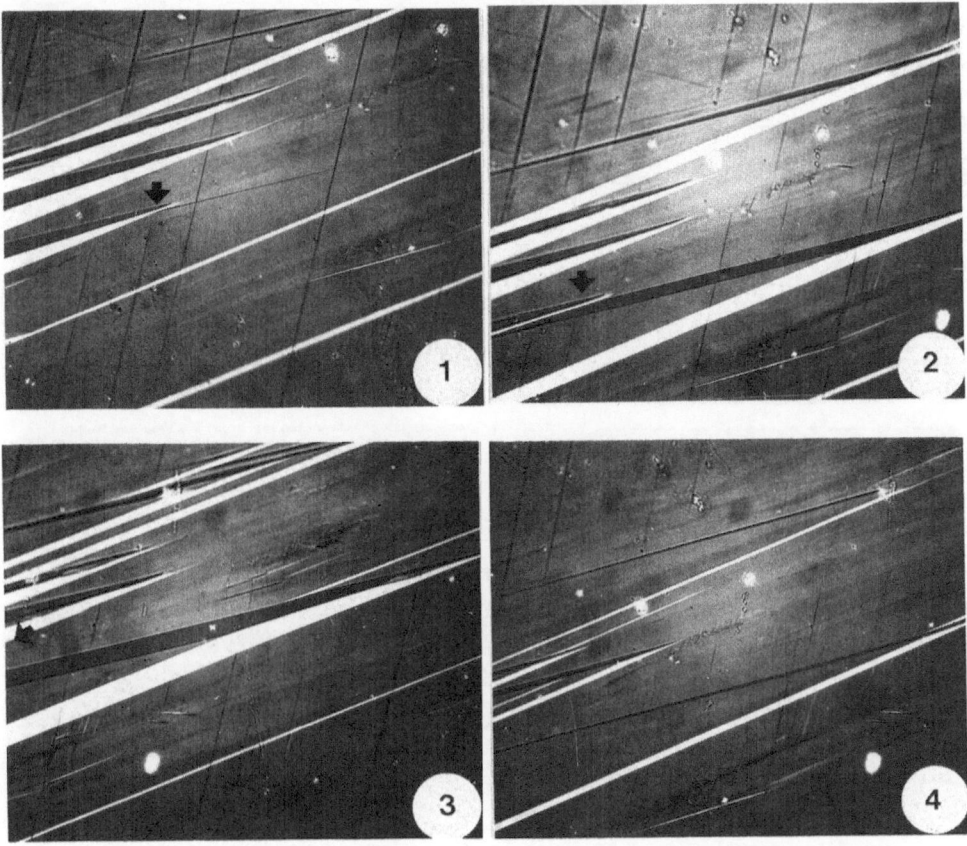

Figure 3 : A series of optical micrographs taken while loading (1,
            2 and 3) and unloading (4) a Cu-Zn-Al β-alloy. Only
            two martensite plate variants are formed (450 x).

thereby decreasing the total elastic stored energy, a band of thin
isolated martensite plates will be formed. These martensite plates
will terminate tapered at their ends as has been frequently observed
also in other alloy systems (3,6). Tapering means a deviation from
the invariant plane, which in its turn means introduction of dislocations. The appearance of these tapered plates has been analysed
in detail by Basinski and Christian for In-Tl alloys (5), for elastic twins by Kosevich and Boĭko (12) and applied later to the β-type
martensite (6). The main conclusion reached in the above cited
works is that the lattice is locally bent by the introduction of tapered martensite plates. The bending of the lattice is clearly visible in figure 4.3, which shows four sets of martensite plates, one
horizontal and three vertical. One set of vertical martensite plates terminates at the vertical position of the upper horizontal

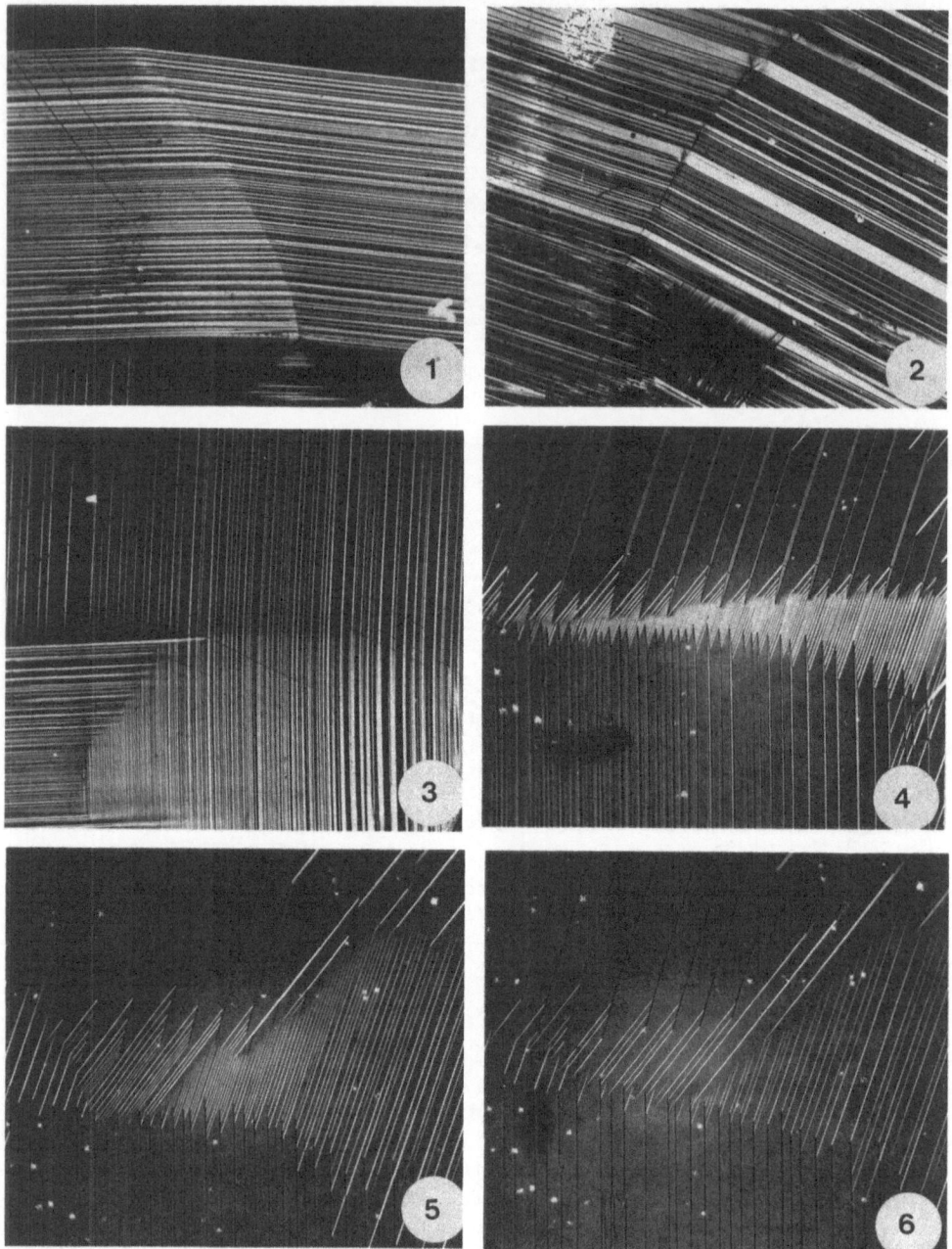

Figure 4 : The stress-induced formation of parallel martensite plates   1 : group of plates of one variant interacting with a β-grain boundary; 2 : group of two variants; 3 : mutual interaction of two groups and bending of the sample; 4, 5 and 6 : interaction between two nearly parallel groups (500 x).

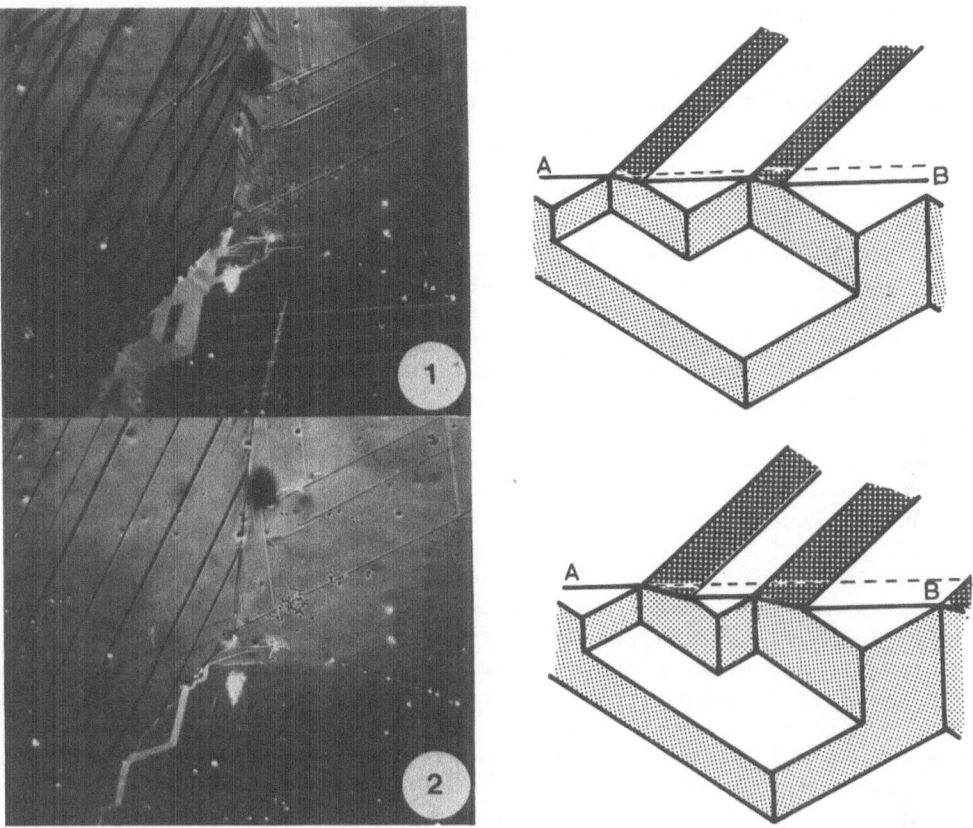

Figure 5 : optical micrographs (450 x) (1 and 2) showing the strain-accommodation at the β-grain boundaries; 3 : schematic representation

plate. The set of horizontal plates introduces a macroscopic shear of the β-lattice along the habit plane. The shear of the lattice is taken over by a bending due to the tapered ends of the martensite plates. The bending is clearly visible by the change in direction of the martensite plates, which extend from the lower part to the upper part of the figure.

Sometimes two variants of martensite plates appear in one deformation band (figure 4,2). The β-grain boundary becomes zig zagged on being crossed by such a group. Increased straining of the sample will then lead to grain-boundary cracking since not much space is available for further accommodation.

Figures 4.4 to 4.6 are a series of micrographs showing the

interaction between two groups of martensite plates which are nearly
parallel. The tapering is taken over by secondary martensite plates
emitted from the ends of the longer plates. The tapered ends are
clearly visible in figures 4.5 and 4.6.

Accommodation at the β-grain boundaries

The orientation of the β-grains on both sides of the boundary
is not always favourable for allowing transgression of the martensite plates. Work is now in progress to analyse the behaviour of
grain-boundaries in bicrystals of the β-phase. However, a semi-
crystallographic analysis of the behaviour of grain boundaries yields
important microstructural information.

Figures 5.1 and 5.2 are two micrographs out of a larger series
of micrographs showing the interaction of a group of thin parallel
martensite plates and a β-grain boundary. Along one of the boundaries a ribbon of V-shaped martensite plates forms separating the
deformed left grain and undeformed lower right grain (fig. 5.2). With
increasing deformation the martensite plates in the left grain
broaden and new plates appear. At the same time further zig-zag
arrangements and newer plates appear. The accommodation across the
grain boundary is schematically represented in figure 5.3, wherein
AB represents the β-grain boundary. The upper figure represents the
situation at a low stress level, whereas the lower figure is valid
for a higher stress level. In the lower part of the figure the β-
grain is divided into regions separated by martensite plates. Along
the grain-boundaries these small isolated β-regions form one rigid
body with the β-regions on the other side of the grain-boundary.
The elastic stresses are thus divided over a larger area and not
any more around the distortions at the β-grain boundaries alone.
Here again the strain-accommodating character of the martensite together with the reciprocal movement of the martensite plate boundaries plays an important role.

Before increasing the stress the zig-zag shaped martensite plates
join until a complete martensite region is formed along the grain-
boundary. Further strain accommodation is taken up by the formation
of new martensite plates crossing the older ones. This mechanism
will be explained below.

Intersection of Martensite Plates

A further increase in stress in order to induce more martensite
plates as obtained by the mechanism explained above also results in
martensite plates of completely other orientations which crossing
the bands of martensite plates and even large single variant regions.
The intersection of martensite plates is reversible as shown by the

series of micrographs in figure 6.1. The micrographs were taken while increasing or decreasing the applied stress. Figure 6.2. shows an enlarged view of the intersection. Two different intersections are shown here, the intersection marked as B which is characterized by a lateral displacement of the martensite plate boundaries and the intersection marked as A where no lateral displacement is visible. Figures 6.3a and 6.3b give a schematic representation of these two intersections. These two models are very similar to the models for the intersection of twins, model 6.3a corresponds with the model of Cahn (13) and model 6.3b with the model of Lin (14).

In model 6.3a the martensite B is the intersected plate and martensite A, the intersecting plate. Directions 2 and 1 are the directions of macroscopic shear corresponding to the variants B and A respectively. Variant C is the product at the intersection of A and B. The area around A, B and C is the β-phase, in contrast to the twinning models where evidently the area around the twins has the same crystal structure as the plates. Direction 1 is parallel to the habitplane H_A as well as H_B.

In model 6.3b the intersecting plate remains undeviated, but the intersected plate B is sheared in the direction 1.

Such intersections occur in completely martensitic samples (ref. 11) whence a much more complicated crystallographic situation arises. In a twinned microstructure the various plates are twin-related to the matrix; the plate C has the same orientation as one of the already existing twins. However, the various martensite plates are related to the matrix by 24 different orientations (the 24 variants), and twinning in one of the plates may transfer area C into an orientation of one of the other variants or into a new, not yet existing orientation.

The Irreversible Plastic Deformation

Irreversible plastic deformation can occur either in the β-phase prior to martensite formation or in the β- and martensite phases while transforming or in the already transformed sample.

Figure 7.1 shows the influence of plastic deformation in the β-phase on the orientation of the stress-induced martensite. The plastic deformation was imparted by a thermomechanical treatment, consisting hot-rolling in the β-phase followed immediately by a water-quench. Most of the β-phase could be retained at room-temperature with the exception of some α-formation along the β-grain boundaries. Then resulted in martensites forming in a fan shape induced by stressing at room-temperature in areas close to the β-grain boundaries indicating severe local plastic deformation.

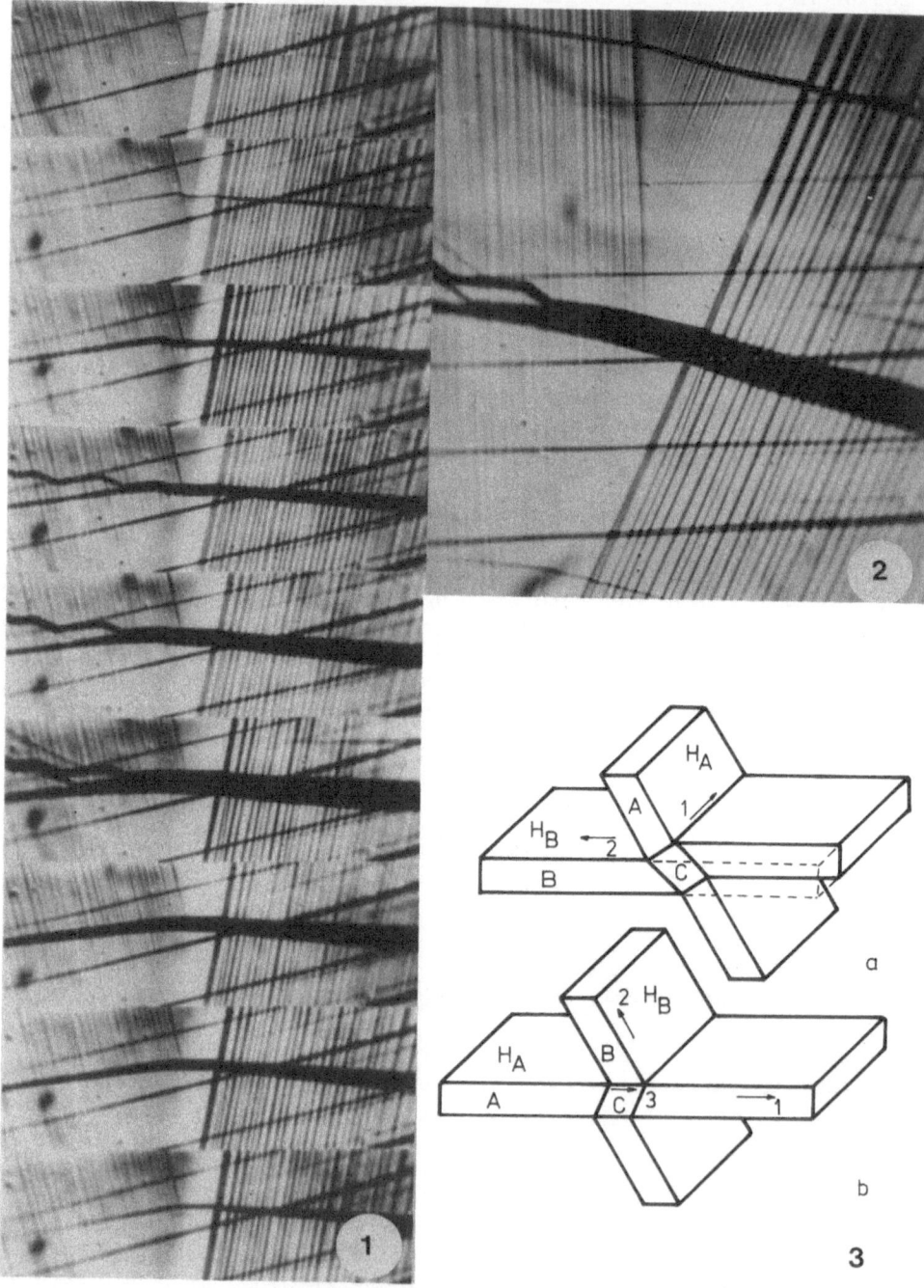

Figure 6 : Intersection of martensite plates. 1 : a series of op-
          tical micrographs taken while loading and unloading the
          sample (450 x); 2 : details of the intersection; 3(a and
          b) : mechanism

Figures 7.2 and 7.3 are taken from samples which have been hotrolled while being water quenched.

The martensite plates shown in figure 7.1 have a pseudoelastic behaviour in that they disappear when the sample is unloaded. The martensite plates shown in the figures 7.2 and 7.3 do disappear only after being heated considerably above the A_s-temperature.

Figure 7.4 is a micrograph out of a series of micrographs, that prove the pseudoelastic reorientation (6). The dark area in the micrograph corresponds to the stress-induced martensite formed in the earlier stages of stressing. On further stressing new mar-

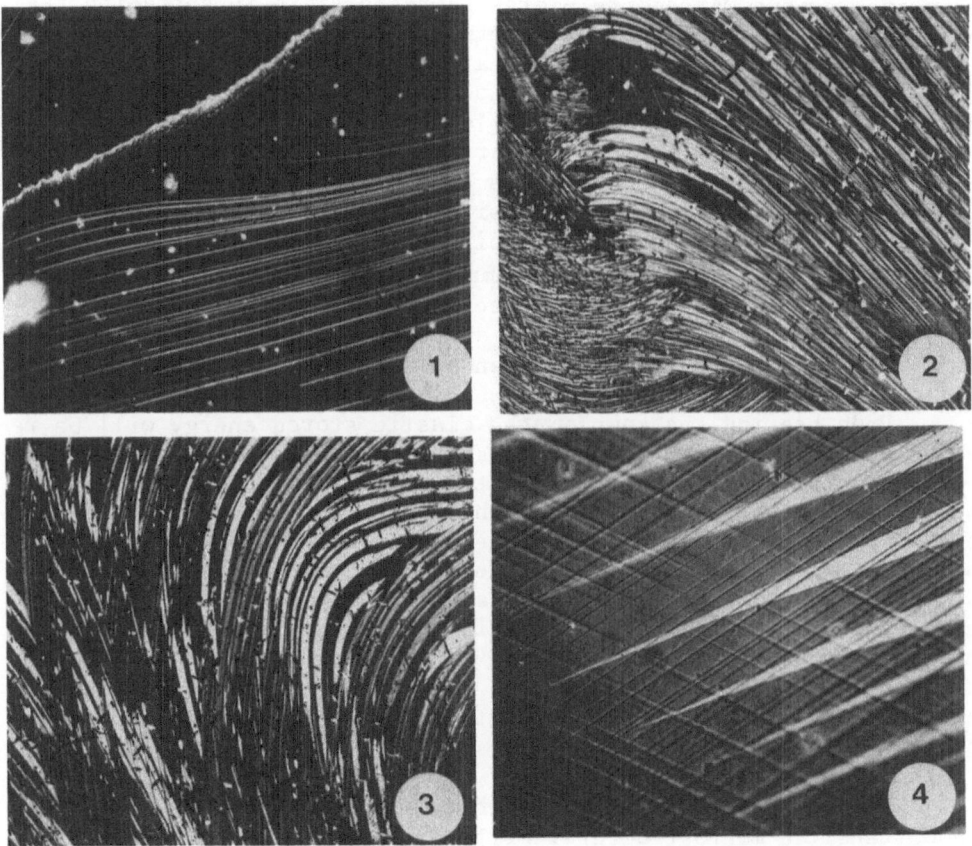

Figure 7 : Microstructural effects due to heavy deformation. 1 : deformation of the β-phase before transformation; 2 and 3 : while transforming; 4 : pseudoelastic reorientation (450 x).

tensite variants appear and these disappear again after the sample
is unloaded.

    Although the martensite formed in samples discussed in this sec-
tion are reversible, i.e. they show the pseudoelastic behaviour, some
irreversible plastic deformation has occurred which can be observed
either by optical microscopy, as presented in this paper, or by electron microscopy as presented in a following paper (15).

Conclusions

    Two important mechanisms determine the microstructure obtained
by stressing a polycristalline material. The mechanical coupling between the various martensite plates is responsible in spreading the
shearing and thus the level of internal stresses over the sample area.
The tapering is responsible for local bending of the lattice. Based
on these mechanisms it will now be possible to carry out quantitative
calculations concerning the crystallography, morphology and mechanical properties of stress-induced martensite. If a sample is stressed
above M_s in order to induce a shape change through the above discussed
stress-induced formation of martensite and subsequently cooled below
M_f in order to retain the macroscopic shape deformation, the microstructure will be frozen in. In samples which exhibit the shape memory effect will thus also be characterized by a high level of elastic stored energy, thus internal stresses exist at the grain-
boundaries and around the tapered ends of the martensite plates.

    Upon heating the sample, the elastic stored energy will be re-
leased and thus not only the martensite-to-β transformation but also
the elastic stored energy should enter as an important factor in the
thermodynamical balance governing the growth and shrinkage of a martensite plate. The first martensite plates which start to shrink
upon heating will be those where the opposing stress during their
formation is the largest. These plates are thus the last that form,
which explains the so many observations that the last formed stress-
induced martensite plates disappear first, either upon releasing the
stress or upon heating. It is thus the internally stressed regions
that help to push back the martensite plates in a sequence opposite
to that of their appearance.

    The existence of internal stresses around inhomogeneities in
the martensitic microstructure, will also have an influence on the
mechanical properties of the sample. Figure 8 shows schematically
the stress-strain curves for the pseudoelastic behaviour (fig. 8 a,b)
and for the shape memory effect (fig. 8 c,d). The curves in figures
8a and b correspond to a temperature $T = T_1 > A_f$ and the martensite
starts to form as soon as the applied stress reaches σ_1. Curve a
shows nearly no strain hardening whereas in curve b the stress to induce more martensite increases with increasing strain. This means

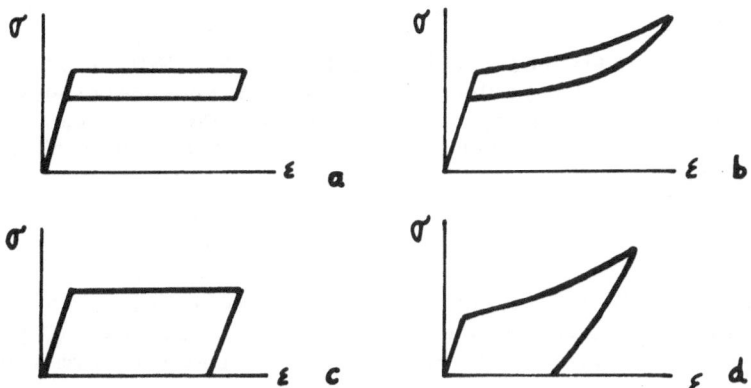

Figure 8 : Schematic stress-strain curves showing the influence of
the internal stresses on the pseudoelastic and shape-
memory behaviour.

that in the sample corresponding to curve b, high internal elastic
stresses are built up, and that due to the strain accommodation in
the sample increasing resistive stresses will oppose the applied
stress. Less internal stresses will be built up on stressing a sample
with a suitable texture or stressing a single crystal, and the curve
will be flatter.

The curves of figure 8 c and d taken at a temperature $A_s < T$
$= T_2 < M_s$ can now be explained using the same arguments.

It is now evident that in samples characterized by a stress-
strain behaviour as represented in figures 8b and 8d, local irreversible plastic deformations in the parent and/or in the martensite
phase will occur at a far more lower macroscopic strain than in
samples characterized by a stress-strain behaviour as represented in
the figures 8a and 8c.

Such local plastic deformations will cause residual stresses
which in their turn may influence the martensite microstructure obtained by cooling the sample during subsequent manipulations. The
spatial distribution and sign of such local residual stress centres
will depend upon the external stress (tensile, compression, bending,
torsion ...). With this in mind calculations such as the ones carried out by Perkins (16) can be performed to explain the two-way
memory effect (reversible shape memory effect). The elastic back
stresses arising from the plastically deformed local regions or arrays
of dislocations may partly help the reversible characteristic of the
shape-memory effect.

Acknowledgements

This work has partially been supported by the N.F.W.O. (Nationaal Fonds voor Wetenschappelijk Onderzoek) and has been continued with the financial aid of I.N.C.R.A. (International Copper Research Association). The authors want to thank G. Smeesters- Dullenkopf for the optical microscopy.

References

1. W. Arneodo and M. Ahlers, Acta Met. 1974, vol. 22, p. 1475.
2. R.V. Krishnan and L.C. Brown, Met. Trans., 1973, vol. 4, p. 423.
3. L. Delaey and H. Warlimont, Proceedings of the Int. Symp. on SME, 1975, this volume.
4. H. Tas, L. Delaey and A. Deruyttere, Met. Trans., 1973, vol. 4, p. 2833.
5. Z.S. Basinski and J.W. Christian, Acta Met., 1954, vol. 2, p. 148.
6. L. Delaey, R.V. Krishnan, H. Tas and H. Warlimont, J. of Mat. Science, 1974, vol. 9, p. 1521, 1536 and 1545.
7. L. Delaey, G. Hummel and J. Thienel, Encyclopaedia Cinematographica, Ed. G. Wolf, Göttingen, to be published Winter 1975.
8. H. Warlimont and L. Delaey, Progress in Mat. Science, 1974, vol. 18, p. 1 - 157.
9. L. Delaey and J. Thienel, Proceedings of the Int. Symp. on SME, 1975, this volume.
10. L. Delaey, H. Tas and A.Q. Khan, ASM-Seminar on copper, ASM-publi. 1972, 062/2.
11. H. Tas, L. Delaey and A. Deruyttere, Z. für Metallk., 1973, vol. 64, p. 855 and 862.
12. A.M. Kosevich and V.S. Boĭko, Soviet Physics Uspekhi, 1971, vol. 14, p. 286.
13. R.W. Cahn, Acta Met., 1953, vol. 1, p. 49.
14. Y.C. Liu, Trans. AIME, 1963, vol. 227, p. 775.
15. R. Rapacioli, H. Chandrasekaran and L. Delaey, Proceedings of the Int. Symp. on SME, 1975, this volume.
16. J. Perkins, Scripta Met., 1974, vol. 8, p. 1469.
17. H. Tas, L. Delaey, A. Deruyttere, J. Less Common Metals, 1972, vol. 28, p. 141.

THE INFLUENCE OF THERMAL HISTORY ON THE PSEUDOELASTICITY OF COPPER-ZINC BASED ALLOYS AND THE MECHANICAL BEHAVIOUR OF MARTENSITE

R. Rapacioli [*], M. Chandrasekaran and L. Delaey

Dept. Metaalk. Kath. Univ. Leuven, Belgium

[*] Esercito Argentino - Investigacion y Desarrollo

Introduction

Alloys exhibiting shape memory and pseudoelastic behaviours have many potential applications. A common feature amongst these alloys is that they all undergo a martensitic transformation either on cooling or under an applied stress. The alloys are thus characterized by certain parameters as M_s, the martensite transformation temperature on cooling, and $\sigma_T^{P \rightleftharpoons M}$, the critical stress needed for the transformation above M_s. The choice of the alloy depending on the specifics of use and the conditions of application are therefore dependent to a very large extent on the above mentioned parameters. The constancy of these parameters in a material under the influence of temperature and stresses during use is essential. Any changes otherwise would seriously alter the shape memory and/or the pseudoelastic recovery properties of the alloy and thereby affect the lifetime of the part or parts being used.

A study of the influence of various thermal and mechanical treatments on the pseudoelastic and shape memory behaviour would thus involve a study of the effect of such treatments on the martensitic transformation itself. To this end influence of such factors as creep (1) has been reported previously.

In the present investigation changes that occur in pseudoelastic behaviour in Cu-Zn-Al alloys as a result of certain thermal treatments are reported. In addition, the mechanical behaviour of martensite has been studied with some interesting effects observed after specific thermal and mechanical treatments. The primary aim of this report is to make the users of such alloys, for potential applica-

tions, aware of certain anomalii that occur and which need be taken
into account before the material is put to use. The study has also
been restricted to property changes in single crystals only, so that
the mechanisms responsible for shape memory effect and pseudoelasticity themselves could be investigated in the process.

Specimen Preparation and Experimental Methods:

Cu-Zn-Al alloys were prepared by melting high purity copper,
aluminium and zinc in sealed quartz tubes. Alloys were selected
based on their martensitic transformation temperatures. Two compositions were prepared corresponding to transformation temperatures
of − 90 °C and + 50 °C (2) and shall henceforth be referred to as
composition (1) and composition (2) respectively. An important
characteristic of these alloys is that they are associated with
minimum betatising temperatures.

Single crystals of the alloys were grown in sealed quartz tubes
by modified Bridgman technique. These crystals were solution treated
at 900 °C for one hour and then quenched into ice water.

The solution treated crystals were spark machined and then
chemically polished to yield round samples from which specimens for
tensile and compressive tests and for dilatometry were prepared.
Samples for the tensile tests were made by spark machining in the
centre part only (15 mm long, 3 mm diameter), so that the thicker
ends could be fitted into grips. The compression and dilatometric
samples were of the same shape and size, 3 mm diameter and 7 mm
long. The tensile and compression tests were carried out on a TT-
DM-L model Ins tron testing machine equipped with a chamber for
testing at temperatures above the room temperature. A theta-dila-
tometer (model Dilatronic III Research) was used for dilatometric
studies.

Optical micrographs were obtained both under normal and pola-
rised light conditions. Samples for electron microscopy were prepared from 0.3 mm discs that were sectioned out of the 3 mm ϕ regions. These discs were first dished by jet polishing and finally
electropolished in a solution of orthophosphoric acid and water.
A JEM 120 electron microscope operating at 120 KV was used in this
investigation.

Experimental Results

1) Variation in σ_T^{P-M} with different thermal treatments
a) single crystals of composition 1 :
Fig. 1a shows schematically the pseudoelastic behaviour exhibited by
a single crystal of β. The stress σ_T^{P-M} needed to induce the martensi-

tic transformation, as shown in this figure, is plotted as a function of test temperature in figure 1b. The entire curve as shown in figure 1a was not traced out in determining these stresses. Instead, the samples were tested only to reveal the onset of the transformation. This was done to avoid any residual deformation in the samples at the end of testing, when the tests were carried out close to temperatures at which the plastic flow stress of β and σ_T^{P-M} are the same (3).

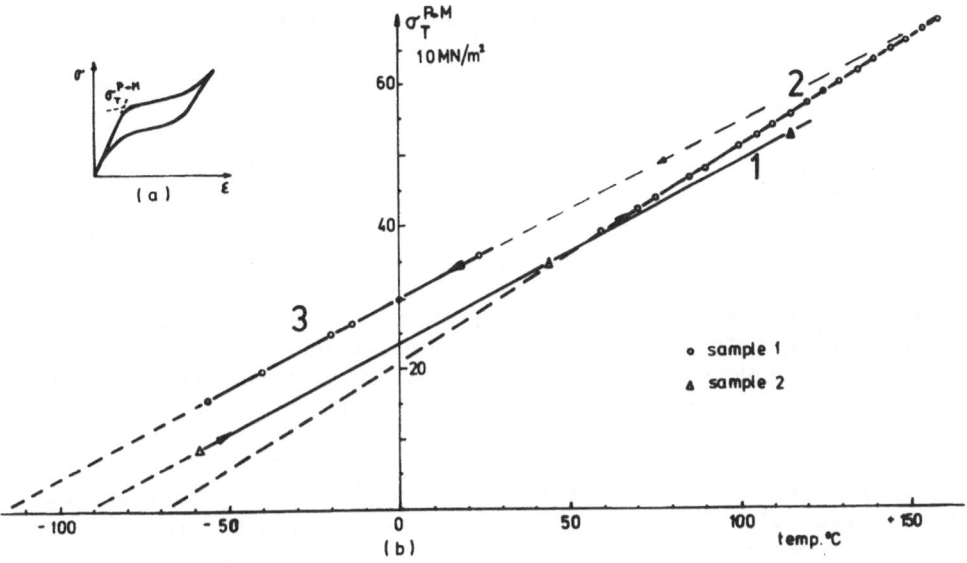

Fig. 1 : a. Schematic stress-strain curve showing pseudoelastic behaviour
         b. Stress, necessary to induce martensite (σ_T^{P-M}) vs temperature.

Three different lines have been drawn through the various data points in figure 1b and these will be presently explained :
Line 1 : The sample was tested in compression at − 60, + 45 and + 100 °C respectively. The sample, in these tests was introduced every time into the testing chamber after the latter had attained the proper temperature and was removed outside (room temperature) immediately upon conclusion of the test at each temperature.
Line 2 : Another sample of the same single crystal was tested at 60 °C and then at higher temperatures till 170 °C. The sample throughout these tests was kept inside the test chamber and not taken out until completion of the final test at 170 °C.
Line 3 : The above tested sample that was taken out of the chamber, was subsequently tested at room temperature and lower temperatures.

     It is observed from the figure that :
a) line 1 intersects line 2 at the starting point of tests corres-

ponding to line 2 (60 °C),

b) a higher stress for transformation is always needed for the samples maintained, during the course of testing, inside the chamber than for the samples that were introduced after the furnace had attained the proper test temperature (line 1),

c) though the sample has been used in further tests at room temperature and lower temperature, the data points (line 3) from these tests do not lie on the continuation of line 2; however, line 3 extrapolates to the last test at 170 °C on line 2;

and d) the sense of the above mentioned discontinuity is such that much larger stresses are required for the onset of the transformation at lower temperatures of testing than is predicted from an extrapolation of line 2 to lower temperature.

Normally, the plot of stress needed for transformation, σ_T^{P-M} vs temperature should result in a continuous straight line which extrapolates to the M_s temperature at zero stress (7). This indeed is observed in line 1 which extrapolates to an M_s temperature of − 90 °C. This also is the M_s temperature of the alloy prior to commencing the tests along line 2. The observed discontinuity from line 2 to line 3 must therefore be a consequence of the mode in which the tests were carried out along line 2, viz., the sample was allowed to remain inside the chamber until completion of the final test at 170 °C.

To get an insight into this anomalous behaviour, further tests were carried out. The σ_T^{P-M} values were recorded after two different treatments were imparted to the quenched sample. The sample after quenching was

(A) subjected to a 1 hour isothermal anneal at 100 °C.

or   (B) flash heated (10 seconds at ∿ 300 °C)

In the former case the σ_T^{P-M} needed to induce martensite at room temperature (σ_{RT}^{P-M}) increased gradually with time of hold at 100 °C reaching a saturated value after 1 hour. This value and the increased σ_{RT}^{P-M} for a flash heated sample are plotted along the ordinate in figure 2.

Changes in σ_{RT}^{P-M} with ageing at temperature for samples subjected to either treatment were also followed and are shown in figure 2. While no detectable change in σ_{RT}^{P-M} had taken place within the time interval of observation for the samples annealed at 100 °C, a decay in the value was definitely observable for the flash heated sample as seen in figure 2.

The effect of flash heating on the plot of σ_T^{P-M} vs temperature is shown in figure 3a. The sample after flash heating produces another straight line (line b) shifted towards a lower temperature but parallel to the untreated sample (line a). The samples were tested to produce complete pseudoelastic curves, as shown in figure 1a, in this set of experiments.

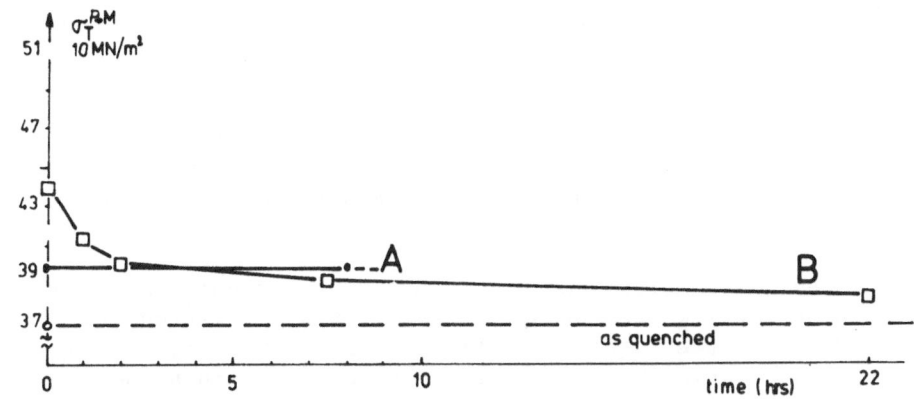

Fig. 2 : σ_T^{P-M} after different thermal treatments and subsequent changes on room temperature ageing.

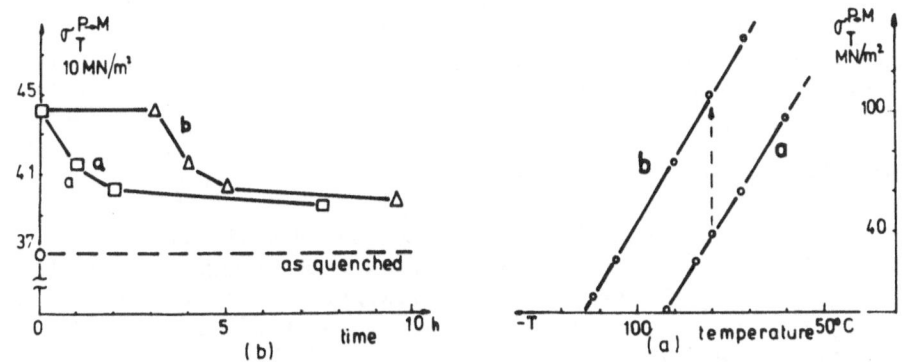

Fig. 3 : a. Effect of flash heating on the σ_T^{P-M} vs temperature plot.
b. changes in σ_T^{P-M} following flash heating.
Line a : decay with room temperature ageing.
Line b : decay with room temperature ageing after initial hold at - 60 °C for 3 hours.

     Ageing temperature after flash heating also has an influence on the decay as was revealed by the following experiments. Two samples were flash heated and σ_T^{P-M} measured before and after flash heating. One sample was allowed to remain at room temperature after the flash heating, and the decay in σ_T^{P-M} was followed with time (line a on figure 3b). The other sample was held, immediately after flash heating at - 60 °C (above the M_s temperature) for 3 hours (line b on figure 3b). Subsequent testing, at room temperature, of this sample revealed no change in σ_T^{P-M} from that measured immediately after flash heating. When the same sample was then annealed at room temperature, the decay in $\sigma_T^{P \rightleftarrows M}$ could be observed.

In order to determine whether any structural changes were associated with the samples subjected to these different thermal treatments, optical and electron microscopic investigation of the samples was performed. No changes could be detected in samples even when annealed for 2 hours at 100 °C.

Flash heated samples when viewed under an optical microscope exhibited boundary like features. However, it was verified from Laue Patterns that the sample was still a single crystal, thereby eliminating the possibility of the features being grain boundaries. Electron microscopy of similar samples often showed the presence of sub-boundaries consisting of an array of dislocations.

The existence of superlattice could be detected, by selected area diffraction, in as quenched sample as well as samples subjected to different thermal treatments. Antiphase domain boundaries could be imaged with the superlattice reflections and the presence of large domains were observed.

b) Single Crystals (in β) of composition 2

A similar set of experiments was performed with samples of composition 2 that were martensitic at room temperature. As explained in the next section, the samples were first treated to exhibit a constant A_s and then pseudoelastic curves were obtained at 90 °C (in β condition) te determine the $\sigma_{90}^{P \rightleftarrows M}$. Also the effect of ageing at 90 °C on the $\sigma_{90}^{P \rightleftarrows M}$ was studied. The results of these tests are reproduced in figure 4a which shows that $\sigma_{90}^{P \rightleftarrows M}$ decreases with ageing. However, the unloading path remains unaltered throughout, indicating that the hysteresis in the pseudoelastic curves reduces with ageing at 90 °C. In order to confirm that this is a true ageing effect and not an effect of the previous stress cycle, the $\sigma_{90}^{P \rightleftarrows M}$ of another sample aged 30 minutes at 90 °C to start with and then with further ageing at 90 °C was determined. The $\sigma_{90}^{P \rightleftarrows M}$ as also the hysteresis were unaltered in these latter tests proving that the decrement observed in figure 4a is indeed an effect of ageing at 90 °C.

When a sample in the as received condition (to exhibit constant A_s) was flash heated, the $\sigma_{90}^{P \rightleftarrows M}$ as well as the hysteresis were found to be unaltered; i.e., the pseudoelastic curve comprised of the loading curve 1 and the unloading path of figure 4a. An interesting result was, however, obtained when a sample aged substantially at 90 °C to exhibit the curve 3 and the minimal pseudoelastic hysteresis of figure 4a was flash heated. As can be observed in fig. 4b the entire pseudoelastic curve is shifted; i.e., $\sigma_{90}^{P \rightleftarrows M}$

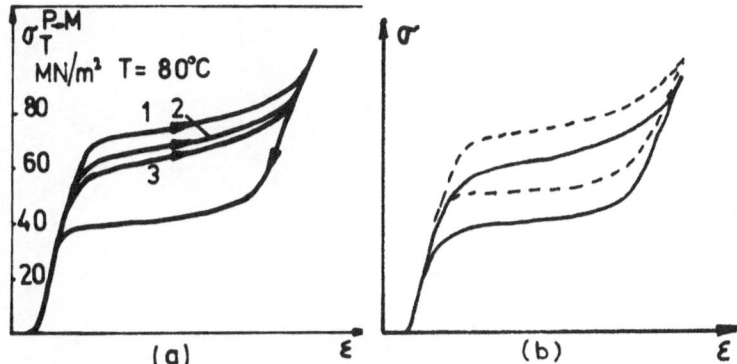

<u>Fig. 4</u> : a) variation of pseudoelasticity and accompanying hyste-
resis at 90 °C for 5 minutes (1), 20 minutes (2) and
45 minutes (3). Note that the unloading curve remains
unaltered while the loading curve changes with ageing.
b) shifting of the entire pseudoelastic curve after flash
heating.

increases without any change in pseudoelastic hysteresis. Subsequent
ageing at 90 °C, however, gradually shifted the pseudoelastic curve
till it is brought down to the original starting condition of figure 4b. This increase in σ_T^{P-M} on flash heating and the subsequent
decay on ageing can be compared to that already reported for single
crystals of composition 1.

Antiphase domains could be observed utilising dark field techniques in transmission electron microscopy, in the untreated and
treated samples. Figures 5a and 5b show the domain sizes in a
sample in the as received condition and in another after it had been
aged at 90 °C for 3 hours. The increased domain size after the
treatment is clearly observable. The domain sizes of samples after
flash heating always remained the same as that prior to flash heating.

2. Mechanical behaviour of martensite and its dependence on prior thermal and mechanical history

Samples of composition 2 when tensile tested at room temperature

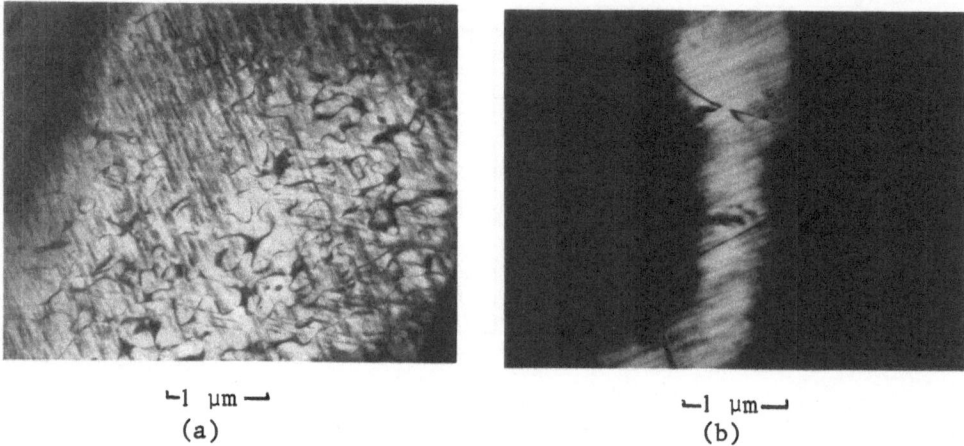

┗1 μm ┛   ┗1 μm┛
(a)   (b)

Fig. 5 : Antiphase domains in untreated sample (fig. 5a) and a
         sample aged 3 hours at 90 °C (fig. 5b).

in the as quenched (quenched from 900 °C into water at room temperature) yielded a stress-strain curve as shown in figure 6a. To
investigate the unusually large amount of pseudoelastic recovery
in these tests, the as quenched samples were studied by dilatometry.
It was observed that the reverse transformation (martensite→β) temperature for the as quenched martensitic sample (200 °C) was much
higher than the transformation temperature for the same sample in
succeeding dilatometric cycles (60 °C). Also, the mechanical behaviour of the same sample at the end of tests in figure 6a and
after it had been subjected to a dilatometric cycle (heating above
200 °C to remove the first martensite and then cooling) varied
exhibiting curves characteristic of shape memory effect as shown in
figure 6b. If the sample immediately after quenching was subjected

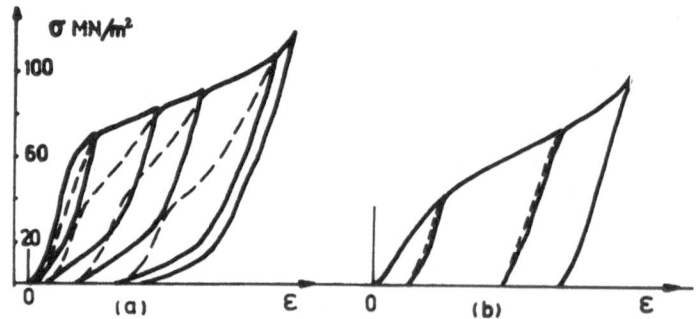

Fig. 6 : Stress-strain curve obtained in tension for as quenched mar-
         tensitic sample (a) and after first heating cycle (b).

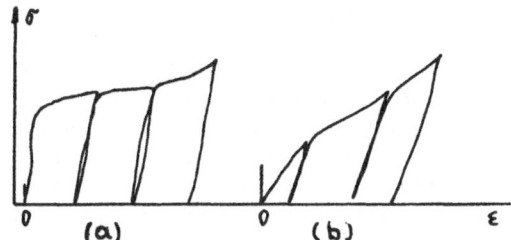

Fig. 7 : Stress-strain curve obtained in tension for a sample subjected to one dilatometric cycle immediately after quenching
(a) and after the next dilatometric cycle (b).

to a dilatometric cycle and then tested, the stress-strain curve
though characteristic of shape memory behaviour differed from that
of figure 6b in as much as a well defined plateau was observed (fig.
7a). Curves similar to figure 6b, however, resulted when the sample
was subsequently subjected to another dilatometric cycle and then
tested (fig. 7b).

That the as quenched martensite has a higher transformation temperature than the martensite after cycling was also confirmed through
tensile tests at 80 °C on an as quenched sample and a sample subjected
to the first transformation cycle after quenching. In the former
case shape memory behaviour was observed whereas the latter sample
yielded a pseudoelastic curve.

The mechanical behaviour was found to be drastically altered
when the as quenched sample was subjected to a number of shape me-

(a)   (b)

Fig. 8 : Mechanical behaviour (a) and microstructure (b) at the end
of shape memory cycles + 900 °C and quenching treatment.

mory cycles followed by heating to 900 °C and water quenching. As
shown in figure 8a, a very low stress level (in tension) is needed
to obtain the same elongation of the martensitic sample after the
above treatment. When the same samples were tested in compression,
complete recovery of the samples for small values of stress could
be obtained on unstressing. It was possible during the course of
our studies to produce single crystals of martensite. These single
crystals also exhibited similar effect on compressing as mentioned
above. The crystals could be subjected to as much as 30 % deformation with complete recovery of this deformation on stress release.
This effect was identified with the "rubberlike" behaviour. Results
pertinent to this effect and their analysis are dealt with separately (4).

     Optical microscopy of the as quenched samples revealed large
martensite plates often intersecting each other with groups of
smaller plates between the larger intersecting plates as shown in
figure 9a. It was observed that subjecting the same sample to the
first dilatometric cycle had resulted in the removal of many intersections and smaller plates so that the microstructure could be
described as consisting of larger plates with many plates belonging
to a preferential variant as shown in figure 9b. This trend was
observed and more markedly in samples subjected to a number of shape
memory cycles and subsequent solution treatment at 900 °C and quenching. As can be seen in figure 8b much larger plates have resulted
due to the treatment. As an extreme case it was possible to obtain
a single crystal of martensite through spontaneous thermal transformation during the final controlled quenching after the sample had
been initially subjected to specific thermal and mechanical treatments.

                    (a)   (b)

Fig. 9 : Optical micrographs showing plate arrangement and sizes
         (a) in a quenched sample
         (b) in the same sample after the first thermal cycle.

The crystal structure of the martensite in all the untreated and treated samples was the same (3R) as verified by electron diffraction.

Conclusions and Discussion

1. Variation in σ_T^{P-M}

The first conclusion that can be drawn from the measured σ_T^{P-M}-values is that stress values as high as 650 MN/mm^2 have been reached without any observable loss in pseudoelastic behaviour. This means that the plastic flow stress of the β-phase corresponding to the composition of the alloys (Alloy 1 : Cu 66.55, Zn 18.9, Al 14.55; Alloy 2 : Cu 70, Zn 12, Al 18 (at %) used in this work is considerably higher than the plastic flow stress of the β-alloys analysed by Pops (3) and Pops and Ridley (5). This is because the upper temperature and stress limit for obtaining the pseudoelasticity is controlled by the balance between the stress to form martensite and the stress to cause plastic flow. Thus, alloys presently used should have a plastic flow higher than 650 MN/mm^2.

Disregarding for the moment the reasons for the observed increase in σ_T^{P-M} with different treatments and hence a lowering of M_s, evidence presented here suggests that such effects are either more or less permanent in nature depending upon the conditions of use. Thus if an alloy of composition 1 were to be used at room temperature after flash heating the effect is only transitory as the σ_{RT}^{P-M} decay s on ageing and tends to return to its original value. However, if the same alloy after a similar treatment were to be used at - 60 °C the effect is more permanent as verified by the negligible or no decay, at this temperature, in σ_T^{P-M} value.

Evaluating the observed effects thermodynamically and therefore referring to the thermodynamics of the transformation (6), the above discussed shifts in M_s temperature should be discussed in terms of a shift in T_o or an increase in ΔG_{Ms}^{P-M}. Assuming that the driving force for nucleation is independent of temperature and applied stress, $(dT_o/d\sigma)$ may be assumed to be equal to $(dM_s^\sigma/d\sigma)$. From the slope of the curves, σ_T^{P-M} vs T, the transformation entropy can be calculated. Allowing a change in the driving force for nucleation, ΔG_{Ms}^{P-M}, with heat treatment, a change in slope of the σ_T^{P-M} vs T curve should be expected. From figure 3a, it can now be concluded that the flash heating only shifts T_o and ΔG_{Ms}^{P-M} can thus be taken as constant. However, line 2 of figure 1 shows an apparent change in transformation entropy with thermal history. Any point on line 2 has now to be considered as representing a β-phase with a lower free energy, the free energy of the β-phase at the temperature of the preceding test which corresponds to a shift of T_o. The point measured at 170 °C lies on the extrapolation of line 3, which has been measured on the same sample but quenched after completing the measurement at 170 °C. Moreover, line 3 is parallel to line 1, proving again a shift in M_s and T_o.

We can now analyse the cause of such changes in σ_T^{P-M} and the shift in T_o. The observed decay at room temperature and its absence at - 60 °C suggests that a diffusional process or processes is/are probably operative. Clark and Brown (7) from a study on β Cu-Zn quenched from temperatures between 25 and 250 °C concluded that dilute but equilibrium disorder could be retained, whence the re-ordering through defect assisted diffusional mechanisms would ex-hibit kinetics characteristic of a decay process. Analysing the present results it seems probable that the increase in σ_T^{P-M} due to thermal treatments and the subsequent decay can be related to such changes in long range order. More quantitative studies are currently in progress.

The antiphase domains observed in samples of composition 2 have been retained at room temperature due to their growth being interrupted by the martensitic transformation in these samples. Though doma in growth has been observed in samples aged at 90 °C, the absence of growth in the flash heated samples may be taken to indicate that a 10 second treatment at about 300 °C (flash heating) is not sufficient to cause any detectable change in domain size as also that no equilibrium domain sizes exist below the critical ordering temperature. The decrease in σ_{90}^{P-M} on ageing samples of composition 2 at 90 °C with attendant decrease in pseudoelastic hysteresis can be attributed to a domain size effect following the explanation given below :

It was often found that the martensite transformation had led to the creation of elongated antiphase boundaries (fig. 10). Extra boundary-curvature at the intersections with the thermal boundaries are also observed. Thus, extra energy is stored in the martensite along those anti-phase domain boundaries. The larger the domain size, the lower will be the number of intersections and the marten-site formation will be less hindered. Upon reversion of the trans-formation, anti-phase domain boundaries can be inherited in the - phase. The dislocations associated with the transformation will now follow preferential paths, thus lowering the hysteresis.

2. Mechanical behavior of martensite :

It has been clearly established that the mechanical behaviour of the first formed martensite after quenching is subject to change

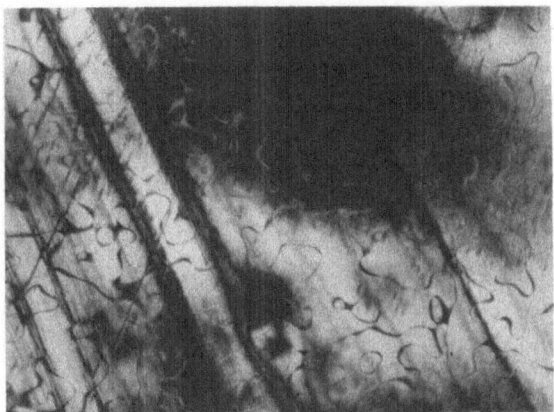

<u>Fig. 10</u> : Thermal and slip induced antiphase domain boundaries in
          Cu-Zn-Al martensite.

under the influence of stress and temperature. The process of acieving a martensite with reproducible transformation temperatures and mechanical behaviour is shown in figures 6 and 7. It is seen that while stress cycles immediately following the quenching and then a dilatometric cycle hastens the process of constancy in behaviour, subjecting the as quenched sample first to a dilatometric cycle only leads to an intermediate stage as shown in figure 7a. Thus the first stress cycle that a martensite is subjected to seems to be a key factor in obtaining the final martensite with reproducible characteristics.

     Another point of interest is the influence of the martensite plate config uration on the stress necessary to deform the sample. If larger isolated plates are formed, high internal stresses have to be overcome which results in a pseudoelastic behaviour of the martensite. Due to the heat treatments, self accommodating plate groups are formed, and because of their ease of mechanical coupling, deformation will proceed at much lower stress level, which explains the mechanical behaviour shown in figure 8.

     Notwithstanding certain questions remain to be answered.
1) Why should the first formed martensite be associated with much higher transformation temperatures ? - This has been attributed in the past as merely due to quenching stresses (8) but in the present system some additional factors such as order parameter may also be responsible.
2) the reasons for the large sized plates with pronounced directionality (fig. 8b) after the shape memory cycles + 900 °C and quenching treatment. When samples of composition 1 (M_s = - 90 °C) were sub-

jected to a similar treatment and examined after the quench in the electron microscope, large domains of mottling were seen rather uniformly distributed in the matrix (6), than the usual continuous mottling on a finer scale that is obtained after the first quench. These results seemingly indicate that some singularities even at as high temperatures as 900 °C remain which are probably responsible for the preferential orientation of the plates.

Finally a general remark can be made concerning the applicability of Cu-Zn-Al alloys for energy conversion purposes (9,10). The alloys possess most of the requirements considered essential for such purposes; viz.,
  i)thermoelastic martensite transformation
 ii)high parent phase yield stress
iii)low T_o and
 iv) large recoverable strain.
Other factors, viz., fatigue life and latent heat of transformation need to be investigated.

Acknowledgements

The present work has been carried out with the financial support of the International Copper Research Association. The authors are grateful to T. Vermaelen for carrying out the tensile tests, Ing. J. Nijs and Ing. J. Mariën for technical support and M. Roosen and J. Mellaerts for alloy preparation.

References

1. S. Miura, S. Maeda and N. Nakanishi, Phil. Mag., 1974, vol. 30, p. 565.
2. M. Ahlers, Scripta Met., 1974, vol. 8, p. 213.
3. H. Pops, Met. Trans., 1970, vol. 1, p. 251.
4. R. Rapacioli, M. Chandrasekaran, M. Ahlers and L. Delaey, Proc. of Int. Symp. on SME, this volume.
5. H. Pops and N. Ridley, Met. Trans., 1970, vol. 1, p. 2653.
6. L. Delaey and H. Warlimont, Proc. of Int. Symp. on SME, this volume.
7. J.S. Clark and N. Brown, Phys. Chem. Solids, 1961, vol. 19, p. 291.
8. I. Cornelis and C.M. Wayman, Scripta Met., 1974, vol. 8, p. 1321.
9. M. Ahlers, Scripta Met., 1975, vol. 9, p. 71.
10. H.C. Tong and C.M. Wayman, Met. Trans., 1975, vol. 6A, p. 29.

THE MARTENSITIC TRANSFORMATION IN β-BRASS AND THE SHAPE MEMORY EFFECT

M. Ahlers [+], R. Rapacioli [++] and W. Arneodo [+++]

Centro Atómico, S.C. de Bariloche, Argentina

[+]Max Planck Inst.f.Eisenforschung, Düsseldorf, Germany, [++]Kath.Univ.Leuven, Belgium, [++]Eserito Argentino, Investigación y desarrollo, [+++]IMAF Cordoba, Argentina

The present discussion of the martensitic transformation and the memory effect (SME) in Cu-Zn will be divided into two sections : A) An analysis of the stability of the different structures which include i) the spontaneous martensite that forms on cooling at the temperature $M_s(0)$, ii) the martensite which is induced by an externally applied stress σ at a temperature $M_s(\sigma) > M_s(0)$ and iii) the structures that form when the martensite is deformed plastically, of main interest being that one which is related to SME. B) A study of the kinetics of the nucleation and growth of the martensite and the discussion of an atomistic model of the martensitic transformation and of SME.

For the experiments single crystals of binary β-brass were used, ranging in composition from 39.2 to 40.6 at % Zn and having an $M_s(0)$ between 70K and 150K. The spontaneous and stress induced martensite was investigated by means of optical and transmission electron microscopy, electrical resistivity and acoustic emission.

A. The Stability of the Martensitic phases :

i) Spontaneous martensite : For binary β-CuZn (1) and ternary alloys based on CuZn (2) the M_s temperature depends strongly on the alloy composition. Although the equilibrium phases occurring in the CuZn system are controlled by the electron concentration e/a (3) no correlation between M_s and e/a had been found (2). Instead M_s is related to the tetragonality c/a of the basic face centered tetra-

gonal structure from which the martensitic structure is derived by
stacking shifts (4) (c/a = 1 for the cubic structure). It is likely,that the tetragonality is due to the order which the martensite
inherits from the bcc β-phase (4), and that therefore the free energy contribution of the long range order essentially determines the
composition dependence of the free energy difference between the β
phase and the martensite.

ii) Stress induced martensite : External stresses can affect
the martensitic transformation in three ways : a) The free energy
difference ΔG is changed by an additional term which is the mechanical work done during the transformation. This leads to a Clausius-
Clapeyron type of equation in which the entropy of transformation
ΔS is related to the transformation stress σ_M at the temperature
$M_s(\sigma_M)$ and the transformation strain ε_M by $\Delta S = \varepsilon_M \, d\sigma_M/dM_s(\sigma_M)$.
Neglecting the small volume change of about 0.2 %, ΔS can also be
expressed by the stress τ_M that acts on the martensite variant and
by the corresponding martensite shear which is calculated to be γ =
0.17 for brass : $\Delta S = \gamma d\tau_M/dM_s(\tau_M)$. b) The stresses can lead to
structural changes of the martensite during the transformation and
thus affect the free energy difference. This would mean that ΔS
or the enthalpy difference ΔH or both were a function of the
stresses and would depend on the orientation of the single crystal.
c) The friction that causes the hysteresis during a transformation-
retransformation cycle may depend on the applied stress.

The experimental results obtained when martensite is induced
in β-brass single crystals by compressive or tensile stresses can
be grouped into the behaviour at the beginning of the transformation and during the progression of the transformation ("transformation hardening"). It is observed (5) by optical microscopy that the
martensite variant with the highest shear stress component is induced and that the critical resolved shear stress τ_M at the beginning
of the transformation depends linearly on $M_s(\tau_M)$. The slope $d\tau_M/dM_s$ is indepent of the orientation of the single crystal, and the
extrapolated transformation temperature $M_s(\tau_M = 0)$ coincides with
the M_s temperature that was measured for spontaneous burst type
martensite in polycrystals (1) (within experimental scatter). The
stress hysteresis is small and hardly depends on the stress or the
temperature (6). From these results it is concluded that the martensitic structure is not changed by applied compressive or tensile
stresses at the beginning of the transformation. The value for ΔS
is $\Delta S = 0.31 \pm 0.01$ cal/molK (5), independent of concentration
(within experimental scatter) for alloys ranging in composition from
39.2 to 40.6 at % Zn.

During the progression of the transformation the stress increases with strain (6). For compressive tests the transformation
hardening coefficient $d\tau/da \equiv \Theta$ has values between 2.5 and 7.5
kg/mm^2 (τ and a are the resolved shear stress and strain for the

most favoured martensite variant). For tensile tests the τ- a curve
depends strongly on the crystal orientation. For most crystals a region of low hardening is followed by a high hardening of about 30
kg/mm^2 after the first few percent of transformation strain. It is
surprising that the stress-transformation strain curves for compressive and tensile transformation differ so much although the orientation of the variants with respect to the sample axis is nearly the
same in both cases for a given orientation and the hysteresis depends only weakly on the transformation strain (6). To explain this
difference it has been argued (6) that the secondary shear system
of the principal martensite variant has only a small stress component in the martensite induced by compression whereas the component
for the martensite formed by tension is considerably higher. It
seems that during the progression of the stress induced transformation structural changes can take place which are caused by shears on
the secondary shear plane.

   iii) The plastic deformation of stress induced martensite : at
a resolved strain of approximately 0.17 the transformation is completed and no markings are left on the surface (7), except for some
faint and diffuse traces occasionally. When the corresponding stress
is sufficiently low, an increase in stress is observed on further
straining until a new deformation mechanism sets in. Plastic deformation induced by tensile stresses in martensite which is formed
in tension recuperates again on unloading and the martensite disappears, i.e. the original shape of the single crystal is restored.
This is the SME for single crystals. By compressive stresses a different type of deformation is induced which does not recover during
the retransformation. These results are in contrast with the deformation behavior of martensite single crystals deformed below the
deformation temperature (14). It has been shown (7) that the SME-
type deformation occurs on the same system on which the secondary
shear during the martensitic transformation takes place, i.e. on a
$\{111\} < 11\bar{2} >_{fcc}$ system of the faulted face centered martensitic lattice. This means that partial dislocations move in such a way that
stacking faults are eliminated (5). The critical resolved shear
stress τ_B does not depend on temperature and is related to the
stacking fault energy Γ by $\tau_B = - \Gamma/b_p$, with b_p the Burgers vector
of the partials (5). Γ is negative, i.e. the structure containing
the stacking faults is more stable and the elimination of a stacking
fault requires energy. Γ depends sensitively on the alloy composition. Γ changes from 10 ergs/cm^2 at 40.6 at % Zn to 6 ergs/cm^2 at
39.6 at % Zn and extrapolates linearily to $\Gamma = 0$ at 38.2 at % Zn.
The reason for the strong concentration dependence of Γ has to be
related to order. Order can affect the bonding between next nearest
neighbours and by its influence on the lattice parameters, especially on c/a, can also change the distance of atoms between the close
packed layers, and consequently the interaction energy. The structure of the martensite is 3R orthorhombic consisting of close packed
planes with the stacking order ABCBCACAB and additional faults (8).

For stability considerations such a structure can not be treated a
priori as a faulted fcc lattice, i.e. from the observation that a
stacking fault in the 3R structure is unstable it can not be concluded in general that the face centered structure is more stable.
It seems however, that to a first approximation this conclusion can
be drawn for the CuZn system since it has been observed that the
structure of the martensite changes from twinned face centered (at
37.7 at % Zn (9)) to the 3R structure (at 38.3 at % Zn (10)) at approximately the same concentration at which Γ is zero.

When martensite induced by compression is deformed plastically
in compression, a different deformation mechanism is favored because
the secondary shear system has an unfavorable orientation with respect to the sample axis and thus has a small shear stress component.
The alternative is a deformation on the $(110)_{fcc}$ plane which contains both the secondary shear direction and the secondary shear
plane normal (7). This type of shear is not recovered when the
retransformation is completed, but leads to a permanent deformation
and thus does not contribute to a shape memory effect. It is important to realize therefore that not all deformation processes in
the martensite recuperate. This aspect can become important in polycrystals.

B. The kinetics of the transformation and the interpretation of
the transformation by a model: an evaluation of the SME includes the
analysis of the hysteresis between the formation and the retransformation of the martensite after the SME type plastic deformation has
been induced. This requires that as a first step the hysteresis associated with the transformation and retransformation of undeformed
martensite has to be studied. Some investigations have been made on
this aspect in single crystals of β-brass by optical microscopy (6,
7,11), electrical resistivity (11), acoustic emission (11) and transmission electron microscopy (12). It had been shown that the martensite grows in two stages : first by the expansion of thin martensite
platelets into the β-matrix and then by the thickening of these
plates. The hysteresis is caused by the friction during the thickening of the plates (6). For stress induced martensite the friction
stress is nearly independent of the amount transformed and thus
likely is caused by the difficulty with which a single martensite
interface grows into the β-phase matrix. In spontaneous martensite,
on the other hand, the friction increases with the progression of
the transformation, due to the additional interaction between different variants of martensite plates. At the beginning of the transformation the friction for spontaneous martensite is the same as for
stress induced martensite. When the crystal is unloaded after the
SME type deformation has been induced, the friction stress is increased (7). Till now no systematic studies of the influence of deformation on the hysteresis have been made.

Although the phenomenological models are able to predict the

geometry of the martensitic transformation it is difficult to explain for example why stress induced martensite retransforms to the
original shape of the crystals even after the habit interfaces are
eliminated, and why the SME occurs.

A different model which accounts for these observations has
been described recently (13). It takes into account the path of
the atoms during the transformation and is able to explain the reversibility and the SME. This model predicts the same geometry as
the phenomenological models, it accounts for the observed secondary
shear plane and removes the equivalence of different paths of the
atoms during the transformation to a single one which is shorter
than other paths leading to the same structure, even after the memory type deformation has been induced in the martensite. In this
model the transformation is described by a transformation shear on
the "primary" $(110)_\beta$ plane which is nearly parallel to the habit
plane, in a direction $[1\bar{1}0]_\beta$ of amount $\gamma_1 = 0.25$. This shear alone
produces close packed planes inclined to the primary plane but whose
stacking order does not yet correspond to the correct stacking of
the martensite. The correct stacking order is obtained by a secondary shear on these planes. In figure 1 the movement of the
atoms during the primary shear is shown and in figure 2 are drawn
the close packed planes, using the same letters for the corresponding atoms as in figure1. The secondary shear means a displacement from D' to D'' or double the distance to D'''. The average
shear which is composed of these two elemental shears is determined
by the condition that an undistorted habit plane exists. The path
of atom D to D'' or D''' is shorter than the path to any other equivalent position after the transformation. It can be expected that
during the retransformation to the β-phase the atoms move on the
shortest possible path as well, which means from D'' or D''' back
to D, and that therefore the original shape is restored. From
this picture the SME is also easy to understand : as has been discussed above the shape memory type deformation occurs on the secondary shear system, this means that D''' is moved to D''. On retransformation the original shape is restored since D'' and D'''
move back to D, independent of whether D' moved to D'' during the
transformation or via D''' after the additional shear. As is seen
this model does not require that the β-phase be ordered.

A second conclusion concerning the stability of the martensite
can be drawn from the present model : the experimental results indicate that by eliminating stacking faults the free energy of the
martensite is increased. Consequently an increase in stacking
fault density is expected to stabilize the martensite. Why then
does the martensite phase not contain more stacking faults than
it actually has ? Within the framework of the model the secondary
shears are restricted to displacements corresponding to D' → D''
or D' → D''' of figure 2. The fraction n_a of planes on which
a shear D' → D'' occurs is fixed by the condition that an undis-

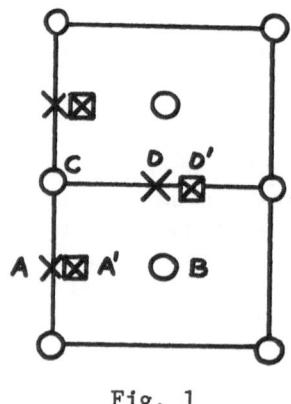

Fig. 1

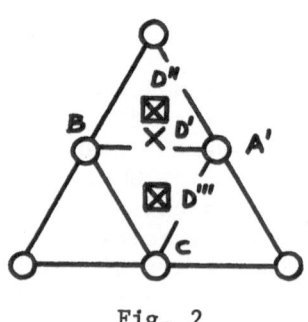

Fig. 2

torted habit exists. An increase in stacking fault density means changing n_a , thus giving rise to a distortion of the habit plane and consequently to an increase in the interface energy which has not been observed. This result shows that the discussion of the stability of the martensitic phases has to take into account the mechanism by which the martensite forms.

References :

1. H. Pops and T.B. Massalski, Trans.AIME 230 (1964) 1662
2. H. Pops, Trans.AIME 236 (1966) 1532
3. C.S. Barrett and T.B. Massalski, Structure of Metals, 1966
4. M. Ahlers, Scripta Met. 8 (1974) 213
5. W. Arneodo and M. Ahlers, Acta Met. 22 (1974) 1475
6. M. Ahlers, R.Pascual, R.Rapacioli, W.Arneodo, to be published
7. W. Arneodo and M. Ahlers, Scripta Met. 7 (1973) 1287
8. H. Warlimont and L. Delaey, Progr.Mat.Sci. 18 (1974) 1
9. I.Cornelis and C.M.Wayman, Acta Met.22 (1974) 301
10. D.Hull, in "Electron Microscopy and Strength of Crystals", ed.
    G. Thomas and J. Washburn. Interscience New York 1963, p. 936
11. R. Pascual, M. Ahlers, R. Rapacioli and W. Arneodo, Scripta
    Met. 9 (1975) 79
12. R. Rapacioli and M. Ahlers, Scripta Met. 7 (1973) 977
13. M. Ahlers, Z. Metallkunde 65 (1974) 636
14. R. Rapacioli, M. Chandrasekaran, M. Ahlers and L. Delaey
    Submitted to Scripta Met. (1975)

THE RUBBERLIKE BEHAVIOR IN CU-ZN-AL MARTENSITE SINGLE CRYSTALS

R. Rapacioli[°],[°°°],   M. Chandrasekaran[°], M. Ahlers[°°] and

L. Delaey[°]

[°]   Departement Metaalkunde, Kath.Univ.Leuven (Belgium)
[°°]   Max-Planck-Institut für Eisenforschung, Düsseldorf
         (Germany)
[°°°]   Escercito Argentino, Investigación y Desarrollo

The answer to the shape memory effect lies in successfully explaining the mechanical behavior of martensite. As a first step, the mechanical behavior of β'-martensite single crystals of different orientations and under different stress conditions should be analysed. During the course of this work a rubberlike effect has been observed within the martensitic structure of a CuZn based alloy. To the authors' knowledge it is the first time that such a behavior has been reported in copper-zinc based alloys.

Single crystals of a CuZnAl alloy (18 at % Al, 12 at % Zn) with a M_s temperature of 50 °C were prepared by growing the single crystals in sealed quartz capsules by the Bridgeman method. The crystals were then annealed at 900 °C and waterquenched. By spark machining the samples were cut into cyclindrical shape of a diameter of 3 mm. The ends had a larger diameter in order to fit the samples into the grips of a tensile instron machine : the samples were then annealed at 800 °C for 1 h and quenched into water to room temperature. A short flash heating to approximately 300 °C followed, in order to remove the mar‑ tensite first formed during quenching (1,2). The crystals were deformed by tensile stresses at 80 °C (i.e. in the β-phase) until the whole sample had transformed into a stress induced martensitic single crystal and by quenching under constant elongation to room temperature the crystal could be retained as a martensitic single crystal. That the sample was indeed a single crystal was confirmed through optical microscopy (polarized light), X-rays and electron microscopy.

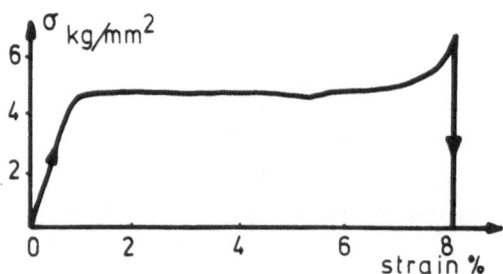

<u>Fig. 1</u> : Transformation stress–strain curve at 80 °C

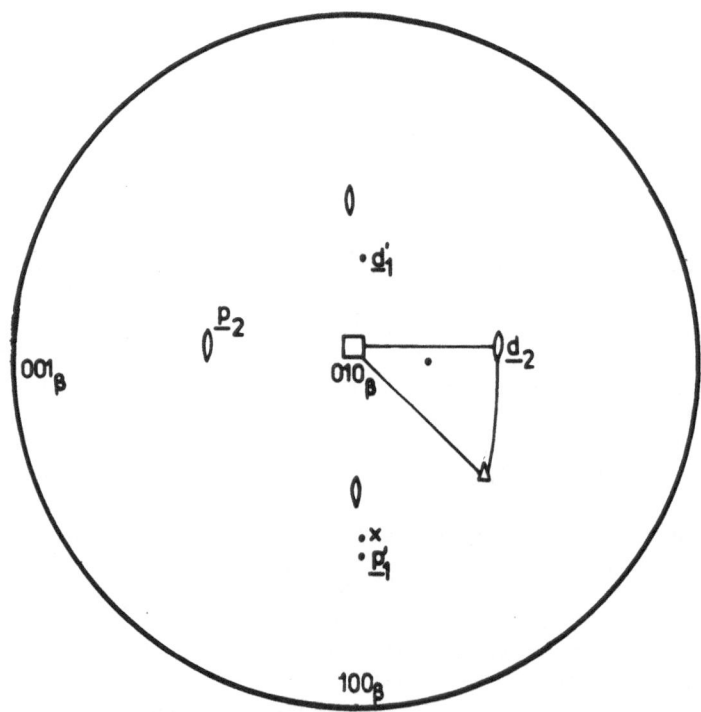

<u>Fig. 2</u> : Stereographic projection showing the stress axis within the
unit triangle, the primary (p_1' and d_1') and secondary (p_2 and
d_2) shear systems for the martensite single crystal and the
normal to the plane of markings obtained on compressing the
single crystal.

(In some cases when a single crystal was not obtained, the sample was flash heated again and the deformation and quenching process repeated). In figure 1 is shown a typical transformation stress-strain curve at 80 °C in obtaining a single crystal. As is seen the transformation was terminated when the stress started to rise steeply. The amount of strain at this point of about 8 % is the value expected for the martensite shear in crystals with a Schmidfactor of $\sim 1/2$. When this martensitic single crystal was subsequently compressed at room temperature, parallel surface markings appeared accompanied with an audible noise. These markings broadened, coalescing into one single band through the entire crystal with increasing stress. The surprising feature now is that on unloading in most cases the crystal regained its original shape which it had before the compressive stresses were applied. It should be emphasized that this effect is a phenomena which occurs entirely in the martensite phase and which does not involve the β-phase. This rubberlike behavior is observed in compression and not in tension, when the martensitic single crystal is induced in tension.

In order to understand the mechanism which is responsible for the rubberlike behavior, the following analyses were carried out : 1) The orientation of the original β-phase single crystal was determined by X-rays. 2) The orientation of the habit plane for the stress induced martensite was determined from a multisurface analysis after partial stress induced transformation and quenching. 3) The plane corresponding to the surface markings that appear on compressing the martensitic single crystals was also analysed. The results are plotted in figure 2 on a stereographic projection. In plotting the pole of the compression induced markings the orientation change resulting from the previous tensile stressing was not taken into consideration. As is seen from the figure the plane of the markings is nearly parallel to the habit plane. Knowing the habit plane normal p'_1, the macroscopic shear direction $\underline{d}'_1$, the secondly shear plane p_2 and the direction $\underline{d}_2$ can be deduced (3) and are also shown in figure 2. This result already permits us to exclude several deformation mechanisms which have been discussed in the literature : 1) A structure change to fcc or hcp by a shear on the secondary shear plane would result in strain markings being observed on the secondary shear plane. 2) As discussed by Tas, et al. (4) the orthorhombic martensite having a stacking sequence ABCBCACAB can transform to a twin variant. Their analysis shows that the twin plane normal for the two existing possibilities lies in the plane that contains both $\underline{p}_2$ and $\underline{d}_2$ and thus is far removed from the plane normal corresponding to the observed markings. 3) When martensite induced by compressive stresses in CuZn alloys was further deformed in compression, deformation markings on a plane that contains both $\underline{p}_2$ and $\underline{d}_2$ were found. This possibility also is excluded in the present case, as is seen from the figure.

The only possibility that seems to remain is a change to a different martensitic variant on compressing the already martensitic
single crystal. The question remains why the deformation should be
associ ated with a large resoring force, as is observed. Experiments
to this end are in progress.

References :

1. C.M. Wayman, Scripta Met. $\underline{8}$ (1974).
2. M. Ahlers, R. Pascual, R. Rapacioli and W. Arneodo, Shape Memory
   Symposium, Toronto 1975.
3. W. Arneodo, M. Ahlers, Acta Met. $\underline{22}$ (1974) 1475.
4. H. Tas, L. Delaey, A. Deruyttere, Z. Metallkde, $\underline{64}$ (1973) 855, 862.
5. W. Arneodo, W. Ahlers, Scripta Met., $\underline{7}$ (1973) 1287.

SUPERELASTICITY AND SHAPE MEMORY EFFECT IN Cu-Sn ALLOYS

S.Miura,Y.Morita and N.Nakanishi*

Department of Mechanical Engineering,Doshisha University,

Kyoto,Japan, *Department of Chemistry,Konan University,

Kobe,Japan

ABSTRACT

The martensitic transformation and the mechanical behavior in β-Cu-Sn alloy,mainly the composition of Cu-15.0 and 15.3 at.%Sn, have been studied by means of electrical resistivity measurements, X-ray diffraction patterns and tensile tests. The martensitic transformation temperature in Cu-Sn alloys decreased with increasing tin contents and the crystal structures of both thermal and stress-induced martensites in Cu-15.3 at.%Sn were found to be orthorhombic. The shape memory effect of thermal martensite in Cu-15.0at.%Sn alloy appears only for a slightly deformation, and the effect is mostly concerned with the stress-induced martensite in Cu-Sn alloy. After ageing at room temperature in β_1 phase, the tensile behavior showed remarkable superelasticity and critical stress to induce the martensite increased with increasing ageing time and deformation temperature. And also,the strain rate dependence of superelastic behavior was remarkable, at the lower strain rate the superelasticity disappeared because of the stabilization of the stress-induced martensite.

1. INTRODUCTION

In Cu-Sn alloys which have disordered bcc β-phase at the electron concentration of about e/a=1.5 (e/a:electron atom ratio), martensitic transformation occurred, but the crystal structure and the alloy composition range for the several martensites have not so far been well clarified. However, it is known that the martensite phases

form as following way with increasing tin concentration [1,2],which
are (i) β_1'-ordered long-range stacking structure, faulted (18R structure). (ii) β_1''-lamellar composite of structures β_1' and γ_1'. (iii)
γ_1'-ordered orthorhombic twined and faulted (2 H structure). (iv) β'-
ordered orthorhombic,faulted (4H structure). Here, the nomenclature
of (i),(ii),(iii) is as according to Warlimont[1] and with regard
to (iv), β' phase named by Nishiyama et al.[3] was used. As pointed
out by Kennon and Miller[2], β' phase which are not formed as part of
the sequence $\beta_1' - \beta_1'' - \gamma_1'$ described above, formed in alloys varied
from 13-16.2at.% Sn quenched from above some minimum temperature.

It can be posited from the results by Nishiyama et al.[3],Warli-
mont[1] and Soejima[4] that in the concentration range from 15.0 to
15.48at.%Sn,the martensites formed during quenching from 720°C to 0℃
and subsequent cooling to -196°C are only γ_1' phase.

Up to now superelasticity and shape memory effect have been reported in various β phase alloys such as Au-Cd[5,6], Ti-Ni[7],
Cu-Al-Ni[8], Cu-Zn[9] and Ni-Al[10] etc.,however, no work has been
reported on β-phase Cu-Sn alloys.

The present authors have noted that in the alloys which exhibit
the shape memory effect, the lattice invariant strains in the martensite phase are twins, therefore, the composition of Cu-15.3at.%-
Sn and Cu-15.0at.%Sn have mainly chosen for the specimen. This paper
will discuss the superelasticity and the shape memory effect in Cu-
Sn alloys.

Recently, Kennon[11] has reported that the volume fraction of
transformation to γ_1' formed during cooling to -196°C in Cu-15.1at%-
Sn aged at 20°C before subzero cooling, decreasing with increasing
the aging time. In the present study same phenomenon was also recognized and its relation to superelasticity behavior is reported.
Furthermore, strain rate dependence of superelastic behavior was
also discussed in terms of the stabilization of stress-induced martensite.

2. EXPERIMENTAL PROCEDURES

The materials used in this experiment are Cu-Sn alloy single
crystals containing 14.8-15.4at.%Sn. The alloys were prepared by
melting at 1100°C pre-weighed amounts of Cu(99.99%) and tin(99.999%)
in sealed quarltz tubes under a partial pressure of argon to prevent
the tin loss due to volatilization. Single crystals grown by the
Tammann method were homogenized at 720°C for 20 hours and then quenched into ice brine in order to get bcc phase. The orientation of
bcc single crystals was determined by the Laue back-reflection method.

The martensitic transformation temperatures of the alloys were determined by the electrical resistivity method. To determine the crys-

tal structure of specimens in as-quenched, subzero-cooled or cold
worked state, the powder X-ray diffraction analysis was made using
Cu-Kα radiation. The specimens which were filed and annealed at 720°
C for 15 min. and then quenched into ice brine at 0°C, were used as
the as-quenched specimen. The X-ray powder diffraction patterns for
as-quenched state were at 20°C and for subzero-cooled state were
taken at -178°C by using the low temperature X-ray apparatus. Also,
annealed powder was ground in the mortar to get the cold worked
specimen.

Tensile tests were performed with an Instron type testing machine
at a strain rate of 5.56x10^{-3}/sec., otherwise noted. Specimens,2.0-
2.5 mm in diameter with 30 mm gauge length, were used for the tensile
testing. The temperature baths used for the tensile testing were
methyl alchohol and iso-pentan refrigerated by liquid nitrogen below 0°C, water from 0°Cto 100°C and silicon oil above 100°C.

3. EXPERIMENTAL RESULTS

3.1 Measurements of Transformation Temperature

Electrical resistivity measurements were carried out to determine
the transformation temperatures of Cu-Sn alloys containing 14.8-15.4
at.%Sn, which have β_1-phase immediately after quenching. Figure 1
shows,for example, the resistivity vs. temperature curve of a Cu-
15.3at.%Sn alloy. It is shown that Ms and Mf, stand respectively
for the starting and finishing temperature of the martensitic transformation on cooling, can be obtained from the figure, but As and Af,
those of on heating,cannot be known because of the absence of a clear
inflection point in this curve. The situation was same for the other
Cu-Sn alloys tested. The Ms temperature determined for various Sn
contents are shown in Fig.2, indicating that Ms decrease from -30°C
to -125°C with increasing Sn contents from 14.8 to 15.4at.%Sn.

3.2 X-ray Diffraction Patterns

X-ray diffraction pattern of Cu-15.3at.%Sn alloy immediately after quenching from 720°C into ice brine is shown in Fig.3(a). The
diffraction pattern was taken at 20°C, and its parameter is a=2.98A.
In this case, any superlattice line has not been found. Then the X-
ray diffraction pattern was taken at -178°C as shown in Fig.3(b).
From the comparison between the observed and calculated values,
analysing the pattern as an orthorhombic structure, its parameters are
a=4.56, b=5.40 and c=4.36 A. Figure 3(c) shows the diffraction pattern after heating the powders used for Fig.3(b), the figure shows

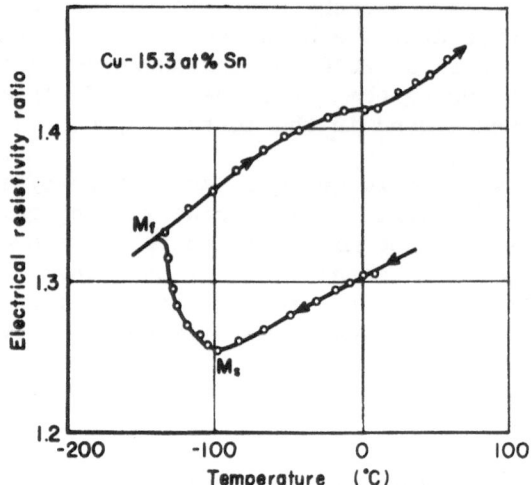

Fig.1   Variation of the electrical resistivity with temperature
        in Cu–15.3 at .%Sn alloy.

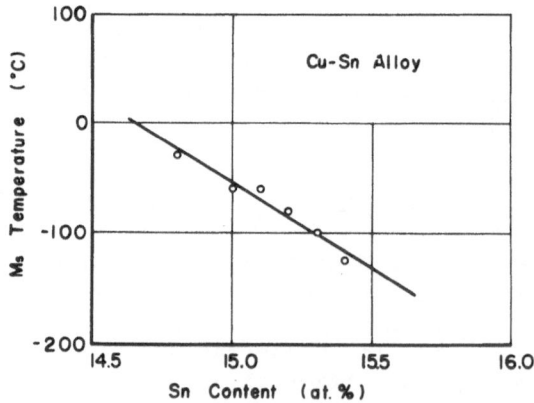

Fig.2   Variation of the Ms temperature with composition in Cu–Sn
        alloy.

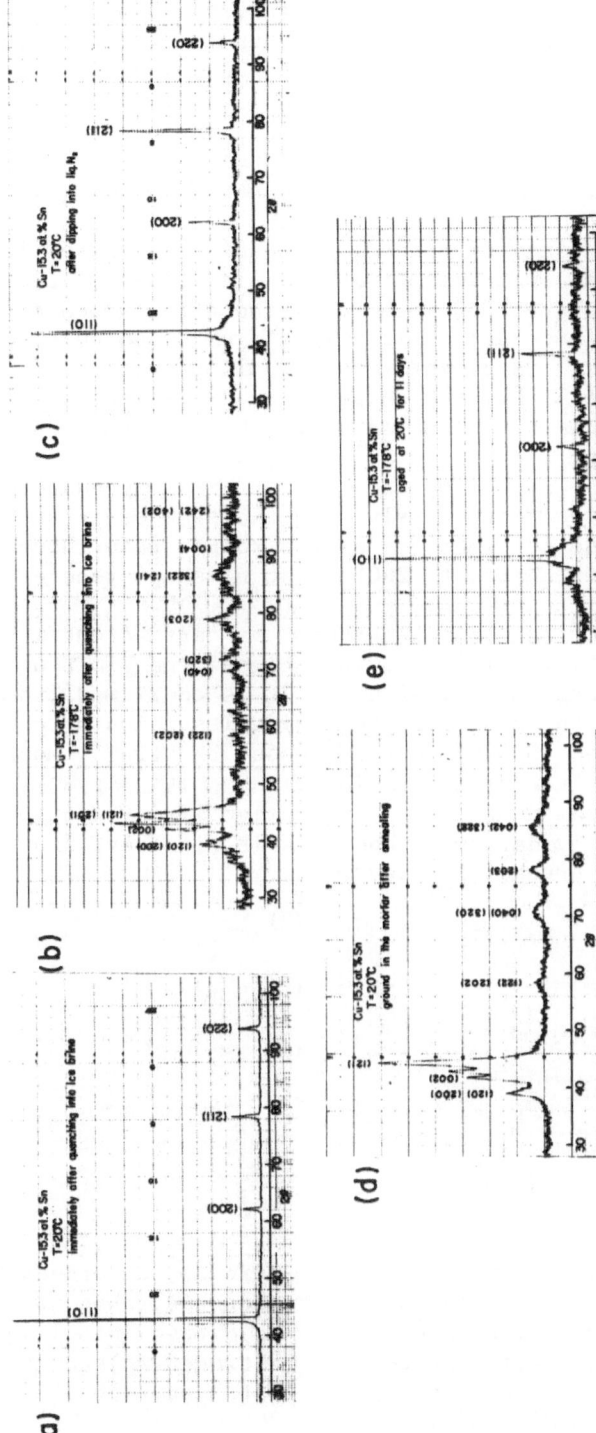

Fig.3   X-ray diffraction patterns of Cu-15.3at.%Sn alloy.   (a) As-quenched, taken at 20°C.
(b) As-quenched, taken at -178°C.   (c) After dipping into liquid nitrogen, taken at
20°C.   (d) Ground in mortor after annealing.   (e) After aged at 20°C for 11 days,
taken at -178°C.

bcc structure. Powder were then ground in the mortar and diffraction pattern was taken at 20°C, as shown in Fig.3(d). Figure 3(e)
shows the diffraction pattern at -178°C, when the quenched powders
were aged at 20°C for 11 days. It is also confirmed the bcc structure as compared with Fig.3(a) and (c).

3.3 Shape Memory Effect in Cu-15.0at.%Sn Alloy

The shape memory effect of thermal martensite (γ_1'phase) in Cu-
15.0at.%Sn alloy is shown in Fig.4(a)-(e). In Fig.4(a), the specimen is in β_1-phase,as-quenched into ice brine after annealed at 720°
C, for 20 hours, and then dipped into liquid nitrogen below Ms(Fig.
4b).   Subsequently it was slightly bent in liuid nitrogen(Fig.4c),
and quickly heated to 100°C, the shape recovery was observed(Fig.4d).
Furthermore, the specimen was heated up to 200°C which is above As
temperature, the shape recovery was completed.
Figure 5(a) and (b) show a specimen which was as-quenched,and dipped into liquid nitrogen respectively with the same orientation as
one used in Fig.4. The specimen was bent stronger than those in Fig.
4(c) (Fig.5(c)), then heated to 200°C, but no shape recovery has
been observed.
The same experiment as mentioned above was carried out in tensile
testing as shown in Fig.6, i.e., to investigate the limit of residual strain on the shape memory effect. It is confirmed from this
figure that for strain about 2%, the residual strains were completely recovered if heated to 200°C,but for about 3% strain, most of
them still remained after heating. Thus,in a Cu-15.0at.%Sn alloy the
shape memory effect was occurred only for slightly deformation but
not at all for strongly deformed, and this is different from the
other thermoelastic martensite alloys.

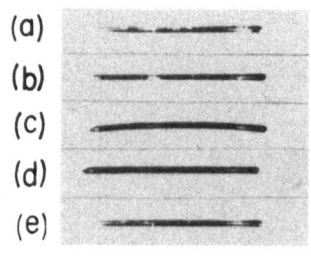

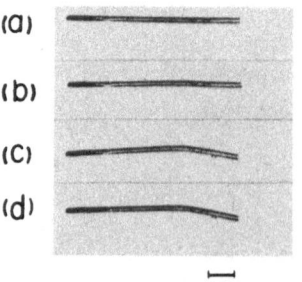

Fig.4 The shape memory effect in
      Cu-15.0at.%Sn alloy slightly
      deformed. (a)as-quenched,
      taken at 20°C,(b) dipped in-
      to liquid nitrogen,(c) bent
      in liquid nitrogen,(d) heat-
      ed to 100°C,(e) heated to
      200°C.(Ms=-50°C,Mf=-70°C)

Fig.5 The shape memory effect
      in Cu-15.0at.%Sn alloy
      strongly deformed. (a)as-
      quenched,taken at 20°C,
      (b)dipped into liquid
      nitrogen,(c)bent in liquid
      nitrogen,(d)heated to
      200°C.

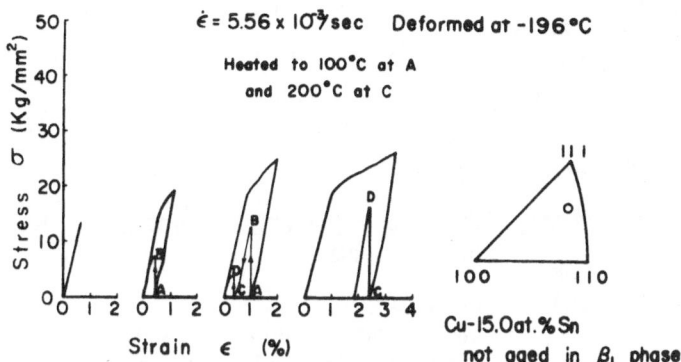

Fig.6 Limit of residual strain for the occurrence of shape memory effect in Cu–15.0at.%Sn γ_1' thermal martensite. Specimen was heated at A(100°C) or C(200°C) and then released from the stress at B or D.

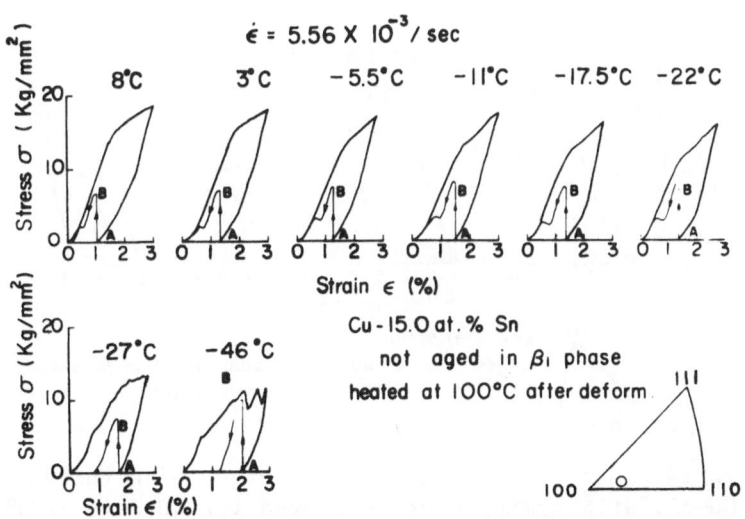

Fig.7 Variation of the stress-strain curves with temperature in a Cu–15.0at.%Sn single crystal which was as-quenched.

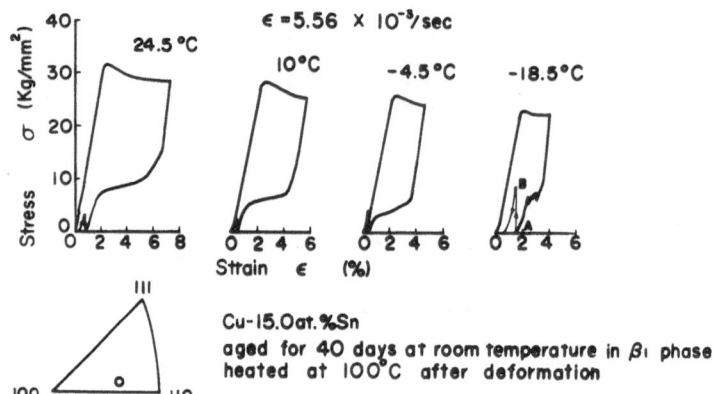

Fig.8   Variation of the stress-strain curves with temperature in a
Cu-15.0at.%Sn single crystal aged for 15 days in β_1 phase.

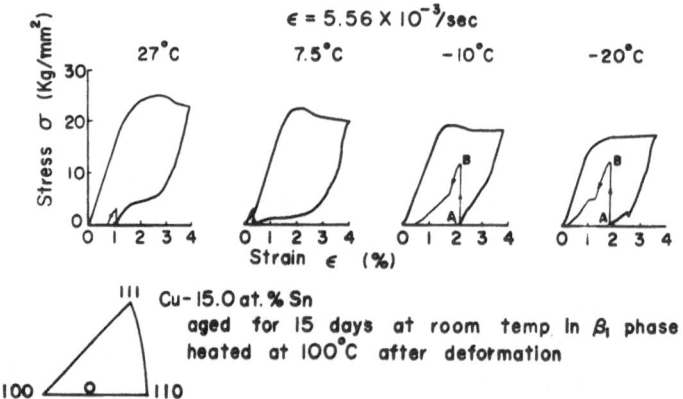

Fig.9   Variation of the stress-strain curves with temperature in a
Cu-15.0at.%Sn single cryatal aged for 40 days in β_1 phase.

3.4 Tensile Behavior

Stress-strain curves of an as-quenched Cu-15.0at.%Sn alloy at various temperatures with a strain rate 5.56x 10^{-3}/sec. are shown in Fig.7. At lower temperature near the Ms, residual strains due to a remaining stress-induced martensite were left upon unloading, but warmed up to 100°C at point A, the stress rises at B because the remaining stress-induced martensite disappeared. Then released from the stress at B again, the residual strains were almost recovered. Namely, this behavior is the shape memory effect of SIM.

Figures 8,9 and 10 show the stress-strain curves aged in β_1 phase at room temperature for 15,40 and 50 days, respectively. From these it is found that even at the temperature range in which the large residual strain has been observed in as-quenched specimen, the stress-strain curves became to show superelastic behavior with increasing aging time. For example, in the specimen aged for 50 days, the stress-strain curves show remarkable superelastic loop almost at temperatures used in this test,and about 11% strains have been recovered upon unloading. The critical stress(the apparent yield stress of SIM formation) obtained from Figs 7,8,9 and 10 were replotted against temperatures as shown in Fig.11. The critical stress increased linearly with increasing temperature as was reported in thermoelastic alloys such as Au-Cd[5,6],Cu-Al-Ni[8],Cu-Zn-Sn[12], Au-Cu-Zn[13].Moreover,the critical stress increased with increasing aging time and in the specimen aged for 50 days, the critical stress increased about three times as high as not aged specimen.

The effect of strain rate on tensile behavior in Cu-15.3at.%Sn alloy is shown in Fig.12. With an order of 10^{-3}/sec., superelastic behavior was observed but decreasing the strain rate, the amount of superelastic recovery was decreased and with the strain rate of $3.7x10^{-4}$/sec.,large residual strain has been observed upon unloading. However, heated up to 90°C and the released from the stress, the residual strain was almost recovered.

The effect of aging during the tensile test is shown in Fig.13. As shown in Fig.13(a), stress-strain curve shows almost superelasticity when deformed at 20°C with the strain rate of $4.7x10^{-3}$/sec. During subsequent straining, the specimen was aged at the deformation temperature for 5 min. under the interruptted load(Fig.13 b) . On further straining, the flow stress was increased but the strain corresponding to the prior deformation before aging remained. In Fig.13(c), it is shown that the strain remained by the aging for 5 min.after the straining.

In order to examine the microstructure of the deformed specimen,

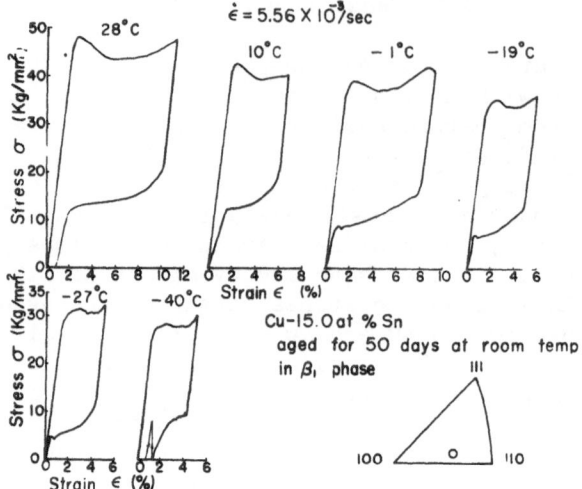

Fig.10   Variation of the stress-strain curves with temperature in a
        Cu-15.0at.%Sn single crystal aged for 50 days in β_1 phase.

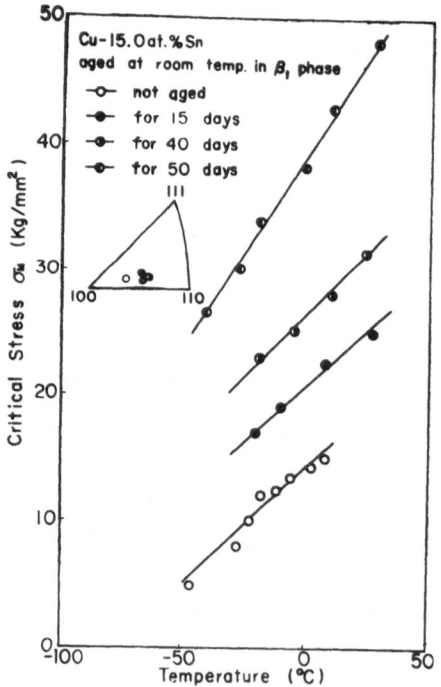

Fig.11   Variation of the critical stress with temperature of
        each aging time.

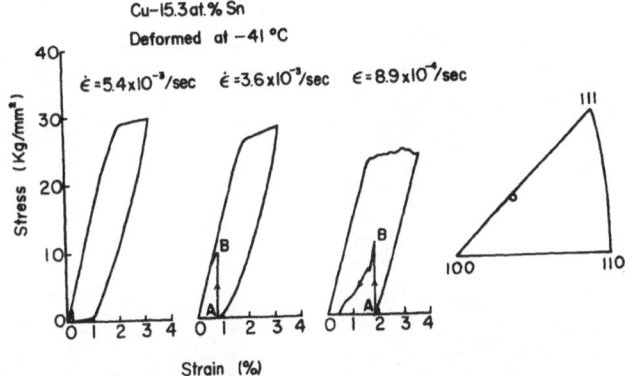

Fig.12 Variation of the tensile behavior with strain rate in a
Cu-15.3at.%Sn single crystal which was quenched into
ice brine, deformed at -41°C.

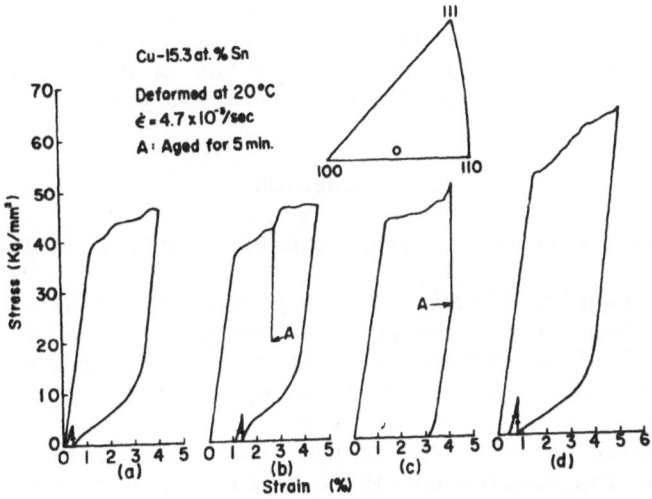

Fig.13 Effect of aging on tensile behavior in Cu-15.3at.%Sn
single crystal which was quenched into ice brine.

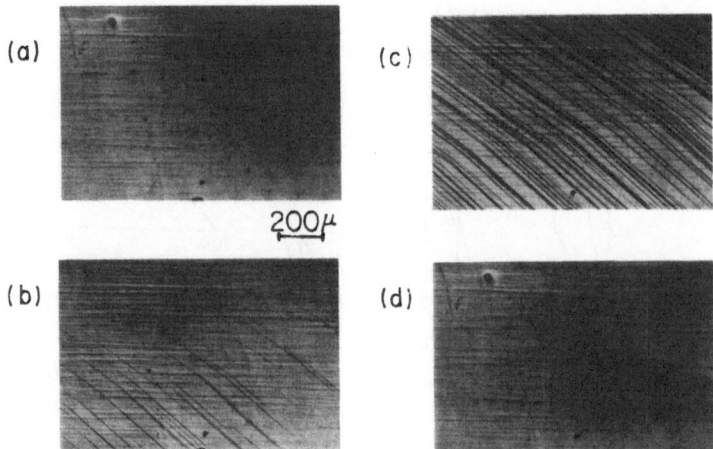

Fig.14 Photomicrographs showing the process of stress-induced
        transformation in a Cu-15.3at.%Sn alloy at 20°C.
        (a) Before straining. (b) Immediately after straining.
        (c) On further straining. (d) After unloading.

photomicrographs were taken at 20°C as shown in Fig.14(a)-(d). Fig.
14(a) shows a micrograph under no load, and the same area immediately after reaching the critical stress is shown in Fig.14(b) indicating the occurrence of the stress-induced martensite. On further
straining, plates of stress-induced martensite increase(Fig.14c)
and disappear on unloading(Fig.14d). In the case of lower strain
rate, it was found that a part of SIM remained upon unloading.

4. DISCUSSION

4.1 Martensitic Transformation of Cu-Sn Alloys

The transformation behavior of bcc β-phase Cu-Sn alloys had been
studied as regards to tin contents and quenching temperatures by
Kennon and Miller[2]. According to them, β_1'martensite was completely obtained at 20°C when Cu-Sn alloys containing 13.0-14.39at.
%Sn were quenched from above 667°C, while complete transformation
to γ_1' martensite was observed in the specimen containing 15.0-
15.48at.%Sn during cooling to -196°C after quenching from above
600°C.

In Cu-Sn alloys, the disordered bcc β-phase must become ordered
to β_1 which is a Fe_3Al type ordered phase, and critical temperature of order-disorder transformation has been reported about 700°
C by Morikawa et al.[14], and just below the melting point by
Soejima[4] respectively. Therefore, in Cu-15.0 and 15.3at.%Sn

alloys, β phase becomes ordered β_1 phase when quenched into ice
brine at 0°C from 720°C and then complete transformation to γ_1' martensite occures during cooling to -196°C.

In the present study, transformation temperatures were determined from the electrical resistivity measurements, but As and Af
could not be determined. However, on a considerable fast heating
rate(about 30°C/min.), As and Af could be determined by the movement of galvanometer in the apparatus for the measurement of the
electrical resistivity and were found to be 90°C and 113°C, respectively in Cu-15.0at.%Sn alloy, although the resistivity vs. temperature curve could not be obtained because of fast heating rate.
These results are in good agreement with the results by Murakami[15]
that Ms and Af in a Cu-15.0at.%Sn alloy measured by differencial
scanning calorimeter are -40° and 90°C respectively.

Recently, Arbuzova et al.[16] have reported the heating rate
dependence of reverse transformation from γ_1' martensite to β_1 phase.
According to them, on a slow heating rate(0.5°C/min.), As and Af do
not appear as a clear inflection point and surface relief due to γ_1'
remains, but on fast heating rate(30°C/min.), they appear distinct-
ly and completely reverse transformation occurs. Also,they have reported that Ms and Af temperatures in a Cu-15.3at.%Sn alloy are
about -100° and -30°C respectively.

Thus, the transformation hysteresis(As-Ms) in Cu-Sn alloys is
large such as 70°-140°C. The transformation hysteresis was found
to be only 20°C for a Au-47.5at.%Cd which undergoes thermoelastic
transformation and about 400°C for Fe-31at.%Ni alloy which undergoes athermoelastic transformation. In this sense, it may be presumably be considered an intermediate between thermoelastic and
athermoelastic alloy.

4.2 Shape Memory Effect in Cu-15.0at.%Sn Alloy

According to Otsuka and Shimizu[17], the conditions of the materials which show the shape memory effect are as follows: (i) The
martensitic transformation proceeds thermoelastically, (ii) the
lattice invariant strains in the martensite phase are not dislocations but twins, (iii) the structure is ordered, (iv) if ordering is
disregarded, the martensite phase has a hcp structure, where (i)
and (ii) are basic conditions, but (iii) and (iv) are supplemental
ones.

Considering these conditions in the case of Cu-15.0at.%Sn alloy,
(i) although it cannot be determined micrographically, the martensites can hardly be considered as a thermoelastic martensite because the transformation hysteresis is large such as 140°C as mentioned in the previous chapter. (ii) According to Morikawa et al.
[14], twin faults exist on (121) plane in the γ_1' martensite(named
β'' in their paper), and stacking faults exist on (121) plane inside
the twin faults. (iii) The superlattice line was not recognized by

powder patterns in this work, but the ordering was found by Morikawa et al.[14]. (iv) The crystal structure is orthorhombic but nearly hcp structure.

As mentioned above, in the case of a Cu-15.0at.%Sn alloy, the conditions (iii) and (iv) are satisfied, but conditions (i) and (ii) are not. Thus, the shape memory effect in a Cu-15.0at.%Sn alloy cannot be explained by the mechanisms which have so far been proposed. However, as regarded to have an intermediate nature of thermoelastic and athermoelastic martensites from the viewpoint of the transformation hysteresis, the coherent nature which is a characteristic of thermoelastic martensite is still maintained on slight deformation, and shape memory effect can be considered to occur on heating. But on strong deformation, it is considered that the coherency at the habit plane is missed and plastic deformation that is irreversible occurs.

4.3 Effect of Aging in β_1 phase and Superelastic Behavior

Kennon[11] has reported that the stabilization of β_1 phase occurred during aging at 20°C after quenching in a Cu-15.1at.%Sn alloy and after aging of about 10^4 min., volume fraction of γ_1' martensite approaches to 0. The same results were obtained in this work, i.e.; γ_1' martensites were observed on the whole surface of Cu-15.0at.%Sn specimen during cooling to -196°C after quenching, but the volume fraction of γ_1' martensites was decreased by aging at 20°C after quenching. These results were caused by a decrease of Ms and Mf points due to aging, i.e. due to stabilization of β_1 phase.

As shown in Fig.11, the increase of critical stress σ_M with increasing aging time is also explained by the stabilization of β_1 phase. To explain the stabilization of β_1 phase, the nucleation theory and the growth theory have been proposed, but the nucleation theory by Kennon[11] in which stabilization occurs by exhaustion of martensite embryos resulting from segregation of tin atoms to form precipitates of a tin-rich phase, can well explain the decrease in Ms and Mf points. On the other hand, the growth theory by Glover and Smith[18] and Morgan and Ko[19] in which dislocations in the matrix are pinned by solute segregate or precipitate particles and resulting matrix-hardening opposes growth of martensite plates, can explain the decrease of Mf, but not the decrease of Ms.

As mentioned above, the increase of critical stress is as follows; because the martensite embryos are exhausted by tin atoms by aging, the higher stress is necessary to activate the exhausted embryos.

Figure 15 shows a relation between superelasticity and shape memory region schematically. In the region A, the stress-induced martensites are unstable in this temperature range, so on unloading the SIM disappears and superelastic behavior can occurs. However, in the region B , as the martensites are stable in this temperature range, SIM can only disappear by heating to above the Af temperature. Now, as mentioned above, Ms temperature decreases as the

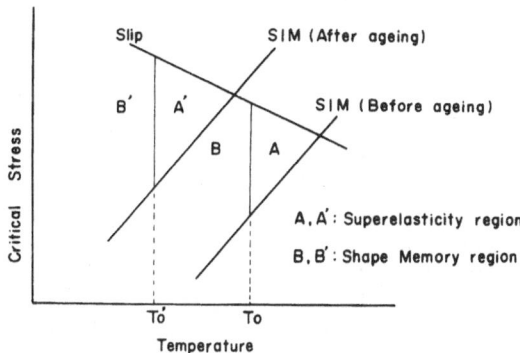

Fig. 15   Schematic illustration of regions of superelasticity and
          shape memory in Cu-Sn alloys before and after aging in
          β_1 phase,respectively.

aging time in β_1 phase increases. Namely, the transformation temperature To(=(As-Ms)/2)displaces to lower temperature site To'.
Accordingly, the superelastic and shape memory regions will displaces to a lower temperature regions A' and B' respectively. Therefore, in the deformation temperature range used in this experiment,
though the memory effect occurs in an as-quenched specimen as shown
in Fig.7, the tensile behavior becmes to show superelasticity with
increasing aging time as shown in Figs.8,9 and 10.

In Au-Cu-Zn[13] and Au-Cd[20] alloys, a band consists of alternate layers of bcc and martensite phase occurs from one end of the
specimen and advances toward the other end. The stress-induced martensite appearing behind the moving band is a single crystal with
an orthorhombic structure(2H), and the strain due to superelasticity
corresponds to the strain existing in a single martensite.

As shown in Fig.14, the formation mode of stress-induced martensite is as follows. Immediately after reaching the critical stress,
plates of martensite begin to form and its amount increases with
the strain. It can be said that the strain due to the superelasticity depends on the total amount of plates. This formation mode is
similar to that of Cu-Zn-Sn alloy[12].

4.4 Strain Rate Dependence of Superelasticity and Aging Effect
     during Tensile Test.

The superelasticity in this alloy depends on strain rate,differ-
ent from in other alloys (this phenomenon has been reported in Cu-
Zn-Si alloy [21].). As shown in Fig.12, as the strain rate decreased, the residual strains were remained on unloading, but they were
almost recovered upon heating to 90°C. It is thought that the

strain rate dependence is due to the locking of martensite by gathe-
ring of tin atoms; the interface of formed streaa-induced martensite are locked due to the accelerated diffusion of tin atoms under
the stress. Upon heating, tin atoms re-diffused and the locking of
interface was released; therefore the residual strains were recovered on removal from the stress.

The effect of aging during the tensile test, as shown in Fig.13
can be explained by the same mechanism. As shown in in Fig.13(b),
the strain corresponds to the prior deformation before aging for 5
min.remained, on further straining after aging since the stress-
induced martensite formed before aging is locked by tin atoms during
aging. Partial recovery is due to the formation of new stress-induced martensite, on further deformation, since these new martensites moved back in the opposite direction upon removal of the stress.

Thus, tin atoms play an important role in Cu-Sn alloys. Kennon
[11] has reported that the actiation energy for diffusion of tin
atom is (0.63 ± 0.07)eV and is probably related to the activation
for diffusion of tin enhanced by a supersaturation of quenched-in
vacancies. However, the diffusional process of tin atoms is considered to be attributed to vacancies due to the deviation from the
stoickiometric composition rather than quenched-in vacancies, considering quenching rates and a size of specimens.

5. CONCLUSIONS

1) The transformation temperature in Cu-Sn alloys determined by
the measurements of electrical resistivity decreases with increasing tin concentration.

2) Crystal structure of the thermal martensite and the stress-
induced martensite are orthorhombic in each case.

3) The shape memory effect of thermal martensite in Cu-15.0at.%
Sn alloy appears only for a slight deformation but not for strong
deformation.

4) The transformation hysteresis in Cu-15.0at.%Snis about 140°C,
and in this sense Cu-Sn alloys has an intermediate nature between
thermoelastic and athermoelastic alloys.

5) critical stress increases with increasing aging time in β_1
phase and stress-strain curves shows remarkable superelastic behav-
or after aged for 50 days.

6) According to the observation of stress-induced transformation,
the formation of martensite plates occurs, immediately after reaching the critical stress and its amount increases with the strain.

7) The superelasticity in a Cu-15.3at.%Sn alloy depends on the
strain rate, and the similer phenomenon is appeared in the tests
of aging under stress. These phenomenon can be explained as follows;
the interface of formed stress-induced martensite are locked due to
the accelerated diffusion of tin atoms under the stress.

ACKNOWLEDGEMENT

The authors are gratefull to Mr.T.Mori of Hitachi Research Laboratory,Hitachi Ltd.,Hitachi, and Mr.S.Maeda of Dainichi Nippon Densen Ltd.,Amagasaki, for their useful advices and discussions.

REFERENCES

1. H.Warlimont: Iron and Steel Inst., Special Rep., 1965, vol. 93 ,p. 58.
2. N.F.Kennon and T.M.Miller: Trans.JIM,1972,vol.13,p.322.
3. Z.Nishiyama,K.Shimizu and H.Morikawa: Trans.JIM,1968,vol.9,Suppl., p.930.
4. T.Soejima: Bull. Univ.of Osaka Prefecture,1967,Series A,vol.16, p.341.
5. N.Nakanishi,T.Mori,S.Miura,Y.Murakami and S.Kachi: Phil.Mag., 1973,vol.28,p.277.
6. H.K.Birnbaum and T.A.Read: Trans.AIME,1960,vol.218,p.662.
7. F.E.Wang,W.J.Buehler and S.J.Pickart: J.Appl.Phys.,1965,vol.36, p.3232.
8. K.Oishi and L.C.Brown: Met.Trans.,1971,vol.2,p.1971.
9. C.M.Wayman: Scripta Met.,1971,vol.5,p.487.
10. K.Enami and S.Nenno: Met.Trans.,1971,vol.2,p.1487.
11. N.F.Kennon: Met. Sci. J.,1972,vol.6,p.64.
12. J.D.Eisenwasser and L.C.Brown: Met. Trans.,1972,vol.3,p.1359.
13. S.Miura, S.Maeda and N.Nakanishi: Phil. Mag.,1974,vol.30,p.565.
14. H.Morikawa, K.Shimizu and Z.Nishiyama: Trans. JIM,1967,vol.8, p.145.
15. Y.Murakami: Private communication.
16. I.A.Arbuzova, Yu.N.Koval', V.V.Martynov, and L.G.Khandros: Fizika Metallov i. Metallovedenie,1973,vol.35,p.1278.
17. K.Otsuka and K.Shimizu: Scripta Met.,1970,vol.4,p.469.
18. G.Glover and T.B.Smith: " The Mechanism of Phase Transformation in Metals", Inst. Metals,1955,Monograph No.18,p.256.
19. E.R.Morgan and T.Ko: Acta Met., 1953,vol.1,p.36.
20. S.Miura, T.Mori and N.Nakanishi:"Mechanical Behavior of Materials"(Proc. of the 1974 Symp. on Mech. Behavior of Materials), The Soc. Materials Sci. Japan,1974,vol.Ⅱ ,p.141.
21. D.V.Wield and E.Gillman: Scripta Met.,1972,vol.6,p.1157.

SHAPE MEMORY EFFECT AND ANELASTICITY ASSOCIATED WITH THE

MARTENSITIC TRANSFORMATION IN THE STOECHIOMETRIC Fe₃Pt ALLOY

M. FOOS, C. FRANTZ and M. GANTOIS

Laboratoire de Génie Métallurgique - Ecole des Mines

Parc de Saurupt - 54 042 NANCY-CEDEX (France)

INTRODUCTION

We have studied the behavior of the iron-platinum alloys of
the Fe₃Pt type subjected to various mechanical conditions before,
during and after the martensitic transformation. Previous papers
(1) (2) and (3) have shown that the characteristics of this martensitic transformation depend strongly on the long range order
state of the parent austenitic phase ; the M_s temperature is
lowered as the degree of order is raised. When the order parameter
S is ≥ 0,60, the transformation is thermoelastic (2) (3), and when
this parameter is < 0,60, the transformation exhibits an appreciable thermal hysterisis. We have shown that the value S = 0,60 defines a limiting order parameter from which the disordering of the
stoechiometric Fe₃Pt alloy becomes a two-phase process. We shall
report in this paper the results obtained in our study of the
Shape Memory Effect (S.M.E.), superelasticity (or pseudo-elasticity), transformation plasticity and the internal friction. Theses
effects are all related to the martensitic transformation and they
will be discussed in connection with the initial order state of the
austenite.

EXPERIMENTAL PROCEDURES

Alloys were prepared in an induction furnace under helium
atmosphere from platinum and electrolytic iron of 99,99 % and
99,95 % purity respectively. The electrolytic iron was previously
melted under vacuum. The alloys were homogenized at 1150°C for 48 h
and brought to the required order state by an appropriate heat-
treatment (2) (3). The kinetics of the martensitic transformation
has been established by using radiocrystallographic analysis : the
samples are platelets of fine compressed powder placed on an X ray

diffraction goniometer working at high and low temperatures. S.M.E. is studied by the torsion of 80 mm length and 1 mm diameter wires with a simple torsion apparatus with which one can specify the applied couple and the torsion angle. This apparatus is also used to study the transformation plasticity in the case of high initial deformations. For the microdeformation and internal friction experiments, we used wires of 70 mm length and 2 mm diameter placed on a classical inverted torsion pendulum which works at frequencies near 1 c/s. Superelasticity of 0,5 mm diameter wires was studied with a tensile micromachine.

SHAPE MEMORY EFFECT AND SUPERELASTICITY

Using the pendulum or the simple torsion apparatus we have quantified S.M.E. for two alloys ; one is partially long range ordered (S = 0,60) and the other is disordered (S = 0). We distinguish two sequences for each test : the first is related to the martensite (α') $\rightarrow$ austenite (γ) transformation, the second to the inverse $\gamma \rightarrow \alpha'$ transformation. The results are expressed in % of the sample initial deformation which does not exceed $1,6.10^{-2}$ and is always applied when the sample was cooled for the first time in the martensitic range ; the other test sequences were always carried out without external applied stress. In the case of the partially ordered alloy, S.M.E. is complete at the first $\alpha' \rightarrow \gamma$ transformation, the sample deforms 30 % and recovers its initial shape, by heating, after the complete reversion of the martensite. Other thermal cycles done in the transformation range lead to identical results.

In the case of the disordered alloy, S.M.E. is 40 % at the first $\alpha' \rightarrow \gamma$ transformation. At the following $\gamma \rightarrow \alpha'$ transformation, the deformation is 20 % and the sample recovers the geometrical shape it had at the end of the first $\alpha' \rightarrow \gamma$ transformation when it is heated above A_f. These results show that, independent of the state of order, Fe_3Pt alloys exhibit S.M.E. which becomes reversible from the second $\gamma \rightarrow \alpha'$ transformation. This S.M.E. reversible component corresponds respectively, for the partially ordered and disordered alloys, to 30 % and 50 % of the initial deformation in the martensitic state. We have verified that in the $\alpha' \rightarrow \gamma$ sense, S.M.E. can overcome an important opposing external stress whereas in the $\gamma \rightarrow \alpha'$ sense it can be easily annihilated by a small opposing stress. S.M.E. depends on the alloy order state and its nature depends upon whether the transformation occurs in the $\alpha' \rightarrow \gamma$ or in the $\gamma \rightarrow \alpha'$ sense : it is associated with the martensitic transformation and is strongly related to the characteristic temperatures M_s, M_f, A_s and A_f. If Tong and Wayman's criteria (4) for the prediction of S.M.E. in some alloys are verified by the partially ordered Fe_3Pt alloys, they do not apply to the disordered alloys which do not exhibit a thermoelastic martensitic transformation. Our results pose again the problem of the true nature of S.M.E. and of the definition of the required conditions for the occurence of this

effect. All that is certain, is that the alloy must exhibit a martensitic type transformation. This criterium is insufficient and we
still have to know the mechanisms which lead to S.M.E. A detailed
study of the plastic deformation modes of the parent and product
phases correlated to a crystallographic and microstructural study
of the martensitic transformation may bring out features leading
to a better understanding of S.M.E.

The superelasticity which appears when a sample is deformed
at temperatures higher than A_f ressembles S.M.E. Indeed, the martensite, formed either by cooling below M_S or by deformation above
M_S, remembers the parent austenitic phase. Figure 1 shows the
stress-strain curves obtained at 25, – 70, – 90 and – 195°C in the
case of a partially ordered alloy (S = 0,60) whose M_S and A_f temperatures are respectively – 120 and – 100°C. The slope of the
$\sigma - \varepsilon$ curves decreases with testing temperature and loops appear.
The latter characterize an anelastic behavior for an unloading
and loading sequence. At 25°C there is no anelasticity ; the alloy
remains austenitic. At – 70 and – 90°C, the initial state is austenitic, the application of a stress induces martensite but, taking
the deformation into account, the alloy is cold-worked, and this
leads to a decrease of order and an increase of the A_f temperature.
The greatest part of induced martensite during the test does not
disappear when the stress is suppressed and we only observe a
tendency of the sample to take its initial shape. At – 195°C, the
alloy is two-phase (about 40 % martensite), the application of a
stress increases the transformed phase volume fraction (we have
been able to confirm this by using optical microscopy) and deforms
the already present martensitic and austenitic phases. On unloading,
a fraction of the martensite formed under stress is reverted. One
will note that, at this testing temperature, the initial shape of
the $\sigma - \varepsilon$ curve is lower than that obtained on unloading. It seems
that there is already martensite formation as soon as the stress is
applied ; on unloading, all the martensite which could be formed at
this stress level and at this temperature has appeared ; on reloading, the value of the stress reached before unloading must be exceeded in order to induce martensite again. In the case of the three
tests done at – 70, – 90 and – 195°C, a reheating of the sample at
room temperature which is far above A_f, transforms all the martensite present in the alloy and shows S.M.E. which, for the tests realized at – 90 and – 195°C, is respectively 40 % and 75 %. These
results show that Fe_3Pt alloy exhibits a tendency to superelasticity though we have not been able to make it as evident as in
Pops's work (5) for other alloys. The use of monocrystals may make
its observation easier.

TRANSFORMATION PLASTICITY AND INTERNAL FRICTION

Their analysis requires first the study of the martensite
transformation kinetics in relation to the initial order state of
the alloys. Figures 2 and 3 represent respectively the variations

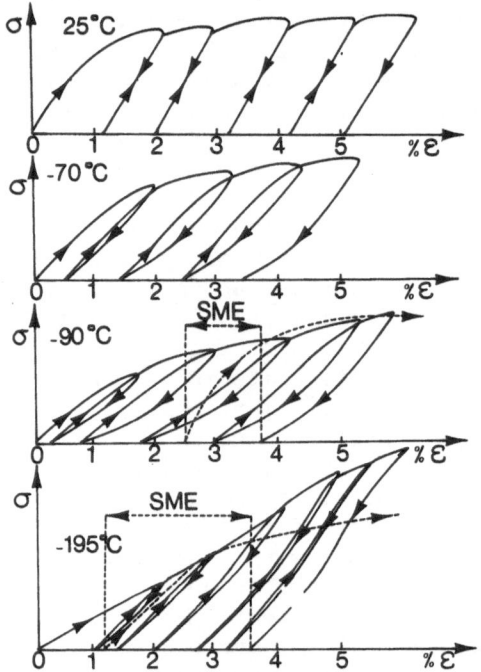

<u>Figure 1</u> : Stress-strain curves at different test temperatures.
For the two last test temperatures (- 90 and - 195°C), dashed curves correspond to a reloading at room temperature after complete
reversion of martensite and show S.M.E.

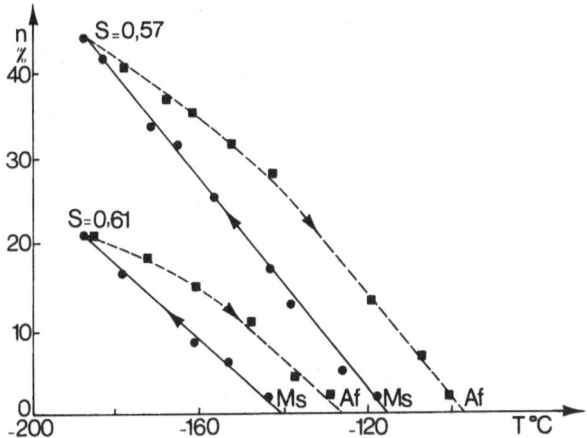

<u>Figure 2</u> : Variations of the martensite volume fraction n with temperature for two partially ordered Fe_3Pt alloys (S=0,61 and S=0,57).

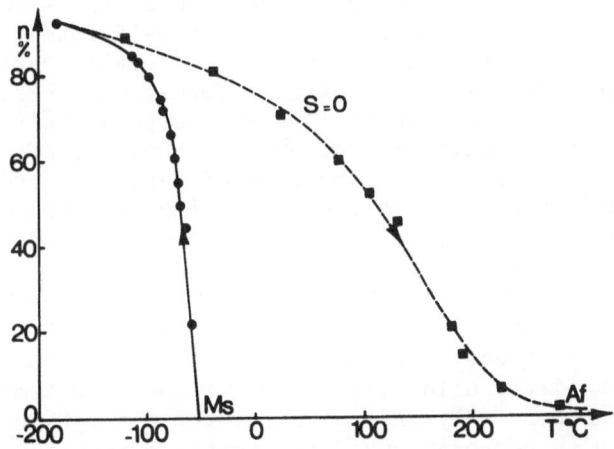

<u>Figure 3</u> : Variations of the martensite volume fraction n with temperature for the disordered Fe_3Pt alloy.

of the martensite volume fraction n with temperature T for the partially ordered (S = 0,61 and S = 0,57) and for the disordered alloys. Table 1 gives the M_S and A_f temperatures, the α initial slopes and the n values at - 195°C for these three alloys. The curves
can be represented by the following expression :

$$n = \frac{\alpha (M_S - T)}{1 + \alpha (M_S - T)}$$

Table 1

S	0	0,57	0,61
M_S °C	- 53	- 117	- 142
A_f °C	300	- 97	- 125
α °C^{-1}	$8,9.10^{-2}$	$0,64.10^{-2}$	$0,44.10^{-2}$
$n_{-195°C}$	90 %	45 %	20 %

The transformation progresses only during cooling and is athermal.
The reversion martensite kinetics, and in particular for the disordered alloy, are different. Indeed, in this particular case, when
the martensite is completely reverted at 300°C, the M_S temperature
on subsequent cooling passes from - 53°C (M_{S1}) to about 40°C (M_{S2}).
This M_S evolution with thermal cycle is due to the transformation
stresses which lead to a plastic deformation of the alloy. The
evolution of the relative variations of the half-intensity width,
$\Delta l/l$, of the (111) austenite fundamental reflection represented in
figure 4 confirms the existence of transformation stresses which
plastically deform the austenite. The M_{S1} temperature (- 53°C) is
again observed if austenite is heated much above A_f after the first
martensitic transformation. In the case of partially ordered alloys,
the $\Delta l/l$ variations are reversible and vanish on heating since the
temperature reaches A_f. In this case, the transformation stresses
are purely elastic and the thermal cycles in the transformation
range do not influence the M_S temperature. One notes that austenite and martensite lattice parameter measurements show that, at the
$\gamma \rightarrow \alpha'$ temperature transformation, the relative volume variation
$\Delta V/V$ of the disordered alloy ($\simeq 3.10^{-3}$) is about 30 times higher
than that of the partially ordered alloy ($\simeq 10^{-4}$). Figure 5 shows
the evolution with temperature of the total deformation γ_T in the
partially ordered (S = 0,60) alloys and in the disordered alloys
tested in torsion and submitted to various initial elastic strains
γ_E in the austenitic range. In all of the alloys considered, no

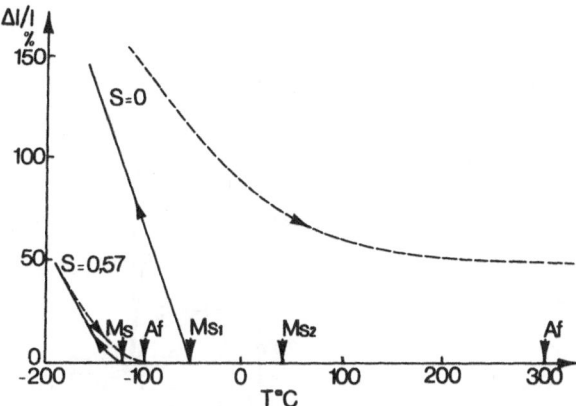

<u>Figure 4</u> : Relative variations with temperature of the half-intensity width, $\Delta\ell/\ell$, of the (111) austenite fundamental reflection in the partially ordered and disordered Fe_3Pt alloys.

transformation plasticity is observed during the $\alpha' \rightarrow \gamma$ transformation ; there is no deformation in the sence of the applied stress and this is a reflection of the S.M.E. Figure 6 shows that the total deformation γ_T is proportional to the initial deformation γ_E and that the plasticity factor

$$h = \frac{\gamma_T - \gamma_E}{\gamma_E} \quad (6)$$

is greater in the disordered alloy than in the partially ordered ones but is independent of the initial deformation γ_E. Also the h/n ratio is the same for the different alloys and is about 11. The capacity for transformation deformation of the Fe_3Pt alloy as measured by the volume fraction of transformed phase, is not very sensitive neither to the austenite initial long range order nor to the character of the martensitic transformation. This fact seems surprising since the Magee's theory (7) may be thought to account for most of the transformation plasticity in partially ordered alloys and the Greenwood-Johnson's theory (8) for the disordered.

Figures 7 and 8 represent the results obtained from internal friction experiments. Figure 7 corresponds to the S = 0,60 ordered

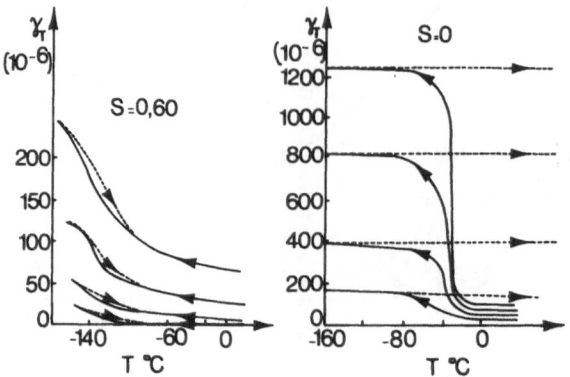

<u>Figure 5</u> : Evolution with temperature of the total deformation γ_T in the partially ordered and disordered alloys submitted to different initial elastic strains in the austenitic range.

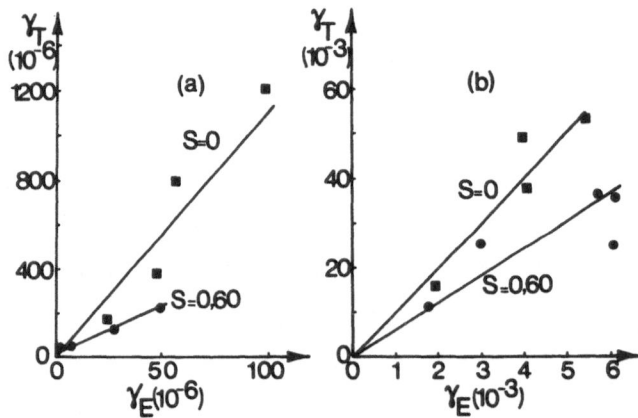

<u>Figure 6</u> : Evolution of the total deformation γ_T with the initial elastic deformation γ_E.
(a) Microdeformations measured with the pendulum at – 160°C
   ($n_{s=0} \simeq 90$ %, $n_{s=0,60} \simeq 28$ %).
(b) Higher deformations measured with the simple torsion apparatus
   at – 195°C ($n_{s=0} \simeq 93$ %, $n_{s=0,60} \simeq 45$ %).

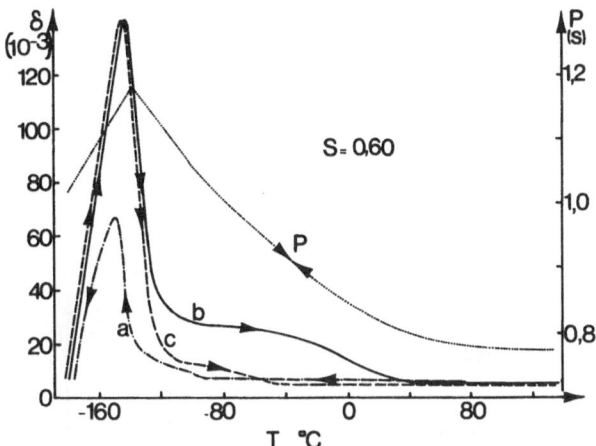

Figure 7 : Evolution with temperature of internal friction δ and
period P in the partially ordered Fe₃Pt alloy (S = 0,60).
Curve a : cooling under external magnetic field   (dT/dt < 2°C/mn)
Curve b : heating under external magnetic field   (dT/dt = 2°C/mn)
Curve c : heating without external magnetic field (dT/dt = 2°C/mn)

alloy. On cooling, from the M_s temperature, internal friction increases, reaches a maximum and then decreases. On heating, during
the martensite reversion, it passes again through a maximum which
is higher than that obtained on cooling because the heating speed
is faster. The background due to magnetic interactions partially
disappears when an external magnetic field of about 300 Oe is applied. It passes from 25.10^{-3} (full curve) to 10^{-2} (dashed curve).
One notes that the abnormal behavior of the period (or modulus)
which starts at the Curie temperature, is not affected by the
applied magnetic field. We show, for the disordered alloy, two
internal friction curves in figure 8 which correspond to different
cooling speeds and different frequencies. The magnetic background
no longer appears since the magnetic and martensitic transitions
are close to each other. However, the period anomaly is appreciable and irreversible. The friction maximum is very visible and
exhibits a plat portion at low temperatures. In all the alloys
(ordered and disordered), the friction peak area is proportional
to the temperature variation speed, to the period and to the mar-

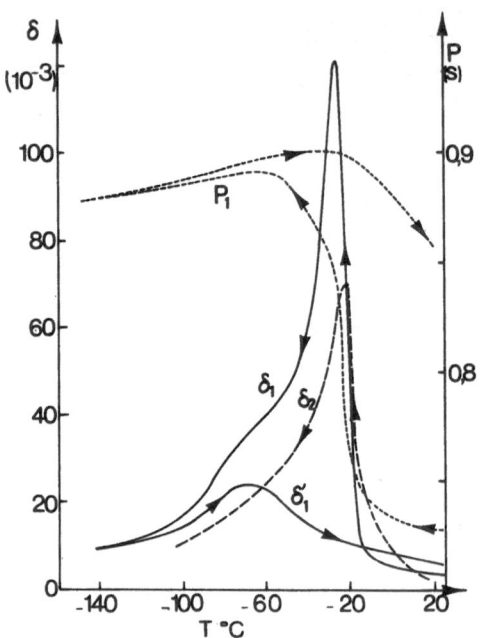

Figure 8 : Evolution with temperature of internal friction and period in the disordered Fe_3Pt alloy.
    Curve δ_1 : cooling at 1,3°C/mn.
    Curve δ_1' : heating at 1,3°C/mn.
    Curve δ_2 : cooling at 1°C/mn.
    Curve P_1 : Period relative to δ_1.

The period relative to δ_2 is between 0,43 s and 0,53 s.

tensite volume fraction. The proportionality factor K, the so-
called normed internal friction (6), in units of volume fraction of
transformed phase n is a constant whose value is about 280 for the
present case. This surprising result agrees however with the transformation plasticity experiments.

CONCLUSION
        Whatever their long range order state, the Fe_3Pt alloys
which are deformed during the martensitic transformation or in the
martensitic range, are able to recover partially or completely
their initial geometrical shape after the total martensite reversion. S.M.E. exhibits a reversible component which seems more important in relative value in the case of the less ordered alloys. In
appropriate conditions, for temperatures higher than M_s (and A_f),
some tendency to superelasticity appears which is related to the
formation of stress induced martensite and to its partial disappearance on unloading. Transformation plasticity experiments have shown
that the plastic deformation is proportional to the initial elastic
deformation and to the martensite volume fraction. The proportionality factor (or plasticity factor) in units of transformed phase
seems constant and independent to the alloy order state and hence,
to the nature of the martensitic transformation. Dynamic internal
friction experiments lead to the same conclusion : the friction
peak area is proportional to the martensite volume fraction, to the
temperature variation speed and to the period. The proportionality
factor in units of transformed phase seems independent to the alloy
order state.

BIBLIOGRAPHIE

(1) D.P. DUNNE and C.M. WAYMAN : Met. Trans., 1973, Vol. 4, p. 137.
(2) C. FRANTZ and M. GANTOIS : International Symposium of Order-
    Disorder Transformations in Alloys, Sept. 1973, Tübingen.
(3) O. HERBEUVAL, C. FRANTZ and M. GANTOIS : Journées d'Automne
    Société Française de Métallurgie, Oct. 1973, Paris ;
    Mém. Sc. Rev. Mét., 1974, LXXI, n° 10, p. 647.
(4) H.C. TONG and C.M. WAYMAN : Scripta Met., 1973, Vol. 7, p. 215.
(5) H. POPS : Met. Trans., 1970, Vol. 1, p. 251.
(6) J.F. DELORME, R. SCHMID, M. ROBIN and P. GOBIN : J. Phys.,
    1971, Supp. n° 7, Vol. 32, p. C - 101.
(7) C.L. MAGEE : Thesis Carnegie Mellon University, 1966.
(8) G.W. GREENWOOD and R.H. JOHNSON : Proc. Roy. Soc., 1965,
    Vol. A 283, p. 403.

COMPRESSION OF CuAlNi CRYSTALS

L. A. Shepard

Materials Sciences Division
Army Materials and Mechanics Research Center
Watertown, Massachusetts 02172

Single crystals of beta Cu-14.1Al-3.0Ni in (100) orientation were reversibly strained in compression to 18%, and 200,000 psi. Above the Martensite Finish temperature (-11°C), the initial luders band deformation produced a single crystal of (010) oriented gamma prime martensite, the stress level linearly dependent upon temperature. Below -11°C, the martensite itself was reoriented to a single crystal.

Beyond 8 to 10% strain, martensite twinning produced an almost linear reversible strain with a temperature independent slope of 1.5 to 2 million psi. No plastic strain was detectable, the crystals being completely brittle, but the total strain was recoverable.

Although the measured Austenite Finish temperature was 20°C, crystals compressed below 50°C remained martensitic upon unloading with a residual 7.8% unrecovered strain. Upon warming to 50°C, the sample instantaneously and completely converted to the beta phase in a burst.

INTRODUCTION

In the two recent reviews of pseudoelasticity and the shape memory effect,[1,2] the martensitic Cu-Al-Ni alloys received prominent attention. Austenitic single crystals of these alloys (near Cu_3Al in composition) have shown high strength and a maximum reversible deformation of 24%.[3] A strongly temperature dependent yield stress decreasing toward the M_s temperature, a deformable martensite below M_s[3] and a sizable SME strain recovery[4] all suggest potential for technological development.

The austenite to martensite transition for alloys containing 14 wt.% Al and 3 wt.% Ni is a transformation from the ordered body centered cubic β to an ordered, twinned orthorhombic γ' martensite. The correspondence relationships are [5,6]

$$(101)_\beta \,||\, (001)_{\gamma'}$$

$$[010]_\beta \,||\, [010]_{\gamma'}$$

and the habit plane is the $(\bar{1}33)$ plane, close to the (011) plane of the β phase. The martensite twins on the $(121)_{\gamma'}$ plane which is parallel to the (011) plane in the β phase. The $\beta-\gamma'$ structure relationships are shown schematically in Figure 1.

A β phase crystal can be compressed along the $[0\bar{1}0]$ direction to a maximum of 8.7% strain to produce a single deformation martensitic crystal. In tension, a maximum 6.6% strain will occur on stressing the β crystal along the $[\bar{1}01]$.

A number of authors[4,7,8,9,10] have reported deformation of several single crystal pseudoelastic materials to single crystal martensites. In general, the strains reported were somewhat less than the theoretical maximums. Oishi and Brown[4] found a 6% maximum tensile strain for Cu-14.1Al-3Ni for example, compared with the 6.6% theoretical.

Two reports[3,8] were concerned with deformation of pseudoelastic materials beyond the point of the full martensite transformation. Pops[8] noted one example in Cu-Zn in which a reversible linear stress-strain behavior was found in this region.

The present work is concerned with the deformation behavior of Cu-14.1Al-3.0Ni alloy single crystals in compression.

EXPERIMENTAL PROCEDURE

A series of alloy compositions, all containing 3.00 wt.% Ni, and with aluminum contents varying from 13.96 to 15.16 wt.%, were prepared as β single crystal rods. The 15.16 wt.% Al alloy was two

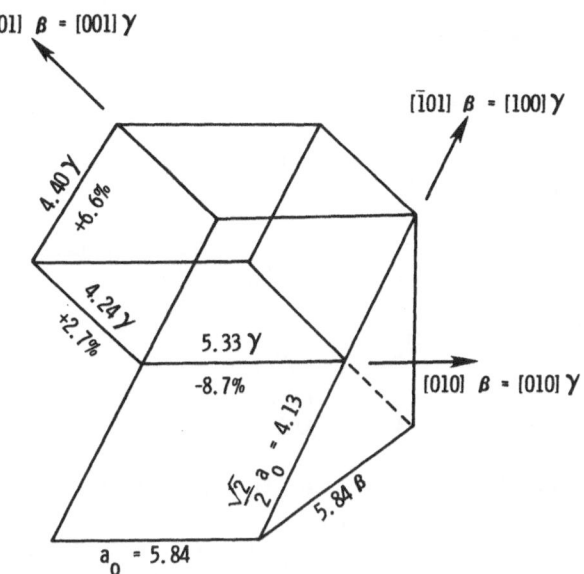

Figure 1.   Dimensional Changes in the β-γ´ Transformation (Cubic to Orthorhombic)
Cu-14Al-3Ni

phased, and the 14.54 wt.% Al alloy was very stiff at room temperature
(A_f at -10°C). The adopted composition was the same as that
used by Oishi and Brown[4], 82.90 Cu, 14.10 Al and 3.0 Ni.

Reproducible strength and transformation characteristics can
be maintained with reasonable precaution against aluminum loss during
processing. Alloy ingots were vacuum induction melted and poured
under argon. Single crystals, rod or strip were cast down and grown
in graphite molds under argon, using a Bridgeman type furnace.
Maximum temperature of the melt was 1160°C, 100° above the melting
point of the alloy. The final heat treatment was 4 minutes at 900°C
in air, and water quench.

Crystals were finished by sanding, chemically polishing and
etching in 2 parts nitric acid, 1 part water and X-raying for orientation.

Compression samples for the present study were cut to half inch
lengths, the ends ground parallel, and tested between hardened
platens in a temperature controlled bath. Deformation rate was
6.6×10^{-4} per second.

EXPERIMENTAL RESULTS

A. Deformation

Transformation temperatures for crystals from the two ingots used in this study are given in Table I.

Table I. Thermoelastic Martensite Transformation
Temperatures (°C)

Ingot	M_s	M_f	A_s	A_f
1	+3	−11	+15	+20
2	+6	−11	+14	+20

The axial orientation of crystals used in these tests lay within 3 degrees of the [100]. Tests were performed over a range of temperatures from −195° to 95°C. Figure 2 shows the compressive properties of a crystal tested to failure at 29.5°C, and is characteristic of loading curves above the M_f temperature. An initial slope, 1 x 10⁶ psi at its steepest, to an upper yield point, a luder strain, ending typically at 9% strain and finally a roughly linear by increasing stress to failure, at 18.1% and 194,000 psi. The curve is distinguished from others shown in the literature for this alloy[3,4] and

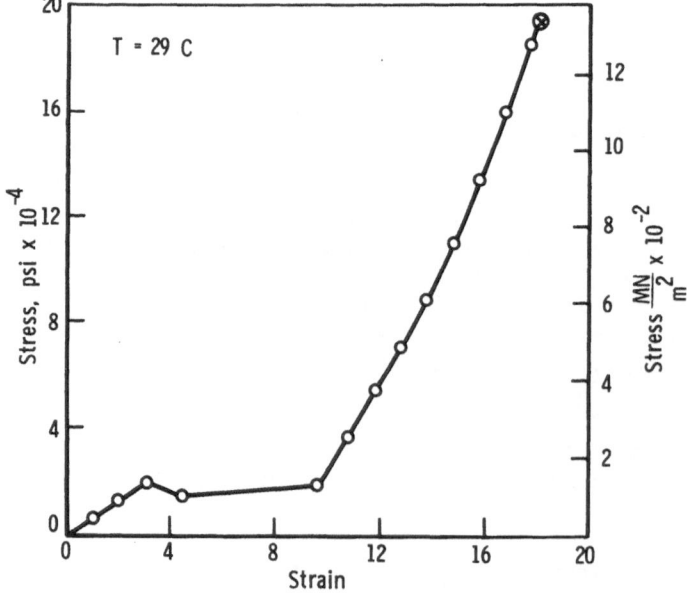

Figure 2. Single Crystal (100) Orientation in Compression
T = 29.5°C

the several thermoelastic materials by the long second slope, and the high fracture stress.

Figure 3 shows a crystal carried through a loading and unloading cycle at 75° and 0°C. The initial slope, the upper yield point and the luders stress are quite temperature dependent.

Strain along the second slope is reversible and the slope is temperature independent. Values of the slope measured over the 290 degree test temperature range are in close agreement, and average 1.46×10^6 psi at 50,000 psi and 1.94×10^6 psi at 100,000 psi, 1.0 and 1.34×10^4 MN/m^2 respectively.

The unloading curve at 75°C drops to a minimum value, then increases in strength with an interesting negative modulus[8] and a reverse luders strain to the initial modulus line. At 50°C, the minimum unloading stress decreases to zero (Figure 4). At all lower temperatures, unloading ends with 7.8% residual strain in the crystal, as shown here for the 0°C curve.

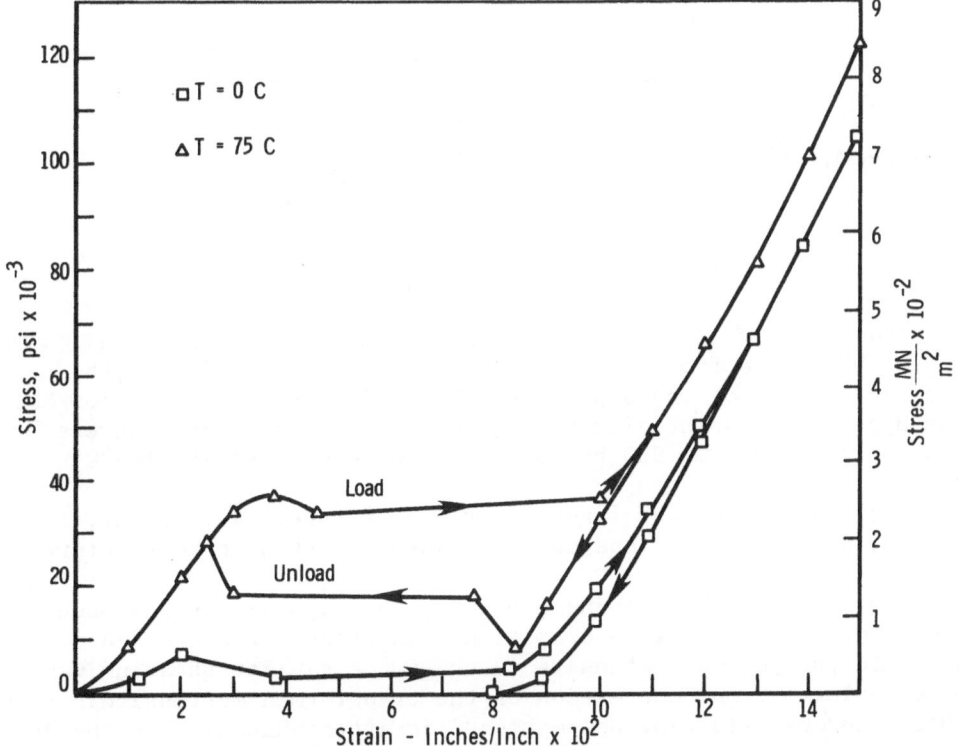

Figure 3. Single Crystal (100) Orientation in Compression

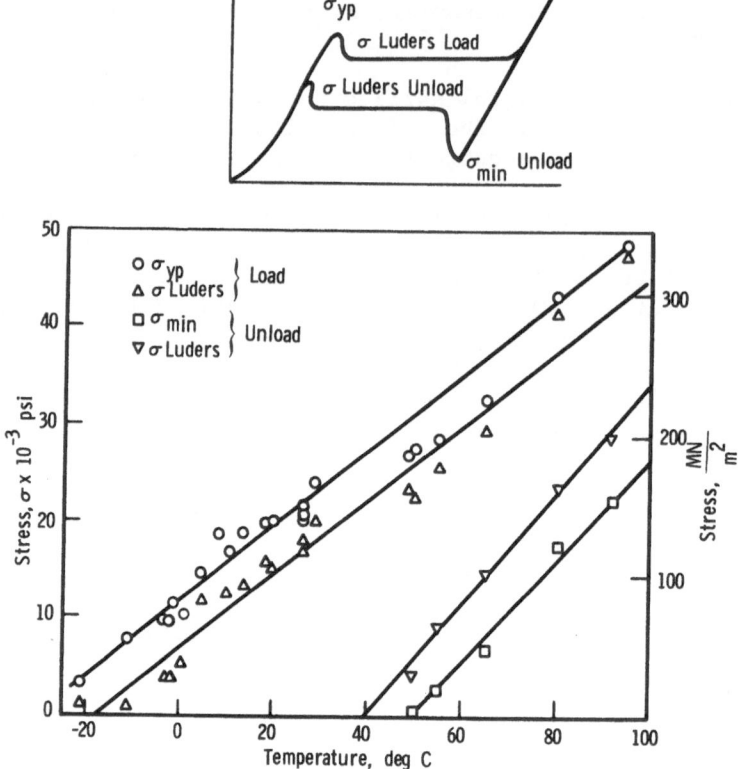

Figure 4. Yield and Luders Stresses Upon Loading and Unloading
vs Temperature (100) Orientation

The shape of the compression curve remains essentially the same
as the 0°C curve from 49°C to the M_f temperature, -11°C, with decreasing values of the upper yield and luders stresses. Below M_f,
the upper yield point disappears, and the average luders stress
rises gradually from 900 psi at -11°C to 16,000 psi at -195°C.

Figure 4 shows the linear decrease of upper yield point, luders
stress, unload minimum and unload luders stress with temperature.

During luders straining above M_s, the crystal visibly sheared,
and transformed by motion of a single interface from the red β
phase to the yellow martensite γ´ phase. A similar shearing was
noted below M_f, on compression of the transformed martensite. Below
50°C, the crystals remained martensitic after unloading if the deformation exceeded the luders strain. Laue photograms of the unloaded martensitic crystals show essentially single crystal γ´.
Actually one or more faint spots occur within 2 degrees of the main

spot indicating small lattice misalignment. There is no evidence, however, of twins in the γ' produced by compression.

B. Shape Recovery

Shape recovery was observed to occur by the transformtion of deformed martensite or deformation martensite to the β phase. For the axially loaded Cu-14.1Al-3Ni alloy crystals in the (100) orientation, shape recovery was always complete, and the specimens returned to their exact original diminsions. This was true for specimens repeatedly strained to 18%, and stresses over 150,000 psi, $10^3 MN/m^2$. The recovery temperature of the deformed martensite was greater than A_f, 20°C, in all cases, and increased with increasing prior deformation, as noted by Oishi and Brown[4].

The greatest amount of shape recovery, 7.8% occurred, as noted above, in specimens strained in excess of the luders strain at any temperature below 50°C. For this special case, shape recovery occurred at 50°C, instantaneously - with a snap! No partial recovery was measured over long periods of time below 50°C.

Crystals deformed to less than the luders strain, either above or below the M_f temperature, recovered their shape more slowly, with a growth of bands of β, and over a few degree temperature range. Specimens compressed at -195°C, 21° and 29°C to 4 and 5.5% (total-elastic) strain all were completely recovered by 29°C. Crystal strained to 7% recovered rather abruptly over a narrow range from 39° to 40°C.

Recovery from other types of deformation, for example, bending, followed this same behavior. However, it was difficult to impose a strain uniformly over the specimen, and unbending started at 20°C, A_f and often finished at 40°C.

C. Other Orientations

Specimens of a series of orientations across the stereographic triangle, Figure 5, were tested; the orientation furthest to the right lies near the ($\bar{1}$33) pole of the martensite habit plane. The initial modulus increased by a factor of 1.75 from the (001) to the ($\bar{1}$33) orientation, from 1 to 1.75 x 10^6 psi, 7 to 12 x 10^3 MN/m^2. The initial yield strength doubled, and apparently does not follow a critical resolved shear stress law.

Beyond yielding, test results for crystals off the (001) axis could not be interpreted, since the specimen axes sheared at an angle to the load axis. However, it is most interesting to observe that all four specimens farthest from the (001) developed a series of kink bands near the ends during deformation. The kink bands accommodated the bending between the specimen ends, which remained parallel

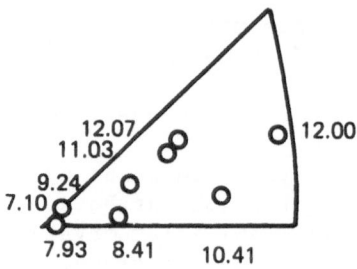

Figure 5.   Initial Elastic Modulus ($MN/m^2 \times 10^{-3}$) as a Function
of Orientation   T = 27°C

to the platens, and the sheared central region. The kinks and the
bending remained when these specimens were reconverted to β.

The two samples oriented 15° and 18° radially from the [001]
recovered after 11% deformation to right cylinders of their original
lengths upon unloading and heating. At larger strains, however,
these specimens also developed permanent kink bands.

DISCUSSION

In reviewing the compression properties of the (100) oriented
Cu-14.1A1-3Ni, there are three observations which make this material
distintive:

1. A strength level considerably higher than previously
reported pseudoelastic alloys[1] and a large recoverable strain
with no plastic deformation.

2. A temperature independent, almost linear pseudoelastic
strain of the order of 10% following luders deformation.

3. An unusual shape memory effect with an instantaneous shape
recovery at 50°C, a temperature 30°C above A_f.

The alloy shows potential for elastic energy storage both as
a spring material, and, in its unloaded state, as a temperature
triggered spring through shape memory. The limitation to application
of the material is that it is brittle, and therefore must be used
in single crystal form. Large compatability stresses between grains
in stressed polycrystalline material cause cracking and failure.

The behavior of this alloy is not greatly different from that

reported for other pseudoelastic materials, for it has been established[2] that the basic processes differ only in detail. Primary
differences are concerned with the martensite pliability. For
easily twinned materials, such as In-Tl[7], the deformation is associated with mechanical twin boundary displacement. In the alloy
considered here, and CuZn, the effect is due to the growth and
shrinkage of different martensite varients. The properties observed
are relatable to earlier work, with minor modification.

The decreasing yield stress with temperature, Figure 4, is in
general agreement with the observations of Pops[8] on CuZn. However,
the Pops curve ends at M_s, a reasonable value, while the curve in
Figure 4 continues to M_f. Since the austenitic and martensite
coexist between M_s and M_f, it is clear that it must be easier to
mechanically convert austenite to martensite in this alloy than to
reorient the existing thermal martensite plates. During luders
straining, the luders stress rises from a lower to a higher level,
suggesting that the two processes, deformation martensite formation
and thermal martensite reorientation occur in succession.

The second slope of the deformation curve, Figures 2 & 3, also
reported by Pops in Cu-Zn is probably due to mechanical twinning.
It has not been possible to verify this mechanism, though all observations are in agreement. Since the (121) twinning plane in the γ
is parallel to the (011) β shear plane for the β-γ transformation[5]
a (100) crystal would remain axial on deformation, as is observed.
Also, the twinning shear (actually determined for transformation
twins rather than mechanical twins[6]) is sufficiently large to account
for the approximately 10% strain observed. Finally, the pseudoelasticity and temperature insensitivity suggest twinning. The
process is probably a kin to the rubbery behavior in In-Tl[7] though
at a higher stress level.

The unusual shape memory effect, the instantaneous 7.8% strain
recovery at 50°C of crystals deformed below that temperature, is
associated with the unload minimum, and the γ'-β retransformation
during the unload luders strain, Figures 3 and 4. The observed 7.8%
strain recovery is only 90% of the potential 8.7% decrease in the
β axis of the unit cell in going from β-γ', Figure 1. However,
Laue photograms of the deformation induced γ' show a somewhat imperfect but twin free γ' single crystal.

An examination of the unloading curve at 75°C, Figure 3, offers
a clue to the residual strain retention. At the unload minimum,
there is a reluctance for the material to transform from γ' to β.
Nevertheless, as seen by the negative modulus at the start of the
reverse transformation, the process proceeds with a considerable
decrease in free energy. Clearly, this SME is due to a nucleation
difficulty, and two possibilities are suggested. It could be simply
a reluctance for the β phase to nucleate martensitically. However,

this is unlikely, since the β phase forms spontaneously in the
thermally transformed γ´ at the A_s temperature. Therefore, the
explanation must lie in a change in the γ´ phase that occurs with
deformation.

Transformation twins in the γ´ phase accomodate the lattice
invarient shear associated with the formation of a β-γ´ interface[2,6]
in thermal martensites. The absence of such twins in the deformed
martensite, and the need to nucleate them before retransformation
may be the cause of this higher temperature shape memory effect.

Although no plastic deformation was observed in (100) orientation specimens, the kink bands formed in crystals of other orientations show that plastic deformation is possible. The bands developed
in these latter crystals are wedge shaped martensite plate groups.
They end within the crystal, and are bounded by dislocations at the
tips of the wedges[1]. Thus complete recovery of large strains not
automatically guaranteed in this, or any other pseudoelastic material.
It is dependent upon the educated choice of crystal orientation
relative to the stress axis.

CONCULSIONS

1.   Single crystal β phase Cu-14.1Al-3Ni in a (100) orientation exhibits both large pseudoelastic strain and high
strength in compression.

2.   At strains larger than the luders deformtion, an almost
linear pseudoelastic strain region occurs with a slope increasing from 1 1/2 to 2 million psi. The slope is temperature
independent, and evidently associated with mechanical twinning.

3.   If strained beyond the luders strain at temperatures less
than 50°C, the crystals exhibit a shape memory effect at that
temperature, recovering 7.8% strain in a burst.

ACKNOWLEDGEMENTS

The author wishes to thank Professor C. M. Wayman of the
University of Illinois, and Drs. R. Beeuwkes, W. J. Croft and
C. B. Walker of AMMRC for helpful discussions in the course of
this work.

REFERENCES

1. L. Delaey, R. V. Krishnam, H. Tas and H. Warlimont; J. Mat.
    Sci., V9, (1974), Pt. 1, p 1521, Pt. 2, p 1536, Pt. 3, p 1545.

2. C. M. Wayman and K. Shimizu; Met. Sci. J., V8 (1972) p 175.

3. R. E. Busch, R. T. Luedeman and P. M. Gross, Final Report Parts
    1 and 2, AMRA CR 65-02/1,2; Feb. 1966.

4. K. Oishi and L. C. Brown; Met. Trans., V2 (1971) p 1971.

5. M. J. Duggin and W. A. Rachinger; Acta Met., V12 (1964) p 529.

6. K. Otsuka and K. Shimizu, Trans. Japan Inst. Met., V15 (1974)
    p 103; p 109.

7. Z. S. Basinski and J. W. Christian; Acta Met., V2 (1974) p 101.

8. H. Pops; Met. Trans. V1 (1970) p 251.

9. H. Tas, L. Lelaey and A. Deruyttere; Scripta Met., V5 (1971)
    p 1117.

10. J. D. Eisenwasser and L. C. Brown; Met. Trans., V3 (1972) p 1359.

TENSILE PROPERTIES OF SUPERCONDUCTING COMPOSITE CONDUCTORS AND

Nb-Ti ALLOYS AT 4.2°K*

D. S. Easton and C. C. Koch

Metals and Ceramics Division, Oak Ridge National

Laboratory, Oak Ridge, Tennessee 37830 USA

Large superconducting magnets will be used for plasma confine-
ment in controlled thermonuclear fusion experiments. The presence
of thermal, mechanical, and magnetic forces in these coils makes
the understanding of mechanical properties of the superconducting
windings important. This paper presents the results of tensile
tests at 300, 77, and 4.2°K on commercial Nb-Ti composites with
both Cu and Cu-Ni matrices. Tests were also conducted on Nb-Ti
filaments produced by chemically removing the matrix of some composites as well as with sub-sized tensile samples of Nb-Ti alloys.

A pseudoelastic strain region was found in the Nb-Ti/Cu (Cu-Ni)
composites and in the Nb-Ti alloys themselves. The anelastic
behavior observed in the composites is typical for composites consisting of strong filaments in a weak ductile matrix. However, the
pseudoelasticity found in the Nb-Ti itself was the first observation
of this behavior in these alloys. Possible explanations are the
formation of a reversible stress-induced martensitic transformation
and/or twinning and de-twinning. The pseudoelastic strain is
accompanied by audible "clicks" upon both stress loading and unloading.

---

*Research sponsored by the Energy Research and Development
Administration under contract with the Union Carbide Corporation.
Funding provided by the Superconducting Magnet Development Program
of the Thermonuclear Division.

The composite conductors and the Nb-Ti alloys exhibited serrated stress-strain curves. The yield-elongation serrations are
apparently the major method of plastic strain in the Nb-Ti alloys
at 4.2°K. Both the pseudoelastic phenomenon and the stress-strain
serrations can produce an energy loss that is important. The well-
known "training" effect in large superconducting magnets may be related to these mechanical effects.

INTRODUCTION

Superconducting magnets have been employed for the past 15
years, mainly as research tools. During this period there have been
many investigations of the effects of microstructure and composition
on superconducting properties; however, little attention has been
paid to the mechanical properties of the material at 4.2°K. Today,
with the proposed extremely large ($\sim$ 6 m ID, $\sim$ 12 m OD) magnets to
be used for plasma containment in controlled thermonuclear fusion
experiments, it is imperative that low temperature mechanical
property data be obtained. Indeed, the ability of the conductor to
tolerate the stresses generated by magnetic forces as well as thermal
contraction forces during cool-down may well be the limiting factor
in the design of these large devices.

The candidate material for the conductor at this time is a
composite consisting of Nb-Ti alloy filaments in a Cu and/or Cu-Ni
matrix.[1-4]

To remain in the superconducting state, the material must
(1) be below a certain critical temperature, T_c, (2) be below a
critical magnetic field, H_{c2}, and (3) not carry an electric current
in excess of a critical current density, J_c.

A poorly understood problem with large superconducting magnets
is their tendency to "train"[5] Training occurs when the amount of
current that can be passed through the conductor is well below the
critical current density, J_c, as measured in a "short sample" test
but increases after a number of superconducting to normal state
cycles. In the use of superconductors in high field magnets thermal
instabilities (flux jumps) make it impossible to use the "short
sample" current carrying capacity of the superconductor. Thus,
stabilization techniques[3,6] such as employing fine filaments of
superconductor in a high conductivity normal matrix (i.e., Cu) are
necessary. The phenomenon of "training" has been observed in short
samples (unstabilized) as well as in magnets. Training has been
rationalized as due to a minimization of induced currents,[7] conductor
movement,[8] and other mechanical effects such as cracking of potting
epoxy.[8] Evans[9] proposed that training is related to the serrated
yielding observed in Nb-Ti alloys at 4.2°K.

Serrated stress-strain curves have been noted in almost all metals when cooled to liquid helium temperatures. Wigley[10] reviewed many of the explanations for this behavior such as: stress-induced phase transformations (e.g., martensite), deformation twinning, burst dislocations, and yielding caused by adiabatic heating. The latter mechanism was shown by Basinski[11],[12] to be generally acceptable for most metals although the other mechanisms may apply in specific cases. For adiabatic heating, the low specific heat capacities of metals at $4.2°K$ combined with poor thermal conductivity can cause localized heat increases at points of stress concentration. If the yield stress is sensitive to temperature, as is generally the case, load drops would then occur. Basinski[12] found temperature increases up to $60°K$ during load drops in Al alloy samples immersed in liquid helium. Erdmann and Jahoda[13],[14] studied Cu-Ni alloys in an isolation calorimeter and found increases in temperature to $35°K$ due to discontinuous slip.

Thermoelasticity, pseudoelasticity, and shape memory effects associated with martensitic transformations, which produce hysteretic stress-strain curves, have been reviewed.[15-17] Richman[18] showed that twinning and de-twinning can occur in Fe_3Be single crystals, and cause pseudoelasticity.

Composite materials consisting of a ductile matrix and strong filaments or fibers (e.g., Cu matrix, W filaments) have been employed for their high strength and toughness for a number of years. There have been extensive studies on their mechanical properties by many investigators.[19-25]

Anelasticity produces thermal energy due to the difference in stored energy during loading and the energy recovered during unloading. Assuming optimized compositions, Young and Boom[26] show an idealized design curve where at 6T a Nb-Ti/Cu composite has a current density, J_c, of 2×10^5 amp/cm^2 at $4.2°K$ and 2×10^4 amp/cm^2 at $6.2°K$. Since the operating temperature for large superconducting magnets may be $5°K$, small changes in temperature could result in drastic effects upon magnet performance.

This paper describes observations of both anelastic stress-strain behavior and discontinuous load drops occurring upon loading and unloading of both conductor composites and Nb-Ti alloys. Using specific heat capacity data at $4.2°K$, we show the maximum temperature increases possible due to such behavior.

MATERIALS

The body-centered cubic Nb-Ti alloys in this work were taken from commercially produced magnet conductors. The compositions were

either 45 or 48 wt pct (62–64 at. pct) Ti. Impurity analysis is
unknown and, since the test samples came from many separate lots,
some small differences in impurity levels and percentage of cold
work might be expected. With the exception of subsized tensile
samples which were annealed at 800°C, all samples were cold worked
(> 99.9 pct) with transverse dislocation cell sizes on the order of
500 Å. Wire samples were obtained from composites by chemically removing the Cu matrix.

EXPERIMENTAL APPARATUS AND TECHNIQUES

Tests were made using a tensile machine with a 4-in.-ID helium
cryostat suspended from the bottom of the movable cross-head. The
lower ends of the test sample were attached to a compression member
and the top ends to a tensile pull rod. Wire samples were wrapped
around 3/4-in.-diam mandrels, rectangular composites were held in
serrated grips, and button-head tensile specimens were held in
machined grips that matched the radii of the head. The grips were
attached to either ball and socket joints or rotatable pins to
ensure alignment along the tensile axis.

Strain measurements were obtained by one (and/or a combination)
of three methods: (1) clip-on extensometer calibrated at the test
temperature by a micrometer capable of reading 20 μin., (2) use of
gage marks where possible and measured using an x-ray film compara-
tor, and (3) cross-head movement. Comparison of the three methods
indicated that the first two agreed quite well for the larger samples
while the third method was only good for testing wire samples.

Reductions in area were measured by the use of a shadowgraph
with a magnification factor of 20x.

Tests on samples of high purity Cu were made as checks on the
experimental technique. The data were in good agreement with that
of Reed and Mikesell.[27]

EXPERIMENTAL RESULTS

The stress-strain curve at 4.2°K of a 0.05 x 0.12-in. (nominal)
Cu matrix conductor containing 18 filaments (0.01-in. diam) of a
Nb–48 wt pct Ti alloy is shown in Fig. 1. The anelastic shape can
be explained by the general behavior of composites consisting of
ductile matrices and strong filaments or fibers. Composites when
stressed to fracture undergo three major stages: (1) both matrix
and filaments deform elastically, (2) the matrix goes plastic while
the filaments remain elastic, and (3) both matrix and filaments deform plastically until fracture. If stage 3 is not reached and the
sample is unloaded the matrix will be in compression with the

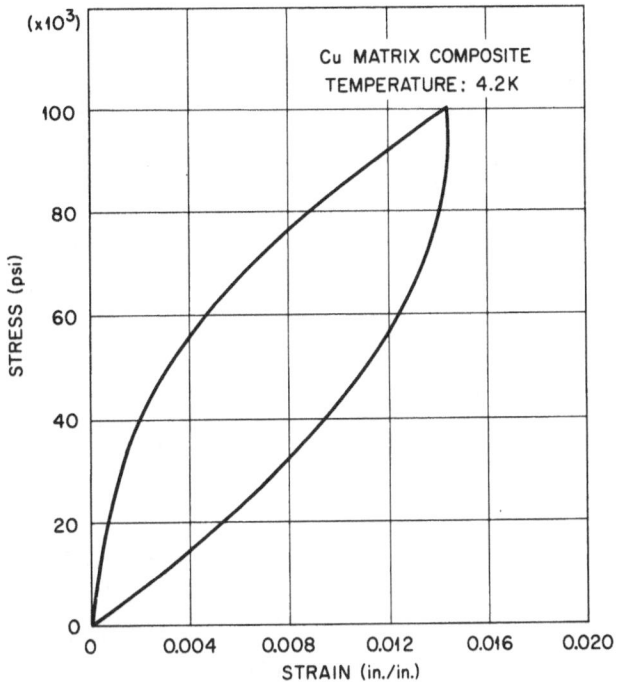

Fig. 1. Stress-strain curve of a Nb-Ti/Cu composite at 4.2°K.

filaments in tension. Such composite behavior has been discussed by many investigators.[20,24]

When samples of Nb-Ti alloys were tested at 4.2°K, they also showed both hysteretic stress-strain curves and discontinuous slip. To ensure that the effects were real, checks were made for possible hysteresis of the extensometer and tests were made on high purity annealed Cu samples of about the same dimensions of the test materials. Hysteresis of the extensometers was so small it could be entirely eliminated, and the tests on Cu showed only very slight deviations on loading and unloading that are normally observed in tensile tests. The elastic modulus of the Cu was measured as 16.9×10^6 psi at 4.2°K and 16.5×10^6 psi at room temperature, in good agreement with Reed and Mikesell.[27]

Figure 2 shows a stress-strain curve (extensometer output) of a tensile sample of Nb-45 wt pct Ti stressed to ~ 210,000 psi where it underwent a discontinuous slip. The sample was unloaded and then cycled four more times to the same stress. On the first cycle there was some plastic strain as seen at zero stress but the second

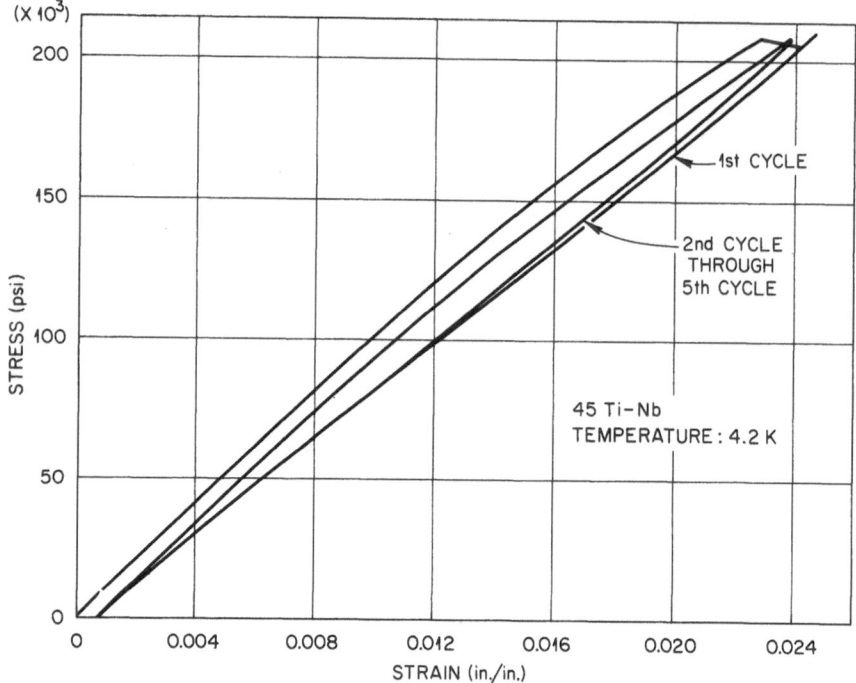

Fig. 2.   Stress-strain curve of a Nb-45 wt pct Ti alloy.

through fifth cycles were pseudoelastic in that they returned to the
same length even though the stress-strain relationship was non-linear.
Some samples were given up to 100 cycles without showing further
change in the stress-strain behavior.   The amount of initial plastic
strain in the first cycle was less than that generated by the discontinuous slip.

Figure 3 shows a load-elongation curve (x-axis timed to match
cross-head travel) of a Nb-48 wt pct Ti wire sample.   The unloading
curves have been reversed as indicated by the dashed lines to indicate
the hysteretic behavior.   Despite the extensive discontinuous yielding,
the first cycle exhibits little plastic strain and the second cycle
shows complete hysteresis (i.e., no plastic strain).   The only linear
stress-strain regions occur during and immediately following some of
the serrations.   Serrations can also be seen on the unloading curves.
This effect is more apparent in Fig. 4 that shows a more sensitive
scale of the unloading curve of the same sample after being reloaded
to ~ 115,000 psi.

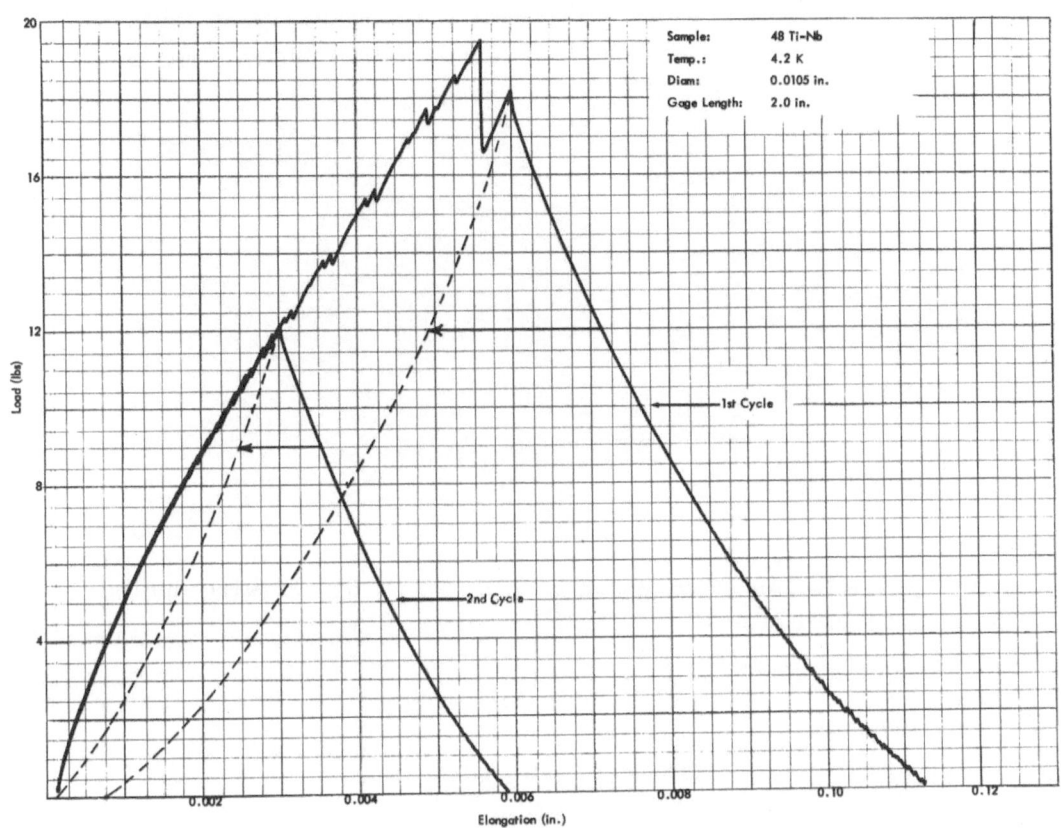

Fig. 3.   Load-elongation curves of a Nb-48 wt pct Ti wire.

A crystal pickup was attached to the pull rod, external to the helium dewar, and the signal fed into an audio amplifier. Almost immediately upon application of stress on samples at 4.2°K "clicks" could be heard. When similar samples were measured at 300 or 77°K no serrations or audible clicks occurred. The frequency of the clicks was greatest at low stresses both on loading and unloading. If a sample was stressed to some value, unloaded, then reloaded to a greater stress, a noticeable increase in both the amplitude and frequency of the clicks occurred at the former stress level. In Fig. 5 a 2-pen recorder shows both load values and the audio signal during a series of loading and unloading cycles on a Nb-45 wt pct Ti tensile sample. The pen recording the audio was offset ∼ 0.2 in. to the right on the chart and should be adjusted to the left to properly interpret the graph. Note the initial amplitude declines until the previous stress level is exceeded upon which there is an increase in

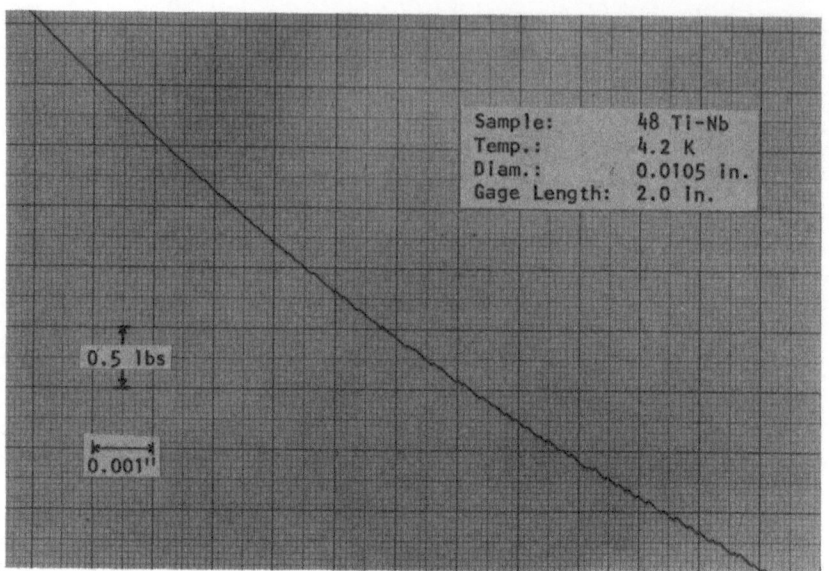

Fig. 4.   Load-elongation curve of a Nb–48 wt pct Ti wire during
          unloading.

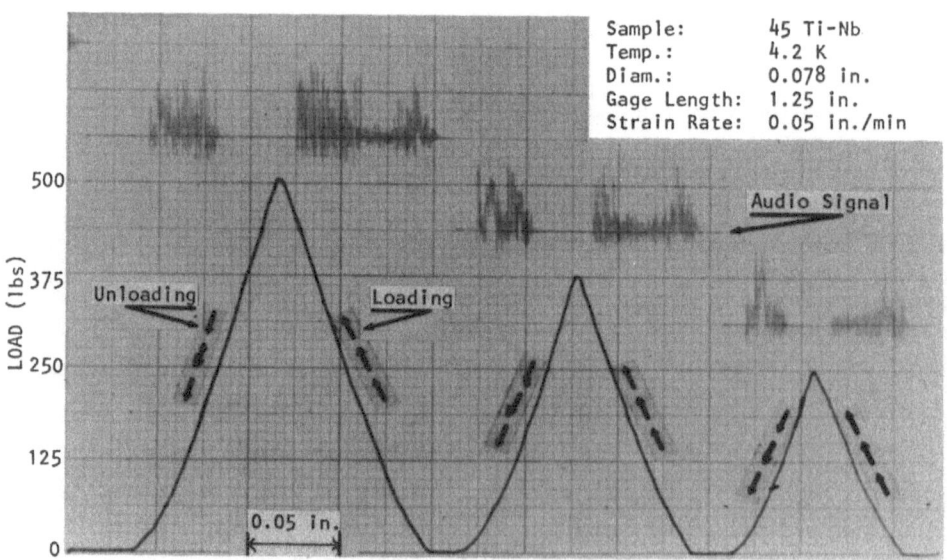

Fig. 5.   Audio signals recorded during loading and unloading of a
          Nb–45 wt pct Ti alloy.

the amplitude. Upon unloading there is no signal until the stress
level is ~ 1/3 of the stress cycle. There also seem to be characteristic "bursts" during unloading.

DISCUSSION OF RESULTS

The cause for the anelastic behavior in Nb-Ti at 4.2°K is not
known. The shape of the stress-strain curves are similar to those
shown by Miura et al.[28] for $Au_{26}-Cu_{28}-Zn_{46}$ for which stress induced
martensite was responsible.

Optical, SEM, and x-ray examination of the samples before and
after testing revealed no differences in structure. There was no
evidence of slip lines, twins, or martensite when the samples were
examined at room temperature. If twins and/or martensite are forming, they are stress and/or thermally reversible. The audible
"clicks" sound like the "cry" of metals during twinning.

Another possible mechanism for the pseudoelastic behavior is
the production of hydrides which are both stress and thermally
activated. Hydride formation at low temperatures has been observed
in niobium.[29-31] These hydrides redissolve when the sample is
returned to room temperature.

In addition to hysteretic stress-strain curves, other evidence
for the reversible nature of the deformation mechanism can be seen
in Table 1. With the exception of the relatively large tensile
samples, the apparent elongation is greater at 4.2°K than at 300°K,
while the reduction in area (which was measured at room temperature)
is lower at 4.2°K. Hence, one has the unusual situation in which
two separate methods of measuring ductility are contradictory. The
first showing a more plastic behavior at low temperature and the
second a more brittle behavior. Our explanation is that the
apparent plastic elongation is actually pseudoelastic in nature and
the total plastic deformation, after a load cycle, is really less at
low temperature than at room temperature.

There was an apparent effect of sample diameter, whereby the
tensile samples (i.e., large diameter) produced only a few load
drops, sometimes only one before fracture, while the wire samples
(i.e., small diameter) underwent a large number of the serrations.
It is possible that in the small diameter wires a discontinuous slip
and/or twin could extend through the entire cross section similar to
Lüders band formation. A dependence upon grain size-thickness ratio
on martensite transformation temperature and the magnitude of
reversible strain in Cu—33.6 wt pct Zn—4 wt pct Sn alloys, has been
reported by Dvorak and Hawbolt.[32] They showed that when the grain
size-thickness ratio exceeded 2, the magnitude of transformational
elasticity (pseudoelasticity) increased markedly.

Table 1. Mechanical Properties at Various Temperatures

Sample	Type	Diameter (in.)	Temp. (°K)	Ultimate Tensile Strength (psi) x 10³	Total Elongation (pct)	Reduction in Area (pct)	Elastic Modulus (psi) x 10⁶
Composite, Cu matrix, 18 filaments Nb-48 wt pct Ti	Rectangular	0.055 x 0.116	300	62	17	16	15
			77	100	26	--	--
			4.2	115	2	14	17
Nb-45 wt pct Ti	Tensile	0.078	300	72	9.6	77	11.9
			77	159	3.3	47	--
			4.2	219	< 0.2	7	12.0
Nb-48 wt pct Ti	Tensile	0.078	300	73	8.8	71	11.9
			4.2	219	< 0.2	11	12.0
Nb-48 wt pct Ti	Wire	0.0125	300	201	4.1	15	--
			4.2	313	4.8	12	--
Nb-48 wt pct Ti	Wire	0.0105	300	150	1.2	20	--
			77	251	2.4	17	--
			4.2	320	4.8	8	--
Nb-48 wt pct Ti	Wire	0.0084	300	161	2.3	22	--
			4.2	314	4.5	7	--
Pure Ti	Tensile	0.078	4.2	190	4.0	20	--
Pure Nb	Tensile	0.078	4.2	188	1.3	5	--

Micro-yielding could be occurring prior to the occurrence of visible load drops on the stress-strain curve. Heat generated by this micro-yielding could be responsible for the "training" effect. A point of stress-concentration in a conductor could result in yielding at some low stress level caused by initial application of a magnetic or mechanical force and thus cause quenching of the superconducting state. Since this particular stress-concentration would then be eliminated, a higher field (or force) could be tolerated on each successive excursion.

Both anelastic behavior and discontinuous slip release thermal energy. In a superconducting device, such as a magnet, a sufficiently large increase in temperature can cause the superconductor to revert to the normal state. These mechanical effects are of course rate dependent. If the stress changes slowly, the energy can be dissipated to the surrounding environment (i.e., liquid or supercritical helium) and does not increase the temperature of the bulk material. However if the thermal conductivity is low and close contact with the cooling medium is not preserved, cyclic stresses could develop (perhaps from pulsing) to the extent that would quench the superconductor.

The magnitude of the temperature increase, ΔT, due to anelastic, hysteretic stress-strain behavior can be calculated from the low temperature specific heat capacity as follows: First, assume that the cycle occurs rapidly enough that all the heat is trapped in the sample. In the case of discontinuous slip, Basinski[11] found that samples immersed in liquid helium responded as though there was an insulating barrier between the sample and liquid (helium vapor) for the time interval corresponding to a single load drop.

$$C_p = \gamma T + \beta T^3 \tag{1}$$

$$\Delta Q = \int_{4.2}^{T_f} C_p \, dT$$

$$= \frac{\gamma T_f^2}{2} - \frac{\gamma (4.2)^2}{2} + \frac{\beta T_f^4}{4} - \frac{\beta (4.2)^4}{4} \tag{2}$$

and

$$\Delta T = T_f - 4.2 . \tag{3}$$

Where ΔQ = area under hysteresis curve in J/mole; $\beta = 12 \, \pi^4 R / 5\theta^3$; $R = 8.317$ J/deg mole, gas constant; θ_D Cu = 342°K; θ_D Nb-Zr = 238 °K; θ_D = Debye temperature; $\gamma_{Cu} = 6.93 \times 10^{-4}$ J/K^2 mole; γ_{Nb-Zr} = 83.0×10^{-4} J/K^2 mole; and γ = electronic coefficient of low temperature specific heat capacity.

The values for θ_D and γ are those of Gschneidner[33] for Cu and Heinger et al.[34] for Nb-Zr. Since no data exist for Nb-Ti in the

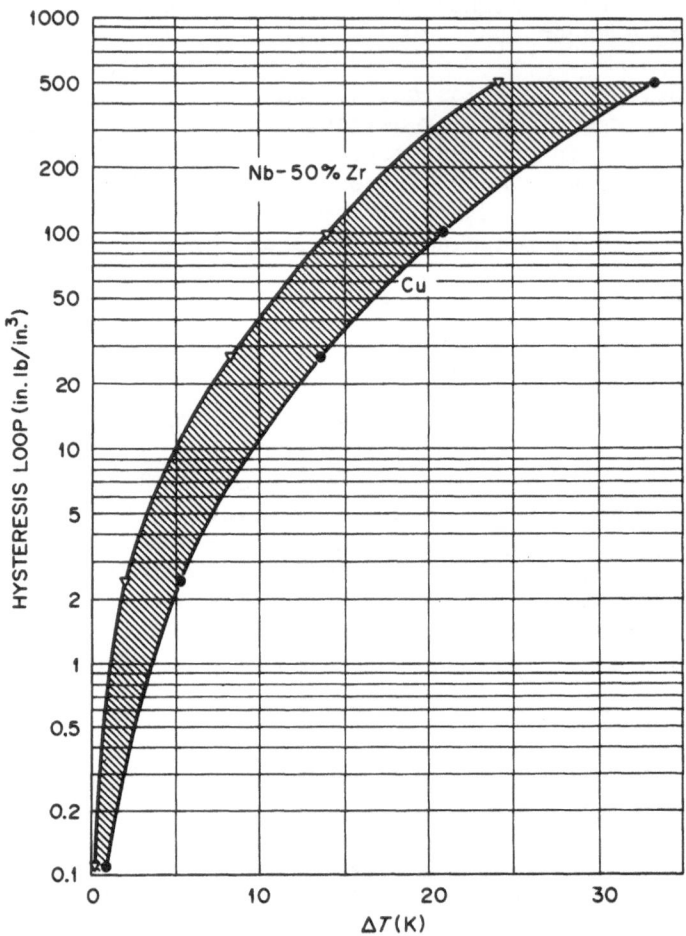

Fig. 6. Increase in temperature, ΔT, as a result of varying
hysteretic stress-strain cycles.

appropriate composition ranges, data on Nb–50 Zr was selected as a reasonable substitute. In the case of the composite curve shown in Fig. 1, ΔQ = 507.5 J/mole, ΔT_{Cu} = 33.2°K, and ΔT_{Nb-Zr} = 24°K. Of course such large values are extreme since a magnet should not normally undergo such large stress cycles. The areas under several experimental hysteresis loops (ΔQ) were measured, and using the calculation shown above, the resultant temperature increases (ΔT) for both Cu and Nb-Zr are shown in Fig. 6. Composite material values should lie between the two curves.

SUMMARY

Anelastic stress-strain behavior at 4.2°K has been found in composites, consisting of a Cu matrix and Nb-Ti filaments, and Nb-Ti alloy samples. If the sample were strained past certain values, deformation proceeded by discontinuous slip. The mechanisms for these phenomena were not determined; however, their presence reveals a source of thermal energy that can be important in the design of large superconducting devices.

The "training" effect in large superconducting magnets may be related to the heat produced by hysteretic stress-strain cycles and/or discontinuous slip. It is proposed that micro-yielding may occur with the application of very small stresses. This micro-plasticity cannot be seen on tensile curves but can be detected by the presence of audible "clicks". During the unloading portion of a stress-strain cycle serrations are sometimes evident at very low stress levels.

REFERENCES

1. J. E. Evetts and P. J. Martin: Metal Sci. J., 1973, vol. 7, pp. 179-184.
2. M. N. Wilson: Composites, 1970, vol. 1, pp. 341-344.
3. B. J. Maddock: Composites, 1969, vol. 1, pp. 104-111.
4. C. W. Fowlkes, P. E. Angerhofer, R. N. Newton, and A. F. Clark: Characterization of a Superconducting Coil Composite, Report No. NBSIR 73-349, Naval Ship Research and Development Center, Annapolis, Md.
5. D. Saint-James, E. J. Thomas, and G. Sarma: Type II Superconductivity, Pergamon Press, New York, 1969, pp. 264-266.
6. Z.J.J. Stekly and J. L. Zar: "Stable Superconducting Coils," IEEE Trans. Nucl. Sci., 1965, vol. 12, pp. 367-372.
7. M.A.R. LeBlanc: IBM J. Res. Develop., 1962, vol. 6, pp. 122-125.
8. V. W. Edwards, C. A. Scott, and M. N. Wilson: "Applied Superconductivity Conference," IEEE Trans. Magnetics, 1975, vol. 11, pp. 532-535.
9. D. Evans: Science Research Council, Report No. RL-73-092, Rutherford Laboratory, Chilton, Didcot, England, 1973.

10. D. A. Wigley: _Mechanical Properties of Materials at Low Temperatures,_ Plenum Press, New York, 1971, pp. 84-90.

11. Z. S. Basinski: _Roy. Soc. London Proc. Ser. A,_ 1957, vol. 240, pp. 229-242.

12. Z. S. Basinski: _Aust. J. Phys.,_ 1959, vol. 13, pp. 354-358.

13. J. C. Erdmann and J. A. Jahoda: _Rev. Sci. Instr.,_ 1963, vol. 34, pp. 172-179.

14. J. C. Erdmann and J. A. Jahoda: _J. Appl. Phys.,_ 1968, vol. 39, pp. 2793-2797.

15. L. Delaey, R. V. Krishnan, H. Tas, and H. Warlimont: _J. Mater. Sci.,_ 1974, vol 9, pp. 1521-1535.

16. R.V. Krishnan, L. Delaey, H. Tas, and H. Warlimont: _J. Mater. Sci.,_ 1974, vol. 9, pp. 1536-1544.

17. H. Warlimont, L. Delaey, R. V. Krishnan, and H. Tas: _J. Mater. Sci.,_ 1974, vol. 9, pp. 1545-1555.

18. R. H. Richman: _Deformation Twinning,_ 1964, vol. 25, Gordon and Breach Science, New York, ed. R. E. Reed-Hill, J. P. Hirth, and H. C. Rogers, pp. 237-270.

19. D. K. Hale and A. Kelly: _Ann. Rev. Mater.Sci.,_ 1972, vol. 2, pp. 405-462.

20. A. Kelly and G. J. Davies: _Met. Rev.,_ 1965, vol. 10, pp. 1-74.

21. D. Crutchley: _Met. Rev.,_ 1965, vol. 10, pp. 79-141.

22. H. R. Piehler: _Trans. Met. Soc. AIME,_ 1965, vol. 233, pp. 12-16.

23. A. Kelly: _Strengthening Methods in Crystals,_ Applied Science, London, 1971, pp. 433-475.

24. A. Lawley and M. J. Koczak: _Role of the Interface on Elastic/ Plastic Composite Behavior,_ Drexel University, ONR Tech. Report No 7, 1972.

25. D. L. McDanels, R. W. Jech, and J. W. Weeton: _Trans. Met. Soc. AIME,_ 1965, vol. 233, pp. 636-642.

26. W. C. Young and R. W. Boom: _Proc. 4th Conf. Magnet Technology,_ Brookhaven, USAEC CONF-720908, ed. Y. Winterbottom, 1972, pp. 244-251.

27. R. P. Reed and R. P. Mikesell: _J. Mater.,_ 1967, vol. 2, pp. 370-392.

28. S. Miura, S. Maeda, and N. Nakanishi: _Phil. Mag.,_ 1974, vol. 30, pp. 565-581.

29. J. S. Abell and I. R. Harris: _J. Less-Common Metals,_ 1972, vol. 29, pp. 104-108.

30. C. Wert, D. O. Thompson, and O. Buck: _J. Phys. Chem. Solids,_ 1970, vol. 31, pp. 1793-1978.

31. O. Buck, D. O. Thompson, and C. A. Wert: _J. Phys. Chem. Solids,_ 1971, vol 32, pp. 2331-2344.

32. I. Dvorak, and E. B. Hawbolt: _Met. Trans.,_ 1975, vol. 6A, pp. 95-99.

33. K. A. Gschneidner, Jr.: _Solid State Phys.,_ 1964, vol. 16, ed. F. Seitz and D. Turnbull, pp. 275-426, Academic Press, New York.

34. F. Heiniger, E. Bucher, and J. Muller: _Phys. Kondens Materie,_ 1966, vol. 5, pp. 243-284.

SUGGESTIONS FOR APPLYING A PHENOMENOLOGICAL APPROACH TO INVESTIGATIONS OF MECHANICAL BEHAVIOR IN SME ALLOYS

Glen Edwards and Jeff Perkins

Materials Group, Department of Mechanical Engineering

Naval Postgraduate School, Monterey, California 93940

Significant progress in understanding SME alloys has been made in recent years, a fact clearly illustrated by the work presented in recent reviews (1-3) as well as by the research presented at this symposium. Most of this work has been quite basic in nature, aimed at microscopic, crystallographic or thermodynamic interpretation of the observed phenomena. A major success of these efforts lies in the emergence of certain generalities regarding potential alloy systems or the anticipated behavior of known alloy systems. It is generally accepted, for example, that SME systems have in common 1) a thermoelastic martensitic transformation and 2) ordering in both parent and martensite phases. Basic research to further delineate such generalities must of course continue.

SME alloy development has progressed to the stage where serious consideration can be given these alloys for utilizing their unique mechanical behavior in actual application. Indeed, certain progressive industrial concerns are already doing so (4,5). A major hindrance to a more general utilization of these useful materials is the scarcity of empirical data which defines the material characteristics important in fabrication and/or service utilization of SME alloys. Perkins et al (6) have cited several such characteristics; namely, thermal stability of reversion stress, optimal strain prior to reversion, magnitude of reversion stress and strain, and optimal magnitude of unconstrained reversion strain. We see a need for the systematic development and analysis of these and other engineering data; moreover, we believe that a search for phenomenological similarities in these data has great potential for advancing the fundamental understanding of SME alloy behavior. The phenomenological approach has proven to be a powerful tool in other metallurgical studies, and many classic examples, familiar to most materials scientists, come to mind. For example, a major success typlifying the approach we are suggesting is the correlation (appropriate for most metallic systems) between self-diffusion activation energy and the activation energy for high-temperature deformation (7). The following paragraphs cite specific areas of SME alloy research which we believe to be fertile ground for the approach we suggest.

An area of great potential for making useful phenomenological correlations in SME alloys is the study of grain size effects. Consider first the qualitative observation that a decrease in prior austenite grain size increases the apparent flow stress in the mar- tensitic phase of a SME alloy (8). The general explanation for deformation below M_f is that intramartensite twin boundaries move in response to an applied stress, (9, 11) and several researchers (8, 12, 13) have concluded that the prior austenite grain boundaries, by impeding the motion of intramartensite twin boundaries, 1) reduce the magnitude of the potential reversion strain, ε_r, and 2) increase the magnitude of the irreversible strain, ε_{irr}. Some of the initial studies of grain size effects have utilized relatively thin, coarse- grained specimens and the data have been interpreted in terms of d/t, the grain size/sample thickness ratio (12, 13). Although the existence of interacting grain size effects and surface effects for d/t ratios greater than 0.5 can not be denied, we believe that this approach does not completely characterize the role played by prior austenite grain boundaries in altering SME behavior. A study em- ploying samples of d/t ratios less than 0.5 would isolate the effects of grain constraint from the interference of surface effects. A first-order attempt to more specifically characterize the grain size dependence would logically be an emulation of the Hall-Petch relation:

$$\sigma_{M \to M_d} = \sigma_o + Kd^n \tag{1}$$

i.e., the flow stress in the martensitic phase can likely be shown to the sum of a friction stress, σ_o, necessary to move intramar- tensite twin boundaries (independent of grain size) and a grain- size dependent stress, Kd^n. (K and n are empirically determined material constants and d is grain diameter).

A second-order characterization of grain size effects could be aimed at relating the magnitude of irreversible strain to the grain size. This might be simply done by taking accurate strain measurements after unconstrained reversion of samples utilized to formulate the stress relation, Eq. 1, and then searching for a sim- ple analytical expression for the relationship between ε_{irr} and d. An even firmer and potentially more useful correlation between ε_{irr}, and d could result from careful electron microscopy. We would hypothesize that ε_{irr} is increased (and ε_r is decreased) when grain size is small because grain boundary constraints impede twin boundary motion, concentrate the stresses at dislocation sources, and increase the dislocation density at relatively low values of total strain. The increase in dislocation density, $\Delta\rho$, creates the irreversible strain, and the relationship between plastic strain and dislocation density is often quite simple; for example, in copper, $\varepsilon = \alpha \Delta\rho^{\frac{1}{2}}$, where α is about 10^{-6} $(cm^2/disl)^{\frac{1}{2}}$ (14). The increase in dislocation density could in turn be related to grain size, the result being:

$$\varepsilon_{irr} = f(\Delta\rho) = g(d) \tag{2}$$

Since the reversion strain, ε_r, is closely approximated by the difference between initial strain, ε_i, and ε_{irr}, and since reversion stress is known to be linearly dependent on reversion strain (15), successful completion of the work outlined above would completely describe the effect of grain size on reversion stress and strain, parameters of considerable technological importance in utilization of SME materials. Moreover, since crystal structures of both parent and martensite phases in many SME alloys are quite similar, a functional relationship determined for a specific alloy and describing such a phenomenon as the effect of grain size on irreversible strain, might well apply to other SME alloys.

Diffusive processes are occasionally important in the mechanical behavior of SME alloys, and the phenomenological Arrhenius relation can often be constructively applied to these cases. An impressive example of how useful this logic can be is given by Lieberman, et al (16), who show that the experimental activation energy for microstructural changes in a Au-Cd alloy correspond to activation energies for short-range diffusion. This empirical data lends considerable credence to their explanation for the ferroelastic behavior of Au-Cd. A second SME alloy system in which diffusive processes affect mechanical behavior is the Cu-Sn system. Miura, et al (17) have observed an aging effect during deformation above M_s of Cu-15 a/o Sn. Aging increased the critical stress for SIM formation and increased the degree of superelasticity, an effect the authors suggest may be associated with diffusion of tin atoms (See Figures 10 and 11, Reference 17). An empirical analysis of the aging kinetics could, of course, be used to investigate that possibility. Samples aged for varying times at varying temperatures above M_s, then deformed to determine the critical stress for SIM formation at constant temperature, would provide the necessary data. By plotting $\sigma_{p \to m}$ versus T_{aging} for constant aging time, a series of lines very similar to those of Figure 11, Reference 17, would evolve. The assumptions that a specific $\sigma_{p \to m}$ is always representative of a specific microstructural condition, and that the reciprocal of aging time, $1/t_{aging}$, is proportional to the controlling diffusion rate, complete the analysis:

$$1/t_{aging} \Big|_{\sigma_{\rho \to m}} = \text{constant } e^{-Q/RT} \tag{3}$$

Reasonable agreement between the experimental activation energy and the activation energy for tin diffusion would lend considerable support to the supposition that tin diffusion stabilizes the β phase in the Cu-Sn system during deformation above M_s.

A final example of the empirical approach applied to the study of SME alloys is taken from our own engineering study of Ti-Ni alloys (18, 19). We have carefully characterized the reversion stress for a particular Ti-Ni alloy as a function of strain below M_f, unconstrained reversion strain, and temperature. We have also experimentally determined for the same alloy the flow stress associated with SIM formation at similar strains and temperatures above M_s. Flow stress measurements were obtained from the simplest of tests; samples were deformed above M_s at constant temperature and strain rate in uniaxial tension. The value of stress for a particular strain and temperature was taken as the flow stress for those conditions. As stated by Perkins et al (6), these flow stresses must define an upper bound for reversion stress. Furthermore, a comparison of flow stress and reversion stress for a specific Ti-Ni alloy indicates that magnitudes of these stresses are quite similar:

$$\sigma_r \Big|_{\varepsilon, T} \simeq 0.9 \, \sigma_{p \to m} \Big|_{\varepsilon, T} \tag{4}$$

The practical significance of Eq. (2) is immediately obvious; stress/strain data at fixed temperatures are easily generated, while accurate measurements of reversion stress are experimentally much more difficult to obtain. Should Eq. (2) prove to be generally applicable to SME alloys, this simple means of characterizing SME potential should facilitate SME alloy development. The theoretical implications of this comparison are likewise intriguing. As an example, these results imply that microstructural changes which normally increase internal stress (e.g., solute strengthening, dispersion strengthening) might also increase the useful reversion stress of SME alloys.

REFERENCES

1. C. M. Wayman and K. Shimizu, Met. Sci. J., 6 (1972) 175.

2. J. Perkins, Met. Trans., 4 (1973) 2709.

3. L. Delaey, R. V. Krishnan, H. Tas, and H. Warlimont, J. Mat. Sci., 9 (1974) 1521.

4. J. D. Harrison and D. E. Hodgson, Raychem Corp., Menlo Park, CA, (See paper this symposium).

5. Horace Pops, Essex International, Ft. Wayne, In, (See paper this symposium).

6. Jeff Perkins, G. R. Edwards, C. R. Such, J. M. Johnson, and R. R. Allen, (See paper this symposium).

7. O. D. Sherby and P. M. Burke, Prog. in Mat. Sci., 13, (1968) 325.

8. C. M. Wayman, (See paper this symposium).

9. Z. S. Basinski and J. W. Christian, Acta. Met., 2, (1954) 101.

10. H. K. Birnbaum and T. A. Read, Trans. AIME, 218, (1960) 662.

11. F. T. Aoyagi and K. Sumino, Phys. Stat. Sol., 33, (1969) 317.

12. C. Rodriguez and L. C. Brown, (See paper this symposium).

13. J. Duorak and E. B. Hawbolt, Met. Trans., 6A, (1975) 95.

14. J. D. Livingston, Acta. Met., 10, (1962) 229.

15. W. B. Cross, A. H. Kariotis and F. J. Stimler, NASA CR-1433, Sept 69.

16. D. S. Lieberman, M. A. Schemerling and R. W. Karz, (See paper this symposium).

17. S. Miura, U. Morita and N. Nakaniski, (See paper this symposium).

18. J. M. Johnson, M. S. Thesis, Naval Postgraduate School, Mar 75.

19. G. R. Edwards, J. Perkins and J. M. Johnson, to be published.

INTERNAL FRICTION MEASUREMENTS ON COPPER-ZINC BASED MARTENSITE

W. De Jonghe, R. De Batist [*], L. Delaey and M. De Bonte

Dept. Met. Univ. Leuven and SCK-Mol[*]

Introduction

The alloys discussed in this symposium are not only characterized by their shape-memory effect and pseudoelasticity but they
show also peculiar damping properties. The internal friction as well
as the extremely high capacity for damping contact noise are linked
to the same metallurgical and physical phenomena that determine the
SME-effects. Also, since measurements of internal friction yield
important information on the mechanism of martensite formation, premartensitic phenomena, structural changes in the martensite phase,
more attention will be paid in the present paper therefore to the
internal friction.

Internal friction measurements have already been carried out
on the γ_1'-type martensite in Cu-Al-Ni alloys (1,2,3) and on the Ti-
Ni martensite (4,5). Intense research on Mn-Cu alloys (6-9) has
also resulted in a technically applicable material (INCRAMUTE) (10).
However, no work has been reported until now on the internal friction and damping capacity associated with the β-to-β_1'-type of martensite formation in copper-based alloys.

The damping is usually represented by a δ vs temperature diagram (figure 1), wherein three areas can be differentiated : a low
damping in the β-phase field, a maximum around the transition temperature and a high damping in the martensite phase. The reason
given for this behaviour can be summarized as follows. The internal friction in the β-phase is caused by lattice defects. The
high internal friction in the martensite phase is caused by the
movement of twinboundaries, martensite plate boundaries ...
The maximum which occurs at the transition temperature is explained

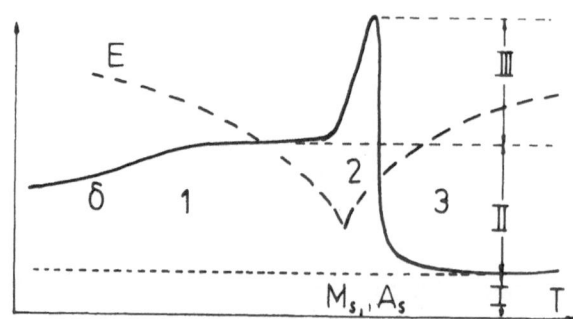

<u>Fig. 1</u> : Schematic diagram of damping vs temperature

by the transformation mechanism itself and is associated with the movement of a β /martensite inter-phase boundary.

Changes in Young's modulus have also been measured and a decrease is observed while approaching the transformation temperature.

The purpose of the present work is to analyse in detail the different phenomena in order to gain more information on the mechanism of martensitic transformation. It is also our aim to contribute to the development of technically useful materials that ought to reduce seriously the emission of contact noise.

Experimental Techniques

1. Apparatus.
The damping measurements, in the present study, can be divided into two categories :
a. Measurements of the internal friction at low amplitudes ("physical measurements"), (i) where $\varepsilon < 10^{-7}$ (ii) where $\varepsilon \sim 10^{-4}$, and
b. Measurements of the audible damping effect : reduction of the sound caused by impact of a hard object ("technological measurements").

Part of the measurements were made using an Elastomat 1015 V (Inst. Dr. Förster). The specimens are rod-shaped (1 = 150 mm, Ø = 10 mm) and can be made to oscillate either by external excitation or by auto-excitation. The exciter- and the receiver-unit can

be either piëzo-electrical (quartz crystal) or magnetic. The oscillations reach a maximum at the natural frequency of the specimen,
or at a higher harmonic. This natural frequency is determined by the
geometry and the dimensions of the specimen, and the nature of the
oscillations (longitudinal, transverse, torsional). In principle,
measurements are possible in the range 800 to 25000 Hz. ε is not
more than 10^{-7}. Difficulties with the measurements were encountered
in the transformation region, particularly when using the piëzo-
electric units. Therefore the magnetic units were tried, by screw-
ing a magnetic part in the heads of the CuZnAl-specimens, which made
the positioning much less critical. The decrement in damping can be
measured either by the method of free decay of the oscillation or by
the half-value method.

     Other measurements were carried out on an apparatus built by
De Batist (11). The specimen is held in the holder as a cantilever,
and the free end is displaced electrostatically by electrodes (order
of magnitude : 10^{-4}). The damping can be determined by measuring
the energy necessary for keeping the vibration stationary. The
samples are very small (10 x 3 x 0,2 mm). The resonant frequency
is determined by Young's modulus and dimensions of the sample, but
lies mostly in the region 300 - 1000 Hz. (This vibration amplitude
is also adjustable and is tuned by F.M.)
     The other series of measurements give information regarding
technological applicability of the material. Measurements were
carried out to determine the sound-level, produced by impact of a
steel hammer or a glass sphere on a plate-shaped material. These
measurements were done with a real-time analyser (analysing time
1/8 sec), that picks up the produced sound, and divides it into
its components (1/3 octaves). From each band, the loudness is
given in dB. Other measurements were performed with a Brüel &
Kjaer - sonometer, provided with an octave band filter.

     2. Specimens
The compositions of the alloys have to lie in a region of the ternary diagram , where the high-temperature phase is β, and where a
martensitic transformation occurs at lower temperatures. The fact
that the alloy is ternary still gives two degrees of freedom, and
thus the possibility to satisfy two requirements e.g.,
- minimum β-temperature (to have the heat-treatment at the lowest
possible temperatures).
- a definite M_s-temperature.
As a first approximation, lines of constant M_s seem to be linear in
a broad composition range, when the compositions are given in atomic
percents (12). Most of the alloys were chosen corresponding to the
intersection of this constant-M_s-line with the line of the β-mini-
mum.

     The measurements discussed in § 1 refer to an alloy with com-
position : Cu , 13.64 at % Zn, 17.06 at % Al and with M_s around 30 °C.

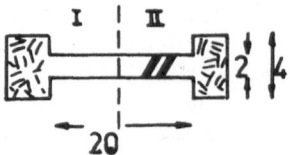

<u>Fig. 2</u> : Sample used for measurements on mono-variant and bi-variant martensite

The alloy made from 99.999 % pure metals was melted and cast under vacuum into a graphite crucible in an induction furnace. The alloy was hot-rolled to 1 mm thickness, and further thinned to 0.2 mm by grinding and electropolishing. Any effects of deformation were removed by a short heat treatment at 700 °C, followed by water-quenching.

For the sample discussed in §.2, a single crystal of β was machined in a spark-cutter, into a tensile sample. This sample was stressed at a temperature   slightly above M_s (≃ 50 °C), to yield a single martensite variant. The sample was then quenched in the deformed state. During grinding another variant was induced. The geometry of the variants is given in fig. 2. This gave us the possibility to measure the internal friction of a mono-variant in region I, and of a bi-variant in region II.

Experimental Results

1. Internal friction and Young's modulus in polycrystalline material.

1.1. Internal friction and Young's modulus as a function of temperature. (fig. 3a)

In the temperature range where the β to martensite transformation occurs, the internal friction behaves in a rather impredictable way. In some occasions, a rather simple damping peak is observed, whereas in other cases only "serrations" of different magnitude occur. Similar observations have been reported by Hasiguti et al. (4) and Sugimoto et al. (9). These serrations were however not reproducible when the sample was cycled through the transformation temperatures. All serrations lay in the region of transformation, where an anomaly in the Young's modulus (natural frequency of vibration) was observed. The Young's modulus followed a smooth curve whose shape and width did not alter appreciably with repeated cycling; the hysteresis (A_s - M_s) however was reduced from 40 °C to 10 °C on repeated cycling.

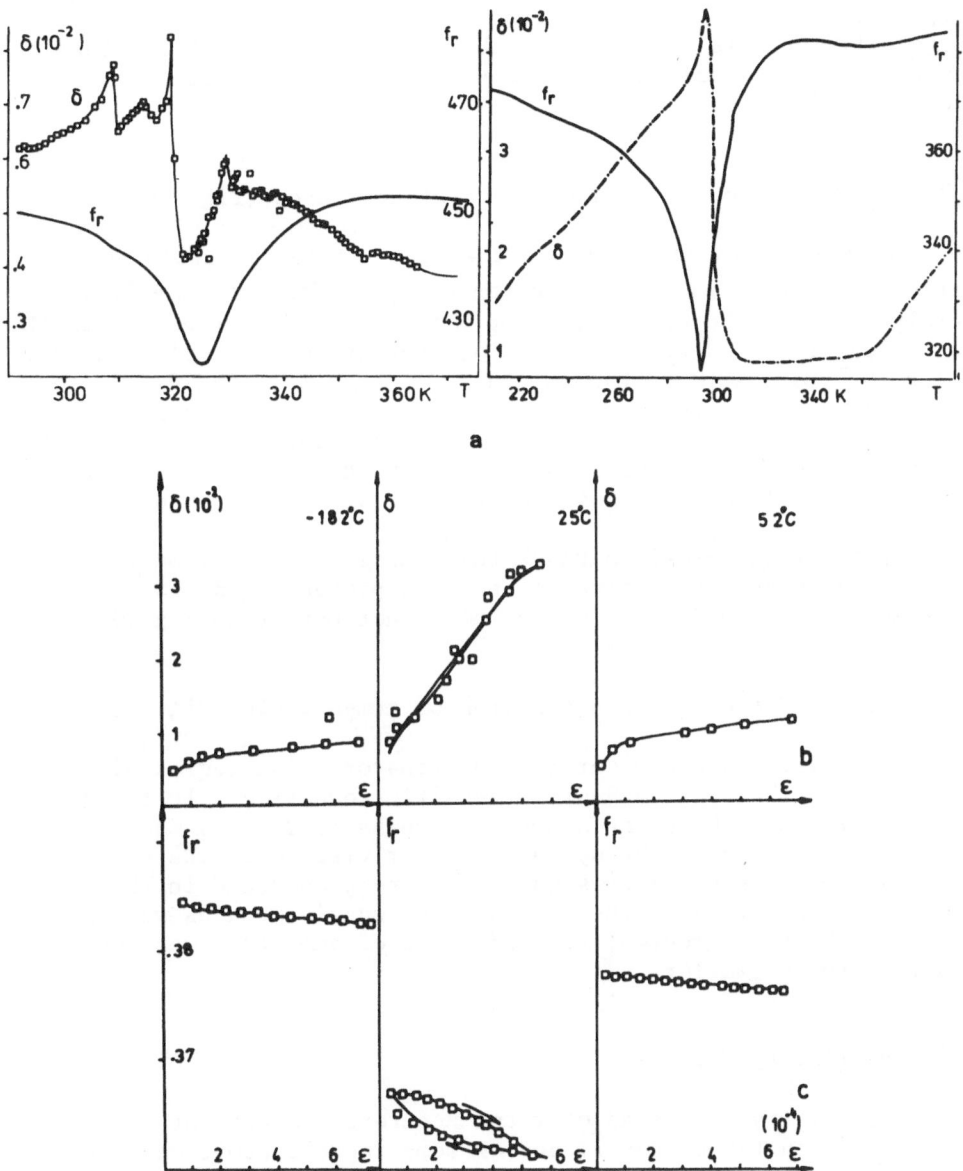

Fig. 3 : Internal friction and Young's modulus in polycrystalline
material as a function of temperature (fig. 3a), measurements on two different samples), internal (fig. 3b) and
natural frequency (fig. 3c) as a function of amplitude;
measurements carried out at three different temperatures).

The internal friction in the martensitic region was not high and was only slightly above that in the β-phase. This is rather surprising. In most other martensite alloys (TiNi, Cu-Al-Ni) a large difference is found, although the magnitude of this effect is dependent on the amplitude. In the present investigation ε was about 3×10^{-4}.

1.2. Internal friction as a function of amplitude (fig. 3b)

It is apparent from fig. 3b that the amplitude exerts a large influence in the transformation range. The effect of amplitude is fairly small in the martensite and β regions. Nevertheless, in these regions it was also found that on increasing the amplitude from 0.2×10^{-4} to 4×10^{-4}, the internal friction is doubled. In some of these measurements, a hysteresis was observed, although the scatter in the results is mostly larger than the width of the hysteresis. The damping level as well as the damping behaviour with respect to amplitude, in β and martensite were not much different.

In the transformation range the sample was partly martensitic (a certain amount of martensitic needles present in β grains). The decrement δ reached a maximum value of not more than 3×10^{-2}.

1.3. Natural frequency as a function of temperature (fig. 3c)

Here also the behaviour in the transformation region differed greatly from that in the β or martensitic condition. In the martensitic region, there was a small decrease in Young's modulus (0,4 %) and also a very slight hysteresis (not visible on the scale used in fig. 3). This hysteresis was much more pronounced in the transformation region. Also the decrease in Young's modulus was greater (1,7 %). In the β-region, the influence of amplitude was again observed to be smaller.

1.4. Calorimetry (fig. 4)

Calorimetry yields an easy and accurate measurement of the transformation temperatures, and of some thermodynamic data (e.g. transformation enthalpy). Only preliminary results will be reported here and discussion will be restricted to the asymmetry of the curve (fig. 4). Independent of the direction in which the temperature changes (heating or cooling) the slope on the martensitic side decreased more slowly than the slope on the β-side, indicating that some changes occur inside the specimen in the region 20-30° below M_s.

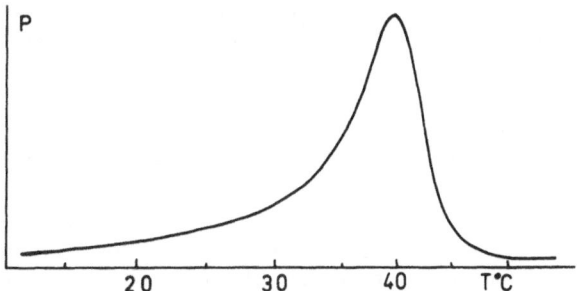

Fig. 4 : Calorimetric curve representing heat transfer during martensitic transformation.

2. Influence of a martensite-plate boundary

2.1. : Measurements on a martensite mono-variant.

    The plot of internal friction as a function of temperature
(fig. 5) is to be compared with these of polycrystals (fig. 3a).
There are significant differences, as well in the internal friction
as in the Young's modulus. The damping peak is much higher (about
25 times) and much narrower than for the polycrystal. Moreover,
it is very smooth, a fact which was never observed in polycrystals.
The differences in the Young's modulus are in the same way : nar-
rowing of the transition region, and more abrupt changes. Another
interesting observation is, that the Young's modulus of the β single
crystal is twice as large as that for the martensite (Young's modulus
= $k.f_r^2$).

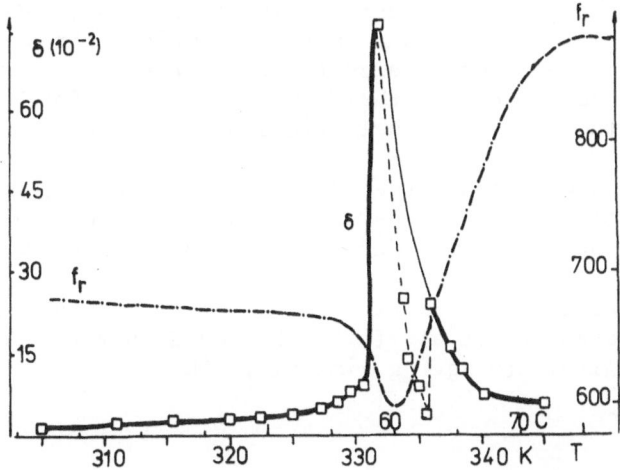

Fig. 5 : Internal friction as a function of temperature for martensite single crystal

It was impossible to go to higher temperatures since due to the shape memory behaviour of the sample, it no longer fitted between the electrodes. This difficulty also was responsible for the temperature stabilisation around A_s. When heating was interrupted, the internal friction fell drastically from 0.76 to 0.03. It rose again, when heating was continued, and seemed to rejoin the original damping peak. There were no anomalies in the Young's modulus.

The behaviour of internal friction as a function of amplitude (fig. 6) was also examined, prior to heating the specimen. The influence of the amplitude was very marked in the first cycle (after mounting), when the damping decrement dropped to 1/3 of its original value. In the following cycles, the effect decreased (cycles 2 ... 5). After the sample had rested for a night, the amplitude dependence practically vanished (cycles 6 - 7), and the damping was found to lie at a lower level.

2.2. : Measurements on a martensite bi-variant

Only the dependence of the decrement vs amplitude was investigated (fig. 7), and it showed a behaviour analogous to fig. 6, (although the over-all level was lower than for the mono-variant). Cycle 3 was measured after a rest of 2 h.

3. Low-amplitude fatigue effect and instabilities.

3.1. : Low-amplitude fatigue effect.

When a martensite specimen is vibrating with a constant amplitude ($\epsilon < 10^{-7}$), it was found that the damping increased gradually (fig. 8). This increment seemed to saturate after some time. Other measurements indicated also that, after vibrating for a certain time at high amplitude, the internal friction was higher.

When the specimen had rested overnight, the internal friction fell to a lower level. The damping then increased again as a function of the vibrating time, and in a more pronounced way.

The lowering in the damping after a rest period has also been observed by Nakanishi et al. (13), who found that the damping decreased gradually with ageing time at 22 °C.

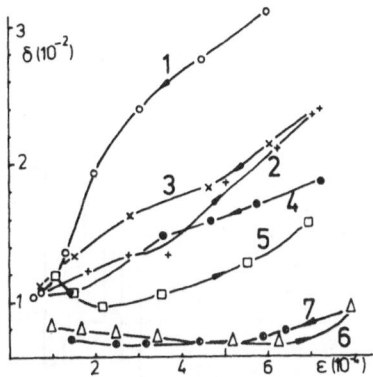

Fig. 6 : Amplitude dependence of internal friction for a martensite
         single crystal.

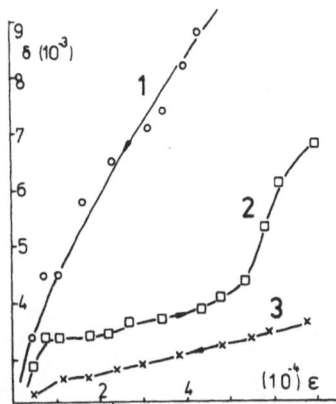

Fig. 7 : Amplitude dependence of internal friction for a martensite
         bi-crystal.

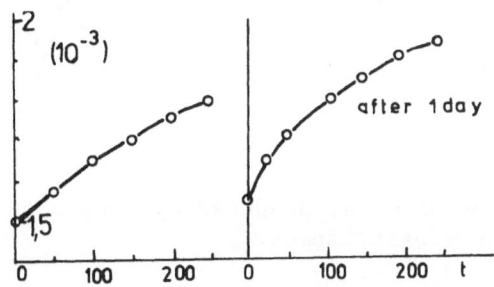

Fig. 8 : Internal friction vs vibration time for a martensitic
         sample.

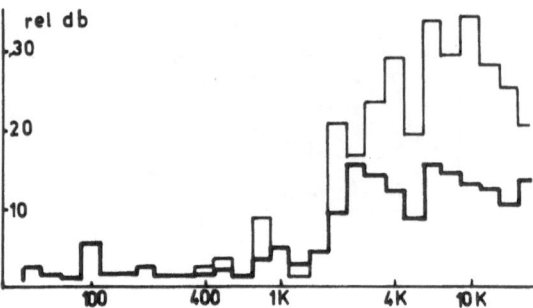

Fig. 9 : Noise produced by low frequency (3 Hz) impact of a steel
         hammer on plate shaped material.

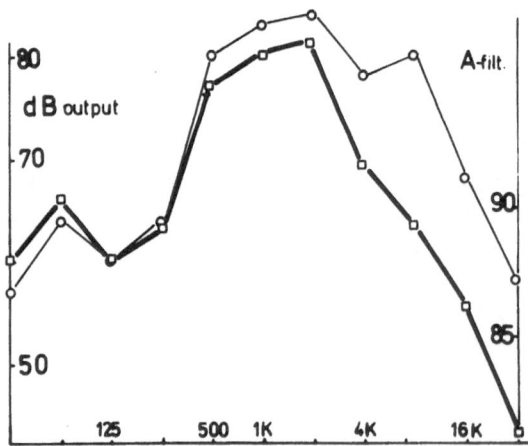

Fig. 10: Comparison of noise produced by damped and undamped ma-
         terial on single impact.

3.2. : Instabilities

Strange effects were observed, while measuring in the transformation region at constant temperature (near room temperature).
After maintaining a constant amplitude of vibration for some time,
(ab.5 min), the amplitude began to oscillate itself, with a certain
frequency, and it became impossible to stabilise the vibration at
the same temperature. Stabilisation however returned on increasing
the temperature a few degrees. But at this (higher) temperature
also, the instability reappeared after a certain time.

This effect is not due to the experimental set-up, but is related with changes in the specimen.

4. Measurements on contact-noise reduction by CuZnAl-martensite.

4.1. Measurements with real-time analyser (fig. 9)

The measured sound was produced by the impact of a steel hammer on a bar (10 x 2 x 0.5 cm). Two non-damping brass samples were
used in these experiments. One was covered with a layer of damping
CuZnAl -martensite.

The sound, produced by impact lay mostly in the frequency region above 1000 Hz. It was observed that the sound reduction
reached 20 dB for some (third-octave) bands.

4.2. : Measurements with an impact sonometer (fig. 10).

In the present experiments, the sound was produced by the impact of a glass ball on a plate (10 x 10 x 0.1 cm). Measurements
were carried out on different CuZnAl-alloys (different composition
and thermo-mechanical treatment), in a "dead chamber" (acoustically
isolated from the environment), and were highly reproducible (within a few dB).

The alloys, which exhibit a "dull sound" confirmed this behaviour during the test. Although there was not much difference in
the A-weighted response (3...5dB) a marked sound-reduction in the
higher frequencies was observed. This confirms the results of (4.1.).

Discussion and Conclusions

From the preceding results it seems that the hypothesis, as proposed in the introduction, is not sufficient to explain all the phenomena. Therefore, it may be interesting to discuss the data in the three different regions (i.e. β, β to martensite and martensite).

1. Internal Friction in the β-region

The internal friction in the β-region does not pose any special problems : the values of the logarithmic decrement reach a normal value, comparable to other similar alloys. There is a slight amplitude dependence of the internal friction and the Young's modulus. A plausible explanation for this can be given by the dislocation dynamics (background-damping). This hypothesis will be tested on alloys with a very low transformation temperature by measurements at low temperatures where the dislocation-influence is less.

The problems lie at the extremities. In some measurements, there were indications that the internal friction was beginning to rise again at higher temperatures (150 - 200 °C). This could be an indication of the formation of Heusler-type ordering, or it can be caused by the onset of a bainitic transformation which becomes important for temperatures above 200 °C.

Further very accurate measurements have to be carried out at temperatures slightly above the transformation temperature in order to collect data on pre-martensitic phenomena. The observed instabilities, close to the transition temperature, might be an indication that premartensitic phenomena interfere with the measurements or the applied constraints.

2. Internal friction in the transformation region

It is very difficult to obtain accurate results in the transformation region (i.e. : at the temperatures where β and martensite coexist), since the damping and resonant frequency vary rapidly with temperature and dimensional changes make stabilisation difficult. This explains the rather substantial amount of scatter in the results.

In polycrystals, mostly different irregular peaks are observed in the internal friction curve, while the Young's modulus shows a very smooth behaviour. These narrow peaks are not reproducible on thermal cycling. On the other hand, the curve of the Young's modulus is fairly reproducible, although it shows a shift with cycling. The fact that the curve of the Young's modulus is smooth, indicates that the

serrations are not due to instabilities. However, this irregular
behaviour of the internal friction is not a general rule. Sometimes,
more regular - but still asymmetric - peaks are observed.

In a single crystal, the peak is much narrower and also much
higher and shows no serrations. (It is also worthwhile to mention
that it reaches the damping values of plastics).

The different widths of the internal friction in the transformation region can perhaps be explained when we take into account
the role of internal stresses on the martensitic and reverse transformation.   In the single crystal, no constraints are imposed on the
transformations, and the material can transform very easily, in a
narrow temperature-interval. In polycrystals, internal stresses
exist, that favour certain variants, and also limits on growth are
imposed by the grain boundaries. Therefore, the martensitic or reverse transformation does not start at the same time in different
places, while some variants appear or disappear earlier, others are
inhibited for a longer time.

The different heights can be explained by taking into account the transformation rate. In other reports (14,15) it was mentioned that the transformation rate is a very important factor, which
determines the magnitude of δ. Belko et al. (15) give a quantitative
formula for CoNi :

$$Q^{-1} = \left(\frac{G.\beta.a^2}{\omega.k.T} \right) \frac{\delta m}{\delta t}$$

wherein G : shear-modulus
β,a parameters, related with the transformation
ω vibration frequency
k   Boltzmann's constant
T   absolute temperature
$\left(\frac{\delta m}{\delta t}\right)$ relative volume of material which undergoes phase transformation in a unit of time.

Although ω and   T vary also, the greatest influence comes
from $\dot{m}$. In a polycrystalline material, the material that transforms
in a certain time, at the same heating rate, is less than in the
single crystal. Therefore the height of the damping peak is much
lower. Perhaps this can also explain the serrations in the internal
friction curve : when a certain amount of material transforms, the
damping rises.

It also explains the drastic fall in the internal friction,
when temperature is stabilised :

$\frac{\delta m}{\delta t} = \frac{\delta m}{\delta T} \cdot \frac{\delta T}{\delta t}$, so when $\frac{\delta T}{\delta t} \rightarrow 0$, δ becomes very low.

This very strong dependence of damping on the amount of transforming material suggests that the internal friction is caused by the transformation martensite $\rightleftarrows$ β. It is premature to decide what mechanism is exactly responsible for damping. It could be the movement of the phase-boundary, the jumps of atoms to other positions, or any other mechanism involved in the transformation.

The shift in the Young's modulus with temperature cycling can be related to the phenomena, described by Rapacioli et al. (16), who found a shift in transformation temperature, depending on the thermal hystory.

3. Internal friction in the martensitic region.

The internal friction in the martensitic region is somewhat higher than in the β-phase. In some measurements, this difference is appreciable, in other ones it is rather small. But, in every case, this damping decrement is not high enough to explain the dull sound, produced by striking the metal. These are clearly two separate problems.

3.1. Relation to structural and microstructural features

The proposed mechanism for the internal friction up to now is the movement of martensite-plate boundaries and /or the movement of (macroscopic) twin boundaries (accommodation of the external strain). Taking into account the results of §2, this hypothesis can not hold any longer. The internal friction in a martensite monovariant is as high, or higher, than in polycrystalline material. Therefore, the reason for the higher level has to lie in the internal structure of the martensite. These mechanisms have to take into account, besides the higher damping, the asymmetric behaviour of the calorimetric curve. The structure and microstructure of the martensite phase itself has now to be analysed in order to find possible mechanisms which may explain the high internal friction. The microstructure of the martensite generally consists of a mixture of two different stacking sequences (17) in the form of thin lamellae. The appearance of the two stacking sequences is didacted by the invariant plane strain condition. However as soon as the β-phase is completely transformed, small rearrangements in stacking can take place. It has been proved that the quench induced stacking sequences do not correspond to the lowest free energy; upon heating the martensite, rearrangements in stacking have been observed (18). The partial dislocations, which have been observed in abundance in martensite, do move already while irradiated by the electron beam (19) and are held by stair-rod dislocations. Movements of partial dislocations will locally alter the stacking sequence. It ought also to be analysed in how far the movement of lamellar boundaries

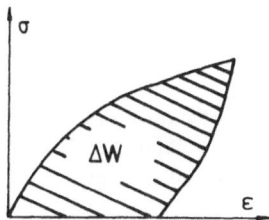

<u>Fig. 11</u> : Stress-strain curve in the martensitic region.

and dislocations, changes the state of long range order. It can be
concluded that the martensite phase does contain structural and
microstructural features, whose movements are associated with changes
in free energy.

3.2. : Damping of contact-noise produced by impact

       Up to now, no direct measurements have been made of the re-
duction in contact-noise produced by impact of a hard object. This
is, however, an important parameter. It gives a quantitative description of what we hear as "a dull sound", and it is an indication
of the usefulness of these materials for technological applications.

       Therefore, measurements have been made of the sound-reduction
(in dB), for the various frequency components. It is very difficult
to compare this sound-reduction with a certain logarithmic decrement.
It is even more difficult to attach a strain-parameter to the impact
(locally very high, but in a very limited region).

       Although comparisons are very difficult, it seems that the
level and the causes for the internal friction, at amplitudes in the
region of 10^{-4}, can not explain the high sound-reduction. There -
fore, a better explanation may be found in pseudo-elastic curves
(fig. 11). The logarithmic decrement for these cases is

$$\delta = 1/2 \ \frac{\Delta W}{W}$$

and reaches values of 0.5 and higher.

       It is thought that the damping behaviour in these cases can be
explained by accommodation in the martensite. It is not yet clear
at what amplitudes there is a transition of the mechanism. Therefore,
measurements will be carried out at higher amplitudes.

Acknowledgments

The authors want to acknowledge L. Eersels (SCK-Mol) for experimental
help, Dr. Rapacioli and Dr. Chandrasekaran for the single crystal and
valuable discussions and Prof. H. Mincke (dept. of physics, KULeuven)
for research facilities. Ir. W. De Jonghe thanks IWONL for a study
grant. This work has received the finantial support of INCRA, for
which the authors are grateful.

References

1. V.A.Teplov, V.A. Pavlov and K.A. Malyshev, Fiz. Metal. Metalloved.,
   1969, vol. 27, n° 2, p. 339.
2. I.A. Arbuzova, V.S. Gavrilyuk and L.G. Khandros, Fiz. Metal.
   Metalloved., 1969, vol. 27, n° 6, p. 1126.
3. K. Sugimoto, T. Mori, K. Otsuka and K. Shimizu, Scripta Met.,
   1974, vol. 8, p. 1341.
4. R.R. Hasiguti and K. Iwasaki, Jl. Appl. Phys., 1968, vol. 39,
   p. 2182.
5. N.G. Pace and G.A. Saunders, Phil. Mag., 1970, vol. 22, n° 175,
   p. 73.
6. D. Birchon, D.E. Bromley and D. Healey, Met. Sc. Journal, 1968,
   vol. 2, p. 41.
7. R.J. Goodwin, Met. Sc. Journal, 1968, vol. 2, p. 121.
8. J.A. Hedley, Met. Sc. Journal, 1968, vol. 2, p. 129.
9. K. Sugimoto, T. Mori, S. Shiode, Met. Sc. Journal, 1973, vol. 7,
   p. 103.
10. Incra-publications about Incramute.
11. Claesen, Wijsmans, R. De Batist, Dissertation.
12. M. Ahlers, Scripta Met., 1974, vol. 8, p. 213.
13. N. Nakanishi, T. Mori, S. Miura, Y. Murakami, S. Kachi, Phil. Mag.,
    1973, vol. 28, n° 2, p. 277.
14. P.F. Gobin, Suppl. au Journal de Physique, tôme 32, fasc. 7,
    p. 62-65.
15. V.N. Belko, B.M. Darinskiy, V.S. Postnikov, I.M. Sharshakov, Fiz.
    Metal. Metalloved., 1969, vol. 27, n° 1, p. 141.
16. R. Rapacioli, M. Chandrasekaran, L. Delaey, Proc. of Int. Symp.on
    SME, this volume.
17. H. Warlimont and L. Delaey, Progr. in Mat. Science, 1974, vol. 18.
18. I. Lefever and L. Delaey, Acta Met. 1972, vol. 20, p. 797, and
    Metall, 1973, vol. 27, p. 1085.
19. Z. Nishiyama and S. Kajiwara, Trans. Jap. Inst. of Metals, 1962,
    vol. 3, p. 127.

A QUANTITATIVE TREATMENT OF THE LATTICE SOFTENING OF

SHAPE MEMORY ALLOYS

N. Rusović and H. Warlimont[+]

Max-Planck-Institut für Metallforschung, 7 Stuttgart 1,

West-Germany;[+] now at: Swiss Aluminium Ltd., Research and

Development, 8212 Neuhausen, Switzerland

The shape memory effect is most commonly observed in β phase alloys which transform martensitically and exhibit the so-called C' anomaly (1,2). This anomalous elastic behaviour is characterized by the following properties of the elastic shear constant C' = $(C_{11}-C_{12})/2$:

(i)   its temperature coefficient becomes positive on approaching the equilibrium temperature T_o from above;

(ii)   its magnitude decreases with changes in composition which raise T_o;

(iii)   its magnitude is of the order of o.1 $\stackrel{<}{\sim}$ C' $\stackrel{<}{\sim}$ o.7 x 10^{11} dynes cm^{-2} (10^{10} Nm^{-2}), i.e. unusually low, at M_s;

(iv)   consequently, the elastic anisotropy factor A = C_{44}/C' is anomalously high, since C_{44} shows normal behaviour.

In other words the lattice becomes increasingly softer against {11o}<11o> shears as the temperature or composition approaches the instability with respect to a martensitic transformation. Fig. 1 shows data of this behaviour for a β_2-NiAl alloy series which will be treated as an example in this paper.

The low absolute value of C' in the transformation range gives rise to a low free energy of nucleation and favourable energetic conditions for comparatively perfect growth which are the bases of SME martensites with a high fraction of recoverable shape change (4).

Many of the β phases possess a CsCl structure. Since the repul-

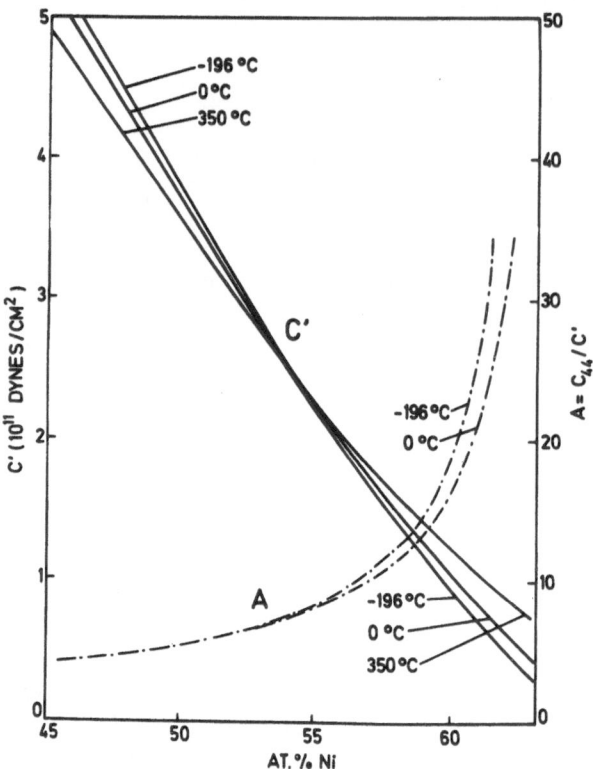

<u>Fig. 1</u> The C' elastic constant and the elastic anisotropy factor A of
β₂-NiAl as a function of Ni concentration at different temperatures (3).

sive forces between the closed ion shells yield a negative contribution to C' in this structure, Zener (5) has attributed the low
values of C' of the β phases to this contribution. Some calculations
were carried out by Nakanishi (6) on the basis of relations derived
by Jones (7) and Leigh (7) but the results did not fully account for
Zener's assumption. In particular, the anomalous temperature dependence of C' could not be derived.

In this paper a quantitative analysis of the elastic constants
of β₂-NiAl is given based on an experimental determination over a
considerable range of temperatures and compositions (3). Even though
the β₂ phase field becomes narrow with decreasing temperature, single
crystals containing 45 to 6o at.% Ni could be obtained which were
essentially single phase and were not transformed martensitically
at room temperautre. The β₂ phase is ordered almost perfectly up to
the melting point. At deviations from the stoichiometric composition
an excess Al atom gives rise to two vacancies on the Ni sublattice
whereas an excess Ni atom will occupy a site on the Al lattice as

an antistructure atom. Since a lattice such as that of β_2-NiAl below
5o at.% Ni containing a high concentration of structural point defects
requires a particular treatment, the present analysis is restricted
to the 5o-6o at.% Ni interval. For alloys in this concentration range
we arrive at the conclusion that the low values of C' and, in particular, their temperature and concentration dependence are predominantly due to the forces arising from the ion core overlap. This
implies that the lattice softening of β_2-NiAl - and probably of
other 3/2 electron compounds - follows from thermal expansion only.
If that is so, ordinary thermal expansion should lead to the same
effects for any CsCl structure if it is associated with the electron
structure specified below and if the ionic radii of the components
are suitably related.

DISPOSITION OF THE CALCULATIONS

We have adopted the usual summation

$$E = E_E + E_F + E_R \tag{1}$$

where E is the total lattice energy, E_E the electrostatic (Ewald)
term, E_F the Fermi energy and E_R the ion repulsion term arising
from the repulsion of the closed ion shells. Since the shear constants $C = C_{44}$ and $C' = (C_{11}-C_{12})/2$ are of central interest here
the calculations were restricted to these two shear elastic constants.

Ewald term. The electrostatic attraction between the positive
ions and the negative electron gas can be computed according to
Ewald (9). The resulting contribution of the electrostatic potential energy to the elastic shear constants of a bcc lattice is
(in 10^{12} dynes cm^{-2}):

$$C_{44}^E = 17.12 \cdot Z_{eff}^2 \cdot a^{-4}, \tag{2a}$$

$$C'^E = 2.3o \cdot Z_{eff}^2 \cdot a^{-4}, \tag{2b}$$

where a is the lattice parameter and Z_{eff} is the effective charge
of the ions. Since 3 electrons of the Al enter the conduction band
in β_2-NiAl and since Ni may be regarded to contribute zero charge,
Z_{eff} can be approximated as

$$Z_{eff} = \begin{cases} 3(1-x) & \text{for } x \geq 0.5 \\ 1.5 & \text{for } x < 0.5 \end{cases} \tag{3a}\tag{3b}$$

where x is the atom fraction of nickel.

Fermi term. Assuming that the Fermi surface touches the {11o} plane of the first Brillouin zone and that no electron states outside that {11o} plane are occupied, Jones (7) has used the following relation

$$E_F = k^2 - k_z^2 + p^2 \cdot f(z) \tag{4}$$

where $p = (2/a)\pi$ is the distance of the {11o} plane of the Brillouin zone from the origin of k space. The coordinates in k space are chosen such that k_z coincides with the normal to one pair of {11o} boundaries; $f(z)$ is a dimensionless function of $z=k_z/p$ and was defined by Jones (7) as

$$f(z) = z^2 - \lambda z^{2/\lambda}. \tag{5}$$

λ is a parameter smaller than unity. In addition, it conforms to some natural restrictions. Since the stabilizing Ewald plus Fermi contribution must not be negative throughout the range of β phase and since, in our case, the Fermi contribution reaches its maximum at approximately $\lambda = 0.5$, the applicable interval of variation of λ is: $0.2 \leq \lambda \leq 0.5$. It is easy to see why the function was chosen in this form; $f(z) \to z^2$ for $z \to 0$ ensures that the energy is a square function of the wave vector near the origin and $f'(1)= 0$ yields the flattening at the Brillouin zone boundary. $f(1)=1 - \lambda$ shows the meaning of λ; from $E_F = k^2 - \lambda p^2$ it may be seen that λp^2 is the amount by which the energy at the Brillouin zone boundary is less than that of the free electrons gas. In equs. (4) and (5) distances are measured in units of the Bohr radius $a_H = o.529 \cdot 10^{-8}$ cm and energies in Rydberg units (Ry) = $21.79 \cdot 10^{12}$ ergs. The contributions to the shear elastic constants are

$$\left(\frac{1}{K}\right) C_{44}^F = -\frac{1}{4}\nu^2 + \frac{4(1-\lambda)}{3(2+\lambda)}\left[(2+3\lambda) - \frac{7(1-\lambda)(8+38\lambda+27\lambda^2)}{3(2+\lambda)(2+3\lambda)(4+\lambda)}\right] \tag{6a}$$

$$\left(\frac{1}{K}\right) C'^{F} = -\frac{5}{8}\nu^2 + \frac{2(1-\lambda)}{3(2+\lambda)}\left[(2+3\lambda) - \frac{11(1-\lambda)(8+38\lambda+27\lambda^2)}{3(2+\lambda)(2+3\lambda)(4+\lambda)}\right] \tag{6b}$$

where $K = \dfrac{p^5}{2\pi^2}\left(\dfrac{Ry}{a_H^3}\right)$, $\nu = \dfrac{2\pi^2}{3}$ and n the number of free electrons

per cell, i.e. $n = 3(1-x)$ for $x \geq o.5$.

The relative contribution of the Fermi energy to C' is considerably lower than to C_{44}. This may be seen from the selected calculated and measured data for room temperature listed in table 1 and from the curves plotted in Fig. 2. The Ewald plus Fermi energy contributions to C' vary in a manner which is qualitatively similar to the concentration dependence of C', fig. 2a.

Table 1. Selected data showing the contributions of the stabili-
sing Ewald and Fermi energy terms to the shear elastic
constants C' and C_{44} of β_2-NiAl (in units of 10^{11} dynes
cm^{-2}) at room temperature; calculated for λ = o.4

composition	C'^F	$C'^E + C'^F$	C'	C_{44}^F	$C_{44}^E + C_{44}^F$	C_{44}
at.% Ni	calc.	calc.	meas.	calc.	calc.	meas.
5o	3.63	3.75	3.56	5.94	11.4o	11.3
6o	2.63	3.12	o.86	3.41	7.o6	12.o

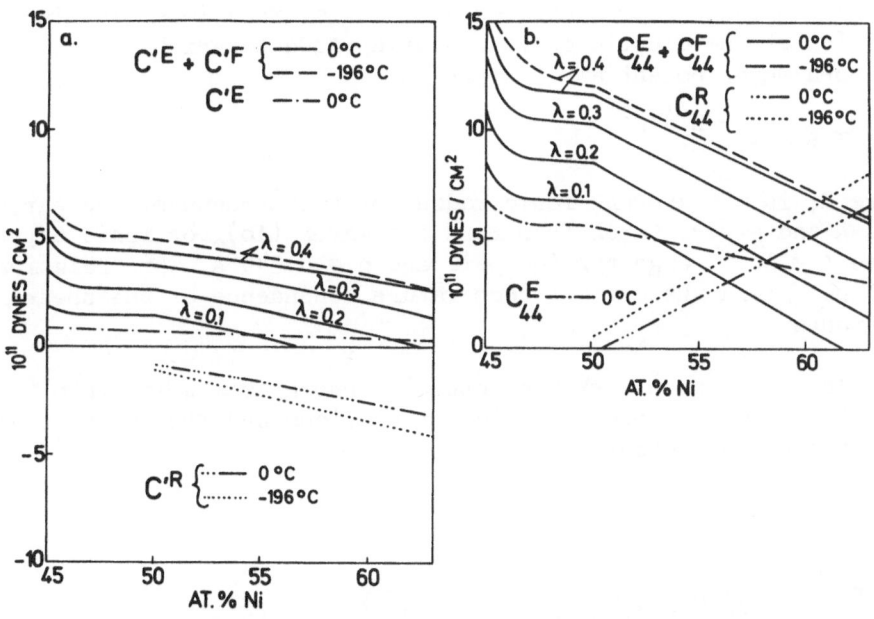

Fig. 2. The Ewald, Fermi and repulsive energy contributions to the
shear elastic constants C' (a.) and C_{44} (b.) of β_2-NiAl
according to the present calculations.

Repulsion term. The repulsive non-coulombic overlap forces between the closed ion shells may be approximated as central forces such that $E_R = w(r)$ is solely a function of interatomic distance. The resulting contributions to the shear elastic constants of the CsCl structure are listed in table 2, where $\Omega = a^3/2 = $ atomic volume, $r_1 = a\sqrt{3}/2$ and $r_2 = a$.

Table 2. Repulsive energy contribution to the shear elastic constants of a CsCl structure.

	nearest neighbours	next nearest neighbours
c'^R	$\frac{1}{\Omega}\ \frac{8}{3}\ r_1 w'(r_1)$	$\frac{1}{\Omega}\left[r_2^2 w''(r_2) + r_2 w'(r_2)\right]$
C_{44}^R	$\frac{1}{\Omega}\left[\frac{8}{9}\ r_1^2 w''(r_1) + \frac{16}{9}\ r_1 w'(r_1)\right]$	$\frac{1}{\Omega}\ 2r_2 w'(r_2)$

The overlap forces are high at short distances and decrease rapidly with increasing distance ("very hard spheres"). Therefore, they are usually represented as exponential functions. In his calculation of the elastic constants of β_2-AuCd Nakanishi (6) employed the Born-Mayer potential

$$w(r) = A\,e^{-(r-2R_i)/\rho} \tag{7}$$

where R_i is the average ionic radius of the structure. The parameters A and ρ were taken from Born and Mayer (1o)$_o$for ionic crystals where $A = 1o^{-12}$ ergs per ion pair and $\rho = o.345$ Å. This relation will not yield the correct temperature dependence of the shear constants.

In the present work the repulsive energy was split into terms resulting from different atomic interactions and from nearest and next nearest neighbours:

$$E_R = \sum_{i,j\geq i,k} f_{ij}^{(k)}\ w_{ij}^{(k)} \tag{8}$$

where

$i,j = 1,2$ $\begin{cases}1 = \text{atom species A}\\ 2 = \text{atom species B}\end{cases}$

$k = 1,2$ $\begin{cases}1 = \text{nearest neighbours}\\ 2 = \text{next nearest neighbours}\end{cases}$

and

$$w_{i,j}^{(k)} = A_{i,j}^{(k)} \, e^{\,(R_i + R_j - r_k)/\rho_{i,j}^{(k)}} = A_{i,j}^{(k)} \, e^{\,-d_{i,j}^{(k)}/\rho_{i,j}^{(k)}}$$

R_i and R_j are the ionic core radii and $d_{i,j}^{(k)} = r_k - R_i - R_j$ is the distance between the ion cores if located on nearest and next nearest lattice sites; $f_{ij}^{(k)}$ are the numbers of the different atomic pairs in an average unit cell, which depend on the concentration and degree of order. Relation (8) permits to take different pairwise interactions of nearest and next nearest neighbours into account such that the repulsive energy contribution of a binary alloy depends on 12 parameters. Fitting the potentials to the experimental data, i.e. the determination of the 12 parameters, is a considerable mathematical problem. If ρ = o.345 Å = const. and A = 10^{-12} ergs per ion pair = const. are used in a first approximation it is found immediately that the temperature dependence of the two shear elastic constants is obtained in close agreement with the dependence observed experimentally whereas the absolute values are incorrect. The lattice shows the normal increase in stiffness with decreasing temperature at the stoichiometric composition; this stiffening decreases with increasing nickel concentration until it turns into lattice softening at 53 at.% Ni.

Since in equ. (8) individual pair potentials are taken into account, the interaction between Ni ions (R_{Ni}(o) = 1.24 Å; R_{Al}(3+) = o.53 Å) and its sensitivity to changes in spacing (derivatives) is more pronounced than between Ni-Al ion pairs. With increasing Ni concentration the average number of Ni-Ni pairs will increase and, thus, the contribution which is sensitive to spacing and, consequently, to temperature.

If the parameters are actually to be fitted to the experimental data the following procedure may be employed. The expressions for C' and C_{44} are determined for 6 different concentrations at constant temperature and are set equal to the corresponding experimental measurements. This yields a system of 12x12 transcendental equations. Since the Jacobian matrix for this system may be set up comparatively easily, the Newton iteration procedure may be employed.

RESULTS

In this work the Ewald and Fermi energy terms were adopted from previous calculations (6) as discussed above. We have further assumed that the temperature dependence of these terms is solely

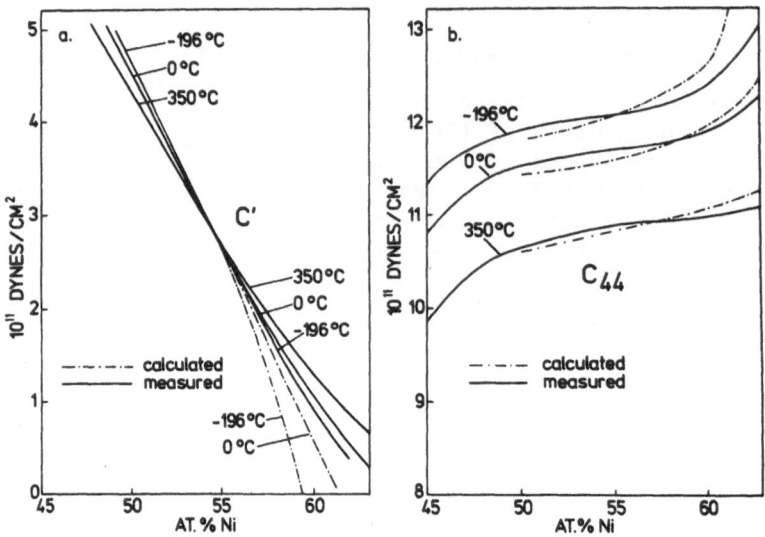

<u>Fig. 3</u> A comparison of measured and calculated shear elastic
        constants C' (a.) and C_{44} (b.) of β_2-NiAl.

<u>Table 3</u>. Values of $A_{i,j}^{(k)}$ [eV] and $\rho_{i,j}^{(k)}$ [Å] according to the present
        calculations:

atom pair	nearest neighbours		next nearest neighbours	
	A	ρ	A	ρ
Ni-Ni	o.o1275	o.11	o.o12	o.15
Ni-Al	o.2885	o.7	o.28	o.7
Al-Al	–	–	o.1	o.7

given by the temperature dependence of the lattice parameter. The results are shown in fig. 2.

Taking these results into account the parameters of the repulsive energy term were fitted to the experimental data of the concentration dependence at constant temperature. Subsequently, the thermal expansion was taken into account which yielded the calculated curves of the shear elastic constants shown in fig. 3. Since the repulsive term gives a positive contribution to C_{44} and a negative contribution to C', relations 8 can reproduce the correct results for both the concentration and the temperature dependence. It should be noted that the calculated elastic behaviour of the β_2-NiAl lattice yields on even more pronounced lattice softening with increasing Ni concentration than actually measured, unless the Born-Mayer parameters would be allowed to vary to an unreasonable extent between nearest and next nearest neighbours. The results presented here are obtained from the set of values listed in table 3. The deviations of the calculated shear constants from the measured ones are probably arising from the onset of decomposition processes in the β_2 phase, invisible in the light microscope, either due to an insufficient rate of cooling of the single crystals after growing or due to premartensitic transformations in analogy to those found other β phase alloys (12).

DISCUSSION

Although the essential contribution of the ion repulsive energy term to the lattice softening of β phases had been pointed out by Zener (5) and has been discussed in relation to martensitic transformations by Nakanishi (2,6) and more recently by Murakami and Kachi (11), only qualitative treatments have been given so far. They do yield the correct concentration dependence of M_S, i.e. an increase of M_S as C' decreases (with the possible exception of AuZn (11) and a sensitivity of this relation to solutes with different ionic radii. But the earlier treatments carried out without splitting the pair-wise interaction into different potentials could not be adapted to the concentration dependence quantitatively nor could they account for the temperature dependence even qualitatively. It should, also, be noted that, neglecting the next nearest neighbours interactions in our calculations, only a few percent of the lattice softening effect are yielded by the calculation. Variations in the attractive potentials beyond those expressed in the relations used in this work appear to be negligible at the present level of accuracy in both experimental and theoretical determination. However, the effects of premartensitic transformations require further study.

It should be noted that the onset of the martensitic transformations is not associated with a vanishing value of C' but rather with a finite, though small value. This indicates that it is the relative stability of two structures which determines the structure formed and the transformation temperature. The present analysis combined with a limited number of data on elastic constants and thermal expansion can be used to predict the temperature and concentration dependence of the shear elastic constants and thus the potential of an alloy series to exhibit the shape memory effect.

REFERENCES

1) L. Delaey, R.V. Krishnan, H. Tas, H. Warlimont,
   J. Mater. Sci. $\underline{9}$ (1974) 1521
2) N. Nakanishi, Mem. Konan Univ.,
   Sci. Ser. no 15, art. 77 (1972), and this Symposium
3) N. Rusović, H. Warlimont, to be published
4) L. Delaey, H. Warlimont, this Symposium
5) C. Zener, Phys. Rev. $\underline{71}$ (1947) 846
6) N. Nakanishi, Trans. Japan Inst. Met. $\underline{6}$ (1965) 222
7) H. Jones, Phil. Mag. $\underline{43}$ (1952) 105
8) R.S. Leigh, Phil. Mag. $\underline{42}$ (1951) 139
9) P.P. Ewald, Ann. Phys. $\underline{64}$ (1921) 253
10) M. Born, M.G. Mayer, Z. Physik $\underline{75}$ (1932) 1
11) Y. Murakami, S. Kachi, Japan J. Appl. Phys. $\underline{13}$ (1974) 1728
12) L. Delaey, A.J. Perkins, T.B. Massalski,
    J. Mater. Sci. $\underline{7}$ (1972) 1197

THE RELATIONSHIP BETWEEN STACKING FAULT ENERGY AND SHAPE MEMORY IN PRIMARY SOLID SOLUTIONS

G.B. Brook*, R.F. Iles[+] and P.L. Brooks[+]

*Fulmer Research Institute, Stoke Poges, Bucks., England

[+]Raychem Corporation Inc., Menlo Park, California, U.S.A.

1. INTRODUCTION

The present activity in research into shape memory and related effects stems from the work of Buehler and his associates on TiNi and similar intermetallic compounds.[1] This has been followed by detailed investigations of the noble metal b.c.c. beta phases which undergo a martensitic transformation to a faulted or twinned martensite on cooling. This has been reviewed thoroughly[2,3] to which reviews reference should be made for further details.

Less attention has been paid to shape memory in primary solid solutions even though it has been shown to occur as a consequence of the martensitic transformations in beta titanium alloys[4], uranium-niobium[5], uranium-molybdenum[6] and uranium-rhenium alloys[6]. In all these systems, the high temperature structure is b.c.c. which, in the case of the uranium alloys, orders before transforming to a characteristically parallel-banded martensite. The maximum shape memory strains achieved with these alloys are comparable to those obtained from TiNi alloys and b.c.c. noble metal beta phases.

However very little examination of shape memory in f.c.c. solid solutions has been made and as a result, as will be shown, some important aspects of shape memory have been missed. A typical f.c.c. solid solution of commercial interest is the austenitic stainless steel based on the metastable iron-nickel-chromium system. Indeed one of the first references to stainless steel was made in the U.S. Patent of Jackson et al[5], in which stainless steel (AISI 304) was used as an example of a material which did

<u>not</u> undergo shape memory when put through the same sequence of
operations which gave rise to shape memory in uranium-niobium
alloys.

2. CHARACTERISTICS OF SHAPE MEMORY
IN AUSTENITIC STAINLESS STEELS

Austenitic stainless steels do indeed undergo shape memory
and detailed results for a very wide range of steels containing
nickel, chromium, manganese, cobalt and other elements have been
given by the present authors[7] and will not be repeated here. Other
authors have shown limited shape memory in the type 304 steel[8]. It
is significant that shape memory in stainless steels is usually
illustrated in terms of the change in the angle of bend in a thin
strip. In this way, a small shape change corresponding to 1%
surface strain can be demonstrated as an angular movement of about
40°. The maximum shape memory strains achieved in tension tests
of stainless steel have not exceeded about 1.6% and this has
militated against industrial exploitation of shape memory in
materials with otherwise attractive properties.

Shape memory in austenitic steels differs in many important
ways from that in intermetallic compounds and the b.c.c. solid
solutions. If Fe-Ni-Cr stainless steels are cooled to below the
M_s temperature, the austenite transforms substantially to alpha
prime martensite, the ferromagnetic acicular martensite which forms
laths or plates depending on composition. There is also another
martensite, epsilon martensite, which is formed in small amounts
on cooling. This is a self accommodating banded martensite which
resembles bundles of stacking faults.

If the steel is allowed to transform completely to martensite
on cooling and is then deformed, it is extremely brittle, and
little shape memory can be achieved. In order to obtain shape
memory, stainless steels must be deformed above the M_s temperature
(and obviously below the M_d temperature) or, less effectively,
cooled through the M_s temperature under stress. On deformation
above the M_s temperature, the metastable austenite transforms to a
mixture of alpha prime and epsilon martensite. The relative
proportions of each depend on the temperature of deformation and
on the composition of the steel.

As might be expected, the shape memory strain recovered on
re-heating a stainless steel is considerably less than the retained
strain after deformation, e.g. it is necessary to deform by 15 to
20% in order to produce a shape memory strain of about 1%.

The amount of the ferromagnetic alpha prime martensite was

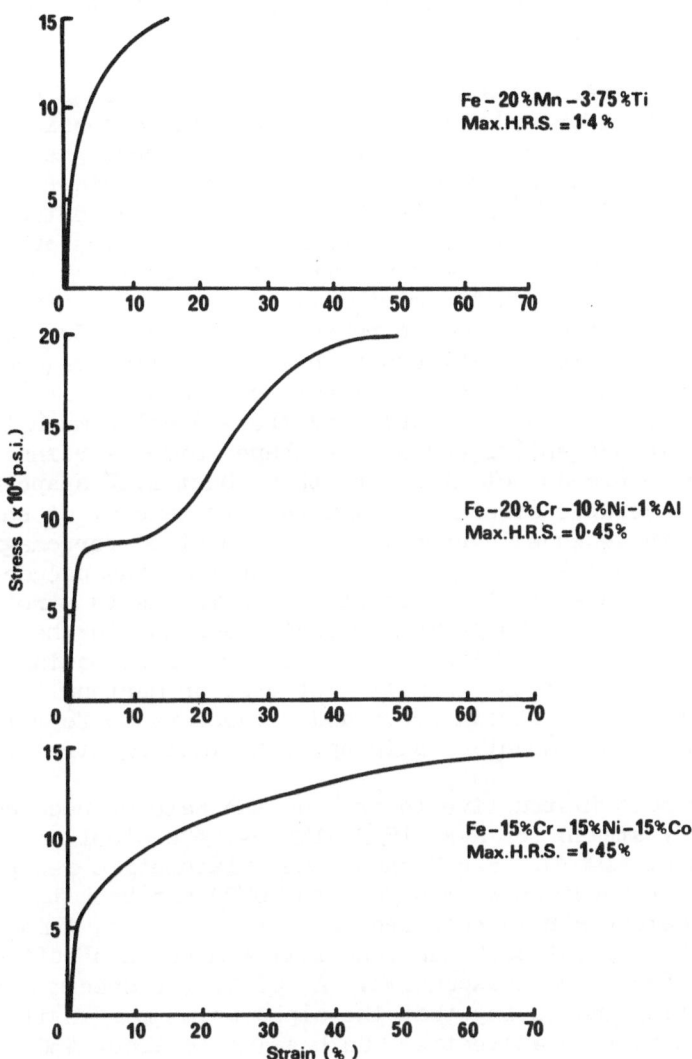

Figure 1. Stress–Strain curves of different heat-recoverable Iron-base alloys, deformed at -196°C.

measured using a simple magnetic balance. The steels were deformed at -196°C by amounts up to 20% and the proportion of ferromagnetic martensite compared with the shape of the stress-strain curve at -196°C and the amount of shape memory strain (or heat recoverable strain, H.R.S.) obtained on reheating to 20°C. The compositions chosen usually contained less than 0.5% ferromagnetic phase on cooling to -196°C.

Typical stress-strain curves for three stainless steels of different behaviour are shown in Fig.1 and it is possible to relate the shape of the curve to the amount of the ferromagnetic alpha prime phase and to the amount of shape memory strain (given as H.R.S. in Fig.1). The stress-strain curve with an extensive pseudo-elastic low strain hardening region (e.g. the curve for Fe-20%Cr-10%Ni-1%Al) was obtained from an alloy which had a high proportion of ferromagnetic martensite after deformation and a low shape memory strain (0.45%) on reheating to 20°C. Such steels are usually high in nickel. Alloys richer in chromium or containing cobalt or manganese have stress-strain curves with little or no pseudo-elastic part of the curve and these develop much less ferromagnetic martensite and higher shape memory strains. However even these steels do not give more than about 1.5% shape memory or heat recoverable strain. In general, the amount of shape memory strain depends inversely on the amount of ferromagnetic martensite. Thus it is possible to infer that the pseudo-elastic part of the stress-strain curve at -196°C is due to stress-induced transformation of alpha prime martensite and not to the epsilon martensite which gives rise to shape memory on reheating. It is impossible to detect the formation of epsilon martensite from the shape of the stress-strain curve and it appears to form in proportion to the amount of slip up to a limiting strain.

It is most instructive to examine the rate of recovery of shape memory strain for a Fe-15%Ni-15%Cr-15%Co steel deformed 20% in tension at -196°C. Fig.2 shows the dilatometric record of reheating the specimen from -196°C to 700°C and back to -196°C. The shape memory strain recovered over the range of temperature from -196°C to about 60°C has been little more than sufficient to compensate for thermal expansion. At 60°C, the specimen was still ferromagnetic, indicating that the alpha prime martensite had not transformed back to austenite. On heating to about 450°C a small contraction of about 0.15% was observed and on cooling to ambient temperature, the ferromagnetism had disappeared. Thus it can be inferred that the reversion of alpha prime martensite was responsible for the further contraction.

The data obtained from ternary Fe-Ni-Cr alloys is summarised in Fig.3. This comprises the metastable phase diagram at 20°C on which has been superimposed the iso-M_S contours for 20°C and -196°C

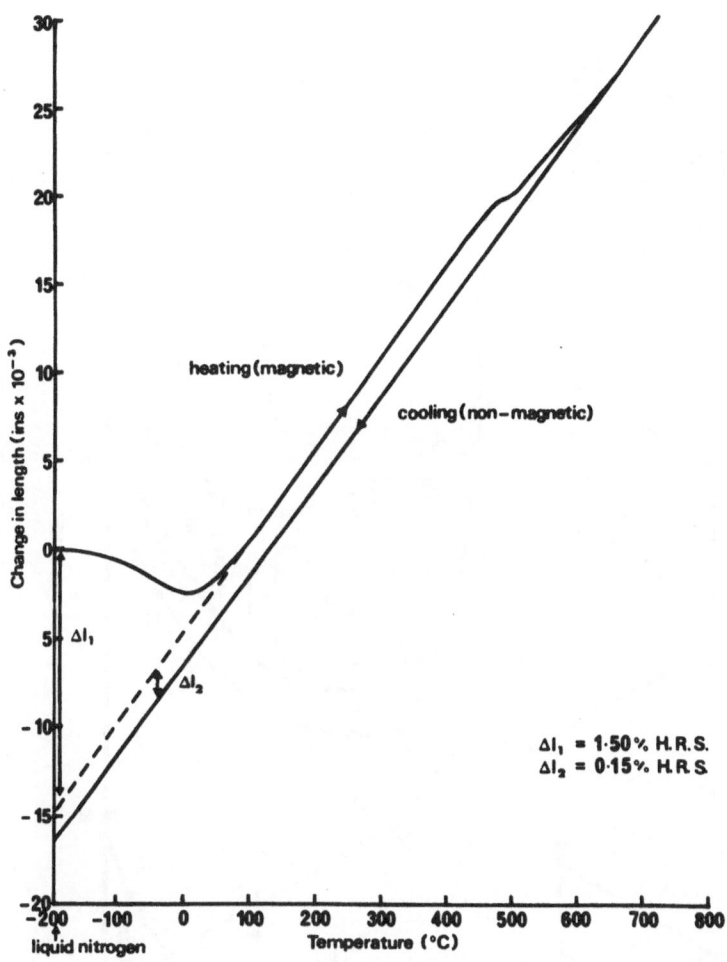

Figure 2. The nature of heat recoverable strain in Fe - 15%Cr - 15% Ni - 15%Co, after straining in tension at -196°C.

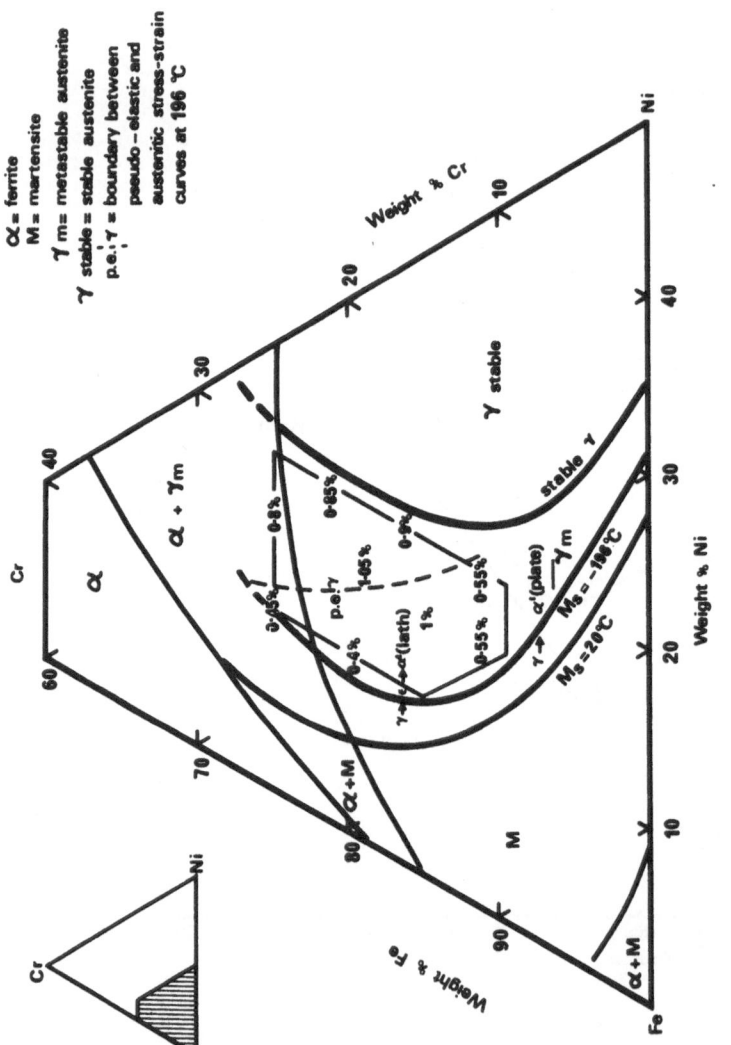

Figure 3. Heat recoverable strains in Fe – Ni – Cr alloys (0·04%C), deformed in tension at –196°C.

together with individual shape memory contractions. The dotted line is the locus of points at which no pseudo-elastic deformation could be detected. As can be seen, the highest shape memory strains were found at compositions between this line and the stable gamma boundary, i.e. in alloys of M_d just above -196°C. This supports the inference that the formation of alpha prime either on cooling through the conventional M_s temperature or by deformation at high temperatures plays little or no part in shape memory in stainless steels. Indeed whether the steels are deformed at -196°C, -80°C or 20°C, the highest shape memory strains are achieved in steels for which the M_s temperature is considerably below the deformation temperature. This suggests that the structure whose reversion is responsible for the shape memory effect is not a true martensite.

Shape memory in austenitic steels is promoted by raising the amounts of elements known to decrease the stacking fault energy of austenite such as chromium, cobalt, manganese and silicon. The influence of nickel is more complex in iron-nickel chromium steels as can be seen by comparing Fig.3 with the data in Fig.4[9]. Raising the chromium content at a constant nickel content increases the maximum shape memory strain up to a critical chromium content beyond which little further change occurs or an actual decrease is noted. This is in agreement with the general trend of the fall in stacking fault energy as chromium is increased at constant nickel content (Fig.4). However if one considers a series of steels of constant chromium content, Fig.4 indicates that as the nickel content is raised, the stacking fault energy rises quite steeply whereas inspection of Fig.3 shows that the shape memory strain rises to a maximum at 15 to 20% nickel. The reason for this anomaly has not been discovered yet. It may arise from the fact that the data on stacking fault energy at ambient temperature are being used to explain observations after deformation at -196°C. However this supposes that it is only the effect of nickel which changes at the lower temperature. Further data on stacking fault energies in chromium-nickel stainless steels are needed. It appears at present that the shape memory strain in these steels is a maximum for steels with more than 15% chromium with 15-20% nickel that their composition bears a relationship to the form of the iso-martensite contours. Nevertheless apart from the nickel anomaly, the correlation between shape memory strain and composition in steels containing manganese and cobalt is satisfactory.

For a specific composition, the amount of shape memory strain is the greater, the lower the temperature of deformation in the range 20°C to -196°C. All these obversations suggest that the structure giving rise to the shape memory effect is the reversion of stacking faults produced by deformation at very low temperatures.

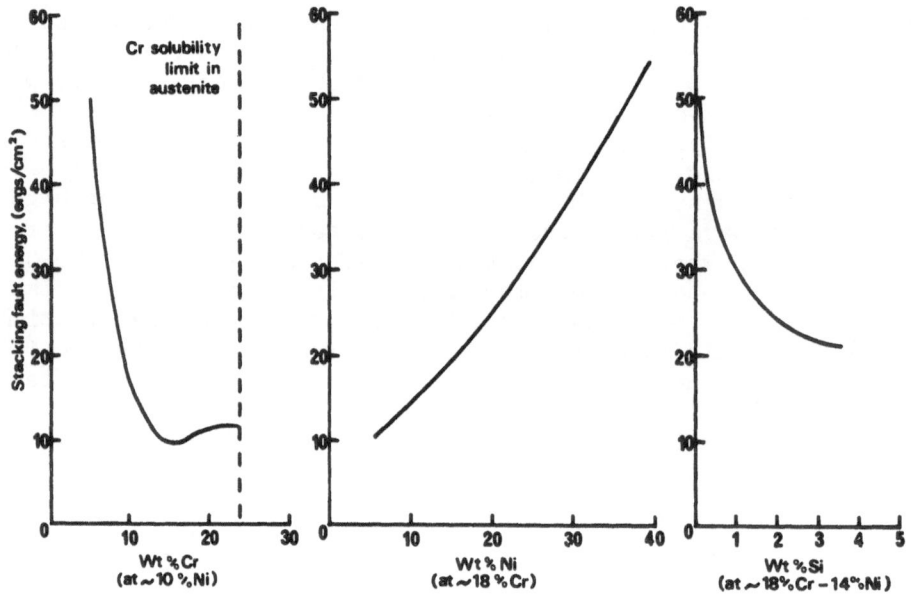

Figure 4. Influence of solute additions on the stacking fault
energy of stainless steels.

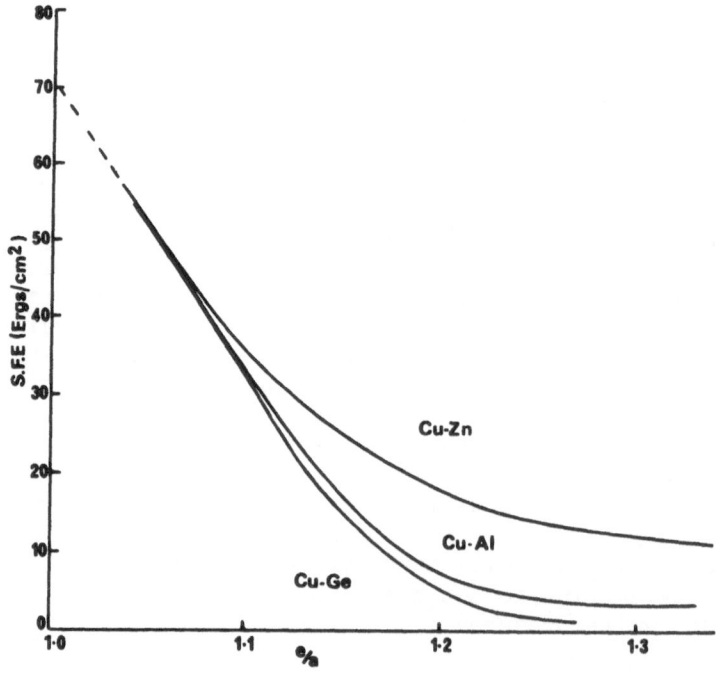

Figure 5. Variation of S.F.E. with $^e/a$ ratio

3. SHAPE MEMORY EFFECTS IN FACE CENTRED
CUBIC SOLID SOLUTIONS

The inference from the research on stainless steels has been
that in face centred cubic solid solutions, a weak shape memory
effect arises from the reversion of a high density of stacking
faults produced by deformation at low temperatures. However in
stainless steel, it is impossible to be certain of distinguishing
the production of stacking faults from that of epsilon martensite,
if indeed it can be held that these are distinct structures. Also
the apparently anomalous effect of nickel casts some doubt on the
argument. If the stacking fault hypothesis of shape memory is
true, it ought to be possible to produce shape memory in solid
solutions of other face centred cubic elements of low stacking
fault energy in which martensitic transformations are thought not
to occur. Copper is such an element and the effect of various
elements on the stacking fault energy of copper is well established[9]
(Fig.5). As saturation of the alpha phase solid solution is
approached, the stacking fault energy reaches a constant and low
value. It is noticeable that even though there may be some doubt
about absolute values of stacking fault energy, the relation effect
of Zn, Al and Ge are in the correct order. It is also known that
the effect of Si is similar to that of Ge. Thus if the hypothesis
advanced above is correct, the addition of silicon and germanium
should produce shape memory. Aluminium should be less effective
and zinc the least effective. Accordingly alloys of Cu-30%Zn,
Cu-8%Al and Cu-4%Si were prepared and were cooled to -196°C and
deformed in tension up to 20%. On heating up to 20°C, there was no
detectable shape memory effect in the Cu-Zn alloy but there was a
contraction of 1.05% and 1.4% respectively in the copper-aluminium
and copper-silicon alloys. This would indicate on the basis of
Fig.5 that shape memory is possible in copper primary solid
solutions at stacking fault energies of less than about 10ergs/cm^2
and that there is now evidence that the shape memory effect does
not <u>necessarily</u> depend on a martensite transformation. However the
martensitic transformation obviously has a role to play in shape
memory: firstly it provides a sharp discontinuity between two
structures with a high and low stacking fault energy (most
martensites are heavily faulted); secondly the density of stacking
faults in martensite is much greater than can be achieved in
worked solid solutions; thirdly it provides a narrow temperature
range over which this change and consequently the shape memory can
take place. In the noble metal intermetallic compounds, shape
memory is complete over a temperature range of about 40°-50°C
whereas in copper and ferrous f.c.c. solid solutions, much smaller
shape memory strains require about 250°C for this recovery. Thus
the evidence at present suggests that whilst a martensitic
transformation is not an essential feature of the shape memory
effect, alloys with martensitic transformations will provide larger

shape memory effects taking place over a narrower temperature range than those without a martensitic phase change. Metallographic confirmation of this is in progress and the results will be published in a later paper.

4.   CONCLUSIONS

Evidence has been presented that in austenitic stainless steels, the shape memory effect is not due to the occurrence of the gamma to alpha prime martensitic transformation. To obtain shape memory, the steel must be deformed above its Ms temperature (which should preferably be below -196°C) and at as low a temperature as possible. Shape memory is also favoured by elements which decrease the stacking fault energy of the steel. Shape memory is attributed to the reversal of stacking faults produced by low temperature deformation. To confirm this hypothesis, it has been demonstrated that shape memory exists in copper f.c.c. solid solutions of low stacking fault energy.

REFERENCES

1.   F.E. Wang, W.J. Buehler and S.J. Pickart. J. Appl. Phys., 1965, 36, 3232.

2.   L. Delaey, R.V. Krishnan, H. Tas and R.H. Warlimont. J. Mat. Sci, 1974, 9, 1521.

3.   H. Warlimont and L. Delaey. Progress in Materials Science, 1974, 18, 1.

4.   C. Baker. Mat. Sci. J., 1971, 5, 92.

5.   R.J. Jackson, J.F. Boland and J.L. Frankeny. U.S. Patent No. 3,567,523.

6.   G.B. Brook and R.F. Iles. U.K. Patent No. 1,315,652.

7.   G.B. Brook and R.F. Iles. U.K. Patent No. 1,346,046.

8.   K. Enamo, S. Nenno and Y. Minato. Scripta Met., 1971, 5, 663.

9.   P.J.C. Gallagher. Metallurgical Transactions, 1970, 1, 2429.

MECHANISMS FOR MARTENSITE FORMATION AND THE SHAPE MEMORY EFFECT

S. Mendelson

The City University of New York Research Foundation
The City College of New York, Steinman Hall, Room 12
138th Street and Convent Avenue, New York, N. Y. 10031

In addition to their technical potential pseudoelasticity and shape memory involve various interrelated effects which manifest themselves in the atomic motions between states. The phenomenological theories of Wechsler, Lieberman and Reed (W-L-R) (1), and Bowles and Mackenzie (B-M) (2), for the crystallography of martensitic transformations were formulated some twenty years ago and have since been applied to various alloys and to studies of pseudoelasticity and the shape memory effect. As is characteristic of most phenomenological theories in physics, discrepancies often exist; well known examples are the findings of different kinds of martensite in the same alloy and sometimes in the same sample, and often with a variable internal structure. In ferrous martensites the habits show wide scatter in the region of {3,10,15}, and habits found near {225} and {111} are not predicted by the theory. Some of the scatter around {3,10,15} are attributed to variations in the twinned structure, while others have been explained by allowing for small dilatations in the product, by changing the elements of the lattice invariant deformation, or by allowing for multiple lattice invariant shears in the product.

While large dilatations of about 1% or more are not likely, and the proposal of a dilated product might appear to be a crutch (ignorance factor (3)) for a weak theory, it may have some validity for small dilatations. One possible source of dilatation is a preferred shear on a low index plane which cannot complete the transformation; another is a magnetically induced thermal expansion in the transformed lattice which may alter the coefficient of thermal expansion sufficiently to leave a dilatation in the product on further cooling. Small dilatations might also develop as a consequence of elastic anomalies and soft mode effects at the transition

temperature. The generally unknown magnitude of thse effects tend
to complicate crystallographic mechanisms. On the other hand, a
valid crystallographic theory of martensite formation would give
insight into electronic effects responsible for the above, since
diffusionless phase transformations are frequently associated with
ferromagnetic transitions, paraelectric to ferroelectric transitions,
semiconductor to metallic transitions, order-disorder changes, ordered antiferroelectric or ferroelectric transitions, and superconductivity transitions. A rationalization of the atomic motions
between states and the criteria which govern them is essential to
clarification of pseudoelasticity and shape memory and, in addition,
can serve as a frame of reference for evaluating the cooperative
displacements associated with soft mode and magnetic and phonon
coupling effects.

TRANSFORMATION THEORIES

In the mathematical representation of the phenomenological
theory (4,5) small vectors $\vec{u}$ of the parent structure are transformed
to $\vec{u}'$ vectors in the product by $\vec{u}'=S\vec{u}$, where S is the lattice deformation matrix and is factorized into a pure deformation P (the Bain
distortion) and a pure rotation R, which gives the final orientation
relationship. Similarly, macroscopic vectors $m\vec{u}$ (where m is large)
of the parent structure are transformed to $m\vec{u}'$ in the product by
$m\vec{u}'=mE\vec{u}$, where E is the shape deformation matrix and is an invariant
plane strain which is factorized into the lattice deformation S and
a lattice invariant deformation G, given by E=SG=RGP. If G represents
a twinning shear, the fraction of which is f, there are two lattice
deformations S_1 and S_2 and the resulting shape deformation is E=
$(1-f)S_1+fS_2$. E=RGP is then solved under the conditions that GP be a
deformation with an undistorted plane, and RGP be a deformation with
an invariant plane. Physically the Bain distortion transforms the
structure into a new phase but which, in general, does not yield an
undistorted habit plane. The latter is proposed to develop by the
subsequent lattice-invariant shear (usually twinning) in the product,
followed by a lattice rotation, introduced to insure that the undistorted habit plane has the same orientation in space as both the
parent and product crystals.

The Bain distortion is one in which a simple homogeneous pure
distortion converts one lattice into another by an expansion or
contraction of the crystallographic axes. For example, in the application to the fcc→bcc transformation it involves one principal strain
of -18.4% and two others of 15.5%; no mechanism has been proposed
for carrying this distortion out. For a single shear transformation
with an invariant plane strain one of the principal strain must be
zero, another increased, and the third reduced; this follows because
there is no dimensional change or mass transfer in the $\vec{S}_m$ direction
normal to the S_m "plane of shear". The principal strain along $\vec{S}_m$ is

zero and the remaining two principal strains must lie in S_m. If a
volume change accompanies the transformation (as is generally the
case) its direction must also lie in the S_m plane. Since the Bain
distortion does not have a zero principal strain it cannot be represented mechanistically by a single shear; the introduction of a
subsequent lattice-invariant shear in the product and a rigid body
rotation does not overcome this difficulty.

The Bain distortion can, however, be accomplished by two lattice-variant shears on systems for which $S_{m2} \perp S_{m1}$ (6,7), or the
transformations can be accomplished by a single lattice-variant
shear (not represented by the Bain distortion) on a system which has
smaller strains than for the Bain distortion. In general, not withstanding the conclusions of others, any Bain type distortion can be
replaced by a smaller shape distortion for which one of the principal strains is zero.

Since martensite forms by shear with a small shear strain it
has characteristics quite similar to deformation twinning (8), Both
occur by plastic shear under an applied stress, chemical instability,
or both. Earlier studies of deformation twinning in various crystal
structures showed that the shear stress is generally much smaller
than that required for twinning by homogeneous shear. The concept
of the zonal dislocation, which allows for atomic shuffling, was
introduced by various authors and later formally defined by
Mendelson (9,10) and applied to various crystal structures (9-12).
The concept of the zonal dislocation was given a still wider application in a partialized dislocation model for the structure of the
dislocations in crystals (9,12,13), and in a more sophisticated
treatment (7) allows for martensite lamellae as well as the more
common stacking faults and twin lamellae at the partialized dislocations.

When impurities interact with the stress field and migrate to
dislocations they lower the distortion energies associated with the
impurities and dislocations. Their presence alters the chemical properties of this region of crystal and thus lowers the energy difference between it an some other crystal structure. Stacking faults
and twins are the more common lamellae of different structure which
forms when alloying elements segregate around dislocations and enhance their dissociation. Other structures are also possible at the
partialized dislocations and , depending on the crystal chemistry,
can be classified as martensite.

In the theory proposed by Mendelson (6,7,14) several mechanisms
were developed with minimum requirements of lattice matching along
$\vec{\eta}_1$ and $\vec{S}_m$ orthogonal directions in the κ_1 shear plane, and the ex-
istance of a zonal dislocation which can transform the lattice with
a small shear strain and simple atomic shuffling; in the more general mechanism the volume change direction is normal to the κ_1 shear
plane, in another it is parallel to the $\vec{\eta}_1$ shear direction, and in
the third a two shear mechanism is developed.

A transformation by zonal dislocations is accomplished by atomic shuffling in the core of the dislocation when atoms on every n^{th} plane undergo homogeneous or pseudohomogeneous shear while those on intermediate planes shuffle into the transformed structure. The zonal dislocations for both deformation twinning and martensite formation are derived from unit dislocations of the parent lattice which meet the martensite or twin plate; these dissociate into transformation partials which glide along the interface plane to transform a lamella of parent to martensite. The residual partials either remain in the interface, dissociate into one or more transformation partials, or into twinning dislocations which move into the lamella.

TRANSFORMATION MECHANISMS

A. Lattice Matching

In the present applications the two phases are distinguished by subscripts b and o, where b represents the bcc phase and o the orthorhombic phase, however they could represent any two phases. $\vec{\eta}_1$, $\vec{\kappa}_1$, and $\vec{S}_m$ are orthogonal vectors with integral indices corresponding to the shear direction, the normal to the κ_1 shear plane, and the normal to the S_m "plane of shear", respectively. Lattice translations and plane spacings in the existing structure are equated to their values in a primative cell multiplied by suitable cell factors which normalize them to their existing structure. $I'\vec{\eta}_1$ is the lattice translation for the $\vec{\eta}_1$ shear direction and I' is the cell factor for that direction. I/κ_1 is the spacing of κ_1 shear planes and I the cell factor for the κ_1 plane. A third cell factor of use is I'', the inverse of the number of atoms in the area bounded by lattice translations along $\vec{\eta}_1$ and $\vec{S}_m$ directions (7). This is given by

$$I'' = \kappa_1 \{\rho_a (I'S_m)(I'\eta_1)I\}^{-1} \qquad (1)$$

Lattice matching along $\vec{S}_m$ is perfect when the ratio of lattice translations $(I_b'S_{mb})/(I_o'S_{mo})$ is equal to the ratio of small integers and q_o/q_b and

$$q_b(I_b'S_{mb}) = q_o(I_o'S_{mo}) \qquad (2)$$

Similarly, lattice matching is perfect along $\vec{\eta}_1$ when $(I_b'\eta_{1b})/(I_o'\eta_{1o})$ is equal to q_o'/q_b' and

$$q_b'(I_b'\eta_{1b}) = q_o'(I_o'\eta_{1o}) \qquad (3)$$

When the volume change is parallel to $\vec{\kappa}_1$ the number of atoms swept up by the dislocation in n_b' planes is equal to that in n_o' planes of product, and when substituted into the above equations we have

$$n_o'q_o'(I_b'I_b''S_{mb}) = n_b'q_b'(I_o'I_o''S_{mo}) \tag{4}$$

$$n_o'q_o(I_b'I_b''\eta_{1b}) = n_b'q_b(I_o'I_o''\eta_{1o}) \tag{5}$$

for mass conservation and lattice matching along $\vec{S}_m$ and $\vec{\eta}_1$ directions respectively.

B. The Zonal Dislocation

The zonal dislocation is defined by its Burgers vector $\vec{b}_s'$, lamella thickness $n'd_{\kappa_1}$, and displacement vector $\vec{b}_n'$, normal to the κ_1 shear plane, and given by components of $\vec{t}$ vectors (generally close packed lattice translations) in the S_m' plane.

$$\vec{b}_s' = [(\vec{t}_b \cdot \vec{\eta}_{1b})/\eta_{1b}{}^2]\vec{\eta}_{1b} - [(\vec{t}_o \cdot \vec{\eta}_{1o})/\eta_{1o}{}^2]\vec{\eta}_{1o} \tag{6}$$

$$\vec{b}_n' = [(\vec{t}_b \cdot \vec{\kappa}_{1b})/\kappa_{1b}^2]\vec{\kappa}_{1b} - [(\vec{t}_o \cdot \vec{\kappa}_{1o})/\kappa_{1o}{}^2]\vec{\kappa}_{1o} \tag{7}$$

where $n'd_{\kappa_1} = [(\vec{t} \cdot \vec{\kappa}_1)/\kappa_1]$ and $b_n' = (n'd_{\kappa_1})_b - (n'd_{\kappa_1})_o$ (8)

If the n' are the minimum number of κ_1 planes in the transformed lamella $\vec{b}_s'$ will transform the structure only if it and $S=b_s'/n'd_{\kappa_1}$ are small. If this is not so a zonal dislocation must be derived with a small shear strain for which n is an integral number of n'. As in deformation twinning the Burgers vector of the zonal dislocation is

$$\vec{b}_s = s\vec{b}_o = sI'\vec{\eta}_1 \tag{9}$$

where s is the fraction of b_o (the lattice translation along $\vec{\eta}_1$) which produces the twinning or transformation shear. The shear is reduced when a multilayer lamella is transformed by zonal dislocations with s, as defined for deformation twinning given by

$$s = ns' - p \tag{10}$$

where s' is the shear fraction for a single layer twin (homogeneous shear) and p is an integer which gives the lowest absolute value of s. The sign of s determines the stress-sense for motion of the zonal dislocation. Similarly for a phase transformation

$$s = (n/n')s' - I'p \tag{11}$$

where s' is the shear fraction for a unit lamella of martensite, formed from n' parent planes and n is the number of planes converted to martensite by the zonal dislocation.

The shear strain is given by $S_s=b_s/nd_{\kappa_1}$, the normal strain by $S_n=b_n/nd_{\kappa_1}$, and the total strain by $S_t=(b_s+b_n)/nd_{\kappa_1}$. If twinning or other shears join the plate they will add to the total shape strain.

The S_m' plane is chosen with close packed directions to define the most favorable zonal dislocations; however the $\vec{S}_m'$ direction (normal to the S_m' plane) will not necessarily be one of perfect matching. Since matching is perfect along $\vec{\eta}_1$, and the Burgers vector of the zonal dislocation is defined for this direction, the $\vec{\eta}_1$ axis is a zone axis for other possible habits of low or high indices, with or without dilatations along $\vec{S}_m$. In general, if $n_b d_{K1b} > n_o d_{K1o}$ then $q_b I_b' S_{mb}' < q_o I_o' S_{mo}'$, and at some angle of rotation about $\vec{\eta}_1$ the dilatation along $\vec{S}_m$ will disappear.

The above is for $\Delta V || \vec{\kappa}_1$ and applies when ΔV and b_n/b_s are small. When ΔV is large the b_n component of the shape strain imposes a torque on the dislocation line which is proportional to b_n/b_s; this tends to rotate the Burgers vector (about the $\vec{S}_m$ zone axis) away from $\vec{\eta}_1$ and makes formation of the first lamellae difficult. The effect of this is to favor transformations for which $\Delta V || \vec{\eta}_1$. This is possible when the zonal dislocation is defined with a slightly different orientation relationship such that $b_n = 0$ and $\Delta V || \vec{\eta}_1$. As the dislocation glides the volume change along $\vec{\eta}_1$ consumes the dislocation core when it moves a distance

$$\vec{L} = \vec{b}_s / [\Omega_b/\Omega_o - 1] \tag{12}$$

where Ω_b and Ω_o are the atomic volumes in the two phases. When part of the volume change occurs along the $\vec{\kappa}_1$ direction the slip distance is between that given by Eq (12) and ∞, and given by

$$\vec{L} = \vec{b}_s / [\Omega_b/\Omega_o - (n d_{K1})_b / (n d_{K1})_o] \tag{13}$$

At this distance the dislocation has effectively climbed (conservatively) by n planes, where it then glides through L on the displaced plane, etc.; a stepped interface results which effectively rotates the habit plane by $\beta° = \tan^{-1}(n d_{K1}/L)$ and the new habit plane is given by

$$\vec{h} = [\vec{L} + n\vec{d}_{K1}] \times \vec{S}_m \tag{14}$$

C. Double Shear

While it is true that the Bain distortion cannot be accomplished by a single lattice-variant shear, it can be accomplished by two homogeneous lattice-variant shears on systems for which $S_{m2} \perp S_{m1}$ (6,7). Since both shears can be homogeneous this type transformation would be favored if the viscosity for atomic shuffling in the single shear mechanism is excessive, or if elastic anomalies or soft mode effects at the transition temperature favor shear on a specific low index system.

General algebraic equations for the two shear mechanism were derived for a general orientation relationship; however these simplify when the low index S_m plane has orthogonal low index lattice translations which remain orthogonal after the shear. We take $\vec{u}_i$ ($i=1,2$, and 3) as orthogonal vectors in the parent, $\vec{u}_i'$ those in the product after the second shear, and $\vec{u}_i''$ those in the product after the second shear. In the first shear $\vec{S}_{m1}||\vec{u}_2$ and $\vec{u}_1$ is transformed to its final dimension ($\vec{u}_1 \to \vec{u}_1' = \vec{u}''$). For the second shear $\vec{S}_{m2}||\vec{u}_1'$ and the remaining two orthogonal vectors are transformed to their final dimensions. It follows for the first shear that

$$R_1{}^2 = u_1{}^2 + u_3{}^2 = u_1{}'^2 + u_3{}'^2 \tag{27}$$

$$(nd_{K11}) = (u_1 u_3)/R_1 \tag{28}$$

$$(nd_{K11})' = (u_1' u_3')/R_1 \tag{29}$$

$$b_{S_1} = (u_1'^2 - u_1{}^2)/R_1 = (u_3{}^2 - u_3'^2)/R_1 \tag{30}$$

$$b_{n1} = (u_1' u_3' - u_1 u_3)/R_1 \tag{31}$$

$$S_{S1} = (u_1'^2 - u_1{}^2)/u_1 u_3 = (u_3{}^2 - u_3'^2)/u_1 u_3 \tag{32}$$

$$S_{n1} = (u_1' u_3' - u_1 u_3)/u_1 u_3 \tag{33}$$

Similarly for the second shear

$$R_2{}^2 = u_2{}'^2 + u_3{}'^2 = u_2{}''^2 + u_3{}''^2 \tag{34}$$

$$(nd_{K12})' = (u_2' u_3')/R_2 \tag{35}$$

$$(nd_{K12})'' = (u_2'' u_3'')/R_2 \tag{36}$$

$$b_{S2} = (u_2'' - u_2')/R_2 = (u_3' - u_3'')/R_2 \tag{37}$$

$$b_{n2} = (u_2'' u_3'' - u_2' u_3')/R_2 \tag{38}$$

$$S_{S2} = (u_2''^2 - u_2'^2)/u_2' u_3' = (u_3'^2 - u_3''^2)/u_2' u_3' \tag{39}$$

$$S_{n2} = (u_2'' u_3'' - u_2' u_3')/u_2' u_3' \tag{40}$$

The equations for the first shear can apply to a single shear transformation when the conditions for orthogonal vectors are satisfied.

CHARACTERISTICS OF MARTENSITE FORMATION
FOR PSEUDOELASTICITY AND SME

The characteristics of martensite formation for pseudoelasticity and SME behavior were reviewed by Wayman and Shimizu (15), Delaey et al. (16) and Perkins (17), and the characteristics of thermoelastic and burst type martensite are discussed by Brown and his associates (11); these are related to current theories of martensite formation. These characteristics are now examined in the light of the present theory and mechanisms.

When cooled below M_s or stressed above M_s the first lamellae of early martensite form by dissociation of the network dislocation lines which are most favorably oriented with respect to the driving force (chemical or mechanical); these can grow into martensite plates when dislocations join the lamellae and dissociate. The dislocations can come from the network when they bow out between their nodes under an increasing driving force, or screw dislocations may cross-slip, if not dissociated, and may multiply by the double cross-slip mechanism at the martensite plate interface.

When cooled below M_s the first thermoelastic martensite plates set up stress fields in regions between them due to the transformation dislocations at the edges of the lenticular plates. These tend to enhance formation of neighboring plates with opposite stress fields and opposite shear in such a way that there are no long range stress fields and no external shape strain develops. In contrast to this behavior, when stress is applied to the untransformed parent above M_s the first lamellae will be thermoelastic, but these will form on planes, generally of a single system, which have the greatest resolved shear stress; long range stress fields may develop in this case and the stress induced thermoelastic martensite will show pseudoelastic behavior, with the lenticular plates shrinking and dissapearing when the transformation dislocations return to their residual partials in the interface as the stress is removed, the driving force for this return being chemical.

If a partially transformed specimen is subsequently deformed below M_s the applied stress will favor the growth of those plates for which the shear is the same, and reversion of those for which the shear is opposite. The applied stress effectively lowers the barrier to martensite formation in the regions between plates and both the elastic and chemical driving forces tend to enhance formation of burst martensite when the resistive energy to initiate a martensite plate exceeds that for its propagation. Burst martensite will be enhanced by lower temperatures below M_s and higher applied stresses, whereas thermoelastic martensite will form just below M_s, and under an applied stress at temperatures above M_s; in the latter case the chemical driving force apposes stress induced martensite formation, and the tendency for burst martensite formation is reduced the higher the temperature is above M_s.

When the temperature is raised or the applied stress is removed

the transformation dislocations reunite with their residuals. Some
of the perfect dislocations may move back to their previous network
positions, but others may not, and those which formed at cross-slip-
ping screw dislocations will stay; some glide polygonization to line
up the dislocations is likely. The finding that the dislocations
line up parallel to the parent slip plane (17) suggests that this
may also be the shear plane for the transformation.

Since reversibility requires return of the transformation dislocations to their residuals, any obsticles or stress fields which
limit this will introduce hysteresis into the stress-strain curve,
and may reduce the amount of pseudoelasticity; the remaining shape
strain which is reversible on heating constitutes the shape memory
effect, and the two effects (pseudoelasticity and shape memory) are
complementary (18). Hysteresis develops when the motion of the
transformation dislocations is restricted, and is a measure of the
relative energy available for formation and reversion of the martensite plates; this difference approaches zero if all the plates are
thermoelastic, and it may include part or all the deformation energy
when the martensite plates are of the burst type. When burst martensite forms by stress below M_s and the stress is removed the restrictions present during the burst may be too large for complete return
and a permanent strain sets in; this is recovered by heating when
the shape memory effect applies.

Transformation dislocations which are tangled with other dislocations and with localized stress fields in burst martensite contribute to the hysteresis and loss of pseudoelasticity. When the
plates extend to grain boundaries and transformation dislocations
pass out during the burst thre may be some loss in SME, in addition
to the loss in pseudoelasticity, this depending on how much of the
burst martensite extends to grain boundaries and how effective the
chemical driving force and back stress from the residuals are in
inducing transformation dislocations to move through steps in the
grain boundaries. In this case pseudoelasticity and SME may not be
completely complementary.

Since the plate thickness depends on the number of dislocations which join the plate it determines the density of residual
dislocations in the interface and consequently, if these are dissociated into twinning partials, the twin fraction in the martensite.
The plates assume the lenticular shape when the density of residuals
is greater in the central region and decreases towards the edges of
the plates, where the transformation partials pile up. If the residuals are dissociated into twinning partials the twin fraction should
be smaller towards the edges. The model therefore requires that the
habit plane at the interface be a function of the number of residuals in the interface, and thus a function of the plate thickness.
If a mid-rib is present it will be parallel to the shear plane,
whereas if twinning partials pass through the mid-rib its orientation will approach that of the interface. The findings of habit
planes in a localized region of the stereographic triangle suggests

that the plates have relatively uniform thickness.

MECHANISMS AND PREREQUISITES FOR PSEUDOELASTICITY AND SME

The suggested phenomenological prerequisite for SME that the elastic limit is not exceeded (17) follows from various studies including those which show only a small increase (about two times) in dislocation density for a transformation cycle. This requires that if dislocations are generated during the transformation they must be eliminated on its reversion. Since dislocation multiplication by the double cross-slip mechanism at the martensite plate interface would significantly increase the dislocation density, this phenomenological prerequisite requires that cross-slip be limited. This would follow if the dislocations are dissociated and suggests that a prerequisite for SME behavior is that the stacking fault energy in these alloys be low. A similar suggestion was made for martensite formation in titanium alloys (14), where the decrease in M_s with alloying and the increased tendency for twinning in the parent phase is attributed to a decrease in stacking fault energy. With the dislocations partialized into twinning dislocations they are less capable of dissociating into transformation dislocations, and the driving force for the transformation is reduced. This defect controlled mechanism is consistent with studies of martensite formation in small highly perfect iron particles in a copper matrix (19).

Unlike double cross-slip, dislocation multiplication by pole mechanisms can be reversible; these were studied and pole mechanisms derived for several cases. However most results are best explained by dissociations of perfect dislocations which join the plate, since a pole mechanism does not leave residuals in the interface and cannot account for twinning or faulting in the martensite.

The prerequisite for SME that the parent and martensite phases be ordered (15) might be related to the special dislocation structure in ordered alloys, where the dislocation density is generally lower and the Burgers vectors are larger than in unordered alloys; these dislocations are more likely to be partialized, making cross-slip difficult.

A general mechanism which follows is that all dislocations in the parent lattice are potential sites for nucleation of the new phase, whether it be martensitic or diffusional. The effect of alloying is then to stabilize the parent phase by lowering the stacking fault energy (or by ordering the structure) and consequently stabilize the dislocations. Dissociation of the dislocations immobilizes them and limits their ability to serve as nucleating sites for diffusional growth at the high temperatures. The stabilized high temperature phase can then transform only under a greater driving force by shear, or at lower temperatures. The lower superelasticity and SME in unordered alloys may be related to the larger number of effective nucleating sites for martensite formation and consequently a finer martensite structure, with plates too small for significant amounts of reversible elastic twinning to occur in.

That the transformations are not isothermal and the M_s-M_f temperature spread is finite, with subsequent lamellae forming at lower temperatures or higher stresses, may be because only those dislocations which are most favorable will serve as nucleating sites at the higher temperatures or lower stresses. Since an applied stress enhances the transformation below M_s it will reduce the M_s-M_f spread and increase the tendency for martensite formation in avalanches or bursts. This suggests that pseudoelasticity and SME will be favored in alloys with a large M_s-M_f spread, and should be designed with this in mind.

The first lamellae of martensite form at the most favorably oriented dislocations and thus require a relatively low driving force; subsequent thickening requires motion of those dislocations which are farther away, and consequently a greater driving force. When thickening occurs at a later interval, the formation of the first lamellae and the rearrangment of network dislocations for their motion to these lamellae would constitute premonitory events. Various studies indicate premonitory effects in martensite formation (20).

Pseudoelasticity and SME in partial or fully martensitic specimens subjected to stress can develop by (i) growth of some plates and reversion of others, (ii) twinning in some plates and detwinning in others, and (iii) a transformation to a new martensitic structure. The first two are likely in a partly transformed specimen in which the martensite is on multiple systems, while the latter is possible in a fully martensitic specimen, with martensite on a single system.

When a crystal, partly transformed by cooling below M_s, is deformed the stress will enhance twinning in some martensite plates by dissociation of the residual interface dislocations and detwinning in others, and will favor growth of those plates for which the shear is the same and reversion of those for which the shear is opposite the applied stress. The relative roles of twinning and plate growth might be indicated by the anelastic studies of Arbuzova et al. (21) for Cu-Al-Ni alloy crystals. These authors find an internal friction peak at 50% martensite which could be attributed to the motion of transformation dislocations at the lenticular edges during growth and reversion of some plates. At 50% martensite the maximum number of plates can grow freely, whereas amounts greater than this would interfere with each other. If the peak were due to twinning within the plates it should show a maximum at 100% martensite; this thus suggests that either the twinning dislocations do not move, or they move with great ease.

When an untransformed crystal is deformed above M_s the stress induced thermoelastic plates generally develop on a single system. This may be accompanied by twinning within the martensite plates if the operating twin system is favorably oriented with respect to the applied stress. If the resolved shear stress on the twin system

is below that required, twinning might occur at a higher stress in
a second pseudoelastic yield on the stress-strain curve; this might
account for the double yield pseudoelasticity in Cu-Al-Ni found by
Bush et al. (22). An alternative interpretation of the second yield
is that it corresponds to a second martensitic transformation in a
prior single martensite structure. This becomes possible when the
additional energy to form the new structure is less than that required for plastic flow; an apparent example of this is seen in a
micrograph by Delaey et al (16). The transformation dislocations
for the second yield arise from the residual dislocations retained
in the first martensite plates, and propagate through the existing
martensite with its own characteristic habit plane, referred to the
firstmartensite structure. If the new martensite does not inheret
twins from the host martensite the new plates should be free of
twins, since the new martensite plates will not have residual dislocations at their boundaries; transmission electron microscopy studies
might be able to establish this.

When multiple variants of the habit are present in a fully
martensitic specimen and stressed, twins will grow in some habits,
while those in others will shrink, and some plates, which have
mobile self accomodating boundaries, will grow into their neighbors.
This can be seen in the micrographs of Delaey et al. (16) for a Cu-
Al-Ni alloy which shows some straight boundaries (probably twin
boundaries) which move parallel to themselves, and new structures
appearing within some plates; the motion of the straight boundaries
appears to be easiest and attributed (16) to reorientation of the
martensite plates by twinning. Ledges along the twin interface
(equivalent to twinning dislocations) can serve this function for
limited growth until the steps are removed. This strain may not be
pseudoelastic, since a smooth twin interface will not revert back
to a stepped (dislocated) interface. The structure within the plates
form at higher stressing and may correspond to twinning: it should
show pseudoelasticity due to the back stress from the twinning dislocations which pile against the interface. This might explain the
double yield behavior recently found by Miura et al. (23) in Au-
Cd-Cu alloys; the second yield is pseudoelastic but the first is
not. When the same specimen was transformed with a single martensite
interface the first yield did not appear and only a single yield,
similar to the second yield in the multiple habit specimen developed, and showed the same pseudoelasticity. Since the stress-strain
behavior is the same for single and multiple variant martensite,
the reorientation mechanism can not apply. Similarly it is not
likely that a new martensitic transformation would show the same
stress-strain behavior in single and multiple variant martensites.
This leaves twinning as the more likely mechanism, as described
above.

APPLICATIONS OF THE THEORY

The initial step in applying the single shear theory is to select a low index S_m' plane in the two phases which contain two low index close packed directions, one of which is parallel in the two phases. The parallel close packed planes and close packed directions define the orientation relationship. For example in the fcc→bcc transformation it corresponds to $(111)_f||(110)_b$ and $[\bar{1}10]_f||[\bar{1}11]_b$, and in bcc→hcp it corresponds to $(0001)_h||(110)_b$ and $[2\bar{1}\bar{1}0]_h||[\bar{1}11]_b$. In fcc→bcc there are two favorable S_m' planes, $(111)_f||(110)_b$ and $(\bar{1}10)_f||(\bar{1}11)_b$. The $\vec{t}$ vectors which define the zonal dislocation are taken as the second close packed (not parallel in the two phases) direction. For example in fcc→bcc with $S_m=(111)_f||(110)_b$

$$\vec{t}_f = \tfrac{1}{2}[0\bar{1}1]_f \quad \text{and} \quad \vec{t}_b = \tfrac{1}{2}[1\bar{1}1]_b \tag{29}$$

The $\kappa_1=(\bar{1}\bar{1}2)_f||(1\bar{1}2)_b$ shear plane results when $a_b/a_f=\sqrt{2/3}$, with $\vec{b}_s = \tfrac{1}{12}[\bar{1}10]_f$. For $a_b/a_f \neq \sqrt{2/3}$ the $\vec{\eta}_1$ shear direction for lattice matching is given by adding a displacement vector Δ to a linear sum of the parallel close packed direction. If we take $\Delta=\vec{t}$ we have for the shear direction

$$\tfrac{1}{2}\rho[\bar{1}10]_f + \tfrac{1}{2}[0\bar{1}1]_f = \tfrac{1}{2}\rho[\bar{1}11]_b + \tfrac{1}{2}[1\bar{1}1]_b \tag{30}$$

where ρ is an integer. Similarly for $S_m'=(\bar{1}10)_f||(\bar{1}11)_b$

$$\vec{t}_f = [001]_f \quad \text{and} \quad \vec{t}_b = [101]_b \tag{31}$$

and the $\vec{\eta}_1$ shear direction is given by

$$\tfrac{1}{2}\rho[33\bar{2}]_f + [001]_f = 2\rho[011]_b + [101]_b \tag{32}$$

Further details of the transformation modes in fcc→bcc will be published elswhere.

For bcc→hcp $S_m'=(0001)_h||(\bar{1}10)_b$

$$\vec{t}_h = \tfrac{1}{6}[\bar{1}\bar{1}20]_h \quad \text{and} \quad \vec{t}_b = \tfrac{1}{2}[111]_b \tag{33}$$

The $\kappa_1=(112)_b||(0\bar{1}10)_h$ shear plane results for $a_b/a_h=2/\sqrt{3}$ with $b_s = \tfrac{1}{12}[111]_b$. For $a_b/a_h \neq 2/\sqrt{3}$ the $\vec{\eta}_1$ shear direction, in analogy with Eq (30), is given by

$$\tfrac{1}{2}\rho[11\bar{1}]_b + \tfrac{1}{2}[111]_b = \tfrac{1}{6}\rho[\bar{2}110]_h + \tfrac{1}{6}[\bar{1}\bar{1}20]_h \tag{34}$$

For the bcc→bco or bcc→fco transformations, which apply to most SME alloys, the orientation relationship follows from that represented by the Bain distortion. For the Ag-45%Cd alloy studied by Krishnan and Brown (18) the lattice parameters are $a_0=4.865Å$,

b_o=4.7536Å and c_o=3.0968Å, with a_b=3.314Å. From the lattice corre-
spondence we take S_m'=$(1\bar{1}0)_b$||$(010)_o$ with lattice matching along $\vec{\eta}_1$
given by

$$[\rho\rho\sigma]_b = [\rho 0\sigma]_o \tag{35}$$

where ρ and σ are positive integers. When $\vec{\eta}_1$=$[111]_b$=$[101]_o$ the
matching has a dilatation of $\delta_{\eta 1}$=1.0047, whereas when $\vec{\eta}_1$=$[9,9,10]_b$=
$[9,0,10]_o$ the dilatation is $\delta_{\eta 1}$=0.9998, or essentially zero. Similarly for the $\vec{S}_m$ direction lattice matching is given by

$$7[(\rho+3\sigma),(\rho-3\sigma),\bar{2}\rho]_b = 3[2\rho,7\sigma,\bar{5}\rho]_o \tag{36}$$

When matching along $\vec{S}_m$ corresponds to $7[2\bar{1}\bar{1}]_b$=$^3/_2[27\bar{5}]_o$ the dilatation is $\delta_{\eta 1}$=1.0023, whereas when matching is along $7[49,\overline{23},\overline{26}]_b$=
$3[26,84,\overline{65}]_o$ the dilatation is $\delta_{\eta 1}$=1.0008.

Taking $\vec{\eta}_1$=$[9,9,10]_b$ and $\vec{S}_m$=$[49,\overline{23},\overline{26}]_b$ for a dilatation free
habit the shear plane is κ_1=$(1,\overline{181},162)_b$ which is near $(0\bar{1}1)_b$,
where many habits are found. The $\vec{t}$ vectors are taken as

$$\vec{t}_b = \tfrac{1}{2}[111]_b \quad \text{and} \quad \vec{t}_o = \tfrac{1}{2}[101]_o \tag{37}$$

and $\quad \vec{b}_s = (^1/_{187})[9,9,10]_b \quad$ with $\quad$ S=0.11 $\tag{38}$

Relating this to reported data, κ_1=$(1,\overline{181},162)_b$ is consistent with
the habit planes but $\vec{\eta}_1$=$[9,9,10]_b$ is not. Firther details of this,
and applications of the two shear theory will be published elsewhere.

DISCUSSION

The theory and mechanisms presented here are valid because they
are based on first principles, however which mechanism applies to a
particular alloy will depend on the prevailing criteria which govern
the transformation. Application of these mechanisms to the various
alloys can lead to the identification of these criteria.

The single and double shear mechanisms for which $\Delta V||\vec{\kappa}_1$ are
applicable to all phase transformations for which the zonal disloc-
cations are defined on low index S_m' planes. The single shear mech-
anism for which $\Delta V||\vec{\eta}_1$ is applicable when $q_b/q_o\neq 1$, and generally
involves more complex atomic shuffling than modes for which q_b/q_o=1.
When q_b/q_o=1, n_b/n_o=1 and the shear will be either homogeneous or
pseudohomogeous. When the shear is homogeneous all the atoms shear
through nb_s' without atomic shuffling, and generally applies when
$d_{\kappa 1}$ for the S_m' plane is the same in the two phases, whereas for
pseudohomogeneous shear part of the atoms on each κ_1 plane also have
a shuffle component parallel to S_m'.

The viscosity associated with atomic shuffling in the dislocation core depends on the nature of the atomic bonding in the alloy,
this being larger the greater the ratio of covalent to ionic type

bonding; it also depends on the dislocation speed. If speed is an
important criterion for SME a high speed homogeneous shear for which
the product is dilated may be favored over a low speed dilatation
free shear with atomic shuffling. A homogeneous dilated shear would
also be favored if it corresponds to a soft mode. In either case
the dilated shear would serve as the first shear in a two shear
transformation. The Bain distortion can be accomplished by two lattice variant shears (without atomic shuffling) on systems for which
$S_{m2} \rfloor S_{m1}$, and if affects like the above apply, may be more favorable
than the single shear with atomic shuffling.

The two shear mechanism is possible for both thermal and stress
induced martensite, but since the two shears depend on their relative orientation with respect to the applied stress, it is likely
that the habit planes and shape strains will differ; various experimental studies report such differences. The two shears can also
account for the lenticular shaped plates and for their limited
lengthwise growth. The dislocations at the edges of the plate transform the lattice into an intermediate structure, and since the second shear does not occur in this region, edge growth will stop when
the incremental elastic distortion of the intermediate product is
equal to the chemical driving energy

Not withstanding the fact that twinning and faulting are found
in most martensites, the question of whether the lattice invariant
deformation is slip, twinning, or faulting (16,17) is not a basic
or primary one; rather the present study shows that the lattice
invariant shear by twinning or faulting is a consequence of the
transformation mechanism which leaves residual dislocations in the
interface, and this will depend on the plate thickness. In general
the habit plane will be independent of the plate thickness only if
the interface has no residuals in it, or if the residuals in the
interface are of random sign, with an average strain of zero. The
former would apply if the transformation dislocations are genrated
by a pole mechanism, or if κ_1 is the slip plane of the host lattice,
and the latter would apply if all the neighboring network dislocations , regardless of their sign, move to the marteniste plate.
This may be possible for thermal martensite but not likely for
stress induced martensite, where only those dislocations which are
favored by the applied stress field will move to the interface.

REFERENCES

1.   M. S. Wechsler, D. S. Lieberman, and T. A. Read, Trans. AIME
     197, 1503 (1953); J. Appl. Phys. 26, 473 (1955).

2.   J. S. Bowles and J. K. Mackenzie, Acta Met. 2, 129, 137 (1954).

3.   D. S. Lieberman, PHASE TRANSFORMATIONS, Amer. Soc. Metals,
     Metals Park Ohio, p. 1 (1970).

4.   B. A. Bilby and J. W. Christian, THE MECHANISM OF PHASE TRANS-
     FORMATIONS IN METALS, Inst. Met. Monographs 18, 121 (1956);
     J. Iron and Steen Inst. Feb. 122 (1961).

5.   J. W. Christian, J. Inst. Met. 386 84, (1955); Comments on
     Solid State Physics 1, 125 (1968).

6.   S. Mendelson, Conference Abstracts, 1971 Fall Meeting of AIME,
     Detroit Mich., Paper No. 302 p. 310.

7.   S. Mendelson, in PHASE TRANSITIONS - 1973, Edited by L. E.
     Cross, Pergamon Press (1973) p. 287.

8.   S. Mendelson, in FUNDAMENTAL ASPECTS OF DISLOCATION THEORY
     p. 550. Ed. by J. O. Simmons, R. deWit and R. Bullough. Nat.
     Bur. Std. (US) Spec. Publ. No. 317 U.S. GPO, Washington DC
     (1970).

9.   S. Mendelson, Proc. Conf. STRENGTH OF METALS AND ALLOYS, Tokyo,
     Japan (1967), Suppl. Japan Inst. Metals 9, 812 (1968).

10. S. Mendelson, Ref. 9, p. 819.

11. S. Mendelson, Mat. Sci. Eng. 4, 231 (1969).

12. S. Mendelson, Ref. 8, p. 495.

13. S. Mendelson, J. Appl. Phys. 43, 2102 (1972).

14. S. Mendelson, in TITANIUM SCIENCE AND TECHNOLOGY, Vol. 3,
     Edited by R. I. Jaffee and H. M. Burte, Plenum Publ. Corp.
     p. 1585 (1973).

15. C. M. Wayman and K. Shimizu, Met. Sci. J. 6 175 (1972).

16. L. Delaey and R. V. Krishnan, H. Tas, and H. Warlimint I, II,
     and III, J. Mat. Sci. 9, 1521, 1536, 1545 (1974).

17. J. Perkins, Met. Trans. 4, 2709 (1973); Scripta Met. 9, 121
     (1975).

18. J. D. Eisenwesser and L. C. Brown, Met. Trans. 3, 1359 (1972);
     K. Oshi and L. C. Brown, Met. Trans. 2, 1971 (1971);
     R. V. Krishnan and L. C. Brown, Met. Trans. 4, 423 (1973).

19. K. E. Easterling and P. R. Swann, in THE MECHANISMS OF PHASE
     TRANSFORMATIONS IN CRYSTALLINE SOLIDS, Inst. Met. Monographs
     33, 152 (1969).

20. J. Perkins, Scripta Met. 8, 31 (1974); 8, 439 (1974).

21. I.A. Arbuzova, V. S. Gavrilyuk and L. G. Khandros, Fiz. Metal
     Metolloved 27, 1126 (1969).

22. R. E. Busch and R. T. Luederman and P. M. Cross, U.S. Army
     Materials Research Reports, AD 629726 (1966).

23. S. Miura, A. Ito and N. Nakanishi, Scripta Met, 9, 247 (1975).

ON ZONAL DISLOCATIONS AND THEIR ORIGIN FOR TWINNING AND MARTENSITE FORMATION

S. Mendelson

The City University of New York Research Foundation
The City College of New York, Steinman Hall, Room 12
138th Street and Convent Avenue, New York, N. Y. 10031

As in the early studies of plastic deformation of crystals a dislocation theory followed to account for the large difference between theoretical and actual strength; the same arguments are valid for martensite formation but, unlike plastic deformation, martensitic phase transformations are complicated by a change in lattice parameters and/or crystal structure.

The zonal dislocation is physically one which transforms a lamella of parent into a new structure (twin or martensite) with a small shear strain when accompanied by atomic shuffling in its core. The name was introduced by Kronberg in 1959 (1), but the concept is inherent in earlier studies by Kiho (2). A formal definition was given by Mendelson (3,4), applied to twinning in various crystal structures (3-7), to a partialized dislocation model for the structure of the dislocation (3,4,7,8), and to martensite formation (9-12). Until this meeting recognization of its potential has been limited by excessive inertia, and some authors have found difficulty in applying them; this may be related to earlier confusion created by applications of inverse logic to Mendelson's definitions, resulting in illogical as well as incorrect definitions (13). Although this was refuted (9,14), the negative effect could not be completely errased.

CHARACTERISTICS OF ZONAL DISLOCATIONS

The physical characteristics of the zonal dislocation can be described as follows: If the lattice translation along the $\vec{\eta}_1$ shear direction is $\vec{b}_0$ and the shear for a lamella (twin plane or martensite lamella) of $n'd_{K_1}$ thickness is $b_s'=s'b_0$ the shear strain will be $S=s'b_0/n'd_{K_1}$. For a lamella of $2n'd_{K_1}$ thickness the homogeneous shear will be $2s'b_0$ and the shear strain is again $S=s'b_0/n'd_{K_1}$;

the same applies for homogeneous shear with n=3,4,---. On the other hand, with zonal dislocations the homogeneous shear of nb_s' is replaced by an inhomogeneous shear of $b_s=nb_s'-pb_o$ for an n-layer lamella of twin or martensite, where p is an integer, usually 1 or 2, chosen to give the smallest absolute value of b_s for a particular n. This follows because when nb_s' approaches the magnitude of b_o, or a multiple of it, one can always accomplish the transformation by a reduced shear, which is the difference between nb_s' and pb_o. Since this shear is no longer homogeneous only the outer plane shears through $\vec{b}_s$, leaving the atoms on intermediate planes to shuffle to their twin or martensite positions. The sign of $\vec{b}_s$ indicates its stress-sence for twinning or martensite formation.

The above physical discription for the Burgers vector of the zonal dislocation is represented algebraically by

$$\vec{b}_s = s\vec{b}_o = sI'\vec{\eta}_1 \tag{1}$$

where $\quad s = ns' - p \tag{2}$

and $\quad \vec{b}_s = ns'\vec{b}_o - p\vec{b}_o \tag{3}$

for defromation twinning, and

$$s = (n/n')s' - p \tag{4}$$

$$\vec{b}_s = (n/n')s'\vec{b}_o - p\vec{b}_o \tag{5}$$

for martensite formation. s' is the homogeneous shear fraction for a single plane of twin, and the shear for a unit lamella of n' planes of martensite. n is the number of planes in the twin lamella and (n/n') is the number of unit lamellae in the martensite lamella which forms when the zonal dislocation glides.

DEFORMATION TWINNING

As an example for deformation twinning we take the martensite in the Ag-45%Cd alloy studied by Krishnan abd Brown (15). The lattice parameters for this martensite are a_o=4.8651Å, b_o=4.7536Å and c_o=3.0968Å and there are six possible {110}<1$\bar{1}$0> type twin systems. The (101)[10$\bar{1}$] and (10$\bar{1}$)[101] twin systems are of the same type and correspond to S_m=(010); the (011)[01$\bar{1}$] and (01$\bar{1}$)[011] similar systems correspond to S_m=(100); while the (110)[1$\bar{1}$0] and (1$\bar{1}$0)[110] twin systems correspond to S_m=(001). The magnitude of the shear strains and Burgers vectors for both systems of each $\vec{S}_m$ zone are the same

If we consider the (101)[10$\bar{1}$] twin system $b_o=\frac{1}{2}$[10$\bar{1}$] and s'= $\frac{3}{7}$ for n=1, giving $b_s'=\frac{3}{7}b_o=\frac{3}{14}$[10$\bar{1}$]. The (101) plane spacing is $d_{K_1}=\frac{1}{14}$[205] and the shear strain is $S=b_s/nd_{K_1}=3$[10$\bar{1}$]/[205]=0.946. This shear strain is too large to operate. For n=2, $b_s=2(\frac{3}{7})b_o-b_o=$ $-\frac{1}{7}b_o$ and the shear strain is $S=-\frac{1}{2}$[10$\bar{1}$]/[205]= -0.158. Finally for n=3, $b_s=3(\frac{3}{7})b_o-b_o=\frac{2}{7}b_o$ and the shear strain is $S=\frac{2}{3}$[10$\bar{1}$]/[205] 0.210. The much smaller shear strains for n=2 and 3 make twinning favorable.

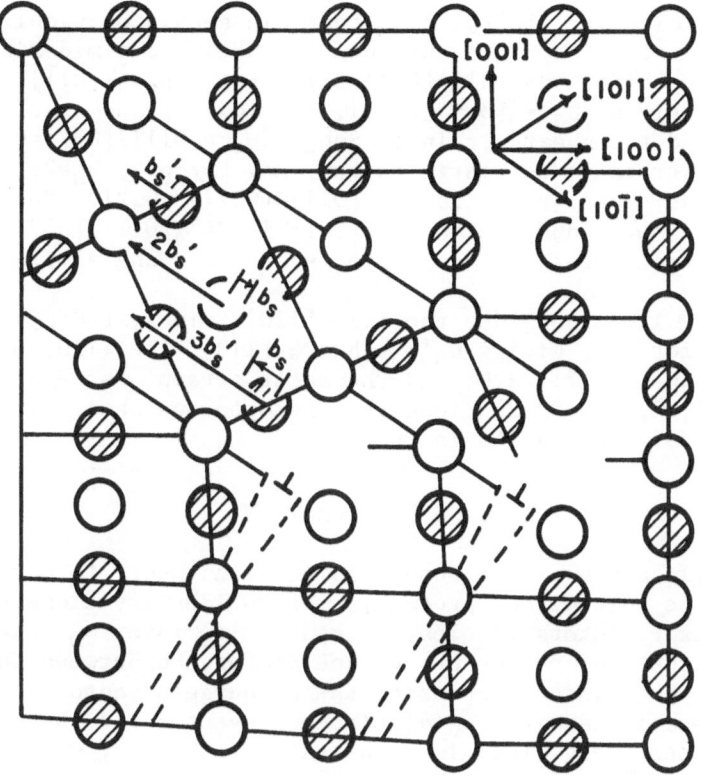

FIG. 1

These shears are illustrated in Fig 1. The projection plane is $S_m = (010)$ for a $(101)[10\bar{1}]$ twin. b_s' $(=\frac{3}{14}[10\bar{1}])$, $2b_s'$, and $3b_s'$ are shown for homogeneous shear corresponding to $n=1,2$, and 3 respectively. b_s for $n=2$ is the difference between b_o and $2b_s'$, with a magnitude of $\frac{1}{3}b_s'$ and a shear strain $\frac{1}{6}$ as large as for homogeneous shear. b_s for $n=3$ is the difference between $3b_s'$ and b_o, with a magnitude of $\frac{2}{3}b_s'$ and a shear strain $\frac{2}{9}$ as large as for homogeneous shear. The zonal dislocations are shown for $n=2$. The possibility of different shears for the same twin system has not been considered in the phenomenological theories, where the twin mode plays an important role. It is also significant that twinning in martensite is possible for both senses of stress. Interface twinning dislocations of both sign can move into thermal martensite, whereas under an applied stress the dislocations of the same stress sense will move into the martensite and the others will move in the opposite direction; both motions contribute to pseudoelasticity. If the twinning partials should polygonize mid-way in the martensite plate they would constitute a mid-rib.

If we consider the $(011)[01\bar{1}]$ twin system $b_o = \frac{1}{2}[01\bar{1}]$ and $s' = \frac{2}{5}$ for $n=1$, giving $b_s' = \frac{2}{5}b_o = \frac{1}{5}[01\bar{1}]$. The (011) plane spacing is $d_{K_1} = \frac{1}{20}[037]$ and the shear strain is $S = b_s/nd_{K_1} = 4[011]/[037] = 0.875$. This shear strain is too large to operate. For $n=2$, $b_s = 2(\frac{2}{5})b_o - b_o = -\frac{1}{5}b_o$ and the shear strain is $S = -[011]/[037] = -0.219$. For $n=3$, $b_s = 3(\frac{2}{5})b_o - b_o = \frac{1}{5}b_o$ and the shear strain is $S = \frac{2}{3}[011]/[037] = 0.146$. The shear strains for $n=2$ and 3 are favorable, and similar atomic models apply.

Finally if we consider the $(110)[1\bar{1}0]$ twin system $b_o = \frac{1}{2}[1\bar{1}0]$ and $s' = \frac{1}{43}$ for $n=1$, giving $b_s' = \frac{1}{43}b_o = \frac{1}{86}[1\bar{1}0]$. The (110) plane spacing is $d_{K_1} = \frac{1}{43}[21,22,0]$ and the shear strain is $S = b_s/nd_{K_1} = \frac{1}{2}[1\bar{1}0]/[21,22,0] = 0.0233$. Since this homogeneous shear strain is small no zonal dislocation is required.

MARTENSITE FORMATION

In application of the zonal dislocation theory to martensite formation the most direct procedure is to identify close packed or low index $\vec{t}$ vectors in close packed or low index S_m' planes of the two phases. The components of the difference between these $\vec{t}$ vectors along $\vec{\eta}_1$ and $\vec{K}_1$ represent the $\vec{b}_s$ Burgers vector for the zonal dislocation and the $\vec{b}_n$ displacement vector for the volume change, respectively. This is illustrated in Fig. 2, with the component vectors given by

$$\vec{b}_s = [(\vec{t}_f \cdot \vec{\eta}_{1f})/\eta_{1f}^2]\vec{\eta}_{1f} - [(\vec{t}_b \cdot \vec{\eta}_{1b})/\eta_{1b}^2]\vec{\eta}_{1b} \qquad (6)$$

$$\vec{b}_n = [(\vec{t}_f \cdot \vec{K}_{1f})/K_{1f}^2]\vec{K}_{1f} - [(\vec{t}_b \cdot \vec{K}_{1b})/K_{1b}^2]\vec{K}_{1b} \qquad (7)$$

If the resulting shear strain $S = b_s'/nd_{K_1}$ is small (≈ 0.2), $\vec{b}_s'$ will serve as the transformation dislocation. If it is large a zonal dislocation for a thicker lamella is derived in the same manner as for deformation twinning.

The above direct procedure is general and applies to most phase transformations because S_m' is a close packed or low index plane in both phases containing close packed or low index $\vec{t}$ vectors

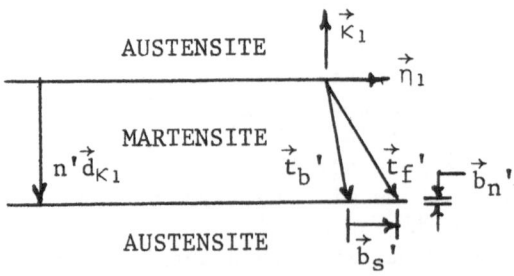

FIG. 2

which can be identified by inspection. If these $\vec{t}$ vectors cannot be identified then $\vec{t}'$ vectors for a unit lamella of martensite can be represented by a linear sum of displacements along $\vec{\eta}_1$ and $\vec{\kappa}_1$ by

$$\vec{t}' = n'\vec{d}_{\kappa_1} + \rho(I/\eta_1{}^2)\vec{\eta}_1 \tag{8}$$

where ρ is an integer and I is the cell factor for the $(hk\ell)$ plane which is perpendicular to $\vec{\eta}_1$. ρ is increased until $\vec{t}'$ is a lattice translation which lies in S_m'. It then follows that

$$\vec{b}_s' = \rho_f(I_f/\eta_{1f}{}^2)\vec{\eta}_{1f} - \rho_b(I_b/\eta_{1b}{}^2)\vec{\eta}_{1b} \tag{9}$$

$$\vec{b}_n' = n_f'\vec{d}_{\kappa_{1f}} - n_b'\vec{d}_{\kappa_{1b}} \tag{10}$$

These $\vec{b}_s'$ can then be substituted for $s'\vec{b}_o$ into Eq (5) to obtain the Burgers vector of a zonal dislocation with a small shear strain.

A. fcc→bcc and fcc→bct Transformations

In applying the theory to the fcc→bcc and fcc→bct transformations two favorable S_m' with the Kurdjumov-sachs orientation relationship are possible; these are

$$S_m' = (111)_f||(110)_b \quad \text{and} \quad S_m' = (\bar{1}10)_f||(\bar{1}11)_b \tag{11}$$

For $S_m'=(111)_f||(110)_b$ matching is perfect along $\vec{S}_m'$ and $\vec{\eta}_1$ when $a_b/a_f=\sqrt{2/3}$, and the shear system is represented by (9)

$$\vec{\eta}_1 = [\bar{1}10]_f||[\bar{1}11]_b \quad \text{and} \quad \vec{\kappa}_1 = (\bar{1}\bar{1}2)_f||(1\bar{1}2)_b \tag{12}$$

The $\vec{t}$ vectors are

$$\vec{t}_f = \tfrac{1}{2}[0\bar{1}1]_f \quad \text{and} \quad \vec{t}_b = \tfrac{1}{2}[1\bar{1}1]_b \tag{13}$$

and corresponds to $n_b/n_f=2/3$ for the $(\bar{1}\bar{1}2)_f||(1\bar{1}2)_b$ shear plane.

Matching along other $\vec{\eta}_1$ directions of the $\vec{S}_m'$ zone for $a_b/a_f < \sqrt{2/3}$ are given by

$$\tfrac{1}{2}\rho[\bar{1}10]_f + \tfrac{1}{2}[0\bar{1}1]_f = \tfrac{1}{2}\rho[\bar{1}11]_b + \tfrac{1}{2}[1\bar{1}1]_b \tag{14}$$

where ρ is an integer. This family of $\vec{\eta}_1$ directions corresponds to $n_b/n_f=1/1$, referred to as pseudohomogeneous shear, and applies for the lattice parameters in steels when $\rho \gtrless 5$, with the Kurdjumov-Sachs orientation relationship within about a $\tfrac{1}{2}°$ deviation from $[\bar{1}10]_f||[\bar{1}11]_b$.

When $\rho=5$ the shear corresponds to

$$\vec{\eta}_1 = [\bar{5}41]_f||[\bar{2}23]_b \quad \text{and} \quad \kappa_1 = (12\bar{3})_f||(3\bar{3}4) \tag{15}$$

Matching along this $\vec{\eta}_1$ applies to high carbon steels, while that for low carbon steels corresponds to higher ρ. The $\vec{t}$ vectors represented by Eq (13), corresponds to a Burgers vector of $\vec{b}_s=5\vec{b}_s'$, where $\vec{b}_s'$ is obtained from

$$t_f' = \tfrac{1}{2}[1\bar{1}0]_f \quad \text{and} \quad t_b' = \tfrac{1}{2}[1\bar{1}\bar{1}]_b \tag{16}$$

The closest atomic approach is maintained in the interface only near the coincident atomic sites; those atoms in the interface between these are distorted, but this is least for an orientation in which the q are small and the shear is pseudohomogeneous; this is so because the atomic displacements for most of the atoms is smaller than for the case when $n_b/n_f \neq 1$. For the above S_m' the coincident interface sites along S_m' are at the levels $2\rho[111]_f = 3\rho[110]_b$ (ρ is an integer), at every sixth S_m' plane. For pseudohomogeneous shear the volume change is parallel to $\vec{\kappa}_1$, with the effect that b_n/b_s imposes a torque on the edge dislocation line which tends to change the orientation relationship so that $b_n = 0$. This is possible for this S_m', but since the shear is no longer pseudohomogeneous the atomic shuffling is less simple and the interface distortions between coincident sites are larger.

An example of shear with $b_n \simeq 0$ corresponds to matching with

$$\vec{n}_1 = [\bar{3}4\bar{1}]_f = [\bar{4}43]_b \quad \text{and} \quad \kappa_1 = (52\bar{7})_f || (3\bar{3}8)_b \qquad (17)$$

Taking the w^c variant of tetragonality (the tetragonal axis is along the w-index) the matching is perfect along $\vec{n}_1$ when $c/a_b = 1.08$ and $a_b/a_f = 0.7823$ in a high carbon steel with 1.78%C. The deviation from $[\bar{1}10]_f || [\bar{1}11]_b$ is 6.5°, within the range reported (16). Although the interface distortions are larger the tendency to reduce b_n may favor it. This shear is also an interesting example because it shows the relationship between the various definitions. The minimum thickness of the transformed lamella of this mode has $n_b'/n_f' = 2/3$ and the t' vectors are

$$t_f' = \frac{1}{2}[2\bar{3}1]_f \quad \text{and} \quad t_b' = \frac{1}{2}[2\bar{2}\bar{1}]_b \qquad (18)$$

This corresponds to

$$b_s' = 0.692a_f \quad \text{and} \quad S = b_s'/3d_{\kappa_1 f} = 4.9 \qquad (19)$$

This large shear strain cannot operate. If we take $(n/n') = 4$ the Burgers vector of the zonal dislocation for the transformation is reduced to

$$b_s = (n/n')b_s' - b_{\circ f} = 0.213a_f \qquad (20)$$

and $\quad S = b_s/12d_{\kappa_1} = 0.307 \qquad (21)$

The same b_s is obtained directly by taking

$$\vec{t}_f = \frac{1}{2}[\bar{1}01]_f \quad \text{and} \quad \vec{t}_b = [001]_b \qquad (22)$$

and shows the relationships between t', t, and Eq (5)

For the $S_m' = (\bar{1}10)_f || (\bar{1}11)_b$ plane of shear with the Kurdjumov-Sachs orientation relationship the atoms are on two $(\bar{1}10)_f$ planes in fcc and on three $(\bar{1}11)_b$ planes in bcc. Since matching along S_m' corresponds to $\frac{1}{2}[\bar{1}10]_f = \frac{1}{2}[\bar{1}11]_b$ when $a_b/a_f = \sqrt{2/3}$ the two fcc S_m' planes are equivalent to three bcc S_m' planes. The shear for planes of this zone is pseudohomogeneous with $n_b'/n_f' = 2/2$. All atoms on each κ_1 shear plane undergo shear through nb_s' with a small shear strain (~ 0.2), but direct shear to their final positions applies to $\frac{1}{3}$ of the atoms on every other κ_1 plane; the remainder have an additional shuffling component parallel to $\vec{S}_m'$. The atomic shuffling for habits of this $\vec{S}_m'$ is more favorable than that for $\vec{S}_m' = (111)_f||(110)_b$ when $\Delta V||\vec{\kappa}_1$; when $\Delta V||\vec{\eta}_1$ the atomic shuffling for this $\vec{S}_m'$ becomes complex and the interface distortions between coincident matching sites very severe.

The $\vec{\eta}_1$ shear directions for this zone with pseudohomogeneous shear and the Kurdjumov-Sachs orientation relationship within about $\frac{1}{2}°$ are given by

$$\tfrac{1}{2}\rho[33\bar{2}]_h + [00\bar{1}]_b = 2\rho[01\bar{1}]_b + [\bar{1}0\bar{1}]_b \tag{23}$$

When $\rho=2$ this corresponds to

$$\vec{\eta}_1 = [11\bar{1}]_f||[\bar{1}4\bar{5}]_b \quad \text{and} \quad \kappa_1 = (112)_f||(321)_b \tag{24}$$

and when $\rho=5$ the matching corresponds to

$$\vec{\eta}_1 = [55\bar{4}]_f||[1,10,11]_b \text{ and } \kappa_1 = (225)_f||(347)_b \tag{25}$$

Both of these and the habits between them apply to steels with dilatations of about $\frac{3}{4}\%$ along $\vec{\eta}_1$ and $\vec{S}_m'$ directions. The $\vec{t}$ and $\vec{t}'$ vectors for the zonal dislocations are

$$\vec{t}_f = [001]_f \quad \text{and} \quad \vec{t}_b = [101]_b \tag{26}$$

$$\vec{t}_f' = \tfrac{1}{2}[33\bar{2}]_f \quad \text{and} \quad \vec{t}_b' = 2[01\bar{1}]_b \tag{27}$$

The dilatations along $\vec{S}_m'$ can be eliminated by changing the $\vec{S}_m$ direction, or by a second lattice variant shear.

B. bcc→hcp Transformations

For the Kurdjumov-Sachs orientation relationship in fcc→bcc, and the Burgers orientation relationship in bcc→hcp, close packed directions and close packed planes are parallel in the two phases. These two orientation relationships are thus equivalent, with the equivalence represented by $\frac{1}{3}[\bar{2}110]_h = \frac{1}{2}[1\bar{1}0]_f$ and $(0001)_h||(111)_f$, and mechanisms for $S_m'=(111)_f||(110)_b$ are equivalent to those for $S_m'=(0001)_h||(110)_b$ in bcc→hcp. The distinction between them is that the stacking of $(111)_f$ planes in fcc is ABCABC---, while that of $(0001)_h$ basal planes in hcp is ABABAB---. Since the latter is the same stacking as for $(110)_b$ planes in bcc the shuffling is more favorable in bcc→hcp than in fcc→bcc for the transformations with the above equivalent S_m'.

The shear system in bcc→hcp, equivalent to that given by Eq (12) for fcc→bcc is

$$\vec{\eta}_1 = [2\bar{1}\bar{1}0]_f||[\bar{1}11]_b \text{ and } \kappa_1 = (0\bar{1}10)_h||(1\bar{1}2)_b \tag{28}$$

This applies with perfect matching along $\vec{\eta}_1$ and $\vec{S}_m'$ directions when $a_h/a_b=\frac{1}{3}\sqrt{3}$. The shear is homogeneosu and no atomic shuffling is required.

Similarly, the equivalent shear system in bcc→hcp to that given by Eq (15) in fcc→bcc is (10)

$$\vec{\eta}_1 = [3\bar{2}\bar{1}0]_f || [\bar{2}23]_b \quad \text{and} \quad \kappa_1 = (14\bar{5}0)_h || (3\bar{3}4)_b \qquad (29)$$

This applies with perfect matching along $\vec{\eta}_1$ in pure titanium. The shear is direct for half of the atoms in each shear plane; the other half undergo a shuffling motion to their martensite positions. The dilatation parameter $\delta_{S_m}=1.009$ for matching along $\vec{S}_m'=[0001]_h$ $||[110]_b$ can be eliminated by a second lattice variant shear.

C. Transformations of B₂ Alloys

In applying the theory to transformations of B₂ alloys we again take, for example, the data of Krishnan and Borwn for Ag-45%Cd. The orientation relationship has $S_m'=(1\bar{1}0)_b||(010)_o$ and the symmetry of the $\vec{S}_m'$ axis is the same in both phases. The matching along $\vec{\eta}_1$ is given by

$$[\rho\rho\sigma]_b = [\rho 0\sigma]_o \qquad (30)$$

where ρ and σ are positive integers. The $\vec{t}$ vectors for the transformation dislocations are

$$\vec{t}_b = \tfrac{1}{2}[11\bar{1}]_b \quad \text{and} \quad \vec{t}_b = \tfrac{1}{2}[10\bar{1}]_o \qquad (31)$$

The matching along $\vec{\eta}_1=[111]_b=[101]_o$ has a dilatation parameter of $\delta_{\eta_1}=1.0047$ and one of $\delta_{\eta_1}=0.9998$ for $\vec{\eta}_1=[9,9,10]_b=[9,0,10]_o$. For $\vec{S}_m'=[1\bar{1}0]_b||[010]_o$ the dilatation parameter is $\delta_{S_m}=1.014$. This dilatation is eliminated by a habit plane rotation about $\vec{\eta}_1$ (12), or by a second lattice variant shear on a system for which $S_{m2}\bot S_{m1}$.

The above exmaples are consistent with some reported data, but the habit plane and shape strain direction are generally not equivalent to the κ_1 shear plane and $\vec{\eta}_1$ shear direction respectively. These depend on the nature of the transformation shear, i.e., on whether $\Delta V||\vec{\kappa}_1$, $\Delta V||\vec{\eta}_1$, it occurs by two shears, or whether the martensite is subjected to other shears from within or without; these depend on the dislocation structure of the interface and on the origin of the dislocations which transform the structure.

ON THE ORIGIN OF DISLOCATIONS FOR MARTENSITE FORMATION

The zonal dislocations necessary for martensite formation can develop either by two-dimensional nucleation during the transformation, or by dissociation and multiplication of those in the host lattice. In the nucleation mechansim very high chemical or mechanical driving forces are required. Since it can only account for growth of existing martensite, a mechanism for formation of the martensite nuclei is required. In the zonal dislocation theory (9-12) martensite formation and growth has its origin in the defect structure of the host lattice; the defects may be as-grown, introduced during the quench, or at the transition temperature.

High densities of small dislocation loops might develop during a rapid quench from a high temperature to form a supersaturated concentration of vacancies; these condense into vacancy disks which

then collapse into dislocation loops. Mukherjee (17) reports a mot
tled microstructure, characteristic of a high density of small
dislocation loops in splat quenched B_2 alloys, however most transformations in these alloys do not undergo so severe a quench from
so high a temperature.

If the length to with ratio of a martensite plate is $L/w = \rho$
the lamella thickness in nd_{κ_1}, and each dislocation dissociates
into n_1 transformation partials the density of dislocaitons required for the transformation is $[n_1 \rho (nd_{\kappa_1})^2]^{-1}$. Substituting
reasonable numbers this calls for a dislocation density of about
$10^{14}/cm^2$, or about 10^6 times the density of network dislocations
in annealed metals.

Dislcoations might be generated in sufficient numbers by multiple cross-slip of screws, as in glide band formation by the double
cross-slip mechanism (18), or by a pole mechanism. In the double
cross-slip mechansim the generated dislocations dissociate into
transformation partials which move to the ends of the martensite
plate, leaving residual dislocations behind in the interface. When
the stress is removed or the temperature is raised reversion
occurs as the transformation partials return to their residuals.
The perfect dislocation loops may then collapse and eliminate
themselves, but a good number will probably be retained as polygon
walls, dipoles, and other debris, and the dislocation density
should increase after each transformation cycle.

A pole mechansim can develop at a dislocation loop or part of
a dislocation loop in the host lattice of Burgers vector $\vec{b}$, which
is oblique to $\vec{\kappa}_1$, when the component of $\vec{b}$, parallel to $\vec{\kappa}_1$, is equal
to the lamella thickness nd_{κ_1}, and the component of $\vec{b}$ parallel to
$\vec{\eta}_1$ is equal or greater than $\vec{b}_s$; the former produces the necessary
displacement (pole strength) and the latter the necessary shear.

POLE MECHANISMS

Since the pole mechanism applies to dislocation loops in the
host lattice when their $\vec{b} = \vec{t}$, and since the $\vec{t}$ vectors for the zonal
dislocation are close packed lattice translations in most crystal
systems the pole mechanism is very favorable. The transformation
partial forms by dissociation of the edge or near edge component
of the dislocation loop and is displaced through nd_{κ_1} by the screw
components when it wraps around them, leaving a transformed loop in
each lamella of the martensite plate. The pole mechanism is reversible and, if the poles are part of the dislocation network, the same
plates will form and disappear at the same positions during cycling.
Various experimental studies report such behavior in various alloys.
In this case the first martensite plates will develop at dislocation
loops which eminate from those network dislocation segments which
are favorably oriented with respect to the driving force (chemical
or mechanical) and bow out with the largest edge component, since

the critical bowing stress is smallest. Subsequent plates will develop at the smaller loops under a greater driving force and, since
this results in higher dislocation velocities, the tendency for
plate penetration and burst phenomena will increase at lower temperatures and higher applied stresses.

A dislocation dissociation is favorable (in the absence of
chemical or mechanical driving forces) when the force between partials is repulsive; this is given by $\gamma_m > 0$ and

$$\gamma_m = (K/2\pi\varepsilon a)\vec{b}_1 \cdot \vec{b}_2 \tag{32}$$

where K is the elastic energy factor for the transformation dislocation at the transition temperature, and $\varepsilon a = r_{12}$ is the critical separation at which linear elasticity applies (10). The dissociation is
spontaneous when $\gamma_m/\gamma > 1$, where γ is the interface energy. When γ_m/γ
<1 the dissociation will occur below M_s or for an applied stress
when $(\gamma_m + \gamma_c)/\gamma > 1$, where γ_c is the chemical or mechanical force.
Since it is not possible to evaluate the interface energy with a
sufficient degree of accuracy it is necessary to show that the
dissociations are energetically favorable in the absence of chemical or mechanical driving forces, as indicated by a positive γ_m.

For fcc→bcc with $S_m' = (111)_f || (110)_b$ two poles are possible
with the Burgers vectors

$$\vec{t}_f = \frac{1}{2}[0\bar{1}1] \quad \text{and} \quad \vec{t}_f = \frac{1}{2}[\bar{1}01] \tag{33}$$

The edge component of these produces the transformation partial,
and the screw component accomplishes the necessary displacement.
When Eq (12) applies for the $(\bar{1}\bar{1}2)_f$ shear plane the Burgers vector
of the transformation dislocation is $b_s = \frac{1}{12}[1\bar{1}0]_f$ and forms ac-
coding to

$$\frac{1}{2}[0\bar{1}1] \rightarrow \frac{1}{12}[1\bar{1}0] + \frac{1}{12}[\bar{1}\bar{5}6] \tag{34}$$

$$\frac{1}{2}[\bar{1}01] \rightarrow \frac{1}{12}[\bar{1}10] + \frac{1}{12}[\bar{5}\bar{1}6] \tag{35}$$

with $\quad \gamma_m = (Ka_f/2\pi\varepsilon)(1/36) \tag{36}$

These correspond to twin related orientations of the martensite
plate on the same $(\bar{1}\bar{1}2)_f$ shear plane

When Eq (15) applies for the $(12\bar{3})_f$ shear plane the Burgers
vector of the transformation dislocation is $0.021[5\bar{4}\bar{1}]_f$ with a_b/a_f
$=0.786$ and n=5, and forms by

$$\frac{1}{2}[0\bar{1}1] \rightarrow 0.021[5\bar{4}\bar{1}] + [\bar{0}.\bar{1}05, \bar{0}.\bar{4}\bar{1}6, 0.521] \tag{37}$$

with $\quad \gamma_m = (Ka_f/2\pi\varepsilon)(1/77) \tag{38}$

These dissociations have their equivalent dissociations in bcc→hcp,
with the same γ_m.

For the transformation of B_2 alloys the poles are the screw
components of $\vec{t}_b = \frac{1}{2}[11\bar{1}]_b$, and the edge component of this dissociates into the transformation dislocation $b_s = \frac{1}{197}[9,9,10]_b$ according
to

$$\frac{1}{2}[11\bar{1}] \rightarrow \frac{1}{197}[9,9,10] + \frac{1}{394}[179,179,\bar{2}\bar{1}7] \tag{39}$$

with $\quad \gamma_m = (Ka_b/2\pi\varepsilon)526/38809 \tag{40}$

The pole mechanism also applies to twinning in martensite. Taking the twin in Fig 1 for Ag-45Cd with n=2 the $\vec{t}=\frac{1}{2}[\bar{1}0\bar{1}]$ vector is the Burgers vector in the martensite and has a pole strength of two twin planes. The twinning partial $\vec{b}_s=\frac{1}{14}[\bar{1}01]$ forms by dissociation of the edge component of the dislocation loop according to

$$\frac{1}{2}[\bar{1}0\bar{1}] \rightarrow \frac{1}{14}[\bar{1}01] + \frac{1}{7}[\bar{3}0\bar{4}] \tag{41}$$

with $\gamma_m = (Ka_0/2\pi\varepsilon)(1/69)$ \hfill (42)

DISCUSSION

The defect controlled model for the origin of the zonal dislocations in consistent with studies which show that highly perfect γ-iron particles in a copper matrix do not transform on quenching, but do under a subsequent applied stress. The martensite was found to nucleate and grow when matrix dislocations pass through the γ-iron precipitates.

The pole mechanism can be related to nucleation and premonitory effects. Since γ_m is positive for the above dissociations of the edge, or near edge, component of the dislocation loop the dissociations are favorable in the absence of chemical or mechanical driving forces. Separation of the partials is not possible at high temperatures because the chemical driving force apposes it. It becomes spontaneous at lower temperatures when $(\gamma_m+\gamma_c)/\gamma>1$ where the γ_c chemical driving force adds to the γ_m elastic repulsion between partials. The first partial should form spontaneously and accelerate until it wraps around the screw poles where it is repelled by the residual partial nd_{K_1} below it.

As the transformation partial bends around the screw poles it cycles through maximum $\gamma_c+\gamma_m'$ and minimum $\gamma_c-\gamma_m'$ driving forces, with

$$\gamma_m' \simeq (Kb_1 \cdot b_2)/2\pi[(\varepsilon a)^2 + (Nnd_{K_1})^2]^{\frac{1}{2}} \tag{43}$$

where N is the number of nd_{K_1} lamellae advanced by the screw poles. Since εa is small, generally similar to nd_{K_1}, γ_m' drops rapidly as N increases, and the difference between maximum and minimum driving forces is significant only for small N, with the transformation dislocation accelerating as N increases.

The effect is that the first lamella will form prior to the martensite avalanche or at a temperature above M_s and constitutes a premonitory effect. The premonitory nucleus is then the first lamella of martensite with a diameter D equal to the spacing of screw poles. If we equate this to the proposed preexisting embrios (20) we can compute the equivalent stress for martensite formation, as the stress required to bow the dislocation out into a semicircle of diameter D, given by

$$\tau = Gb/D \tag{44}$$

Pseudoelasticity and shape memory are enhanced when D is large, and consequently when the dislocation density of the host phase is low.

This equation also applies for dislocation multiplication by the Frank-Read mechanism, but since the critical stress is directly proportional to the Burgers vector the pole mechanism with $b_s \simeq \frac{1}{6}b$ will require an effective stress of about $\frac{1}{6}$ as large, consistent with the phenomenological prerequisite for SME (19). Taking $G = 6 \times 10^{11}$ dynes/cm^2, $D = 1200$Å and $b_s = \frac{1}{12}[\bar{1}10]_f = 0.424$Å ($a_f = 3.60$Å) for the Fe-29.2Ni-0.2Mn alloy studied by Raghavan and Cohen (20) we find $\tau = 2.1 \times 10^8$ dynes/cm$^2 = 3,080$psi as the minimum equivalent stress required to form the first martensite plate. This can be compared with $\tau = 18,500$psi for plastic flow and with $\tau = 10,000-15,000$psi for studies of stress induced martensite.

Since the pole mechanism for martensite formation does not leave residuals in the interface it cannot account for twinning in the martensite and thus might be rejected (9). On the other hand if other pole mechanisms accomplish the twinning the pole mechanism is most favorable and has a lot going for it: (i) it involves a simple dislocation loop which can eminate from the network, (ii) it eliminates the need for high dislocation densities, (iii) it accounts for the low transformation stress, (iv) it can account for both marteniste formation and twinning, (v) it applies to single and multiple shear transformations, (vi) it is a very high velocity mechanism, (vii) it can account for the formation of the same plates in the same places after each transformation cycle, (viii) it can account for both pseudoelasticity and shape memory, (ix) it can account for reversibility without large increases in dislocation density (x) it can explain burst type martensite, (xi) it can account for both the temperature and stress dependence on the amount of martensite formed, (xii) it can account for nucleation and premonitory effects, and finally since $\vec{t} = \vec{b}$ in most transformations, (xiii) it complements the definition of the zonal dislocation.

If the transformation occurs by cross-slip it must be limited to certain screw dislocations, since if many screws can cross-slip, either plastic flow will occur and compete with the transformation, or a very fine martensite structure will develop. In either case the shape memory will be reduced. Thus the phenomemological prerequisite for SME that the elastic limit not be exceeded (19) requires that cross-slip be limited; this follows if the stacking fault energy is low and suggests that this may be a prerequisite for martensite formation. Since multiplication by cross-slip is similar to the operation of a Frank-Read source (18) the stress will be close to the yield stress tending to violet the above phenomenological prerequisite for SME.

REFERENCES

1. M. L. Kronberg, J. Nucl. Mater. $\underline{1}$, 85 (1959).

2. H. Kiho, J. Phys. Soc. Japan $\underline{9}$, 739 (1954); $\underline{13}$, 269 (1958).

3. S. Mendelson, Proc. Conf. STRENGTH OF METALS AND ALLOYS, Tokyo, Japan (1967), Suppl. Japan Inst. Metals $\underline{9}$, 812 (1968).

4. S. Mendelson, Ref. 3, p. 819.

5. S. Mendelson, Mat. Sci. Eng. $\underline{4}$, 231 (1969).

6. S. Mendelson, J. Appl. Phys. $\underline{41}$, 1893 (1970).

7. S. Mendelson, in FUNDAMENTAL ASPECTS OF DISLOCATION THEORY, Edited by J. O. Simmons, R. deWit, and R. Bullough, Nat Bur. Std., Spec. Publ. No. 317 U.S. GPO, Washington D.C. (1970). p. 495.

8. S. Mendelson, J. Appl. Phys. $\underline{43}$, 2102 (1972).

9. S. Mendelson, Ref. 7, p. 550.

10. S. Mendlson, in TITANIUM SCIENCE AND TECHNOLOGY, Vol. 3, Edited by R. I. Jaffee and H. M. Burte, Plenum Publ. Corp. New York, p. 1585 (1973).

11. S. Mendelson, in PHASE TRANSITIONS - 1973, Edited by L. E. Cross, Pergamon press (1973) p. 287.

12. S. Mendelson, Mechanisms For Martensite Formation And The Shape Memory Effect, this symposium.

13. M. H. Yoo, Trans AIME $\underline{245}$, 2051 (1969); Ref. 7, p. 479; Scripta Met. $\underline{4}$, 9 (1970).

14. S. Mendelson, Scripta Met. $\underline{4}$, 5 (1970).

15. R. V. Krishnan and L. C. Brown, Met. Trans. $\underline{2}$, 1971 (1971).

16. C. M. Wayman, Adv. in Mat. Sci. $\underline{3}$, 147 (1968).

17. K. Mukherjee, in these proceedings.

18. S. Mendelson, J. Appl. Phys. $\underline{33}$, 2175 (1962); Phil. Mag. $\underline{8}$, 1633 (1963); J. Appl. Phys. $\underline{43}$, 2113 (1972).

19. J. Perkins, Met. Trans. $\underline{4}$, 2709 (1973); Scripta Met. $\underline{9}$, 121 (1975).

20. V. Raghavan and M. Cohen, Acta Met. $\underline{20}$, 333 (1972).

USE OF TiNi IN MECHANICAL AND ELECTRICAL CONNECTORS

J.D. Harrison and D.E. Hodgson

Raychem Corporation

300 Constitution Drive

CRYOFIT® COUPLINGS

Development of two products based on the shape memory effect will be the topic of this presentation. The products use two different TiNi alloys which are permutations of the Nitinol alloys pioneered by Wm. J. Buehler and his co-workers at the U.S. Naval Ordance Laboratory.

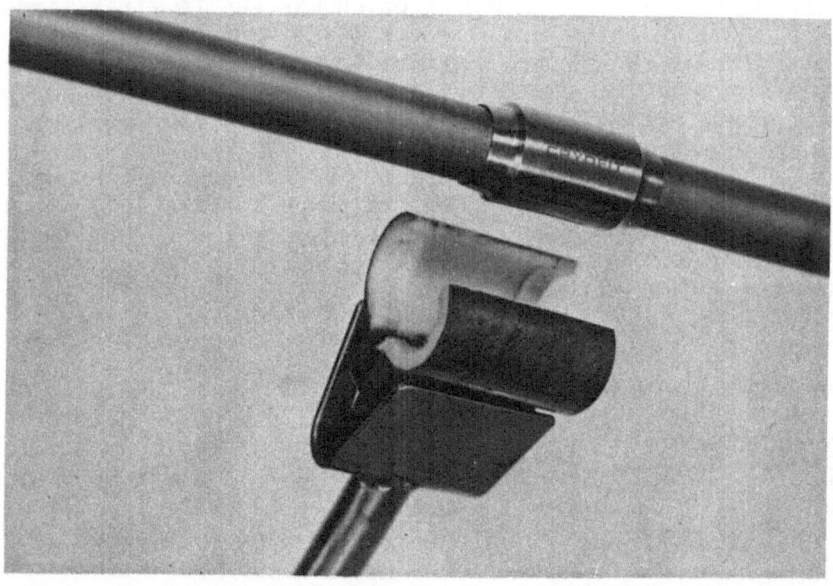

Figure 1. A Cryofit coupling and the installation tool

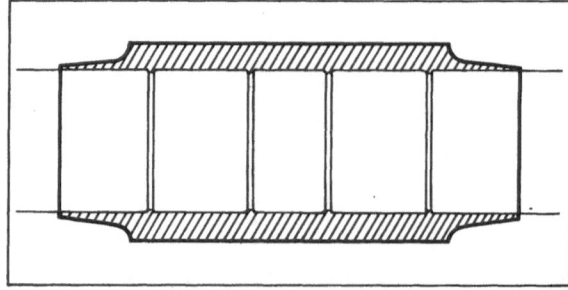

Figure 2.   Sectional view of a cryofit coupling

     The first product is the Cryofit line of permanent couplings for
aircraft hydraulic tubing, Fig. 1.   In terms of conceptual design,
a Cryofit union is a hollow cylinder which has been expanded in its
martensitic range and which shrinks down on the tubes to be joined
as it is heated through its A_s-A_f range.   Installation of these couplings is straightforward: a pre-expanded part is removed from the
liquid nitrogen in which it has been shipped and stored, then the
ends of the hydraulic tubing to be joined are inserted.   As the assembly heats to room temperature, the coupling contracts and effects
a permanent tube union.   Reduction to practice, however, required
simultaneous evolution in mechanical design and alloy development.
Although contraction of the coupling during heating is the essence
of the design, avoiding expansion of the coupling as it is cooling
back through the M_s is the prime consideration for alloy selection.
Stress increases the temperature at which martensite first forms on
cooling.   The alloy in an installed coupling experiences unit stress
in the vicinity of 70 ksi but must not begin to transform to martensite as it is cooled to -54 C, the lowest service temperature specified in aircraft design.   This condition plus a safety margin demands
M_s to be in the vicinity of -150 C.   In the TiNi alloys, the M_s and
A_s occur at about the same temperature.   Liquid nitrogen becomes the
practical refrigerant for expansion, shipping and storage.   Requirements for this low M_s, a 65-ksi minimum room temperature yield
strength and sufficient ductility over the service temperature range
from -54 C to +254 C led to development of an alloy containing 49-50
at.%Ti, 3-4 at.%Fe, balance Ni.

     In addition to the considerations above, the mechanical design
of the coupling, illustrated in Fig. 2, had to provide sufficient
hoop strength to avoid burst at 12,000 psi hydraulic pressure, leak-
tight grip on the tube with sufficient force to avoid tube blow out
and a stress distribution which would assure adequate flex life.
The burst requirement is easily met by providing sufficient coupling
wall thickness.   Leak tightness and tube blow-out resistance are provided for by annular ridges on the inside diameter of the coupling.

These cause rings of local plastic deformation which wipe out imper-
fections in the tube surface assuring leak tightness and anchor themselves in the tube providing resistance to blow out force. Flex life
is assured by the progressive thinning of the coupling wall at its
ends. This provides for a smooth decrease in stress but sufficient
grip is maintained to avoid fretting between the tube and coupling.

The couplings are machined at room temperature, that is in the
austenitic condition, to have an inside diameter 4% smaller than the
tube nominal outside diameter. After being cooled to liquid nitrogen
temperature, the couplings are expanded such that the inside diameter
is 4% larger than the tube outside diameter. The expanded couplings
must be shipped and stored in liquid nitrogen to maintain them in the
martensitic condition. At the site of tube assembly, the couplings
are removed from the liquid nitrogen using tools which have high
thermal inertia, Fig. 1, and which have also been cooled to liquid
nitrogen temperature. The ends of the tubes which are to be joined
are inserted into the still expanded coupling. Within seconds after
the handling tool has been removed, the coupling warms sufficiently
to transform to austenite and contract toward its original shape.

Although structurally similar to mechanically swaged couplings,
shape memory effect couplings differ in a crucial respect. Swaged
coupling are installed by being driven beyond the compressive yield
stress, but as the swaging force is removed, the coupling goes back
through zero stress, then in service, operates under tensile stress.
Shape memory couplings, however, develop tensile stresses as they
contract, locally go beyond the tensile yield stress, but never go
back through zero stress. In addition, the rather low Young's modulus of TiNi provides good compliance to accommodate difference in
coefficient of thermal expansion between coupling and tubing. The
effectiveness of the Cryofit couplings is demonstrated by the ease
with which they meet the performance requirements of MIL-F-18280, but
even more by their performance in service. Since 1970, over 100,000
couplings have been installed in U.S. Navy F-14 fighter planes without a single coupling failure in service.

A number of advantages have resulted from use of shape memory
for couplings. Both the couplings and the installation tools have
low profile which allows high tube density and installation in confined spaces. The installation tools are inexpensive, no special
tube preparation is required and no special skills are required for
installation personnel. Assembly time is short. No heat (other than
ambient) is used, thus the tubes are not degraded by a heat-affected
zone and no flame hazard exists.

CRYOCON^{TM} DEVICES

The second application to be discussed is an electrical pin-and-

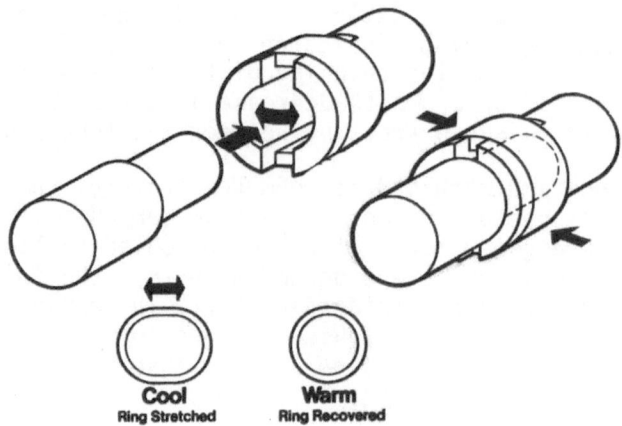

Figure 3.   The Cryocon pin-and-socket contact

socket contact illustrated in Fig. 3.   When the device is cooled, the pin can be inserted or withdrawn with zero force but when the contact is at or above room temperature the pin cannot be removed.

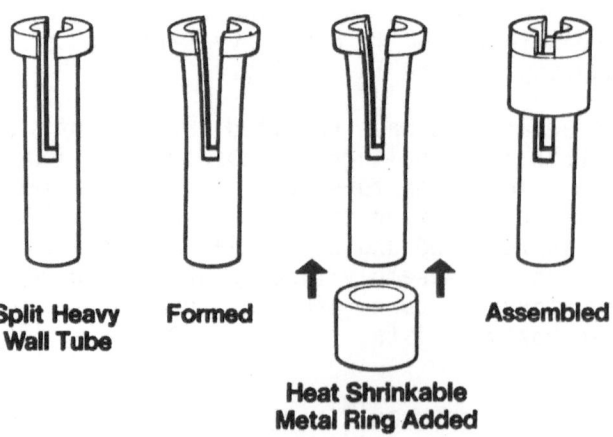

Figure 4.   The socket of a Cryocon contact

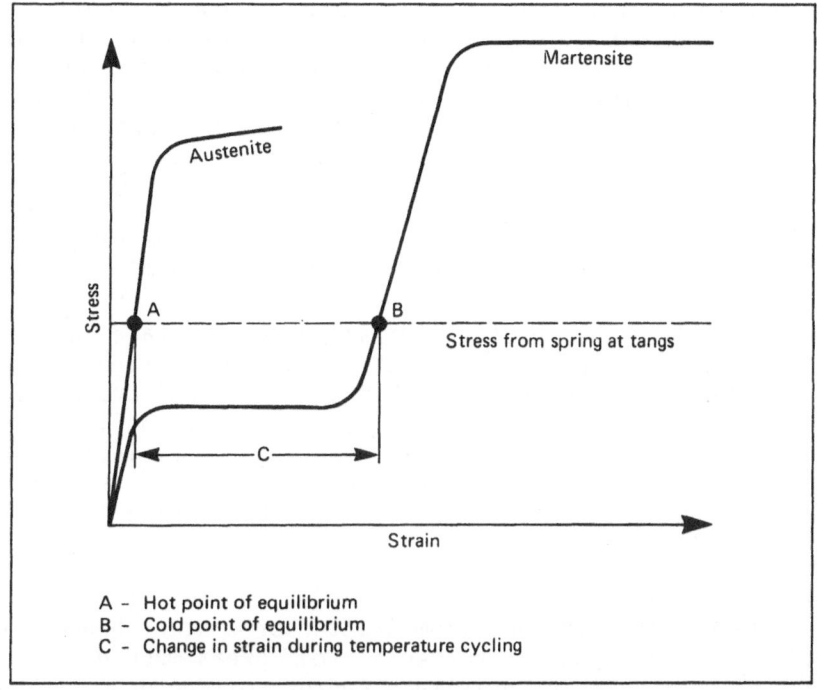

Figure 5.   The stress-strain relationship in Cryocon devices

These products are called Cryocon devices. As in the Cryofit coupling, the shape memory member is a hollow cylinder of TiNi alloy which contracts while it transforms during heating. In traditional pin-and-socket contacts, the socket tangs are bowed inward so the effective inside diameter of the socket is smaller than the pin outside diameter. The spring action of the socket tangs effects the connection. As shown in Fig. 4, Cryocon sockets have the tangs bowed outward and a TiNi ring is slipped around the outside of the tangs. The dimensions are such that the tangs exert a constant outward force on the ring. If the socket is cooled through its M_s-M_f range, the force exerted by the tangs causes the TiNi ring to expand, which permits insertion or retraction of the pin with zero force. As the assembly heats through the A_s-A_f range, the TiNi ring recovers toward its original diameter causing a firm gripping of the pin by the socket. Thus the condition which is to be avoided in the Cryofit coupling is vital to the action of the Cryocon device: cooling a TiNi ring through its martensitic transformation range in the presence of outward stress causes the annulus to expand.

The most important material consideration is selecting a TiNi alloy whose transformation occurs within an appropriate temperature range; A_f must be at or below the normal service temperature and M_f must be at or above the coolant temperature. One Cryocon connect-

or is designed for room temperature service. Freon 22, boiling point
-40 C, is a convenient coolant because it is widely available in
aerosol cans. The TiNi ring must transform to martensite during
cooling from room temperature to -40 C under the stress exerted by
the tangs. Likewise the ring must transform back to austenite during
heating to room temperature against the spring stress of the tangs.
The dimensions of the tangs and ring must give a force balance:
below the M_s the spring force of the tangs must be stronger than the
ring; above the A_s the ring must be stronger than the tangs. A
schematic of the stress-strain relationship is shown in Fig. 5.

The stress developed by the TiNi alloy on shrinkage causes the
tangs of the socket to be held against the pin with extremely high
pressures. Typically, forces greater than 10,000 psi are generated
at that interface. These forces cause small amounts of local deformation of the mating surfaces that results in very intimate contact of the two as illustrated in Fig. 6. This gives a series of
particularly important characteristics when this device is viewed
as an electrical contact.

The characteristics the devices have are these: (a) gas tight
contact surfaces, (b) freedom from chatter or contact bounce in
severe shock or vibration environments, (c) great mechanical
strength, (d) very stable resistance and/or voltage drop, (e) zero
insertion force. Except for (e), these properties are like those of
permanent connections rather than those of classic spring contacts.

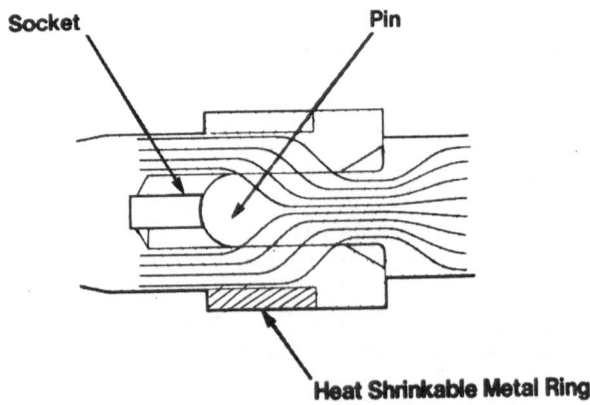

Figure 6. Sectional view of a Cryocon contact

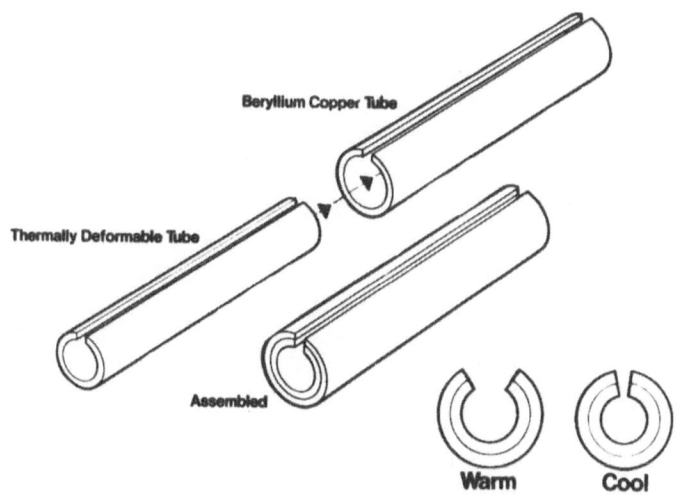

Figure 7. Schematic of a Cryocon roll pin device

The ability to open on cooling makes the device a make-break contact. Cryocon devices are effectively semi-permanent connections that have the desirable properties of permanent joints, e.g. solder, weld, or crimps, yet can be taken apart by simply cooling the TiNi element.

When the imposed strain in a TiNi member is simple tension or compression, motion is limited to about 8 1/2 percent. A way to provide greater motion is to arrange the geometry so the TiNi member recovers in a bending mode. A configuration using this approach is the roll pin device illustrated in Fig. 7, which has many of the desirable properties of the ring-tang device.

Shape memory in alloys has given an opportunity to extend applications from the base provided by heat shrinkable polymers. Success has depended on a combination of materials, technology, and market oriented design.

MANUFACTURE OF AN INTEGRATED CIRCUIT PACKAGE

Horace Pops

Essex International, Inc.

1601 Wall Street - Fort Wayne, Indiana 46804

I. INTRODUCTION

Many techniques have been used to produce integrated circuit packages. All of these assembly methods, such as beam lead, spider, flip chip and others require a large number of separate and distinct processing steps, frequently causing high costs and poor reliability (1). In the chip and wire process, for example, a large number of very fine aluminum lead wires are bonded by ultrasonic cold welding to pads on the semiconductor chip, and in turn to tips of leads on the metal lead frame. Failure of the devise will occur if any of these bonds are defective.

Historically, metals used for lead frames have been selected from conventional nickel or copper based alloys (2). The type of package described herein utilizes thermally induced transformations of copper-zinc based alloys containing additions of silicon and tin, and eliminates some of the more costly processing operations. The only reported application of shape-memory effects for electronic devices is an unsoldering method in dual inline packages (3). One of the lead wires is made from TiNi, which is soldered to a second electrical lead. Upon subsequent heating, the leads will separate as the heat-recoverable material moves away from the connection. Thus, it is possible to replace or repair the device from a printed circuit board without damage to either component. So far as it is known, no material or process has been devised utilizing the shape-memory effect for manufacturing of integrated circuit packages. This paper is directed to such a product.

II. PROPERTIES OF TERNARY ALLOYS

A. Pseudoelasticity

Polycrystalline alloys were prepared in strip form from high purity materials (> 99.99%) and details of specimen preparation and heat treatment are described elsewhere (4). These β phase alloys show pseudoelasticity, i.e., very large recoverable strains resulting from reversible stress-induced martensitic transformation at temperatures above the M_S. Elastic strains as large as 15 percent were measured in single crystals and coarse-grain (1 to 5mm) polycrystalline speciments (5). For wrought alloys having finer grain size, the maximum attainable elastic strains were approximately 8 percent. Furthermore, a small "permanent" set remained in the test sample after removal of tensile stress. This set becomes progressively larger with increasing total strain, as illustrated in Figure 1. Upon heating above the A_F temperature, the "permanent" set associated with stress-induced martensite is erased, and the sample returns to its original shape by the shape-memory process, i.e., by means of the transformation, martensite ⟶ β .

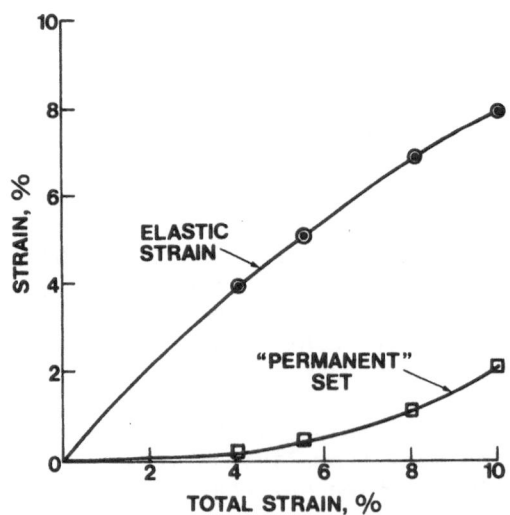

Figure 1. Effect of total strain on pseudoelastic strain and "permanent" set in a Cu Zn Si alloy. Samples deformed in tension at 24°C. M_S = -70°C.

B. Shape-Memory

Strain recovery data during heating were obtained for the same alloys having partly or completely martensitic structures (6). Tests were performed by bending specimens 180° around mandrels of varying radii so that the effects of different strain levels on the outer surface could be evaluated. Following deformation at -60°C (in the β + martensite region), partial recovery of the original flat shape was effected by reheating the bent strips for approximately one minute in different constant temperature baths at successively higher temperatures. As shown in Figure 2, recovery begins near the A_S, but is not complete until the temperature is well above the A_F. It can also be seen that increasing the strain raises the temperature for recovery, for example, completion of the S.M.E. occurs at -10°C and 20°C for samples deformed 4% and 6% respectively. Since residual stress-induced martensite was observed at room temperature, it is also likely that the A_F has shifted to higher temperatures. It was also observed that upon subsequent cooling from ambient temperature (point a) to -60°C (point b), deformed specimens return approximately 20 percent to their bent shape, as is shown in Figure 2 by the curve a-b. Reheating from -60°C causes specimen

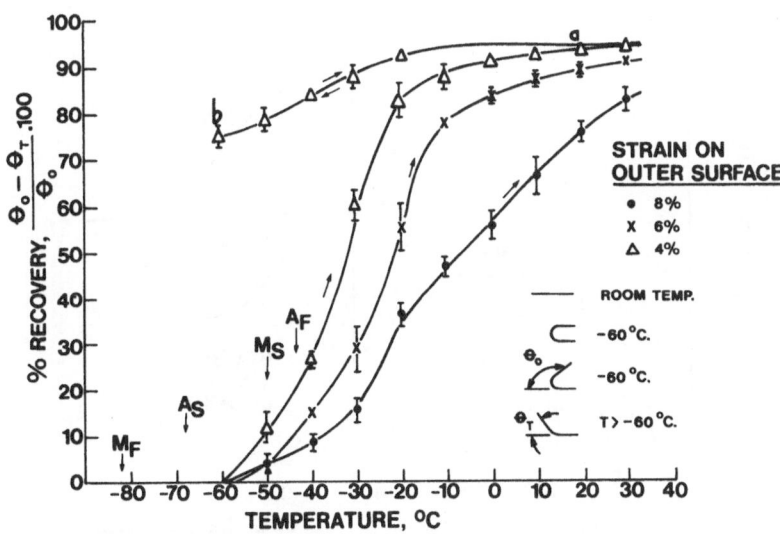

Figure 2. Shape-memory effect in ternary alloy deformed at -60°C in the β + martensite region. Transformation temperatures during cooling and heating refer to no-applied stress. Strain on outer surface is produced by bending 180° around mandrels having different size radii.

recovery, which follows the same path in the direction b-a. This reversible change on transformation during cycling has also been observed by Wasilewski for TiNi alloys (7). Presumably, the residual martensite, after transforming to the β phase, acts as preferred sites for the transformation during recooling below M_S.

It was of interest to determine whether or not stress-induced martensite, in particular that which does not disappear upon unloading, affects shape-memory behavior. To answer this question, specimens were deformed by bending to a 4 percent strain level in liquid nitrogen, because no retained β phase was observed by optical cold stage microcopy at -195°C; almost 100 percent recovery of strain was obtained upon heating through the reverse transformation to room temperature. If stress-induced martensite contributes appreciably to the shape-memory effect, one would expect higher recovery for deformation at -60°C rather than at liquid nitrogen temperature. It follows, therefore, that shape-memory is associated primarily with deformation of thermal martensite. Assuming that deformation at these low temperatures does not produce the

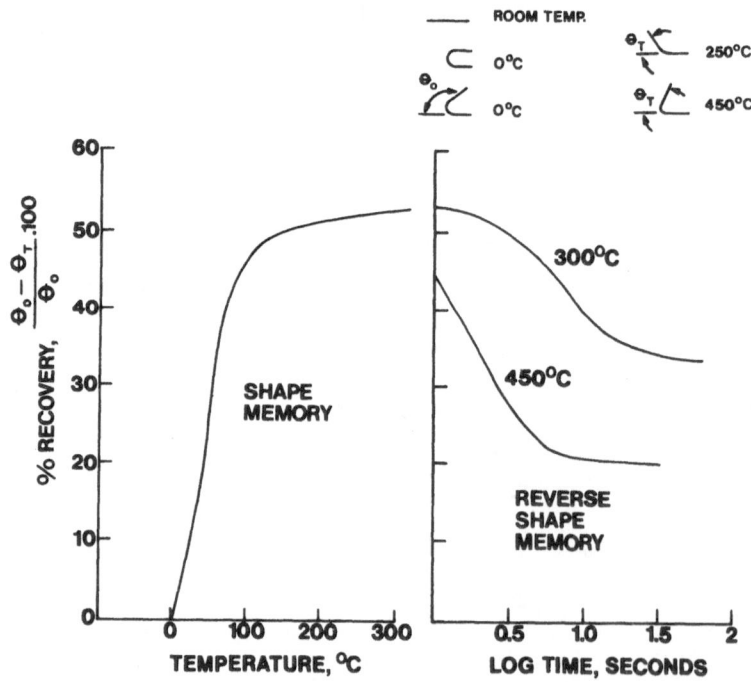

Figure 3. Percentage of recovery by the shape-memory and reverse shape-memory process, illustrating the importance of temperature and holding time. All polycrystalline samples were bent to a 10 percent strain level at 0°C. M_S = +13°C.

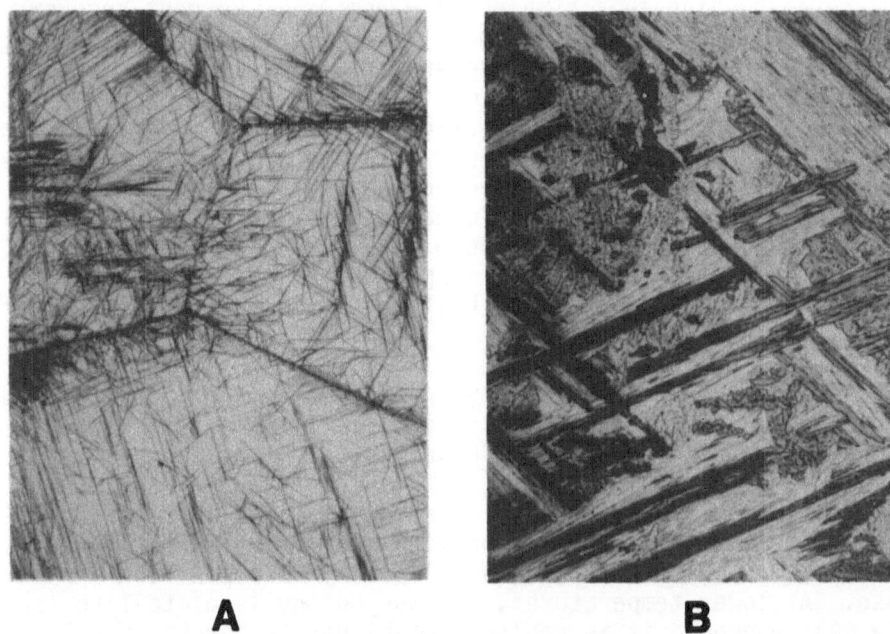

A   B

Figure 4. Microstructures formed by the bainitic transformation
in ternary β phase alloys: (A) Typical plate-like phase occurring
at lower temperatures. 39.6% zinc alloy heated for 4 seconds at
350°C. (B) Rod-like morphology characteristic of higher temperature
reactions. 36.8% zinc alloy tempered for 6 seconds at 504°C.

reverse transformation, thermal martensite ⟶ β, it seems
likely that shape-memory in ternary beta brass alloys is associated with reversible motion of martensitic twin boundaries (8).

C. Reverse Shape-Memory

A new type of transformation has been observed recently in
these alloys, and is called Reverse Shape-Memory (9). The reverse
shape-memory effect (R.S.M.E.) is akin to the shape-memory effect
in that stress is applied below the M_S temperature or slightly
above, and deformation induced martensite reverts when heated above
some reversion temperature. In contrast, however, specimens do not
return to their original (pre deformation) shape, but move in the
opposite direction, i.e., toward the direction of applied stress.
Whereas shape-memory changes occur instantly at temperature, the
R.S.M.E. is time dependent. An example of both effects is shown in
Figure 3. Strips of the ternary silicon brass having an M_S temperature of +13°C were bent 180° around a mandrel at 0°C, the bend

radius corresponding to a strain of 10 percent. Percent recovery plotted versus temperature in Figure 3 shows the expected shape-memory behavior, i.e., movement toward the original flat position. Under the above test conditions, maximum recovery was 50 percent at 250°C. Upon heating to 300°C, a slight but visible movement in the direction of bending begins in about 5 seconds, and is essentially complete in 30 seconds. This movement constitutes the reverse shape-memory effect. Heating of other samples to 450°C resulted in a more pronounced R.S.M.E., as shown in Figure 3. As illustrated in the next section, the reverse shape change may be greater than the S.M.E.. resulting in a final heat treated bend that is sharper than the original bend angle.

Reversion occurs isothermally in the duplex α + β phase field by a diffusion controlled precipitation reaction, a typical temperature range being between 230°C and 550°C.

This process in the absence of stress is similar to the bainitic transformation reported by previous workers (10) for copper based alloys. Two distinct products form by decomposition of the β phase. At lower temperatures, the morphology is plate-like (see Figure 4A), whereas, it is replaced at higher temperatures by a rod-like phase (Figure 4B). Incubation periods for visible precipitation decrease with increasing temperature and decreasing zinc content, as illustrated in Figure 5. Although the kinetics of reverse shape-memory are associated with the isothermal decomposition of deformation martensite, the detailed mechanism of this process has not been established.

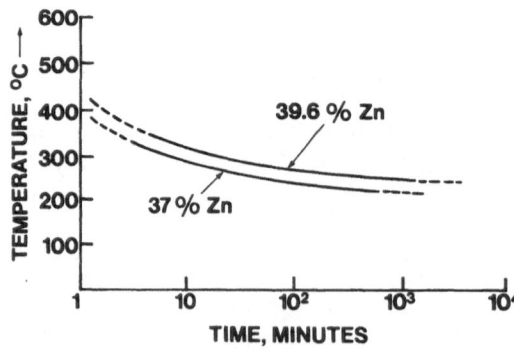

Figure 5. Isothermal time/temperature diagram showing the incubation period for bainitic precipitation of α phase in two different ternary β phase alloys.

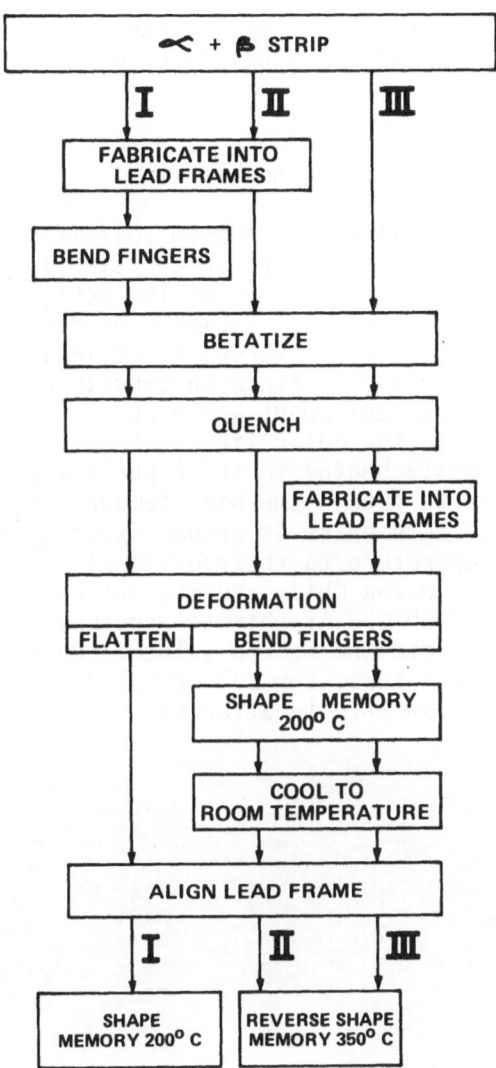

Figure 6. Processing flow chart of the steps for fabrication of an integrated circuit assembly.

III. APPLICATIONS

Three approaches have been used to assemble an integrated circuit package. As illustrated by the flow chart in Figure 6, all three methods utilize the shape-memory effect, but in addition, Methods II and III make use of the reverse process. Examples of these techniques using an alloy having an M_S of -70°C are as follows:

Method I

Since the as-cast alloys have been solidified through a two phase region, all of the rolled strip contains a duplex mixture of alpha and beta phases. Lead frames of the design shown in Figure 7 are fabricated by a conventional stamping or photochemical etching process. This design is not intended to replace the commercial frame member, but merely represents an experimental model. The fingers are bent 90 degrees about a mandrel corresponding to a strain of 7 percent on the outer fiber. Each of the bent lead frames are subsequently heated to the β phase field (~ 830°C), and quenched into water to retain the high temperature β phase. Deformation of the martensite phase is accomplished by flattening the fingers at room temperature to their original shape. The frames are then aligned above the chip, and the entire assembly is heated to a temperature of 200°C. At this temperature the S.M.E. causes the fingers to move into the molten solder. During subsequent cooling to room temperature, the solder solidifies, and the chip and brass fingers become firmly attached.

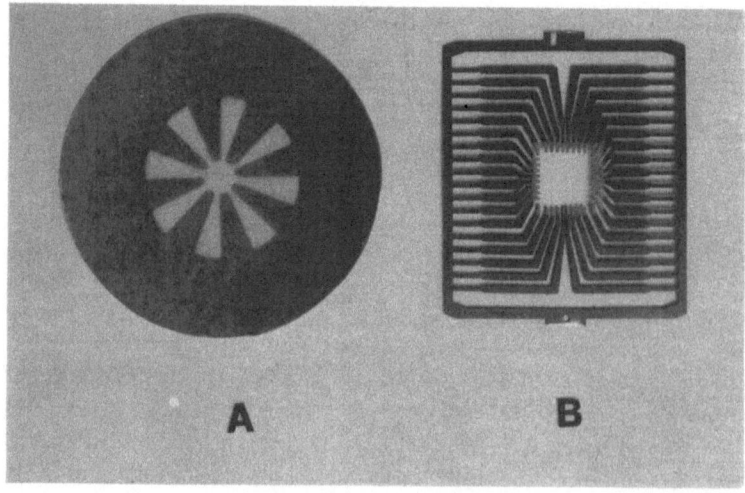

Figure 7. Plan view of typical configuration for shape recoverable alloys (A) and conventional lead frame (B).

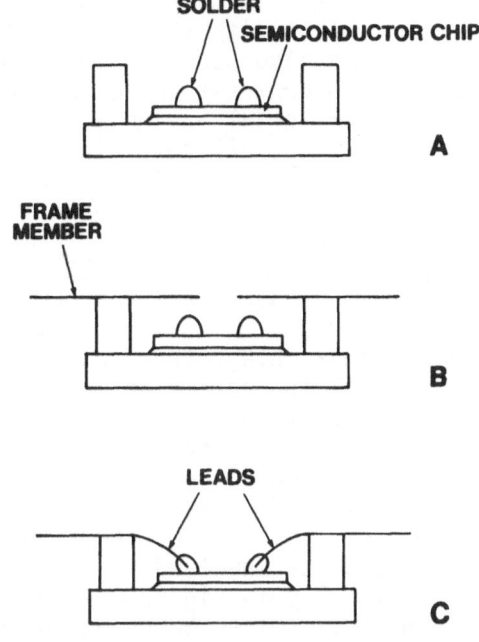

Figure 8. Schematic flow diagram illustrating the methods of assembly.

Method II

The α + β strip is fabricated into lead frames, betatized at 830°C, and quenched in a similar manner as described above. An identical amount of bending (7 percent strain) is used on the fingers, but in this case it is applied to a β + martensite or β microstructure. Heating to 200°C produces shape-memory effects, and tends to flatten the fingers. After cooling to room temperature, the (nearly) flat lead frame is appropriately positioned above the chip, and the package is placed in a furnace at 350°C. Reverse shape-memory is completed in less than two minutes, during which time the fingers move the required minimum distance (10 mils) in a downward direction, and make contact with the molten solder.

Method III

Distortion of the lead frame member was sometimes evident in the above two methods, particularly during quenching of thin specimens. To eliminate this problem, strip was heated to the betatizing temperature, discharged at high rate of speed from the

furnace, and quenched continuously using cold steel rolls of a
rolling mill that is in contact with the strip and a coolant spray.
A tension take-up device was also used to prevent distortion during
the quenching operation. In contrast to Methods I and II, lead
frames are produced from a material having a β or β + martensite
structure. Following this step, all remaining operations are identical to Method II. For example, the fingers are bent 7 percent at
room temperature. Flattening occurs by a shape-memory process, and
is accomplished by heating the deformed lead frame to 200°C. The
reverse shape-memory process (2 minutes at 350°C) makes the fingers
move downward, and helps to produce satisfactory bonds.

In order to produce this integrated circuit assembly, it was
necessary to closely control the direction and magnitude of sample
movement. Previous sections of this paper have dealt with the
importance of such variable as applied stress, transformation temperature, and final annealing conditions. In each of the three
methods, lead frames are aligned above the semiconductor chip, and
heated to cause simultaneous movement of the fingers into reflowed
solder bumps. Basic procedures involved in the soldering operation
are illustrated diagramatically in Figure 8. In step A, lead-tin
solder is reflowed into small solder bumps on the semiconductor.
Step B involves positioning of the heat recoverable brass lead
frame above this chip. The last step (C) involves heating of the
entire assembly, not only to melt the solder, but also to initiate
movement of leads by the reverse shape-memory or shape-memory process. In either case, the fingers extending from the frame member
must exert sufficient pressure to overcome surface tension, and
penetrate the molten solder. Although forces are produced during
subsequent cooling by the reversible change or transformation, they
are quite small and do not pull the recoverable member away from
the solder joint. Thus, bonds obtained by the shape-memory effect
remain intact at ambient temperature.

Since considerable simplicity is achieved if the leads are
deformed at room temperature (rather than below), it was desirable
to raise the martensitic transformation range. Consequently,
several different ternary alloys were prepared on the basis of the
following formula (4):

$$M_S \; (°K) \quad \approx 3280 \quad -80Zn \quad -175Sn \quad -120Si$$

where compositions are expressed in atomic percent. Electrical
resistivity measurements indicated that the temperatures were in
the range -70°C to +25°C.

Angular movement of leads was measured for different strain
levels and M_S temperatures, typical results being summarized in
Figure 9. Recovery is highest for the alloy's transforming

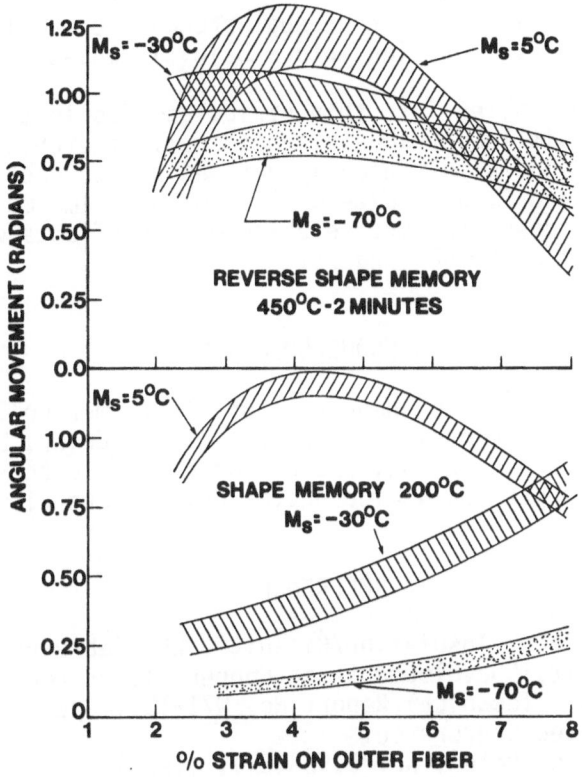

Figure 9. Angular movement versus strain indicating the magnitude of shape-memory and reverse shape-memory for alloys having different M_S temperatures. Samples bent at room temperature.

martensitically near room temperature, since they have the highest volume fraction of stress-induced martensite. In addition, recovery of the R.S.M.E. is larger than movement by the S.M.E., except in the case of high strain levels.

IV. SUMMARY

Pseudoelasticity, shape-memory, and reverse shape-memory effects have been observed in the same wrought polycrystalline copper-zinc based ternary alloys. Reverse shape-memory involves movement of the material towards the direction of original strain upon application of heat when strain is applied at a temperature below the M_S or slightly above. Following deformation, the material is heated to a higher temperature range than is normally used to produce shape-memory effects, typically, between about

230°C and 550°C. This rather unusual process occurs isothermally, and it involves decomposition into a bainitic phase.

Alloys were rolled into thin strip, and fabricated into integrated circuit lead frames by photochemical etching. Multiple bonds were produced simultaneously between solder pads on a semiconductor chip and fingers extending from the frame member. The shape-memory process and reverse shape-memory effects were used to control movement of these fingers.

ACKNOWLEDGEMENTS

The author wishes to thank some of his former colleagues for assistance with the experimental work, in particular, Messrs. Amado Cabo, Richard Barkalow, and Jay Grossman. Discussions with Barry Johnson concerning semiconductor devices are also appreciated.

REFERENCES

1.   J. M. Salzer, Insulation/Circuits, 29, February (1975).
2.   S. H. Butt, "Developments in Copper Alloys for Semiconductor Packages", Technical Report No. W71-12.1, paper presented at 1971 Westec Conference.
3.   R. F. Otte, Patent No. 3,588,618, (1971).
4.   H. Pops, Trans., AIME 236, 1532, (1966).
5.   H. Pops, Met. Trans., 1, 251, (1970).
6.   A. Cabo and H. Pops, unpublished results, report to Essex International, Inc., (1971).
7.   R. J. Wasilewski, Met. Trans., 2, 2973, (1971).
8.   J. Perkins, Met. Trans., 4, 2709, (1973).
9.   H. Pops and B. Johnson, Patent No. 3,854,200, (1974).
10. R. D. Garwood, Special Report 93, The Iron and Steel Institute, Scarborough Conference, 90, (1965).

NITINOL HEAT ENGINES

Ridgway Banks

Lawrence Berkeley Laboratory
University of California
Berkeley, California 94720 U.S.A.

Thermally-related shape memory properties of certain Nickel-Titanium alloys were first observed at the Naval Ordnance Laboratory in the early 1960's, and their application to thermo-mechanical energy conversion systems recognized.[1] The first successful application of these materials to a continuously operating heat engine was demonstrated at the Lawrence Berkeley Laboratory in August 1973.[2] Since that time parallel programs in materials study, thermodynamic analysis, prototype concepts and design of Ni-Ti heat engines have been initiated at LBL principally under a grant from the National Science Foundation/RANN. As the specific mechanisms of the Ni-Ti phase transformation and the thermodynamic analysis of its energy conversion effeciency will be presented in detail elsewhere, this paper will deal with practical aspects of the use of the shape-memory material in low-temperature heat engines, and their application to energy conversion.

BACKGROUND OF THE RESEARCH

There are, throughout the world, abundant resources of low-grade thermal energy which are not being economically utilized at the present time. The recovery of this energy through conversion to mechanical or other useful forms has become the subject of growing interest as energy research gains importance as a matter of technological concern. In some cases conversion of low-grade heat may have positive environmental benefits as well, as in the case of industrial waste-heat and the thermal effluent from nuclear power generating facilities. Despite the limitations imposed on heat engines working at low temperatures across a small ΔT, sources of low-grade heat are so widespread and available that economical conversion of a fractional percent could have a signif-

icant impact on the world energy supply. Such resources as the
oceanic thermal gradient or geothermal hot springs might become
important if a suitable conversion technology for their exploitation could be demonstrated. As all systems operating under these
conditions will suffer from the same thermodynamic limitations,
conversion efficiency may not be the sole criterion for their
competition. Simplicity of the system as a whole as well as installation and maintenance requirements will probably also be important considerations especially in parts of the world where
energy needs cannot be met by conventional sources of power. Among
the small-scale applications envisioned for such engines are the
operating of solar-powered refrigerating units, agricultural irri-
gation pumps, and auxiliary electrical power generating components.

       The project leading to development of a solid-state heat
engine originated at the Lawrence Berkeley Laboratory under aus-
pices of the Solar Energy Program. The many attractive aspects of
the enormous solar resource are widely recognized, but its conversion to useful power, as in the generation of electricity has been
hampered by lack of a technology which would be economically com-
petitive with other available sources. The basic problem lies in
the dilute concentration of solar thermal energy. To achieve temperatures suitable for operation of conventional turbine systems,
the sunlight must be focused by reflecting surfaces converging on
a central boiler. Concentrating collectors have certain inherent
limitations, a serious one being that they must follow the sun
through the sky which involves sophisticated and costly tracking
mechanisms. Moreover, concentrating collectors will only operate
during daylight hours. The reflective surfaces must be resistant
to erosion, and must be regularly maintained to avoid loss through
deposits of atmospheric contaminants. With these considerations,
and the problem of energy storage for an intermittent system, it
may be seen that economical solar thermal energy conversion is
normally thought of as being limited to quite large installations
such as solar "farms" located in desert regions.

       Another form of solar thermal energy -- the oceanic thermal
gradient -- is presently being explored, again in terms of multi-
megawatt installations floating off-shore. These plants are typically designed around turbines which use low boiling-point organic
fluids such as freon as their working fluid. Their closed systems
include massive heat-exchangers as boiler and condensers, with
highly sophisticated pumping and regulatory systems. As in the
case of solar farms the capital requirements for such systems are
so great that they must be scaled for very high output to be considered as economical alternatives. Thus, both approaches are
planned in a way that will interface easily with the existing power

distribution grid. As fossil fuels become more scarce, maintenance
of the centralized power distribution system is, of course, a top
priority. But the centralized system ·itself has some drawbacks: at
the present time 10% of the energy and more than 50% of the unit
price of electric power delivered residentially represents the cost
of system maintenance and distribution.

In contrast to the constraints and limitations of conventional approaches to large-scale solar thermal energy conversion,
solar heat has been exploited in many parts of the world for many
years. Temperatures required for residential heating and hot-
water supply are relatively easily and economically achieved with
stationary flat-plate solar collectors. Unlike the focusing, concentrating, type of collector this approach requires no tracking
mechanism, minimum maintenance and has the ability to collect
both diffuse and specular solar radiation, making them functional
in overcast weather. If a flatplate collector is used to heat
water, energy storage is easily achieved with an insulated tank.
Thus thermo-mechanical conversion of hot water can be a continuous
process if storage volume and collector area are matched to the
demands of the consumer and regional climatic conditions.

At the present time, the only developed technology for the
conversion of low-temperature heat to mechanical energy are
turbine systems of the sort described in connection with conversion across the ocean thermal gradient. A particularly highly re-
fined turbine of this sort, the Ormat Turbine [3] has been manu-
factured and marketed in Israel for use with a solar collector. As
of February, 1974 not one of these turbines in use at that time was
actually powered by sunlight, but rather they were linked to gas-
fired boilers [4]. While this fact certainly reflects the still
relatively low cost of natural gas, it suggests that the requirements of the overall system -- including thermal losses across the
heat exchangers -- may be fairly high. Despite the high efficiency
of the turbine, the conversion efficiency of the system as a whole
must be taken into account.

This is the background for interest in the possibility of
converting low-temperature thermal energy across a small temperature gradient (ΔT) to mechanical work. Much work is being done to
bring the cost of flatplate solar collectors down, as well as to
increase their efficiency. Interest in residential designs capable
of supplying at least part of their energy requirements is growing
in many parts of the world. To compliment existing technology with
the capacity to generate electric power from hot water would address this current interest in an important way.

DEVELOPMENT OF THE LBL PROTOTYPE

The first approach to low-temperature energy conversion at
LBL was a prototype heat engine design exploiting the differential
expansion of bimetallic strips. A multiplicity of bimetallic
springs were mounted on the periphery of a wheel rotating in the
vertical plane. The lower half of the wheel was immersed in a hot
water bath. Cooling of the springs was by convection during the
upper half cycle. Power from the springs was communicated to an
eccentric crankshaft by light-weight connecting rods.

This prototype was never built, but served as the basis for
discussion of the feasibility of low-temperature energy conversion.
On the basis of the conversion efficiency of bimetallic working
elements, the outlook was far from optimistic, being around .07%
absolute conversion of heat in to work out. Consideration of the
simplicity of the machine and the attractiveness of a "rooftop"
generator for modest amounts of power kept the idea from an early
death, however, and in the course of further discussions the author
was given a sample of "55 Nitinol" (5) from a commerical experi-
menter's kit. With nothing more to go on than the instructions
supplied with the sample and the empirical evidence of simple tests
it was nonetheless clear that the potential of this material was
much greater than the bimetallic strips, and the earlier prototype
was modified to employ Nitinol wires, and to permit immersion in
water for both hot and cold half cycles to optimize the heat transfer rate.

The LBL prototype Nitinol heat engine employs twenty 15 cm
loops of Nitinol wire, 1.2 mm in diameter, as its working elements.
The loops are supported by spokes radiating horizontally from a
crankshaft outward through the periphery of a 35 cm wheel. The
centers of rotation of the spokes and of the wheel are offset by
2.5 cm so that the spokes work back and forth in reciprocation relative to the wheel as the system revolves.

The Nitinol wire loops hang between the wheel and fixed
stops on the spokes. Rotation around the stationary crankshaft
results in opening and closing of the distance between these stops
and the rim of the wheel as well as advancing the position of the
wire loops relative to two stationary waterbaths immediately below
the mechanism. (Fig. 1) The semi-circular baths contain water at
about 48° and 24°C respectively.

The Nitinol wires were annealed as straight lengths, and
this is the shape to which they tend to return on heating. Heating is accomplished by immersion in the hot bath, and the mechanism is adjusted so that during this half of the cycle the distance
between the stops and the wheel is increasing. In practice, force-

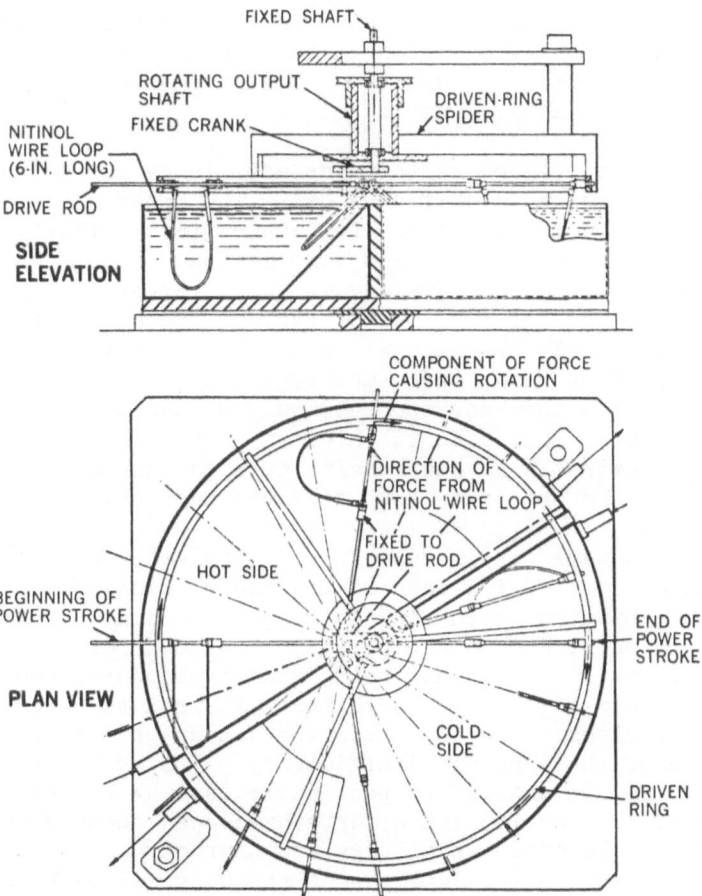

Fig. 1. Schematic view of the original prototype Nitinol engine.

able opening of the wire loop increases this distance, which provides the rotational component (working half) of the engine cycle. At the most "open" point of the cycle the wire loops are transferred to the cold bath, and the closure of the distance between stops and wheel reshapes the wires into a closed loop for the next working half-rotation.

Since its first demonstration in 1973 the device has run consistently at 60-80 rpm (depending on fluctuations in water temperatures) and no deterioriation in the performance of the Nitinol wire elements has been noted after > 17×10^6 revolutions. Outputs have been measured at $\geq$ .2 watts.

Although optimized neither for efficiency nor output, this prototype has produced significant data with regard to application of the Ni-Ti shape memory effect to heat engines. First, we are able to observe that at quite low deformation, perhaps as low as 1/2%, recovery of the shape-memory is quite repeatable at least to the order of tens of millions of cycles. Secondly, after having run the engine for a few hours we were able to observe the development of an apparent cold "shape memory". As the wire loops dropped into the cold bath they now spontaneously closed without the application of stress. This effect, which we have called "double training" is most desirable in heat engine applications as it produces an increase in the net power output of the machine.

The prototype has proven extremely valuable as a demonstration and as a test bed for determining the working lifetime of the wire. The machine has been used to generate enough electricity to light a small bulb, and this experiment has been performed using solar-heated water. It is now being run continuously under load to evaluate possible effects on the wires. It is not, however, a design which promises to be the basis for a high-output heat engine both because of the inherent inefficiency of using Nitinol wire in the mode of flexure and in its poor power density potential. For these reasons LBL development efforts have been focused on new machine design concepts, metallurgy and thermodynamic studies rather than an exhaustive analysis of the original machine and its efficiency.

The LBL prototype was not, as it turns out, the first attempt to apply Nitinol to energy conversion in heat engines, although it seems to be the first device to withstand continuous operation of long duration. Study of photographs showing other prototype concepts shows certain basic similarities between them and the LBL model. Typically, the eccentric configuration of the wheel and crankshaft is common to most of the mechanisms. In many cases, however, operation of other prototypes appears to have permanently deformed the Nitinol working elements. Until the results of

Fig. 2. First conversion of solar heat to electricity by means of
the Nitinol engine at LBL, Nov. 1973.

systematic experimentation are available it is only possible to
speculate why the LBL machine has run continuously without any apparent degradation in performance. The following remarks must
therefore be taken only as the personal opinion of the author.

      In any eccentric system of constant angular velocity, the
mode of convergence and divergence between corresponding points is
sinusoidal. The leverage (mechanical advantage) of a bicycle crank,
for example, is constantly increasing from top dead center until it
reaches a maximum at the angle of 90°, roughly parallel to the
ground, from which point leverage is constantly diminishing until
it reaches an effective value of zero at bottom dead center. Moreover the rate of change in increase or decrease of leverage is not
linear -- it is in fact sinusoidal.

      Nitinol wires in linear tension have been shown to withstand
tens of thousands of cycles without permanent elongation [6] when
they are allowed to recover at their own rate. They will do work

in the course of this recovery, provided that there is an advantageous leverage to do that work. The speed of recovery is, moreover, very fast. When wires in linear tension are incorporated into a mechanism that does not provide an advantageous leverage within the time required for the phase transformation to take place, the stresses developed at that point in the cycle may, over the course of many repetitions have the effect of permanently elongating the wire by a process equivalent to mechanical forging.

In the LBL prototype, the mode of opening of the mechanism is also sinusoidal, but by virtue of using the wires in flexure rather than linear tension, the rate of shape recovery of the wires is independent of the leverage available at any particular point in the mechanical system. The wire loops, in other words, function both as the working elements and as springs which can store mechanical energy until the point of greatest advantage in the leverage of the system. Despite the inefficiency of the flexural mode, the element of compliance it provides may be critical in matching the shape recovery of the working elements to the kinetics of the mechanical cycle.

Much work remains to be done in the study of Nitinol -- or other shape memory materials -- in heat engine applications. Closer observation of the dynamic macroscopic behavior is indicated, as well as study of the microcrystalline features of the phase transformation with a view to developing processing techniques that will optimize thermodynamic efficiencies in realizable materials. Practical and efficient engines have yet to be designed, and the ultimate question of economic competitiveness must be addressed. The very great variety of applications that await practical and compact low-temperature energy conversion units, however, make the resolution of these questions a task of great relevance and interest.

This work is supported by the U.S. Atomic Energy Commission and by the National Science Foundation.

FOOTNOTES AND REFERENCES

1. U. S. Patent No. 3, 403, 238; Conversion of Heat Energy to Mechanical Energy, W. J. Buehler et al., assignors to the United States of America.

2. Proposal for Application of Solid State Energy Conversion to Cooling of Buildings, Lawrence Berkeley Laboratory, Energy and Environment Division, submitted to the National Science Foundation (RANN), Nov. 1973. NSF Grant AG-550, Proposal No. P416452.

3. H. Tabor and L. Bronicki, "Small Turbine for Solar Energy Power Package"; Proceedings of the United Nations Conference on New Sources of Energy; Rome, Italy; Vol. IV pp. 68-79; 1961.

4. H. Tabor, private communication to the author, Feb. 1974.

5. "55 Nitinol" is the generic name given to the nominally equiatomic Nickel-Titanium intermetallic compounds (∿55% Ni, balance Ti by weight) exhibiting the shape memory effect. The name is derived from Nickel, Titanium and Naval Ordnance Laboratory (Silver Spring, Maryland) where the memory effect was first observed.

6. R. Banks, experiment in progress, April 1975.

INTERNAL VIBRATION ABSORPTION IN POTENTIAL STRUCTURAL MATERIALS*

L. Kaufman, S. A. Kulin and P. Neshe

ManLabs, Inc.
Cambridge, Massachusetts   02139

R. Salzbrenner

Massachusetts Institute of Technology
Cambridge, Massachusetts   02139

INTRODUCTION

Machinery contains parts which are subjected to periodic stresses that result in acoustic waves which are set in motion within the solid components. When these waves reach a free surface, audible noise is produced. Since the energy in the waves is generally a small fraction of the total energy output of a machine, conversion of this energy to heat becomes an effective method of noise reduction with an insignificant accompanying temperature rise. Damping materials with good structural properties are required to fulfill this function.

Historically, the discovery of new damping alloys has been rather accidental. The ability to tailor an alloy to a specific engineering need requires understanding of the atomic mechanisms which provide the enhanced damping capacity.

The present study is aimed at rectifying this situation by providing insight into mechanisms of acoustic absorption in alloys. The ultimate goal of the study is identification of alloys which can be employed to reduce vibration and noise associated with the

---

*This research is supported by the Advanced Research Projects Agency of the Department of Defense and monitored by the Army Materials and Mechanics Research Center under Contract Number DAAG46-74-C-0048.

operation of devices containing rapidly moving components.

The initial atomic transformations selected for study as potential internal noise absorption mechanisms were the thermoelastic transformations which occur in 55 w/o Ni-45 w/o Ti and 83 w/o Cu-14 w/o Al-3 w/o Ni alloys. Subsequent sections of this paper summarize the results of coordinated studies of the transformation, damping and mechanical properties of these materials.

METHODS EMPLOYED FOR MEASURING THE DAMPING CAPACITY OF THE SELECTED ALLOYS

The damping capacity of the selected alloys was measured by the resonance dwell technique which has been applied extensively at Bolt, Beranek and Newman, Inc. of Cambridge, Massachusetts, by Heine, Ungar and coworkers (1,2).* This method was selected because it provides an accurate measure of the damping capacity in the audible range at various frequencies and applied stress levels. Accordingly, frequencies in the 150-250 cycles/sec (Hertz) range were used in all of the measurements. This range represents an important spectrum of audible frequencies. The resonance dwell technique is a forced vibration method of indirectly determining the loss factor of a simple structural element by measuring its response to excitation at a modal frequency. The simple structural element used in these tests is the cantilever beam shown in Figure 1.

The method for exciting the beam at a modal frequency consists of fastening it to an electromagnetic shaker which is at one end of the base of the apparatus (near the beam) while the other end of the shaker (opposite the beam) is fastened to the heavy base. An accelerometer is connected to the sample near the beam. The response of the cantilever beam to the electromagnetic shaker is measured by means of a microscope which is focused on the tip of the cantilever beam. The latter is illuminated by a stroboscope that is set to flash at nearly the same frequency as the natural frequency of mechanical oscillation of the tip of the beam. The fundamental natural frequency, f, of a cantilever beam of length L (inch) and thickness h (inch) is

$$f = \frac{1}{2\pi} \left(\frac{1.8751}{L}\right)^2 h \left(\frac{32E}{\rho}\right)^{1/2} \left(\frac{cycles}{sec.}\right) \tag{1}$$

where ρ is the density of the sample in (lb/in^3) and E is the dynamical Young's Modulus (psi). Thus visual observation of the tip

---

*Underscored numbers in parentheses denote references.

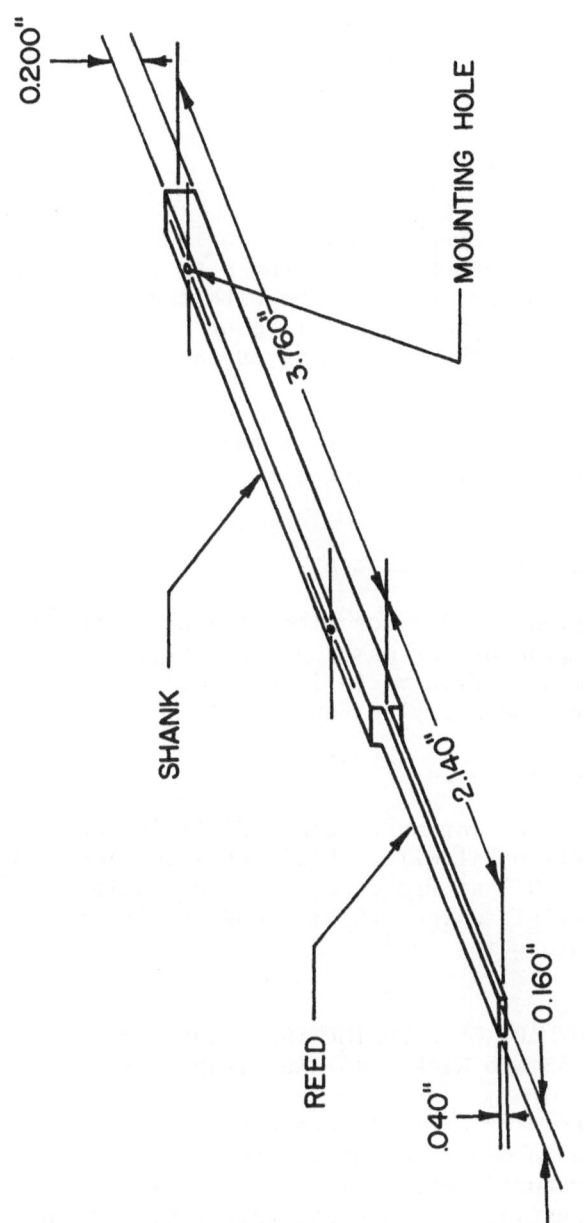

Figure 1. Specimen design for measuring damping capacity.

vibration with the microscope and the stroboscopic light is used to fix the natural frequency, f. The latter is used to compute the dynamical Young's Modulus, E.

For cantilever beams in their fundamental mode of vibration the tip amplitude as a function of peak stress and specimen natural frequency is given by Equation 2.

$$y = 3.63 \ (\sigma/f)(E\rho)^{-\frac{1}{2}} \tag{2}$$

where y is the peak to peak amplitude in inches and σ is the maximum stress in psi. Measurement of the peak to peak amplitude (which is done visually when coupled with the measurement of the natural frequency, f, and dynamical Young's Modulus, E, as described above) yields the maximum stress at the root of the cantilever blade.

The final direct measurement is the damping capacity or loss factor, Q^{-1}. The latter is related to the logarithm decrement by the relation

$$\pi Q^{-1} = \text{logarithmic decrement} \tag{3}$$

Thus the product πQ^{-1} describes the decay in amplitude per cycle. The loss factor (or damping capacity), Q^{-1}, is obtained by measuring the acceleration a_o in inches/sec^2 at the frequency f for the deflection y, and applying Equation 4.

$$Q^{-1} = 0.083(1+0.2L)a_o f^{-2} y^{-1} \tag{4}$$

The loss factor for most metals falls in the range 10^{-5} to 10^{-3}. A good bell metal will exhibit very low loss factors in the range of 10^{-4} to 10^{-5} at frequencies in the audible range. On the other hand, materials which exhibit good damping capacity will have loss factors approaching 10^{-2}.

SUMMARY OF EXPERIMENTAL RESULTS ON THE
55 w/o Nickel-45 w/o Titanium ALLOY

Several heats of Nitinol were purchased from Titanium Corporation of America in order to obtain samples with a range of transformation temperatures near ambient. The alloys were heat treated initially by annealing at 790°C for thirty minutes and quenching in water. Subsequently, the electrical resistance was measured as a function of temperature between -80°C and +80°C on a precision double Kelvin Bridge. The results were employed to fix the M_s temperature by establishing the temperature where a peak in the resistance versus temperature curve occurs. This temperature represents the M_S at which the first observable surface relief forms on cooling (3-5). In keeping with the above mentioned results, it was found that the resistance peak became more pronounced with increasing number of temperatures cycles. Accordingly, multiple cycling was adopted as a standard procedure and each sample was cycled fifty to eighty times between -80°C and +80°C prior to testing. The final cycle terminated by heating from -80°C to +25°C.

The results obtained for the 55 w/o Ni-45 w/o Ti alloy in the annealed (and cycled) condition as well as those obtained after cold working (and cycling) are shown in Figures 2-6. Figure 2 displays the results of loss factor, dynamic Young's Modulus, resistance and heat evolution measurements on a single set of samples in order to establish the M_S temperature for a given heat. In the present case (for heat 4609), M_S was observed at about +7°C. At this temperature the resistance versus temperature curve and the heat evolved on cooling showed pronounced peaks. The loss factor (Q^{-1}) was found to increase on cooling at temperatures substantially higher than M_S and remain fairly high (i.e. near $Q^{-1} = 0.01$) to even higher temperatures on heating. These damping tests were conducted at frequencies in the range 150-200 Hertz (cycles/sec) and at stresses between 2000 and 5000 psi. In order to insure that these stresses applied during the damping measurements did not in themselves alter the M_S, electrical resistance versus temperature was measured for a series of samples of heat 4609 under conditions of uniaxial tension at tensile stresses of 1275 and 5475 psi respectively. These tests disclosed no measurable change in the M_S with applied stress in this range. The heat evolution measurements shown in the bottom portion of Figure 2 were obtained by means of a Perkin-Elmer Scanning Differential Calorimeter. The peak in the heat evolution curve observed on cooling coincides with the M_S temperature. The maximum in the heat absorption curve on heating lies nearly forth degrees centigrade above M_S. It is not clear if this temperature corresponds to A_S or A_f. Nevertheless, the temperature difference between the extremal values shows that substantial hysteresis occurs in this transformation.

The measurements of the damping factor (Q^{-1}), relative electrical resistance and 0.2% offset yield strength shown in Figures 3-6 were performed on heat 4609 and a second heat of 55Ni45Ti designated 4866. The electrical resistance and damping measurements on this alloy indicated an M_S near +17°C. This alloy was used for all of the yield strength and damping capacity measurements shown in Figures 4-6.

Examination of the damping capacity data for the 55Ni45Ti alloy in Figures 3-6 shows that in the annealed condition the loss

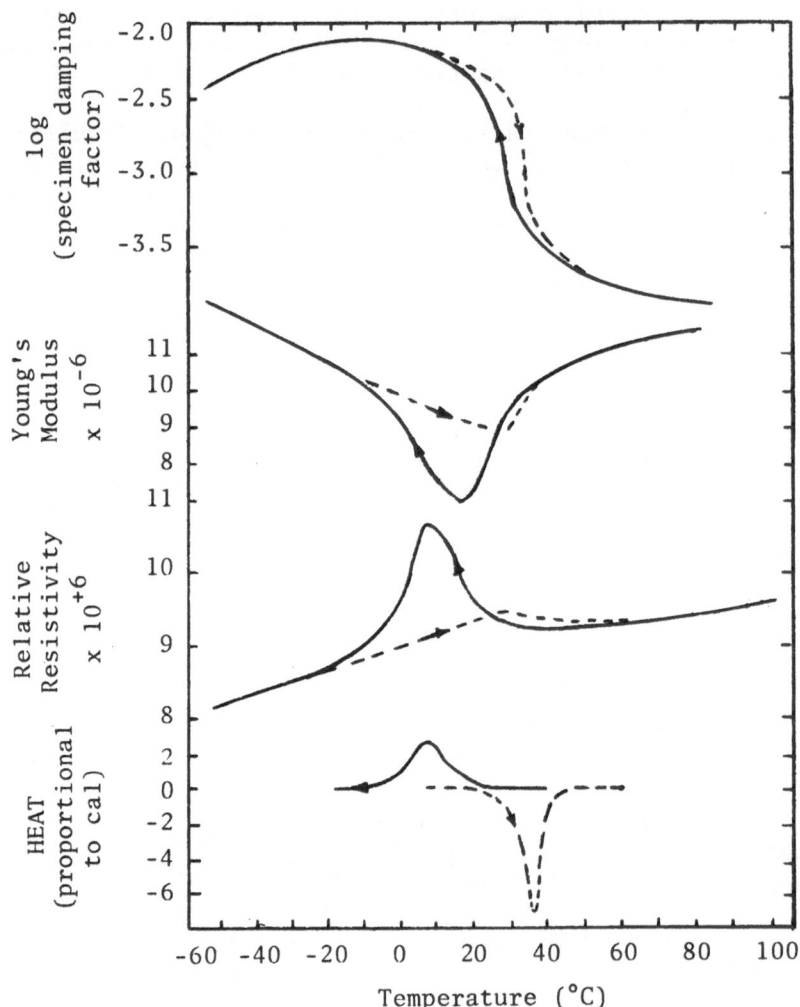

Figure 2. Specimen damping factor, Young's
modulus, relative resistivity and heat
evolved as a function of temperature for the
same series of 55w/oNi-45w/oTi.

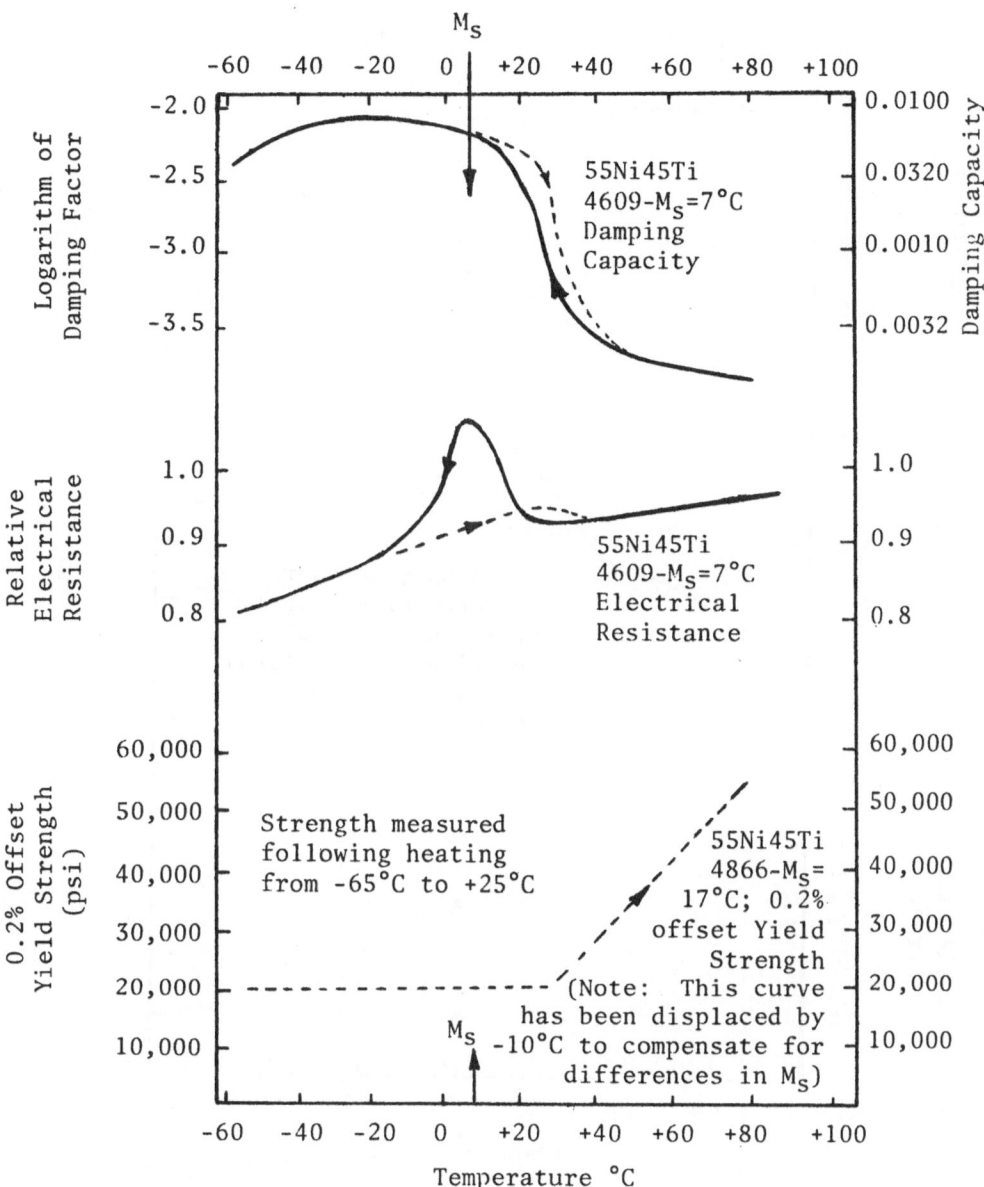

Figure 3. Damping Capacity, Electrical Resistance and 0.2% Offset Yield Strength for 55Ni45Ti as a function of temperature after Multiple Cycling Treatment.

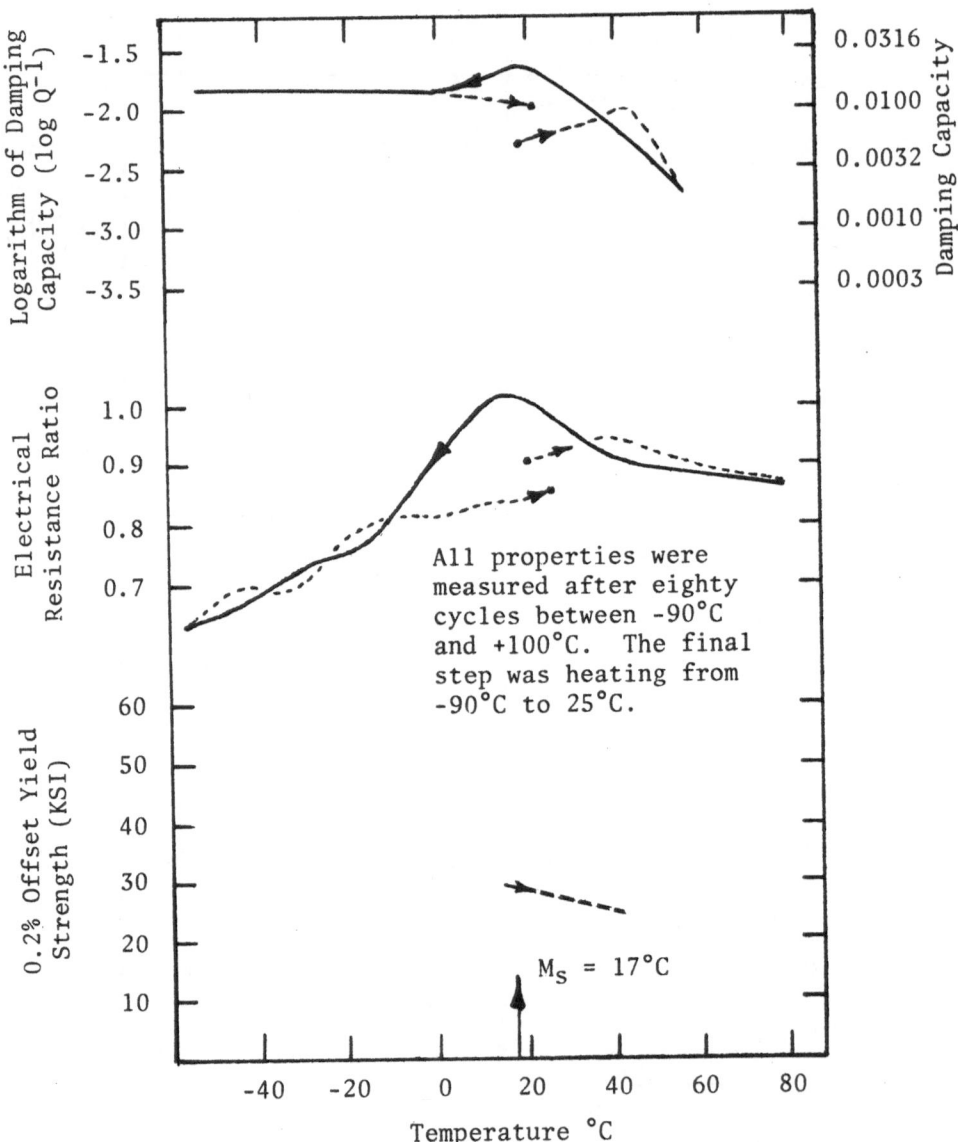

All properties were
measured after eighty
cycles between -90°C
and +100°C. The final
step was heating from
-90°C to 25°C.

M_S = 17°C

Figure 4. Damping Capacity, Electrical Resis-
tance and 0.2% Offset Yield Strength for 55 w/o
Ni-45 w/o Ti versus Temperature after 3% Cold
Reduction and Cycling Treatment.

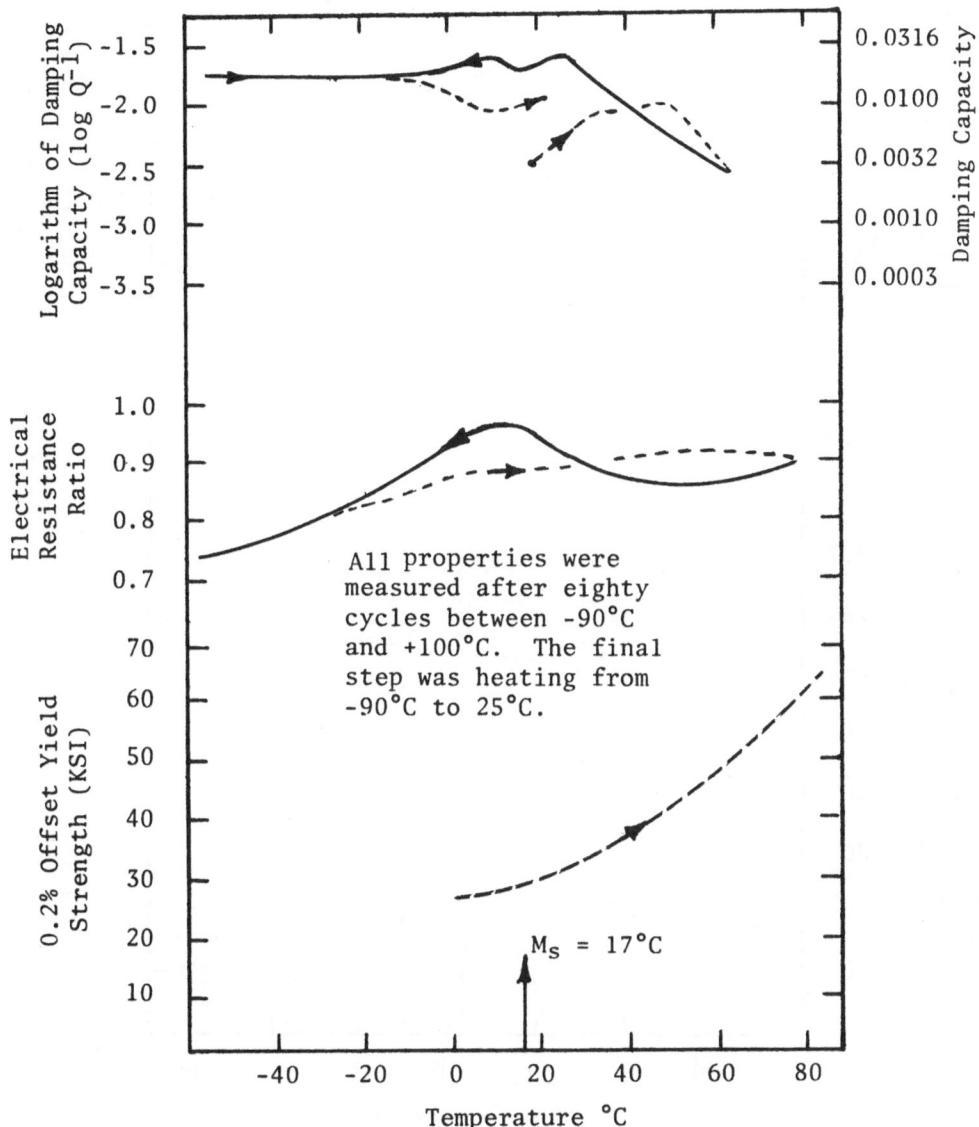

Figure 5. Damping Capacity, Electrical Resistance and 0.2% Offset Yield Strength for 55 w/o Ni-45 w/o Ti versus Temperature after 7% Cold Reduction and Cycling Treatment.

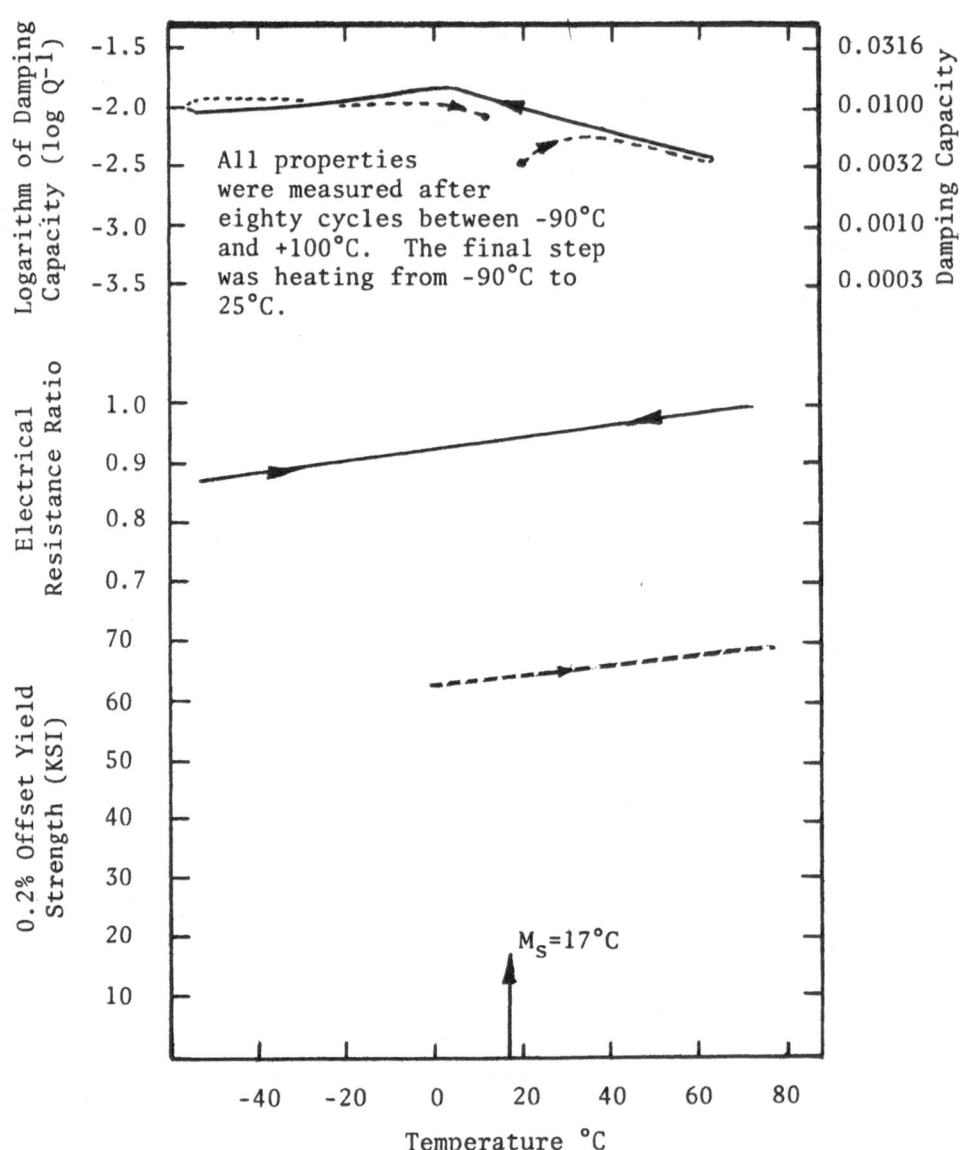

All properties
were measured after
eighty cycles between -90°C
and +100°C. The final step
was heating from -90°C to
25°C.

$M_s = 17°C$

Figure 6. Damping Capacity, Electrical Resistance and 0.2% Offset Yield Strength for
55 w/o Ni-45 w/o Ti versus Temperature after
15% Cold Reduction and Cycling Treatment.

factor increased markedly on cooling at least 25°C above the M_S temperature. In this temperature range, the damping factor increases by more than an order of magnitude. This behavior seems to be caused by some sort of premartensitic phenomenon. However, the exact cause is not clear at present. At temperatures between M_S and M_f the primary cause of damping appears to be associated with the transformation itself. At temperatures below M_f (i.e. below -17°C) the enhanced damping may be due to changes in the internal structure of the martensite itself.

On heating, the damping remains fairly high up to 28°C, indicating that there is some hysteresis in the damping behavior with temperature cycling. The effects of cold work on the damping behavior displayed in Figures 4-6 show that the rapid variations as a function of temperature in the annealed material are reduced by cold working. The cold reduction in thickness was performed at room temperature by rolling and was followed by thermal cycling prior to measurement. The changes produced by 3% reduction in thickness were small. However, readily measurable alteration in behavior can be seen following the 7% and 15% reductions. In particular the 15% cold reduction leads to a damping factor versus temperature which shows only a threefold change over the temperature range from -60°C to +60°C.

The measured electrical resistance versus temperature curves shown in Figures 3-6 for varying degrees of cold work parallel the results disclosed by the damping measurements. The annealed material shows a distinct peak in Figure 3. However, the resistance peak is progressively broadened with increasing cold reduction in Figures 4 and 5, and disappears entirely at 15% in Figure 6.

The 0.2% offset yield strength of the annealed sample is quite low at temperatures below M_S. This is apparently due to the transformation itself. However, the low yield strength is also observed 20°C above the M_S temperature where the damping begins to increase. The yield strength is observed to increase with temperature in the annealed material reaching a value of 55,000 psi near 80°C. The effects of cold work at the 3% and 7% levels did not materially alter this behavior as is indicated in Figures 4 and 5. However, a substantial change is observed at the 15% reduction level. In this case, yield strengths between 60,000 and 70,000 psi are observed over the temperature range from 0°C to 80°C. These strength values can be combined with damping capacity factors near 0.01 in the temperature range from 0°C to 30°C. At high temperatures the loss factor decreases to a value near 0.003 at 80°C.

Thus, introduction of cold work to the extent of 15% reduction in thickness results in a combination of strength and damping capacity which may be attractive for special applications.

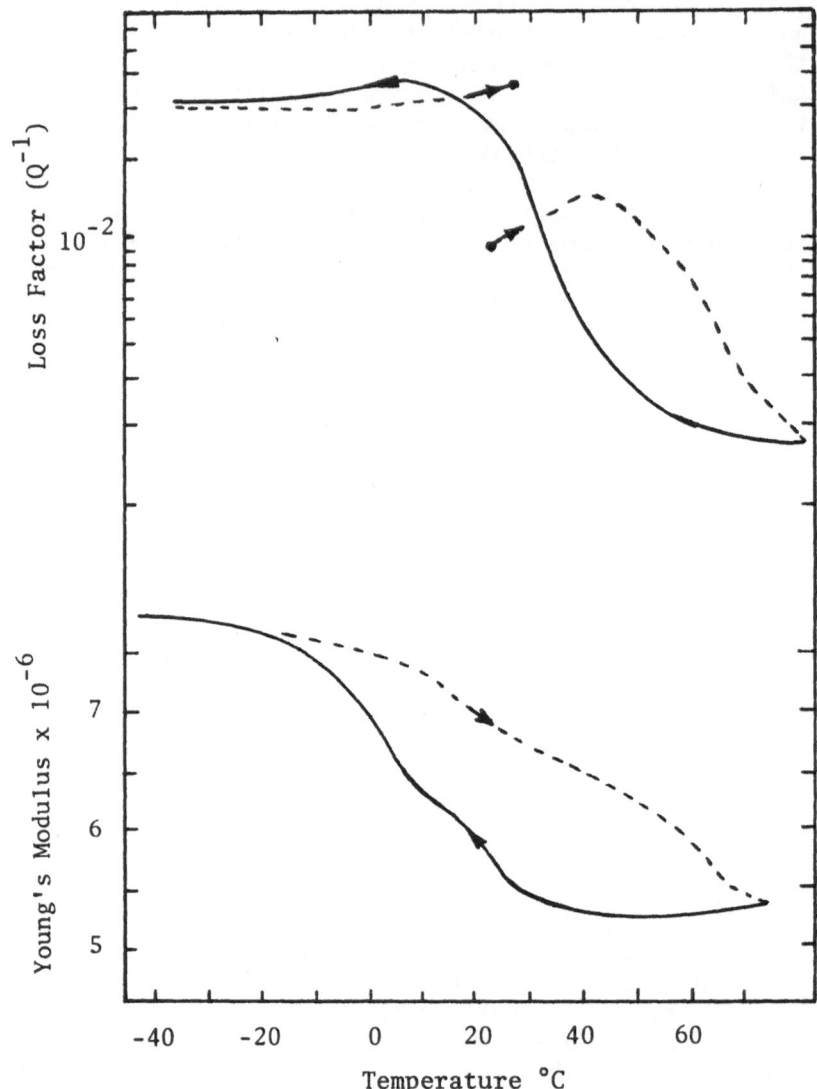

Figure 7. Loss Factor and Dynamical Young's
Modulus for a Single Crystal Sample of Cu-14.0
w/o Al-3.0 w/o Ni measured at 115-150 Hertz.
M_S=+10°C.

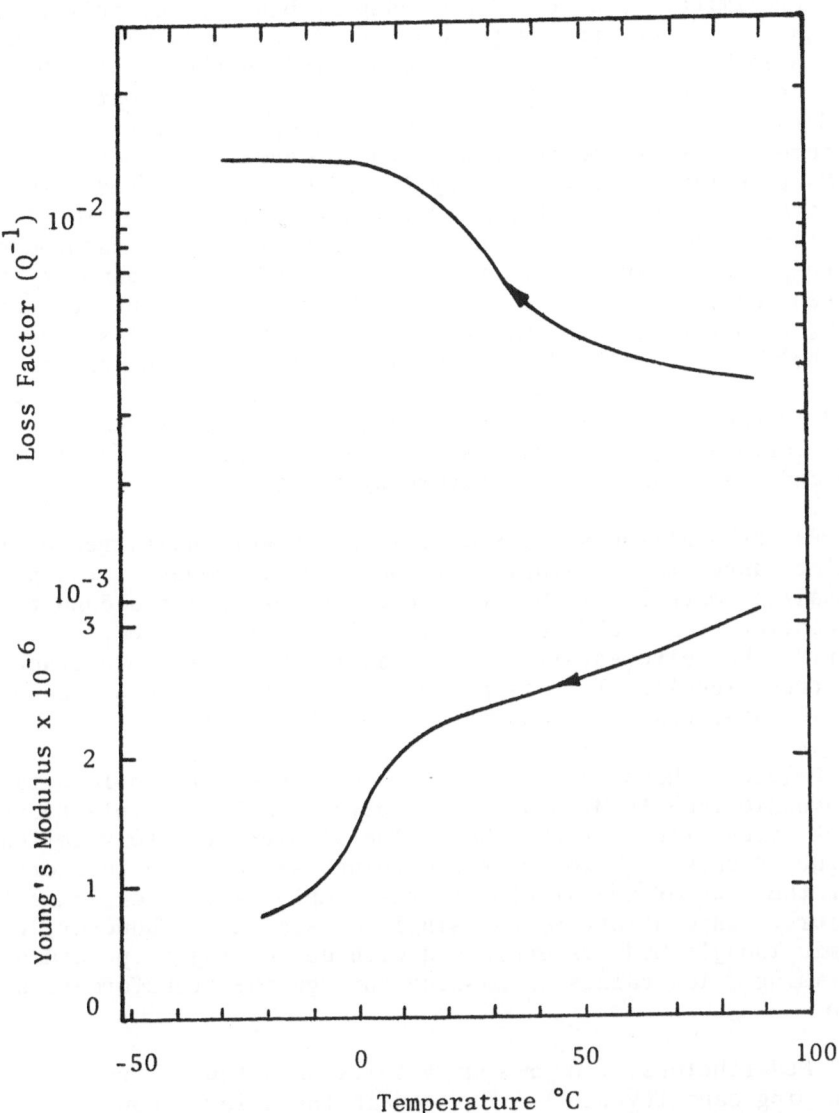

Figure 8.   Loss Factor and Dynamic Young's Modulus
for Polycrystalline Cu-14.0 w/o Al-3.0 w/o Ni
Alloy as a Function of Temperature on Cooling
at Frequencies in the 115-150 Hertz Range.

SUMMARY OF EXPERIMENTAL RESULTS ON THE
83 w/o Copper-14 w/o Aluminum-3 w/o Nickel ALLOY

Preliminary studies of the damping behavior of polycrystalline and single crystal samples of an 83 w/o Cu-14 w/o Al-3 w/o Ni
alloy have been carried out at frequencies in the 115-150 Hertz
range at stresses near 1000 psi. The samples were heat treated at
950°C for thirty minutes, followed by water quenching to room temperature. This procedure retains the ordered (DO$_3$) beta phase.
The M_s temperature of this alloy is +10°C as determined visually by
cold stage microscopy. Figure 7 shows the loss factor (Q^{-1}) and
dynamic Young's Modulus for a single crystal sample measured over
the temperature range between -50°C and +70°C. The damping increases substantially on cooling. The increase begins well above
the M_s temperature. The martensitic phase exhibits loss factors of
0.03 below +10°C. The damping measurements conducted on the cooling cycle disclose a substantial hysteresis similar to that encountered in the Ni-Ti alloy. The dynamic Young's measurements for the
single crystal show an increase in modulus with dropping temperature and a substantial temperature hysteresis.

Polycrystalline samples of this alloy were heat treated in a
similar manner to the single crystal sample. However, the polycrystalline material is brittle in the water quenched condition.
Air cooling from 950°C does not alleviate this problem. Consequently, the polycrystalline material is susceptible to cracking at
low stress levels. This is presumably due to precipitation of a
brittle intermetallic compound at grain boundaries.

Figure 8 shows the temperature dependence of the loss factor
and dynamic Young's Modulus for a polycrystalline sample tested at
a peak stress level of 1000 psi. The sample ultimately failed by
fatigue cracking. However, the damping was observed to increase,
as in the case of the single crystal sample, with decreasing temperature. In contrast to the single crystal data, however, the
dynamic Young's Modulus decreased with decreasing temperature to
surprisingly low values on passing through the transformation
range.

Nevertheless, the present results show that attractive levels
of damping capacity can be achieved in the alloy. However, the
tendency of polycrystalline material to develop brittle characteristics must be overcome if this alloy is to be used in practice.

REFERENCES

1. J. C. Heine, "The Stress and Frequency Dependence of Material
    Damping on Some Engineering Alloys," Ph.D. Dissertation,
    Massachusetts Institute of Technology (1966).

2. L. Cremer, M. Heckl and E. E. Ungar, Structure-Borne Sound,
    Springer-Verlag, New York, Chapter III (1973).

3. G. D. Sandrock, Met. Tr. (1974) 5 299.

4. G. D. Sandrock, A. J. Perkins and R. F. Heheman, Met. Tr.
    (1971) 2 2769.

5. K. Otsuka, T. Sawamura, K. Shimuzu and C. M. Wayman, Met. Tr.
    (1971) 2 2583.

A PROPOSED MEDICAL APPLICATION OF THE SHAPE MEMORY EFFECT:

A NiTi HARRINGTON ROD FOR THE TREATMENT OF SCOLIOSIS.

M.A. Schmerling*, M.A. Wilkov*, A.E. Sanders, M.D.[+]
and J.E. Woosley, P.E.[≠]
*Material Science and Engineering, 433 ENS, University
of Texas at Austin, Austin, Texas 78712, [+]Orthopedic
Surgery, Nix Professional Bldg., San Antonio, Texas 78205
[≠]Woosley Engineering, P.O. Box 18131, San Antonio, Texas
78218

INTRODUCTION

Scoliosis is a deformity which develops in the growing spine.
It consists of either a single or combination of sideways curvatures
leaving a buckled configuration which can be disfiguring and pain-
ful, and can cause interference with internal organs. Figure 1a
shows an X-ray of an untreated curvature in a teenage child.
Harrington instrumentation for the internal correction of scoliosis
accompanied by fusion is currently the surgical technique most
widely used[1]. A Harrington distraction rod with purchasing hooks
is shown in figure 2a. During surgery the hooks are attached to
transverse processes of vertebrae above and below the curve to be
corrected. The spine is distracted (straightened under tension)
by an external device and the rod inserted and made to apply the
correcting force by moving the hook along the ratchets closer to
its end of the rod. Figure 1b shows the X-ray of the same patient
as in figure 1a after instrumentation. This distraction with tension on the concave side is used universally often in conjunction
with a compression rod on the convex side to prevent excessive
stretching of the spinal column.

As the spine is distracted, each additional increment of distraction requires a larger increment of force and produces a
smaller increment of correction[2]. The possibility of break-out
of the purchasing hooks and neurological damage places a restric-
tion on the axial force that can be applied by a Harrington rod.
For these reasons the upper tolerance limit for the distracting
force applied during surgical correction usually does not exceed
thirty to forty kiloponds.

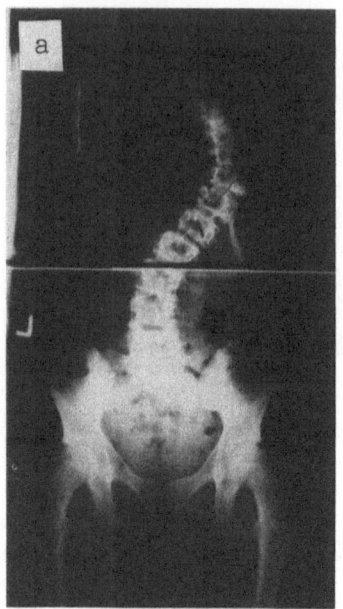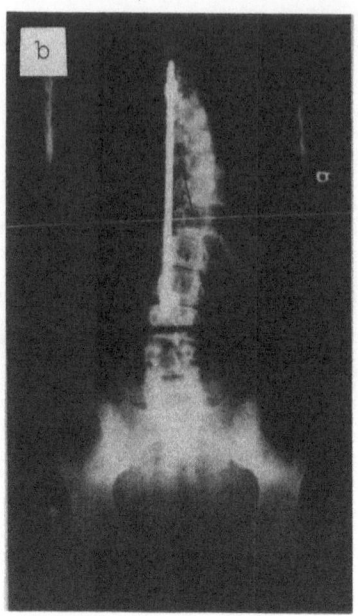

Figure 1: X-rays of scoliotic spine. (a) Uncorrected curvature.
(b) After Harrington instrumentation and fusion.

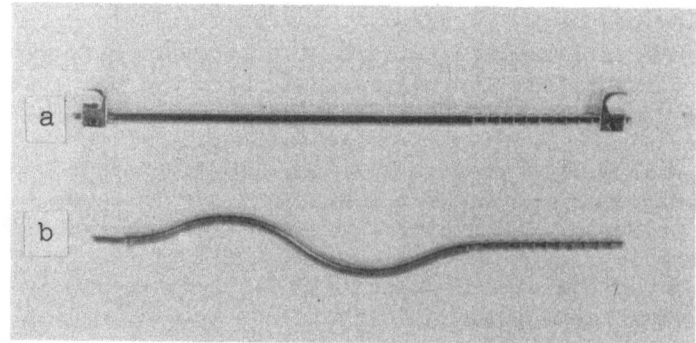

Figure 2: Harrington rods. (a) Stainless steel Harrington rod with
purchasing hooks attached. (b) NiTi Harrington rod
shortened by bending.

When biological tissues are stretched and held in an extended position, the holding force gradually relaxes. Waugh[3] and Nachemson and Elfström[4] have demonstrated using intravital telemetry that the axial force in the Harrington rod relaxes both during the operation as well as in the post operative period. Telemetry measurements by Nachemson and Elfström showed that the correcting force falls about 20% in the first 20 minutes and then slowly decreases, stabilizing at approximately 30% of the initial distraction force after 10 to 15 days. To reestablish the distraction force and gain 5 to 10 degrees more correction, a second operation is sometimes performed at the end of this period.

In this paper, research is described which is aimed towards the application of NiTi to the problem of scoliosis. One of the results will be a NiTi Harrington rod, shortened initially by deformation (figure 2b), inserted and distraction force applied following the normal procedure used for the first operation. After 10 to 15 days, heat (about 5°C above body temperature) will be applied externally causing the rod to regain its undeformed longer shape and reestablish the initial distraction force without the need for further surgical procedure. Only a shape memory effect alloy can accomplish so simply this added correction without a new operation.

MATERIALS AND METHODS

NiTi is initially soft and ductile in the low temperature martensitic phase but after it is deformed past the strain which is recoverable by shape memory effect behavior (i.e. heating above A_f), it can withstand stresses greater than the stainless steel now used in surgical procedures. Work hardening by using up the recoverable strain is an important property of NiTi.

Only two investigations on tissue compatibility of NiTi have been reported. NiTi bone plates implanted in dogs indicated no corrosion after 12 months[5]. Tissue reaction appeared normal with no clinical evidence of rejection. A second investigation[6] observed subcutaneously implanted NiTi wire in rats over a period of nine weeks. There appeared no difference in reaction compared to stainless steel similarly implanted for the same time period. As part of our research tests are underway in Rhesus monkeys. NiTi discs have been inserted in the maxillary region just under the mucoperiosteum and in grooves surgically created in the body of the mandible. After three and six month periods no clinical evidence of foreign body rejection or inflamation was seen.

For use as a Harrington rod a commercial alloy (Titanium

Metals Corp.) was obtained having an A_f just a few degrees
(approx. 6°C) above body temperature when the material was allowed
to transform unconstrained. The as received .76 cm rod was then
machined to a duplicate of a standard Harrington distraction rod.
From the data of Nachemson and Elfström[4] it is clear that to be
beneficial in this surgical treatment the deformed NiTi rod would
have to withstand an initial distraction load of at least 40 kiloponds. After a 10 to 15 day period, when the load has relaxed
to as low as 8 kiloponds, shape recovery should allow the rod to
return to its initial length approximately 1 cm longer, and
reestablish something close to the initial axial force. The
recovery should occur at a few degrees above body temperature even
while exerting this force.

    Before the NiTi rod is inserted at body temperature, the
initial recoverable strain must be exceeded to allow the necessary
initial load to be withstood. In this research both a single
curve and a double or "S" shaped curve, both composed of 8 cm
diameter radius of curvature elements, producing 8% outer fiber
strain for the .63 cm diameter rod, were used. For consistency
a simple set of bending dies (figure 3) was constructed and the
initial deformation was accomplished by squeezing the rod between
the dies while submerged in cold water. It is necessary to cool
the rod while bending since the heat created during deformation is
enough to raise the rod above A_s.

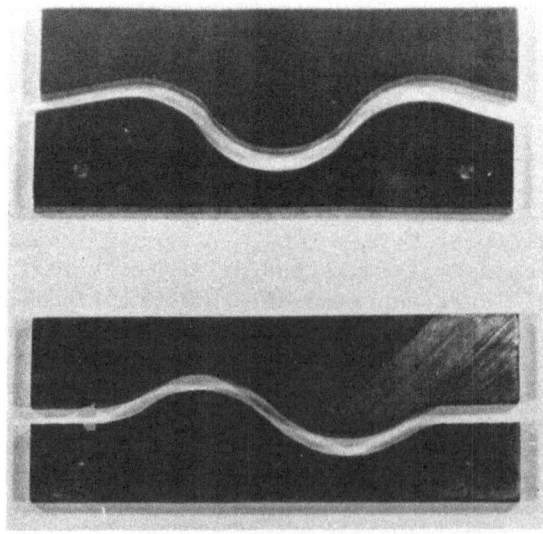

Figure 3: Bending dies for NiTi rods.

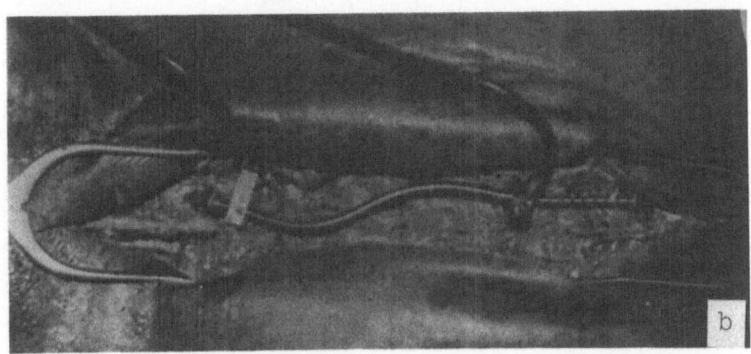

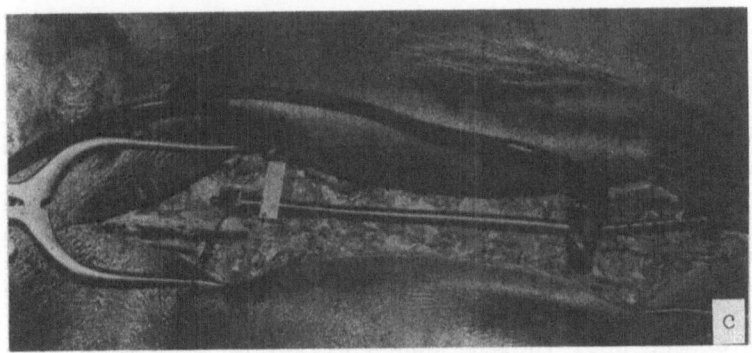

Figure 4: Trial implantation in unscoliotic cadaver.
(a) Outrigger used to stretch spine. (b) "S" curve
NiTi Harrington rod distracted to ~ 40 kp. (c) Recovered NiTi Harrington rod imparting a reverse
curvature to the spine.

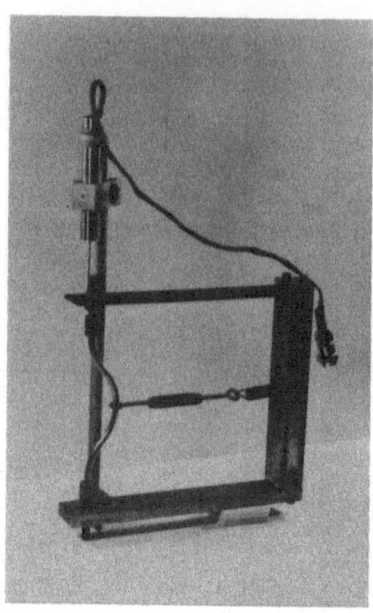

Figure 5: Test fixture to simulate transformation of NiTi
rod under spinal constraints.

        Trial implantation to observe shape recovery against spinal
load was performed on a 37 year old female unscoliotic cadaver
(figures 4a, b and c). Figure 4a shows the normal procedure of
stretching the spine using an outrigger. In figure 4b a NiTi
Harrington rod deformed into an "S" curve has been inserted and
distracted to a load of approximately 40 kiloponds. Shape
recovery was accomplished by direct electrical resistance heating
through the contact wires shown in the figure. Figure 4c shows
the Harrington rod in its recovered shape where it has applied
sufficient force to impart a reverse curvature to the unscoliotic
spine. The bend remaining in the notched segment near the end was
a result of the poor thermal conductivity of NiTi constraining the
shape recovery to only that portion between the electrodes.

EXPERIMENTS AND RESULTS

        The requirement for shape recovery to occur solely by
external heat application requires that the transition temperature
under load be predictable quite accurately. To simulate transformation under spinal constraints, a test fixture was constructed
capable of applying both axial and lateral forces to a .63 cm NiTi

rod 18 cm long during transformation (Figure 5). Holes in the
upper spring plate and bottom rigid plate allowed the specimen
"grips" to be positioned at different distances from the fixed end
of the spring plate. This allowed the spring constant k to be
varied. The "grips" were threaded so that they could be brought
closer together thus applying an initial load. At the top of the
fixture was a DCDT (direct current differential transformation)
for measuring the deflection of the spring plate. This deflection
(and its associated electrical signal) was converted into force
by calibrating it against a known load using an Instron mechanical
testing machine. Two types of specimen grips were used. One set
was "free end" which allowed the ends of the Niti rod to pivot
freely as it expanded. The other set was "fixed end" which kept
the ends of the rod approximately parallel during expansion.

For tests the apparatus, except for the DCDT, was submerged
in a variable temperature water bath. The bath was well stirred
to keep the temperature gradient across the rod at less than $0.5^\circ C$
and the temperature during the experiment was changed at $\sim 1^\circ C/min$.
The temperature was measured by a thermocouple.

The axial load vs temperature during heating of the Niti
rod for two different spring constants, k, in the "free end" grips
are shown in figure 6. In all the figures of load vs. temperature
the vertical dotted line is at body temperature, $37^\circ C$.

It can be seen that as the temperature increases through the
transformation region, the load increases to some final value
which is predetermined by the initial load and spring constant k
of the test fixture and the length change of the rod (~ 1 cm).
These tests indicate that A_f is almost exclusively a function of
final load; that is, a higher temperature is required to do more
work. The dip in the curves before A_f is due to the instability
of the "S" curve as the rod begins to recover with free ends. The
dip occurs when the double curve becomes unstable and a lower
energy single curve occurs.

As an alternate mode of shortening the rod, compressive strain
could be used. Four percent, the maximum compressive strain that
can be totally recovered, shifts the transformation temperature to
a higher value than a similar deformation produced by bending
figure 7. A shift in the transition temperature to an even higher
value is produced by the combination of 4% compression and bending,
to obtain the desired length change of approximately 1 cm.

Figure 8 shows the results of a single curved rod in the
"fixed end" grips. Here the rod was transforming against both an
axial and a lateral load. The lateral load varied from 8 to 20
kiloponds at the same time as the axial load increased from 20 to
40 kiloponds. The limit to the force which can be applied appears

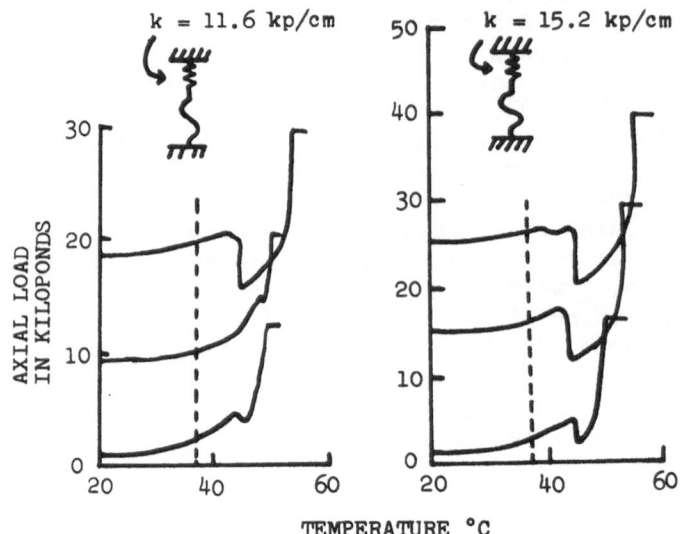

Figure 6: Load versus temperature during heating of an
"S" curve NiTi rod under axial load for two
spring constants and various initial loads.

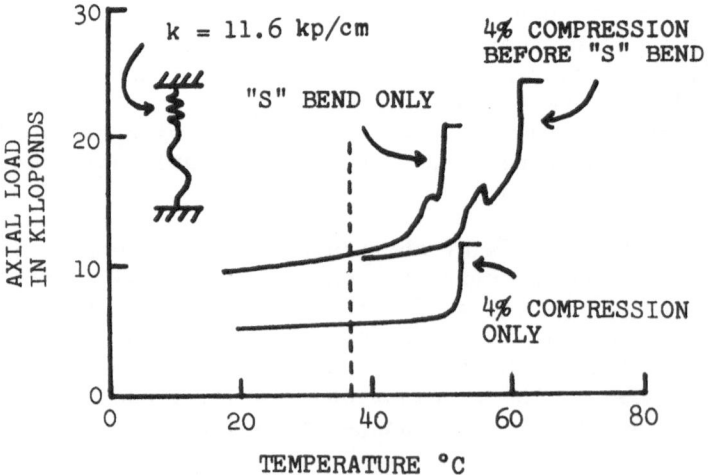

Figure 7: Load versus temperature during heating for
combinations of compression and bending on a
NiTi bar.

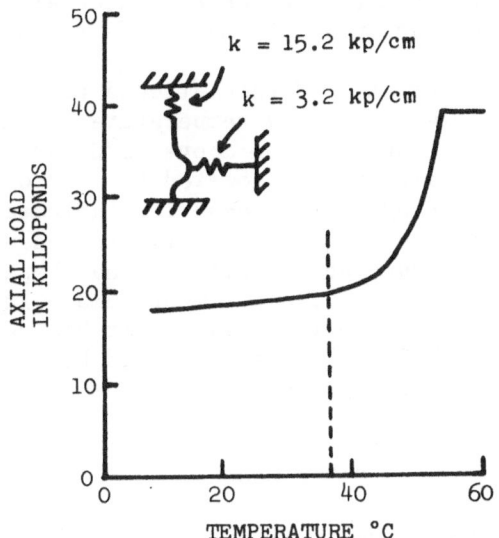

Figure 8: Load versus temperature during heating of a
single curve NiTi rod against both axial and
lateral load. Initial lateral load: 3.2 kp.
Lateral load after heating: 8.0 kp.

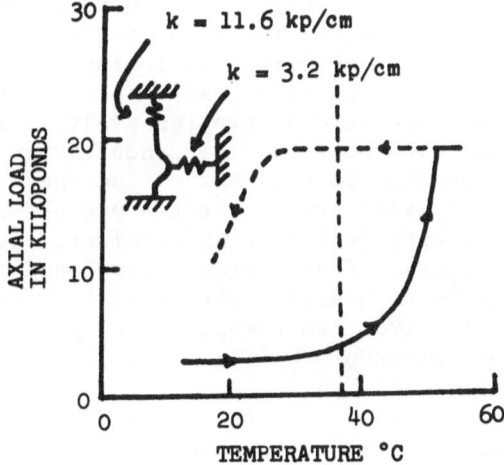

Figure 9: Load versus temperature during heating and
cooling of a single curve NiTi rod under both
axial and lateral load. Initial lateral load:
3.2 kp. Final lateral load: 8.0 kp.

to be determined by the point at which the material plastically
deforms and the highest acceptable temperature reached.

A rod was heated past A_f and then allowed to cool to a
temperature considerably below body temperature, Figure 9. The
rod straightened on heat application above 52°C and remained
straight until the temperature was lowered to ~ 27°C, at which
time it rebuckled. Such hysteresis behavior indicates that the
material remains in its strong high temperature phase at temperatures far below those required to initiate the shape change.
This effect is important since the rod must remain in its final
shape when cooled to body temperature after it has been heated.
If it recovered reversibly, the material would become soft and
ductile, as the low temperature phase is, before its ~ 8% deformation.

DISCUSSION

The use of the shape memory effect in NiTi in the Harrington
instrumentation would offer two distinct advantages in the treatment of scoliosis. First it would enable the surgeon to restore
any relaxed corrective force, post operative, simply by the application of heat. Second, because the material during transformation
is capable of exerting either large forces or moments it could be
used initially at the time of operation to apply a more proper set
of corrective force.

The first application appears to be a more straightforward
use of the shape memory effect. The loss of corrective forces
that occurs in the fifteen day period following the Harrington
procedure could be restored by simply straightening the initially
bent NiTi rod. This could be done progressively, a little at a
time, or all at once. The poor thermal conductivity of NiTi would
also allow selected heating of sections of the rod to avoid tissue
or neuron damage in critical areas. At the present writing the
mode of external heat application being considered is diathermy
(radio frequency heating). Eddy currents developed in the NiTi
rod should enable it to be heated 3-5°C above body temperature
without damage or discomfort to the patient. Correction could be
monitored during this procedure by fluoroscopy.

Proper control of the final force exerted and thus the transition temperature would require some spinal column relaxation data
for each individual. This could be obtained by use of a suitably
instrumented outrigger distracter during the initial insertion
and by post operative x-ray examination of the loss in curvature
correction. Once this data is established, it could be used to
select a rod with a shape recovery transformation temperature which
will be close to the 3-5°C range above body temperature. Error in
estimation at this time would only raise the transformation

temperature a few degrees, and not have severe consequences.

Little data is available for the discussion of the second mode of application. The existence of a geometrical limit on the amount of correction obtainable with the Harrington procedure has been argued by Schultz and Hirsch[2]. As the curve straightens, the moment arm through which the distraction force acts is reduced and thus the rod's corrective effectiveness is decreased. Adding a compression rod to the concave side, converts some of the axial force to a couple (pure moment) but has the disadvantage of complexity and interference with the fusion mass[7]. Physiological limitations of pressure, bone erosion and nerve damage also limit the amount of correction available with the current procedure.

Based on mechanical considerations alone, a lateral force or rotational moment should provide a considerably better corrective procedure than an axial force. Present methods such as the surcingle and Risser or Cotrel localizer cast which attempt to correct the scoliotic curve by external application of both axial and lateral pressure do not appear to be capable of providing sufficient amounts of lateral correction. In Figure 9, it was shown that a .63 cm rod of NiTi is able to apply both axial and lateral forces limited only by the yield stress of the material and the increase in transformation temperatures. A NiTi rod which is bent to conform to the uncorrected spine and attached to the vertebrae at some intermediate points should be able to exert a more proper combination of corrective forces during recovery to its initial shape. In this case recovery would occur at the time of the operation with the NiTi rod serving as both the auxiliary distraction apparatus and the distraction rod itself.

The research reported here is intended only as a feasibility study. Considerably more effort is required in the areas of precise prediction of the transformation temperature, its long term compatability with living tissue, and the problems involved in intravital heating before widespread clinical use is attempted.

SUMMARY

The shape memory effect in near equiatomic nickel-titanium has been used to advantage in the Harrington rod treatment of scoliosis. A NiTi Harrington rod, originally straight but with an imparted curvature, has the ability of straightening to its original shape and post operatively reestablishing the relaxed corrective force. This can be accomplished by external heat application.

A program using simulated spine forces in a test fixture and implantation in a cadaver was undertaken to show the feasibility of modifying the existing surgical treatment.

REFERENCES

1. P. R. Harrington, J. Bone and Joint Surg., <u>44-A</u>, 591 (1962).

2. A. B. Schultz and C. Hirsch, Clin. Orthop., <u>100</u>, 63 (1974).

3. T. R. Waugh, Acta Orthop. Scandenavica, Supplementum <u>93</u> (1966).

4. A. Nachemson and G. Elfström, J. Bone and Joint Surg. <u>53-A</u>, 445 (1971).

5. L. S. Castleman, The 5th Annual Biomaterials Symposium, April 1973, Clemson, S. C. and Private Communication.

6. D. E. Cutright, S. N. Bhaskar, B. Perez, R. M. Johnson, G. S. M. Cowan, J. Oral Surg. <u>35</u>, No. 4 (1973).

7. P. R. Harrington (Private Communication).

WRITTEN DISCUSSION

L. Delaey (Department Metaalkunde, Katholieke Universiteit,
        Leuven, Belgium)

Discussion to Paper 1:

   In analyzing the influence of β-grain boundaries on the shape
of the stress-strain curve of stressed martensitic polycrystalline
samples two very important factors should not be omitted. As
shown in a later contribution (1), the stress-strain curves are
very sensitive to the thermal history. It should therefore be
quoted with each test whether or not the samples have been subjected to a previous deformation step and/or heated to a temperature above A$_s$. The other point that should be analyzed is the
crystallographic orientation difference between two neighboring
grains. It has been pointed out (2) that the high elastic aniso-
trophy of the β-phase is to a great part responsible for the
martensite plate morphology along the former grain boundaries.
Depending upon the orientation of the two grains and of the orientation of the grain boundary weak or strong strain accommodation
is needed.

C. M. Wayman (Department of Metallurgy and Mining Engineering and
        Materials Research Laboratory, University of Illinois
        at Urbana-Champaign, Urbana, Illinois 61801)

Discussion to Paper 8:

   Professor Lieberman has drawn attention to the necessity of
ordering and a basis to explain the rubberlike condition found in
the martensitic state of Au-47.5 at % Cd alloys. These requirements may not be universally applicable since rubberlike martensite
is also found in In-Tl alloys in which case neither ordering nor a
basis are evidently involved.

---

(1) R. Rapacioli, M. Chandrasekaran and L. Delaey, this symposium.
(2) L. Delaey and R. Devrinck, to be published.

S. Mendelson (The City University of New York Research Foundation, Steinham Hall, Room 12, 138th Street and Convent Avenue, New York, New York 10031)

Discussion to Paper 8:

Dr. Lieberman proposed that pseudoelasticity or ferroelasticity in bent AuCd occurs by a lattice deformation in the martensite which results in a structure which is not a twin, but becomes one during aging under stress, when atomic dumbbells rotate by atomic diffusion.

The most likely reason why this kind of a mechanism cannot apply to a close packed alloy like AuCd is that the energy of this structure is enormous. Since the distortions occur on every plane of the structure (not just in the interface), it is effectively a second martensitic transformation to a high energy structure. One is left with the question of why this structure should form under stress when other, also unlikely, structures but of lower energy, do not form, or why the low energy twin should not form directly by zonal dislocations. That the proposed shear is the same as that for deformation twinning has no bearing on this if the lattice can not tolerate the distortions.

There are other mechanisms which could account for the pseudo-elasticity and aging effects under stress; these have been considered by various authors and include (i) twinning and detwinning (real twins), (ii) growth of some martensite plates and reversion of others, and (iii) a second martensitic transformation to a new structure of similar energy. Since all creep phenomena involve some dislocation motion and diffusion, it is more reasonable to seek an explanation for the loss of memory on aging to changes in the forces on dislocations and their morphology.

In studies by Delaey et al. (1) some martensite plates with self accommodating twin boundaries in Cu-Al-Ni alloys grow into their neighbors. Ledges along the twin interface (equivalent to twinning dislocations) can serve this function for limited growth until the steps are removed. This strain may not be pseudoelastic since a smooth twin interface will not revert back to a stepped (dislocated) interface. Diffusion could enhance the passing of the twinning dislocations through the ends of the interface and thus eliminate the back stress responsible for pseudoelasticity and shape memory. If the martensite or twins form by pole mechanisms (2) the diffusion during aging might alter the structure of the

(1) L. Delaey, R. V. Krisnan, H. Tas, and H. Warlimont, J. Mat. Sci. 9, 1521 (1974).
(2) S. Mendelson, this symposium.

transformation or twinning partials and prevent their reversed motion.

L. Delaey and J. Smeesters (Department Metaalkunde, Katholieke
                          Universiteit, Leuven, Belgium)

Discussion to Paper 31:

A SME-heat engine has been constructed (Figure 1), using the same mechanical principles and concepts as developed by R. Banks and O. Weres. Instead of a Nitinol-material, strips of a Cu-Zn-Al alloy were used for the vital parts of the engine. The Cu-Zn-Al alloy showed the β_1'-type of martensite and has a M_s-temperature of about 50°C. The strips were previously taught to exhibit the two-way memory effect; the expanded strips return back to their bent state as soon as they come in contact with the cold water. Using the two-way memory effect it is thus avoided that the strips exert a back-stress on the expanding strips.

By this experiment it is shown that SME-heat engines equally can be built with a copper-based alloy exhibiting the β to β_1'-type of transformation.

Figure 1: SME-heat engine using Cu-Zu-Al alloy strips.

G. Alefeld (Physik Department der Technischen Universitat, Munchen,
           8046 Garching)

Discussion (on reversibility of strain):

   Reversibility of strain is known for elastic or anelastic
deformation of solids. In these cases the strains are small.
During this meeting conditions for the existence of shape memory,
i.e., the reversibility of large-scale strain, have been discussed
by several authors, e.g. by Wayman. The striking aspect of these
phenomena is the fact that each atom knows exactly where it originally came from. The memory for large-scale deformation certainly
would not be possible if slip occurs by a translation vector of the
unit cell of the lattice (= Burgers vector of a dislocation).
Therefore a necessary requirement is that slip must occur by a
vector which is not a translation vector of the lattice (1). Thus
each atom comes into a position which is non-equivalent to the preceding position so that a memory for the direction and distance of
motion in the back-reaction is built in. In general this slip will
be smaller than a translation vector of the lattice. Futhermore,
to preserve the memory effect slip should occur in one plane only
once, or at least not so often that the total slip vector adds up
to a translation vector of the lattice. The addition or subtrac-
tion of two of the possible equivalent slip vectors should not add
up to a translation vector of the lattice. Otherwise the original
shape of the specimen would not necessarily be reproduced in the
back-reaction.

   From the preceding statements it is obvious that ordering an
alloy and thus creating a longer translation vector is a helpful
although not a necessary condition for shape memory. Furthermore,
homogeneous slip by a vector not identical with a translation vector of the lattice produces a new crystal structure. Shape memory
therefore is necessarily connected with a phase transition.

Discussion (on soft modes):

   The dislocation concept has been invented to remove the discrepancy between ideal shear strength and experimentally observed
shear strength. It should be pointed out that the existence of a
soft-acoustic shear mode ($C_{11}-C_{12} \rightarrow 0$) implies that the ideal
shear strength approaches zero. The crystal can shear homogene-
ously over a total plane without moving dislocations. No state-
ment can be made a priori at which shear vector the lattice stabilizes, except that it would not stabilize for a shear by a translation vector. Furthermore, the energy of those dislocations, for
which the self-energy is proportional to ($C_{11}-C_{12}$), approaches
zero, or at least reduces to the core energy. Therefore it may be
possible to observe stacking faults and/or dislocations in thermal
equilibrium close to such a martensitic transformation, which is
accompanied by a soft mode.

SUBJECT INDEX